D1628739

Multidimensional Quantum Dynamics

Edited by

Hans-Dieter Meyer, Fabien Gatti, and Graham A. Worth

Related Titles

Fried, JR

Computational Chemistry and Molecular Simulation

2009

ISBN: 978-0-471-46244-6

Rode, B. M., Hofer, T., Kugler, M.

The Basics of Theoretical and Computational Chemistry

2007

ISBN: 978-3-527-31773-8

Höltje, H.-D., Sippl, W., Rognan, D., Folkers, G.

Molecular Modeling

Basic Principles and Applications

Third, Revised and Expanded Edition

2008

ISBN: 978-3-527-31568-0

Dronskowski, R.

Computational Chemistry of Solid State Materials

A Guide for Materials Scientists, Chemists, Physicists and others

2005

ISBN: 978-3-527-31410-2

Matta, C. F., Boyd, R. J. (eds.)

The Quantum Theory of Atoms in Molecules

From Solid State to DNA and Drug Design

2007

ISBN: 978-3-527-30748-7

Cramer, C. J.

Essentials of Computational Chemistry

Theories and Models

2004

ISBN: 978-0-470-09182-1

Multidimensional Quantum Dynamics

MCTDH Theory and Applications

Edited by

Hans-Dieter Meyer, Fabien Gatti, and Graham A. Worth

WILEY-VCH Verlag GmbH & Co. KGaA

The Editors

Prof. Hans-Dieter Meyer
Ruprecht-Karls-Universität
Physikalisch-Chemisches Institut
Theoretische Chemie
Im Neuenheimer Feld 229
69120 Heidelberg
Germany

Dr. Fabien Gatti
Université Montpellier II
CTMM, Institut Charles Gerhardt
UMR 5253, CC 1501
34095 Montpellier Cedex 05
France

Dr. Graham A. Worth
University of Birmingham
School of Chemistry
Birmingham B15 2TT
United Kingdom

Library of Congress Card No.:
applied for

British Library Cataloguing-in-Publication Data
A catalogue record for this book is available from the British Library.

Bibliographic information published by the Deutsche Nationalbibliothek
The Deutsche Nationalbibliothek lists this publication in the Deutsche Nationalbibliografie; detailed bibliographic data are available on the Internet at http://dnb.d-nb.de.

Typesetting Uwe Krieg, Berlin
Printing Strauss GmbH, Mörlenbach
Binding Litges & Dopf GmbH, Heppenheim

Printed in the Federal Republic of Germany
Printed on acid-free paper

ISBN: 978-3-527-32018-9

Contents

Multidimensional Quantum Dynamics: MCTDH Theory and Applications.
Edited by Hans-Dieter Meyer, Fabien Gatti, and Graham A. Worth

ISBN: 978-3-527-32018-9

Preface

This book describes a powerful and general approach to solve the time-dependent Schrödinger equation. Generally referred to by its acronym, MCTDH, the multi-configuration time-dependent Hartree method has been used to solve a range of problems from interpreting gas-phase spectroscopy to studying surface-molecule interactions at the molecular level.

The problems of molecular quantum dynamics are very challenging tasks numerically. Hence the demand for efficient methods is high. In 1989 Meyer, Manthe, and Cederbaum started to discuss a new approach, later named MCTDH, and the first paper appeared in 1990. Since then a number of people have been involved in the work that has taken MCTDH from an idea to a practical method: understanding the properties and characteristics of the method, developing numerical tools to efficiently solve the equations of motion, writing software to implement the algorithm, and using and testing that software. The editors would like to take this opportunity to thank all involved. In addition to the thanks due to all contributors in this volume, special mention must be made of the following people.

Firstly Uwe Manthe who has been involved in all aspects of the development of MCTDH in general, including the initial development. Andreas Jäckle and Michael Beck who played a major role in the initial development of the Heidelberg MCTDH package, which is the most widely used software implementation of the method. Masahiro Ehara, Marie-Catherine Heitz, and Suren Sukiasyan who were the first people to take up and test the embryonic Heidelberg package and made important contributions to the development of the code. Andreas Raab who pushed the development of the density-matrix formulation of MCTDH and made the 24-mode pyrazine absorption spectrum calculation that amply demonstrated the abilities of the method. This study, published in 1999, can be considered as the first real breakthrough of the MCTDH method. Christoph Cattarius, Stephan Wefing, Frank Otto, Mathias Nest, and Oriol Vendrell who used the now more mature Heidelberg package to study new exciting problems and/or made contributions to the develop-

Multidimensional Quantum Dynamics: MCTDH Theory and Applications.
Edited by Hans-Dieter Meyer, Fabien Gatti, and Graham A. Worth

ISBN: 978-3-527-32018-9

ment of the code. Finally, Michael R. Brill who started the parallelization of the Heidelberg MCTDH code, a very important issue.

Turning to MCTDH in general we would like to thank Haobin Wang who, together with Michael Thoss, has developed and implemented (in his own code) the multi-layer formulation of the MCTDH algorithm. This very important and exciting development has the potential to extend the range of MCTDH to considerably larger systems. It is planned to implement this complicated extension into the Heidelberg MCTDH package in near future.

We are also grateful to the external users of the Heidelberg MCTDH package. They made MCTDH more widely known, noted several bugs, and initiated new developments. Let us mention only a few of them, Oliver Kühn, C. William McCurdy and coworkers, and Jan-Geert Kroes, as they are among the first and most active external users.

Last, but not least, we would like to thank Lenz Cederbaum for his continued interest in and support of the MCTDH project.

April 2009 *The Editors*

List of Contributors

Ofir E. Alon
Ruprecht-Karls-Universität
Physikalisch-Chemisches Institut
Theoretische Chemie
Im Neuenheimer Feld 229
69120 Heidelberg
Germany

Michael Brill
Ruprecht-Karls-Universität
Physikalisch-Chemisches Institut
Theoretische Chemie
Im Neuenheimer Feld 229
69120 Heidelberg
Germany

Lorenz S. Cederbaum
Ruprecht-Karls-Universität
Physikalisch-Chemisches Institut
Theoretische Chemie
Im Neuenheimer Feld 229
69120 Heidelberg
Germany

Cédric Crespos
Institut des Sciences Moleculaire
CNRS UMR 5255
Université Bordeaux I
351 Cours de la Libération
33405 Talence Cedex
France

Fabien Gatti
Université Montpellier
CTMM, Institut Charles Gerhard, UMR
5253, CC 1501
Place Eugène Bataillon
43095 Montpellier
France

Etienne Gindensperger
CNRS, Université de Strasbourg
Institut de Chimie UMR 7177
Laboratoire de Chimie Quantique
4, rue Blaise Pascal, B.P. 1032
67000 Strasbourg Cedex
France

Rob van Harrevelt
Institute for Molecules and Materials
Radboud University Nijmegen
Toernooiveld 1
6525 ED Nijmegen
The Netherlands

Daniel J. Haxton
JILA and Department of Physics
University of Colorado
440 UCB
Boulder, CO 80309
USA

Fermín Huarte-Larrañaga
Computer Simulation and Modeling
Laboratory
Parc Cientifíc de Barcelona
Baldiri Reixach 13
08028 Barcelona
Spain

Multidimensional Quantum Dynamics: MCTDH Theory and Applications.
Edited by Hans-Dieter Meyer, Fabien Gatti, and Graham A. Worth

ISBN: 978-3-527-32018-9

Christophe Iung
Université Montpellier
CTMM, Institut Charles Gerhard, UMR 5253, CC 1501
Place Eugène Bataillon
43095 Montpellier
France

Gerald Jordan
Vienna University of Technology
Photonics Institute
Gußhausstraße 27/387
1040 Vienna
Austria

Horst Koeppel
Ruprecht-Karls-Universität
Physikalisch-Chemisches Institut
Theoretische Chemie
Im Neuenheimer Feld 229
69120 Heidelberg
Germany

Geert-Jan Kroes
LIC, Gorlaeus Laboratoria
Universiteit Leiden
Postbus 9502
2300 RA LEIDEN
The Netherlands

Oliver Kühn
Universität Rostock
Institut für Physik
Universitätsplatz 3
18051 Rostock
Germany

Benjamin Lasorne
Université Montpellier
CTMM, Institut Charles Gerhard, UMR 5253, CC 1501
Place Eugène Bataillon
43095 Montpellier
France

Uwe Manthe
Theoretische Chemie
Fakultät für Chemie
Postfach 100131
33501 Bielefeld
Germany

C. William McCurdy
Departments of Applied Science and Chemistry
University of California
Davis, CA 95616
USA

Hans-Dieter Meyer
Ruprecht-Karls-Universität
Physikalisch Chemisches Institut
Theoretische Chemie
Im Neuenheimer Feld 229
69120 Heidelberg
Germany

Mathias Nest
University Potsdam
Theoretical Chemistry
Karl-Liebknecht-Str. 25
14476 Potsdam
Germany

Thomas N. Rescigno
Chemical Sciences Division
Lawrence Berkeley National Laboratory
Berkeley, CA 94720
USA

Peter Schmelcher
Physikalisches Institut
Universität Heidelberg
Philosophenweg 12
69120 Heidelberg
Germany

Armin Scrinzi
Vienna University of Technology
Photonics Institute
Gußhausstraße 27/387
1040 Vienna
Austria

Alexej I. Streltsov
Ruprecht-Karls-Universität
Physikalisch Chemisches Institut
Theoretische Chemie
Im Neuenheimer Feld 229
69120 Heidelberg
Germany

Michael Thoss
Friedrich-Alexander-Universität
Erlangen-NÃijrnberg
Institut für Theoretische Physik
Staudterstraße 7/82
91058 Erlangen
Germany

Oriol Vendrell
Ruprecht-Karls-Universität
Physikalisch Chemisches Institut
Theoretische Chemie
Im Neuenheimer Feld 229
69120 Heidelberg
Germany

Haobin Wang
Department of Chem. & Biochem.
MSC 3 C, New M. State Univ.
Las Cruces, NM 88003
USA

Graham A. Worth
University of Birmingham
School of Chemistry
Edgbaston, Birmingham
B15 2TT
United Kingdom

Sascha Zöllner
Ruprecht-Karls-Universität
Physikalisch Chemisches Institut
Theoretische Chemie
Im Neuenheimer Feld 229
69120 Heidelberg
Germany

Acronyms and Symbols

Acronyms

BS	Burlisch–Stoer (integrator)
CAP	Complex absorbing potential
CDVR	Correlation discrete variable representation
CMF	Constant mean field
DOF(s)	Degree(s) of freedom
DVR	Discrete variable representation
FBR	Finite basis-set representation
FD	Filter diagonalization
FFT	Fast Fourier transform
FWHM	Full width at half-maximum
KEO	Kinetic energy operator
MCTDH	Multiconfiguration time-dependent Hartree
MCTDHB	MCTDH for bosons
MCTDHF	MCTDH for fermions
MC-TDSCF	Multiconfiguration time-dependent self-consistent field
ML-MCTDH	Multi-layer MCTDH
PEO	Potential energy operator
PES(s)	Potential energy surface(s)
RK	Runge–Kutta (integrator)
rms	root-mean-square
SIL	Short iterative Lanczos
SPF(s)	Single-particle function(s)
TDDVR	Time-dependent DVR
TDH	Time-dependent Hartree
TDSCF	Time-dependent self-consistent field
VMF	Variable mean-field

Multidimensional Quantum Dynamics: MCTDH Theory and Applications.
Edited by Hans-Dieter Meyer, Fabien Gatti, and Graham A. Worth

ISBN: 978-3-527-32018-9

Symbols

Full multidimensional wavefunction	Ψ
Number of degrees of freedom	f
Number of particles (combined modes)	p
Index running over particles	κ $(1 \leqslant \kappa \leqslant p)$
Index running over degrees of freedom	ν $(1 \leqslant \nu \leqslant f)$
Number of primitive functions (or grid points) for DOF ν	N_ν
Number of primitive functions (or grid points) for particle κ	$\tilde{N}_\kappa$
Number of single-particle functions for κth particle (and αth state)	$n,\ n_\kappa,\ n_{\alpha,\kappa}$
Nuclear coordinate for DOF ν	q_ν
(Multidimensional) coordinate for particle κ	Q_κ
jth primitive basis function for DOF ν	$\chi_j^{(\nu)}$
jth single-particle function for the κth particle (and αth state)	$\varphi_j^{(\kappa)},\ \varphi_j^{(\alpha,\kappa)}$
MCTDH expansion coefficient	$A_{j_1 \dots j_p},\ A_J,\ A_J^{(\alpha)}$
MCTDH configuration	$\Phi_J,\ \Phi_J^{(\alpha)},\ \Phi_J = \prod_{\kappa=1}^{p} \varphi_{j_\kappa}^{(\kappa)}$
Projector on the space spanned by the SPFs for κth particle (and αth state)	$P^{(\kappa)},\ P^{(\alpha,\kappa)}$
lth single-hole function for the κth particle (and αth state)	$\Psi_l^{(\kappa)},\ \Psi_l^{(\alpha,\kappa)}$
Composite index J	$J = (j_1, j_2, \dots, j_p)$
Composite index J with the κth entry set at l	$J_l^\kappa = (j_1, \dots, j_{\kappa-1}, l, j_{\kappa+1}, \dots, j_p)$
Sum over the indices for all particles excluding the κth	$\sum_J^\kappa$
Constraint operator for the κth particle	$g^{(\kappa)},\ g^{(\alpha,\kappa)}$
Mean-field matrix elements for the κth particle	$\mathcal{H}_{jl}^{(\kappa)} = \langle \Psi_j^{(\kappa)} \vert H \vert \Psi_l^{(\kappa)} \rangle$
Matrix of mean-field operators	$\boldsymbol{\mathcal{H}}^{(\kappa)}, \boldsymbol{\mathcal{H}}^{(\alpha,\kappa)}$
Density matrix elements for the κth particle (and αth state)	$\rho_{jl}^{(\kappa)},\ \rho_{jl}^{(\alpha,\kappa)}$
$n_\kappa \times n_\kappa$ unit matrix	$\mathbf{1}_{n_\kappa}$
Separable term appearing in the Hamiltonian operator and acting on the kth particle only	$h^{(\kappa)}$

Residual part of the Hamiltonian operator	H_R, where $H = \sum_{\kappa=1}^{p} h^{(\kappa)} + H_R$
Operator acting on the kth particle appearing in the rth term in the product representation of the Hamiltonian	$h_r^{(\kappa)}$, where $H_R = \sum_{r=1}^{s} c_r \prod_{\kappa=1}^{p} h_r^{(\kappa)}$
Hamiltonian matrix	$\mathcal{K}, \mathcal{K}_{JL} = \langle \Phi_J \vert H \vert \Phi_L \rangle$
Number of electronic states	σ
Set of electronic states	$\{\vert\alpha\rangle\}$
κth combined mode	$Q_\kappa = (q_i, q_j, \ldots)$
Complex absorbing potential acting on DOF q	$-\mathrm{i}W(q) = -\mathrm{i}\eta(q - q_c)^b \theta(q - q_c)$
Overlap matrix of the single-particle functions	$\mathcal{O}_{jl}^{(\kappa)} = \langle \varphi_j^{(\kappa)} \vert \varphi_l^{(\kappa)} \rangle$
Heaviside step function	$\theta(q - q_c)$
Autocorrelation function	$c(t)$
Associated Legendre functions	$P_l^m(\cos\theta)$
L^2 normalized associated Legendre functions	$\tilde{P}_l^m(\theta)$
Spherical harmonics	$Y_{jm}(\theta, \phi)$

Symbols for *potfit*

Potential energy value on product grid	$V_{i_1 \ldots i_f}$
Number of natural potentials for κth particle (combined mode)	m_κ
Potential density matrices acting on the kth particle	$\varrho^{(\kappa)}$
with components	$\varrho_{ij}^{(\kappa)}$
Natural potential (ith grid point of jth potential	$v_{ij}^{(\kappa)}$
Natural potential populations	$\lambda_j^{(\kappa)}$
Reduced natural potential populations	$\tilde{\lambda}_j^{(\kappa),\mathrm{red}}$
Expansion coefficients	$C_{j_1 \ldots j_p}$
Contracted expansion functions	$D_{j_1 \ldots j_{\kappa-1} i_\kappa j_{\kappa+1} \ldots j_f} = \sum_{j_\kappa=1}^{m_\kappa} C_{j_1 \ldots j_p} v_{i_\kappa j_\kappa}^{(\kappa)}$
Weight function	w

1
Introduction

Hans-Dieter Meyer, Fabien Gatti and Graham A. Worth

Quantum dynamics simulations are now established as an essential tool for understanding experiments probing the nature of matter at the molecular level and on fundamental time-scales. This is a relatively recent development and for many years the methods of choice were based on time-independent calculations, describing a system in terms of its eigenfunctions and eigenvalues. This book is about the multiconfiguration time-dependent Hartree method, commonly known by its acronym MCTDH, a method that has played a significant role in the upsurge of interest in time-dependent treatments by extending the range of applicability of what are often called wavepacket dynamics simulations.

The book will cover the theory of the method (Part 1), highlighting the features that enable it to treat systems not accessible to other methods. Details in particular will be given on the implementation strategy required. In Part 2, chapters will then detail extensions of the basic method to show how the theory provides a framework to treat systems outside the original aims and to combine different methods. In the final part (Part 3), examples of calculations are given. As the method is completely general, and has been applied to a range of problems, the result is a snapshot of the state of the art in the study of molecular dynamics.

To describe a (non-relativistic) molecular system one needs to solve the Schrödinger equation, which in its time-dependent form reads

$$i\hbar \frac{\partial \Psi}{\partial t} = \hat{H}\Psi \tag{1.1}$$

Unfortunately, this equation is impossible to solve for more than two particles, that is, the hydrogen atom and the field of theoretical chemistry is dominated by developments of methods, numerical and approximate, that provide solutions that can be used to treat atoms and molecules.

A key development in making this problem tractable is the separation of nuclear and electronic motion through the Born–Oppenheimer approximation [1–3]. This allows us to imagine our molecular system as a set of nuclei represented as point masses moving over a potential energy surface (PES)

Multidimensional Quantum Dynamics: MCTDH Theory and Applications.
Edited by Hans-Dieter Meyer, Fabien Gatti, and Graham A. Worth

ISBN: 978-3-527-32018-9

provided by the electrons: the electrons follow the much heavier nuclei, adjusting instantaneously as the nuclei change conformation. This approximation works extremely well. It does, however, break down in certain situations when two electronic configurations strongly mix with dependence on the nuclear conformation [4]. The nuclei then must be imagined as moving in a manifold of coupled electronic states, each associated with a potential energy surface.

The PESs are obtained by solving the time-independent form of the Schrödinger equation applied only to the electrons and treating the nuclei as static point charges. This is the field of electronic structure theory, or quantum chemistry as it is often called. Quantum chemistry is a mature field of research with a number of general-purpose computer programs available, such as GAUSSIAN [5] and MOLPRO [6]. These programs are able to solve the electronic structure problem for a nuclear configuration using a range of methods. They can also provide an analysis of molecular properties at that molecular geometry.

Much of chemistry can be described by an analysis of the critical points on the PES: minima represent stable nuclear configurations and saddle points represent transition states connecting them. In the field of photochemistry, features such as conical intersections and avoided crossings where neighbouring states interact are also important. The nuclear geometries and relative energies of these points can then be used to build up a picture of a reaction in molecular terms. Local analysis of these critical points can further provide information. For example, frequencies related to vibrational spectra can be calculated from the Hessian matrix. These can all be provided by the quantum chemistry programs.

The field of molecular dynamics studies the motion between these critical points. This is chemistry at its fundamental level. How do nuclei move during molecular collisions and reactions, or in response to the absorption of a photon?

Molecular quantum dynamics, aiming to solve the time-dependent Schrödinger equation, is difficult and numerically demanding. Comparing it to quantum chemistry, one may wonder why quantum dynamics cannot solve problems of similar size. Accurate computations of the electronic structure of molecules with more than 100 electrons are feasible, whereas accurate dynamic calculations including 100 atoms are not (except for special simple cases).

One difference is that quantum chemistry is characterized by low quantum numbers. One is usually interested in the electronic ground state, or in the two or three lowest excited states. The molecular orbitals have a simple smooth form and can be well represented by Gaussian basis sets. In quantum dynamics, however, the wavefunction is much more structured and may contain highly oscillatory terms. Moreover, the density of states is rather high. There may be hundreds of eigenstates lying below a fundamental C–H or O–H stretch of a small polyatomic molecule (four or five atoms, say). This high density of states, which is a consequence of the fact that nuclear masses are much larger than the electron mass, is one of the main sources of difficulties.

A second difference is that quantum chemistry is governed by the Coulomb potential, a rather structureless two-body interaction. Quantum dynamics, on the other hand, has to cope with complicated many-body potentials, which are not general but specific for the problem under investigation. The strong repulsion suffered by atoms that come close to each other may lead to a very strongly correlated motion. These differences explain why the techniques used in quantum chemistry and quantum dynamics are often similar in spirit but very different in detail.

The development of quantum dynamics simulations has been driven by research probing the fundamental properties of molecules. Historically, the first field of importance was scattering experiments using molecular beams. To study elementary reactions, it is necessary to enable molecules to collide with known initial states under controlled conditions and to measure the products of the collision [7]. One way to achieve this experimentally is in scattering experiments using beams of molecules. In particular, crossed molecular beam experiments, pioneered by Lee, Herschbach, Toennies and others, have provided a wealth of data in this field.

More recently, time-resolved spectroscopy driven by the development of pulsed lasers has also become important. The work of Zewail was key in the development of techniques to produce and apply pulses of the order of femtosecond duration [8]. This allows molecules to be followed on the timescale in which bonds vibrate and break. Early work studied bond breaking in molecules such as ICN and NaI, and bond vibration in I_2. The motion was evaluated in terms of a nuclear wavepacket moving over the potential surfaces. These 'femtochemistry' experiments have now been applied to a wide variety of systems. A recent example of the detail produced by these techniques include a study of the retinal chromophore in the rhodopsin protein showing *cis–trans* isomerization taking place over a picosecond [9].

Initial theoretical research focused on time-independent methods. Early research focused on understanding line spectra, for which the states must be known. The time-independent equation is easy to adapt to numerical solution using matrix diagonalization methods, and, unless the Hamiltonian is explicitly time-dependent, even ostensibly dynamical problems can be described using the eigenfunctions and eigenvalues of the system.

Despite the early seminal work of McCullough and Wyatt [10, 11], which describes the $H + H_2$ exchange reaction in a time-dependent picture, time-dependent methods have really only become common in the last two decades, and only recently has a textbook on quantum mechanics been published that focuses on a time-dependent perspective [12]. These methods do, however, have advantages over time-independent ones. They are more intuitive, connecting directly with the motion of the system. They are able to treat continuum states in a more natural way, important in unbound systems. They are able to treat time-dependent Hamiltonians, important for including the effect of a laser pulse. Finally, they also provide a better starting point to approximate solutions.

The MCTDH method is one such approximate method. Its power lies in the fact that it uses a variational principle to derive equations of motion for a flexible wavefunction *Ansatz*. The resulting evolution of the time-dependent basis functions means that the basis set remains optimally small, that is, the wavefunction representation is very compact. Cheap, qualitative calculations are possible with a small number of functions, while increasing the basis set until convergence is reached with respect to the property of interest results in the numerically exact answer.

The present importance of quantum chemistry calculations in supporting general chemistry is in no small part due to the availability of computer programs usable by non-experts. Only a small handful of codes have been written implementing the MCTDH method. The main ones are the code developed by Manthe and co-workers in Bielefeld, the Las Cruces code developed by Wang, and the Heidelberg code developed by Meyer, Worth and co-workers [13]. The Heidelberg code in particular has the aim of being general and user friendly. It is by no means yet possible for a non-expert to run quantum dynamics calculations, but it is now possible without the extensive coding for each new problem traditionally required.

MCTDH is, of course, only one method, if a very successful one, in the field, and this book aims to be of interest to a wider audience than just the MCTDH community. The method does have limitations, and for some calculations other approaches are to be preferred.[1] Many of the ideas developed here and the systems looked at are of general interest, and we hope that some of them will be picked up by other communities.

In Part 1 of the book, the MCTDH theory will be reviewed. The background theory is dealt with briefly in Chapter 2, before the MCTDH method is looked at in detail in Chapter 3, focusing on the special features of the method. The various issues associated with quantum dynamics are then looked at. For example, these include how to set up the initial wavepacket (Chapter 5) and how to analyse the results of a propagation in relation to experiments (Chapter 6). The choice of coordinates for a study plays a large role in how easy a calculation is and what information can be obtained. The subject of coordinates and

1) For example, see the recent review [14] on the calculation of the vibrational energies of polyatomic molecules.

obtaining the kinetic energy operator is addressed in Chapter 12. The efficient integration of the equations of motion (Chapter 4) and evaluation of potential matrix elements (Chapters 10 and 11) are also treated.

The MCTDH method is able to do more than just represent and propagate a nuclear wavefunction. It is also able to include thermal effects and environments using density operators. This is described in Chapters 7 and 22. One can also obtain eigenvalues and eigenvectors of an operator taking advantage of the compact form of the MCTDH wavefunction. This can be done either by propagation in imaginary time (termed *relaxation*) or by iterative diagonalization of the matrix representation of the operator. These methods are dealt with in Chapters 8 and 9. If only the eigenvalues are required, the filter diagonalization method presented in Section 6.2.2 is also possible. This uses the full power of the time-dependent formalism. If the operator is the Hamiltonian, the result is a solution of the time-independent Schrödinger equation. Another operator relevant for molecular dynamics studies is the thermal flux operator required in the calculation of rate constants. This topic is treated further in Chapter 19.

In Part 2, extensions to the basic method are looked at. These are all exciting developments that are moving quantum dynamics into new directions. Despite its power compared to standard grid-based quantum dynamics methods, MCTDH calculations are still unable to treat more than a few atoms explicitly: calculations with more than 20 degrees of freedom quickly become intractable. The multilayer approach of Chapter 14 promises to be able to treat hundreds of particles in the MCTDH framework. Parallelization of the MCTDH algorithm to take advantage of modern computer architectures is also a must (Chapter 15).

A separate bottleneck to treating large systems is the need for a potential energy surface. Obtaining this function also becomes prohibitive for many-atom systems. One approach is to use a model, as done in the vibronic coupling approach of Chapter 18, or the n-mode representation used in Chapter 23. Another is to use *direct dynamics* in which the PES is calculated 'on the fly' by quantum chemistry calculations. This approach means that the PES is only generated where the system goes rather than globally, thus saving a huge effort. Its implementation is described in Chapter 13. Direct dynamics provides restrictions on the evolving basis functions, as the PES is only known locally where it is calculated. A formulation of the MCTDH method, termed G-MCTDH, uses a Gaussian wavepacket basis that has the desired properties (Section 3.5). G-MCTDH also provides a framework to describe mixed methods such as quantum–semiclassical dynamics in which part of the system (the bath) is treated at a lower level of theory to again allow larger calculations.

The original MCTDH method, like the vast majority of nuclear dynamics calculations, does not take into account the symmetry of particle exchange: it is assumed implicitly that all nuclei are distinguishable. The imposition of the correct symmetry for fermions leads to MCTDHF in Chapter 16. The method can now be used to describe electrons, and examples are given in Section 16.5 of the dynamics of these particles after the application of ultrashort, intense laser pulses. The bosonic version, MCTDHB, is discussed in Chapter 17, as is the formalism for the mixed case, MCTDH-BF. The resulting theory here thus makes the MCTDH method complete.

In the final part of the book (Part 3), a number of applications are presented. These demonstrate the generality of the method and highlight systems of interest to quantum dynamics studies. Chapter 18 looks at calculating absorption spectra for polyatomic molecules, treating the non-adiabatic coupling between electronic states. The importance of being able to include enough modes is exemplified by calculations on the allene photoelectron spectrum. By including all 15 modes, the assignment of the vibrational peaks was changed from previous work that had used only four modes.

Chapter 19 looks at calculating reaction rates. A convenient and efficient way to calculate the rate constant directly is offered by the flux correlation approach of Miller and co-workers [15]. Combined with MCTDH, Manthe has been able to calculate accurate rate constants for a range of systems such as $H + CH_4$ in full dimensionality. Discrepancies with experimental data are due to errors in the potential surfaces, and these calculations provide a tough test of these functions.

The topic of surface scattering is covered in Chapter 20, where it is shown that MCTDH is able to treat systems such as CH_4 and CH_3I absorbed onto a solid. The inclusion of all relevant modes is shown to be important for accurate results, as reduced-dimensionality studies can introduce artefacts by preventing certain motions. One approach that is used to include the huge number of modes of the substrate in these calculations is the density-matrix formalism. How to treat general open systems using density matrices within the MCTDH framework is then further detailed in Chapter 22.

The topic of intramolecular vibrational energy redistribution is looked at in Chapter 21. Understanding the flow of energy through a molecular system is a fundamental problem that naturally involves the coupling between many modes. The MCTDH method has been used to obtain detailed results in polyatomic systems such as HFCO, H_2CS and HONO.

Proton transfer, another ubiquitous mechanism in chemistry and biochemistry, is treated in Chapter 23. The results presented include the transfer of a proton along a 'wire', and a full 15-dimensional simulation in which the infrared spectrum of the protonated water dimer – the Zundel cation – is assigned and explained.

The effects of a laser field are studied in Chapter 24, where the topic of quantum control is treated. Here, this is combined with the development of a general Cartesian reaction path Hamiltonian to treat a range of systems, such as laser-driven proton transfer and ladder climbing in CO bound to a haem molecule, and controlling predissociation of a diatomic molecule in a rare-gas matrix.

The process of dissociative electron attachment in the water molecule is looked at in Chapter 25. This introduces the problems of complex potential energy surfaces and multiple product channels, and, despite being only a triatomic system, is a hard numerical problem to solve. The last chapter, Chapter 26, reports calculations on ultracold systems where quantum effects become very important for nuclei. Here, the MCTDH method is being applied to a new area in quantum dynamics, away from the traditional molecular dynamics for which it was conceived.

Part 1 Theory

Multidimensional Quantum Dynamics: MCTDH Theory and Applications.
Edited by Hans-Dieter Meyer, Fabien Gatti, and Graham A. Worth

ISBN: 978-3-527-32018-9

2
The Road to MCTDH

Hans-Dieter Meyer, Fabien Gatti and Graham A. Worth

The development of methods to study molecular dynamics followed a number of different paths. Early calculations mainly used methods based on classical mechanics, representing the evolving wavepacket by a swarm of trajectories. To recover some of the error this approximation makes while retaining the simplicity and ease of interpretation of trajectory based approaches, semiclassical methods were also developed. Of these, probably the most influential were the Gaussian wavepacket methods pioneered by Heller in the 1970s [16]. Not only did this develop the idea of an evolving wavepacket, but some of the machinery used in modern quantum dynamics calculations, such as the calculation of a spectrum, was developed as a result.

The first full quantum dynamics simulation reported was probably the study of McCullough and Wyatt in 1969 on the $H + H_2$ exchange reaction [10, 11]. The introduction of grid-based methods, particularly the fast Fourier transform (FFT) method of Kosloff [17] and the discrete variable representation (DVR) method of Light [18–20] in the 1980s, provided exact simulations with an efficient, accurate machinery for general calculations. This we refer to as the *standard method.* It is now fair to say that, given an accurate potential function, grid-based quantum dynamics simulations can match, if not exceed, the accuracy of experiments on small systems.

The standard method, however, suffers from a natural exponential scaling with the number of degrees of freedom (DOFs) of the system. Even with state-of-the-art computing facilities, it is very hard to go beyond systems with six DOFs. This provides an effective limit of four-atom systems, ignoring the vast majority of systems of interest.

A separate strand of research examined how to obtain approximate solutions to the time-dependent Schrödinger equation by applying a variational principle to a guess form of the wavefunction. McLachlan [21] used a wavefunction *Ansatz* with a simple product form, a Hartree product, with one basis function for each DOF of the system. He then derived equations of motion for the one-dimensional basis functions so that their evolution optimally represented the true time evolution of the wavepacket.

Multidimensional Quantum Dynamics: MCTDH Theory and Applications.
Edited by Hans-Dieter Meyer, Fabien Gatti, and Graham A. Worth

ISBN: 978-3-527-32018-9

This, the time-dependent Hartree (TDH) method, has been extensively used by Gerber and co-workers [22–25]. Unfortunately, while TDH is a conceptually simple theory, the performance of TDH is often very poor. This contrasts strongly with quantum chemistry calculations, as the self-consistent field (SCF) method has been extremely successful in describing molecules. The difference is that the potentials in nuclear dynamics provide strong coupling between the modes. The error, as will be shown below, is due to the lack of correlation between degrees of freedom.

It is a rather obvious idea to try to recover the missing correlation by turning to a multiconfigurational form of TDH. The non-trivial task is to make this idea work. First investigations on the propagation of multiconfigurational wavefunctions were undertaken in 1987 by Makri and Miller [26] and by Kosloff, Hammerich and Ratner [27, 28]. In both cases the wavefunction was written as a sum of Hartree products without coefficients. The resulting equations of motion are very complicated and difficult to solve, and the resulting methods are not competitive. It is also worth mentioning the work of Jackson [29], which, however, is strongly focused on phonon excitation in a gas–surface scattering event. All these approaches, though different, run under the name *multiconfiguration time-dependent self-consistent field* (MC-TDSCF).

In 1990, Meyer, Manthe and Cederbaum [30] introduced a new *Ansatz* for the wavefunction that contains coefficients, the so-called A-vector (see Equation (3.1)). This representation of the wavefunction is redundant: a linear transformation of the Hartree functions can be compensated by a reverse linear transformation of the coefficients. To remove the redundancy, constraints must be introduced that make the equations of motion unique but do not narrow the variational space. The constraints are chosen such that the equations of motion assume a most simple form. The resulting method is called *multiconfiguration time-dependent Hartree* (MCTDH).

In the following two sections, the standard method and TDH are reviewed to provide the background to MCTDH. Both are found to be limits to MCTDH, and thus the different starting points of the two theories help to illuminate MCTDH.

2.1 The Standard Method

The most direct way to solve the time-dependent Schrödinger equation is to expand the wavefunction into a direct-product basis and to solve the resulting equations of motion. An f-dimensional wavefunction is hence expanded as

$$\Psi(q_1, \ldots, q_f, t) = \sum_{j_1=1}^{N_1} \cdots \sum_{j_p=1}^{N_f} C_{j_1,\ldots,j_f}(t)\, \chi_{j_1}^{(1)}(q_1) \cdots \chi_{j_f}^{(f)}(q_f) \tag{2.1}$$

Here N_κ denotes the number of basis functions employed for the κth DOF. For technical reasons one mostly uses a DVR [18–20, 31] and represents the wavefunction on a product grid rather than on a basis, but we may ignore this technical point here.

Plugging this *Ansatz* into the time-dependent Schrödinger equation, or (mathematically more sound) applying the Dirac–Frenkel variational principle [32, 33], yields the equations of motion

$$\mathrm{i}\,\dot{C}_{j_1,\ldots,j_f} = \sum_{\ell_1,\ldots,\ell_f} \langle \chi_{j_1}^{(1)} \cdots \chi_{j_f}^{(f)} | H | \chi_{\ell_1}^{(1)} \cdots \chi_{\ell_f}^{(f)} \rangle \, C_{\ell_1,\ldots,\ell_f} \tag{2.2}$$

The *Ansatz*, Equation (2.1), and the resulting method, Equation (2.2), which may be called the *standard method*, is easy to implement and works well. However, the standard method suffers from a strong exponential increase with the number of DOFs of both the computation time and the memory requirements. If there are $f = 6$ DOFs and $N = 20$ functions per DOF, then the basis sets consist of 64×10^6 functions. The coefficient vector C, which defines the wavefunction, takes 1 GB of memory, and one usually needs to keep three coefficient vectors in memory during the propagation. Such a calculation is feasible on today's computers, but investigations using the standard method are rare for $f > 6$. For larger systems, more approximate methods are unavoidable.

2.2 Time-Dependent Hartree

One of the simplest propagation methods is the time-dependent Hartree (TDH) approach. Here the wavefunction is written as a simple product of one-dimensional functions, which, however, are time dependent:

$$\Psi(q_1, \ldots, q_f, t) = a(t)\, \varphi^{(1)}(q_1, t) \cdots \varphi^{(f)}(q_f, t) \tag{2.3}$$

This representation is not unique. One may multiply one function by a factor and divide another function by the same factor without changing the total wavefunction. We have added the redundant factor $a(t)$ to the *Ansatz* to give the same amount of arbitrariness to all functions. To ensure unique equations of motion, one has to introduce constraints that fix the time evolution of these free factors. Note that the constraints do not narrow the functional space of the TDH *Ansatz*. As constraints we choose

$$\mathrm{i}\langle \varphi^{(\kappa)} | \dot{\varphi}^{(\kappa)} \rangle = g^{(\kappa)}(t) \tag{2.4}$$

for $\kappa = 1, \ldots, f$ and – for the time being – arbitrary constraints $g^{(\kappa)}$. By a convenient choice of the constraints, one may finally bring the equations of motion to the most suitable form. Differentiating $\|\varphi^{(\kappa)}\|^2$ with respect to time yields $2\,\mathrm{Re}\langle\varphi^{(\kappa)}|\dot{\varphi}^{(\kappa)}\rangle$, which, according to Equation (2.4), equals $-2\,\mathrm{Im}\,g^{(\kappa)}$. Hence, the norm of the functions $\varphi^{(\kappa)}$ is conserved if the constraints $g^{(\kappa)}$ are real. In the following we will assume that the constraints are real and the functions normalized.

Using the Dirac–Frenkel variational principle, $\langle\delta\Psi|H - \mathrm{i}\partial/\partial t|\Psi\rangle = 0$, to derive the equations of motion yields [34]

$$\mathrm{i}\dot{a} = a\left(E - \sum_{\kappa=1}^{f} g^{(\kappa)}\right) \tag{2.5}$$

$$\mathrm{i}\dot{\varphi}^{(\kappa)} = \mathcal{H}^{(\kappa)}\varphi^{(\kappa)} + (g^{(\kappa)} - E)\varphi^{(\kappa)} \tag{2.6}$$

where

$$\mathcal{H}^{(\kappa)} = \langle\varphi^{(1)}\cdots\varphi^{(\kappa-1)}\varphi^{(\kappa+1)}\cdots\varphi^{(f)}|H|\varphi^{(1)}\cdots\varphi^{(\kappa-1)}\varphi^{(\kappa+1)}\cdots\varphi^{(f)}\rangle \tag{2.7}$$

denotes a mean field that operates on the κth DOF only. Here E is the expectation value of the Hamiltonian

$$E = \frac{\langle\Psi|H|\Psi\rangle}{\langle\Psi|\Psi\rangle} = \langle\varphi^{(1)}\cdots\varphi^{(f)}|H|\varphi^{(1)}\cdots\varphi^{(f)}\rangle \tag{2.8}$$

Note that E is real and time independent for Hermitian Hamiltonians but may become time dependent and complex for non-Hermitian ones.

There are several convenient choices for the constraints. The two most obvious ones are $g^{(\kappa)} = E$ and $g^{(\kappa)} = 0$. The former choice is possible only for Hermitian Hamiltonians and the latter yields

$$a(t) = a(0)\exp\left(-\mathrm{i}\int_0^t E(t')\,dt'\right) \tag{2.9}$$

$$\begin{aligned}\mathrm{i}\dot{\varphi}^{(\kappa)} &= [\mathcal{H}^{(\kappa)} - E(t)]\varphi^{(\kappa)} \\ &= (1 - |\varphi^{(\kappa)}\rangle\langle\varphi^{(\kappa)}|)\mathcal{H}^{(\kappa)}\varphi^{(\kappa)}\end{aligned} \tag{2.10}$$

The second form of the last equation holds because $\langle\varphi^{(\kappa)}|\mathcal{H}^{(\kappa)}|\varphi^{(\kappa)}\rangle = E$. The TDH approximation turns the solution of an f-dimensional differential equation into the solution of f one-dimensional differential equations. This is, of course, an enormous simplification.

The TDH solution can be interpreted as the exact solution of an effective Hamiltonian H_{eff}, that is, $\mathrm{i}\dot{\Psi} = H_{\mathrm{eff}}\Psi$. A short calculation yields

$$H_{\mathrm{eff}} = \sum_{\kappa=1}^{f} \mathcal{H}^{(\kappa)} - (f-1)E \tag{2.11}$$

Let us split the Hamiltonian into separable and non-separable parts,

$$H = \sum_{\kappa=1}^{f} h^{(\kappa)} + V \tag{2.12}$$

where $h^{(\kappa)}$ operates on the κth DOF only. The effective Hamiltonian may now be written as

$$H_{\text{eff}} = \sum_{\kappa=1}^{f} (h^{(\kappa)} + \mathcal{V}^{(\kappa)}) - (f-1)E_{\text{corr}} \tag{2.13}$$

where $\mathcal{V}^{(\kappa)}$ denotes the mean field of V and $E_{\text{corr}} = \langle\Psi|V|\Psi\rangle / \langle\Psi|\Psi\rangle$. Now it is clear that TDH is exact if the Hamiltonian is separable, that is, $H_{\text{eff}} = H$ if $V = 0$.

Next we assume that the non-separable potential V is of product form, that is,

$$V(q_1, \ldots, q_f) = v_1(q_1) \cdots v_f(q_f) \tag{2.14}$$

This form is convenient, as it readily yields the mean field

$$\mathcal{V}^{(\kappa)} = \langle v_1 \rangle \cdots \langle v_{\kappa-1} \rangle \, v_\kappa(q_\kappa) \, \langle v_{\kappa+1} \rangle \cdots \langle v_f \rangle \tag{2.15}$$

where $\langle v_\kappa \rangle = \langle \varphi^{(\kappa)} | v_\kappa | \varphi^{(\kappa)} \rangle$. The product form is, on the other hand, no restriction, because every potential can be written as a sum of products (see Chapter 11). To simplify the calculation further, we assume that $f = 2$ and obtain for the difference between the exact and effective Hamiltonians [34]

$$H - H_{\text{eff}} = (v_1 - \langle v_1 \rangle)\,(v_2 - \langle v_2 \rangle) \tag{2.16}$$

This is a simple but quite illuminating result. TDH will be accurate if the potential terms v_κ do not vary appreciably over the width of the wavepacket. TDH becomes exact in the classical limit when the wavepacket degenerates to a delta function. As a time-dependent wavefunction, a wavepacket, is usually more localized than an eigenfunction, TDH provides in general better results than time-independent Hartree, which is also known as vibrational SCF (VSCF) [35,36]. Moreover, TDH (and similarly VSCF) works reasonably well if the non-separable part of the potential varies slowly and smoothly. TDH and VSCF may fail if there are abrupt changes in the non-separable potential, for example, hard repulsions.

3
Basic MCTDH Theory

Hans-Dieter Meyer, Fabien Gatti and Graham A. Worth

The MCTDH algorithm was introduced in 1990 by Meyer, Manthe and Cederbaum [30]. A first comprehensive description of the method – together with the first non-trivial application (the photodissociation of NOCl) – appeared two years later [37]. The basic theory of MCTDH has been discussed in great detail in two review articles [34, 38]. Hence, in the following, only a brief overview of MCTDH theory is given. The important features that give the method its power are highlighted, as are the various formulations that provide the flexibility required to treat different situations. Further details and derivations are to be found in the review articles [34, 38].

3.1
Wavefunction *Ansatz* and Equations of Motion

The basis of the MCTDH method is the use of the following wavefunction *Ansatz* to solve the time-dependent Schrödinger equation for a physical system with f degrees of freedom (DOFs) described by coordinates $q_1, \ldots, q_f$:

$$\begin{aligned}\Psi(q_1, \ldots, q_f, t) &= \Psi(Q_1, \ldots, Q_p, t) \\ &= \sum_{j_1=1}^{n_1} \cdots \sum_{j_p=1}^{n_p} A_{j_1 \ldots j_p}(t) \varphi_{j_1}^{(1)}(Q_1, t) \cdots \varphi_{j_p}^{(p)}(Q_p, t) \qquad (3.1) \\ &= \sum_J A_J \Phi_J \qquad (3.2)\end{aligned}$$

Equation (3.1) is a direct-product expansion of p sets of orthonormal time-dependent basis functions $\{\varphi^{(\kappa)}\}$, known as *single-particle functions* (SPFs). The coordinate for each set of n_κ functions is a composite coordinate of one or more system coordinates

$$Q_\kappa = (q_a, q_b, \ldots) \qquad (3.3)$$

Thus the basis functions are d-dimensional, where d is the number of system coordinates that have been combined together and treated as one 'particle'.

Multidimensional Quantum Dynamics: MCTDH Theory and Applications.
Edited by Hans-Dieter Meyer, Fabien Gatti, and Graham A. Worth

ISBN: 978-3-527-32018-9

(Typically $d = 1, 2$, or 3). The second line, Equation (3.2), defines the composite index $J = j_1 \ldots j_p$ and the *Hartree product* Φ_J. The *Ansatz* looks similar to the standard wavepacket expansion, Equation (2.1) [39–41], except that the SPFs provide a time-dependent basis set.

Using this *Ansatz*, a variational solution to the time-dependent Schrödinger equation is provided by a coupled set of equations, one for the expansion coefficients

$$\mathrm{i}\dot{\mathbf{A}} = \boldsymbol{\mathcal{K}}\mathbf{A} \tag{3.4}$$

and one for each set of SPFs

$$\mathrm{i}\dot{\boldsymbol{\varphi}}^{(\kappa)} = (1 - P^{(\kappa)})(\boldsymbol{\rho}^{(\kappa)})^{-1}\boldsymbol{\mathcal{H}}^{(\kappa)}\boldsymbol{\varphi}^{(\kappa)} \tag{3.5}$$

A matrix notation has been used with the A coefficients and SPFs written as vectors, that is, $\boldsymbol{\varphi}^{(\kappa)} = (\varphi_1^{(\kappa)}, \cdots, \varphi_{n_\kappa}^{(\kappa)})^{\mathrm{T}}$.

The matrix $\boldsymbol{\mathcal{K}}$ is the Hamiltonian operator represented in the basis of Hartree products

$$\mathcal{K}_{JL} = \langle \Phi_J | H | \Phi_L \rangle \tag{3.6}$$

Thus Equation (3.4) has the same form as the equations of motion for standard wavepacket propagation, Equation (2.2). The difference is that the Hamiltonian matrix is time dependent due to the time dependence of the SPFs.

The equations of motion for the SPFs contain three new entities. The first is the projector onto the space spanned by the SPFs:

$$P^{(\kappa)} = \sum_j |\varphi_j^{(\kappa)}\rangle \langle \varphi_j^{(\kappa)}| \tag{3.7}$$

The operator $(1 - P^{(\kappa)})$ ensures that the time derivative of the SPFs is orthogonal to the space spanned by the functions. Thus any changes cover new regions. When the basis set is complete, the SPFs become time independent and the equations of motion are identical to the standard method. If the SPFs do not provide a complete basis set, then they move so as to provide the best possible basis for the description of the evolving wavepacket. This optimal description is ensured by the variational method used for the derivation.

For the other two new entities, it is useful to introduce the *single-hole function*, $\Psi_a^{(\kappa)}$, which is the wavefunction associated with the jth SPF of the κth particle. As the total wavefunction lies in the space spanned by the SPFs, one can make use of the completeness relation and write

$$\Psi = \sum_a |\varphi_a^{(\kappa)}\rangle \langle \varphi_a^{(\kappa)} | \Psi \rangle = \sum_a \varphi_a^{(\kappa)} \Psi_a^{(\kappa)} \tag{3.8}$$

To make this clear, the single-hole function for the first particle is

$$\Psi_a^{(1)} = \sum_{j_2=1}^{n_1} \cdots \sum_{j_p=1}^{n_p} A_{aj_2 \ldots j_p} \varphi_{j_2}^{(2)} \ldots \varphi_{j_p}^{(p)} \tag{3.9}$$

The *single-hole index*, J_a^κ, is also useful to keep the notation compact. The index can take any values except for the κth position, which has a value a. In the same way, J^κ is a composite index similar to J but with the κth entry removed. Thus there is the single-hole coefficient $A_{J_a^\kappa}$ and single-hole Hartree product Φ_{J^κ}, which denotes a configuration with the κth SPF removed. These definitions allow the single-hole function to be written as

$$\Psi_a^{(\kappa)} = \sum_{J^\kappa} A_{J_a^\kappa} \Phi_{J^\kappa} \tag{3.10}$$

Using this new notation, the *mean-field operator matrix*, $\boldsymbol{\mathcal{H}}^{(\kappa)}$, can be easily written as

$$\mathcal{H}_{ab}^{(\kappa)} = \langle \Psi_a^{(\kappa)} | H | \Psi_b^{(\kappa)} \rangle \tag{3.11}$$

The integration in the brackets is over all particles except κ. This operator on the κth particle correlates the motion between the different sets of SPFs.

Finally, the *density matrix* $\boldsymbol{\rho}^{(\kappa)}$ is given by

$$\rho_{ab}^{(\kappa)} = \langle \Psi_a^{(\kappa)} | \Psi_b^{(\kappa)} \rangle \tag{3.12}$$

$$= \sum_{J^\kappa} A_{J_a^\kappa}^* A_{J_b^\kappa} \tag{3.13}$$

The density matrices, which enter the equations of motion for the SPFs, Equation (3.5), as its inverse, can be used to provide a useful measure of the quality of the calculation. In an analogous way to the use of density matrices in electronic structure theory, the eigenfunctions of this matrix are termed *natural orbitals* and the eigenvalues provide populations for these functions. The lower the population, the less important is the function. As the space spanned by the natural orbitals is equivalent to that of the original SPFs, if the population of the highest natural orbital is such that the function is effectively not required for an accurate description of the evolving wavepacket, the MCTDH wavefunction is of a good quality. As a rule of thumb, averaged quantities such as expectation values and spectra are converged when the highest natural orbitals have a population less than 10^{-3}. Other quantities such as cross-sections are more sensitive to errors in the wavefunction and the populations have to drop below 10^{-6} for converged results.

Although small natural populations are desired because they signal convergence, vanishing natural populations are problematic as they lead to a singular density matrix, which cannot be inverted. Zero natural populations frequently occur at the beginning of a propagation, because one often takes a

Hartree product as initial wavefunction. In this case all eigenvalues of $\boldsymbol{\rho}^{(\kappa)}$ vanish except the first one. To arrive at equations of motion that can be tackled numerically, one must regularize the density matrix. To this end we replace $\boldsymbol{\rho}^{(\kappa)}$ by

$$\boldsymbol{\rho}_{\text{reg}}^{(\kappa)} = \boldsymbol{\rho}^{(\kappa)} + \varepsilon \exp(-\boldsymbol{\rho}^{(\kappa)}/\varepsilon) \tag{3.14}$$

when building the inverse. Here ε is a small number ($\varepsilon \approx 10^{-8}$).

Note that the regularization changes only the time evolution of those natural orbitals that are very weakly populated. The time evolution of the natural orbitals important for the description of the wavefunction remains unchanged.

The MCTDH equations conserve the norm and, for time-independent Hamiltonians, the total energy. MCTDH simplifies to time-dependent Hartree when setting all $n_\kappa = 1$. Increasing the n_κ recovers more and more correlation, until finally, when n_κ equals the number of primitive functions N_κ, the standard method (that is, propagating the wavepacket on the primitive basis) is used. Hence with MCTDH one can almost continuously switch from a cheap but less accurate calculation to a highly accurate but expensive one. We emphasize again that MCTDH uses variationally optimal SPFs, because this ensures early convergence.

3.2 The Constraint Operator

In MCTDH both the expansion coefficients (A-vector) and the SPFs are time dependent, which creates an ambiguity in the time propagation of the wavefunction. The same change of the wavefunction may be accomplished by changing the coefficients or the SPFs. To lift this ambiguity, a constraint is added when deriving the equations of motion. It reads

$$\mathrm{i}\langle \varphi_i^{(\kappa)} | \dot{\varphi}_j^{(\kappa)} \rangle = \langle \varphi_i^{(\kappa)} | \hat{g}^{(\kappa)} | \varphi_j^{(\kappa)} \rangle \tag{3.15}$$

where $\hat{g}^{(\kappa)}$ is an arbitrary operator acting on the κth particle. The constraint operators are related to the time derivative of the overlap matrix,

$$S_{ij}^{(\kappa)} = \langle \varphi_i^{(\kappa)} | \varphi_j^{(\kappa)} \rangle \tag{3.16}$$

$$\dot{\mathbf{S}}^{(\kappa)} = \mathrm{i}[(\mathbf{g}^{(\kappa)})^\dagger - \mathbf{g}^{(\kappa)}] \tag{3.17}$$

where the dagger denotes the adjoint matrix and where $\mathbf{g}^{(\kappa)}$ is the matrix representation of $\hat{g}^{(\kappa)}$, that is, $g_{ij}^{(\kappa)} = \langle \varphi_i^{(\kappa)} | \hat{g}^{(\kappa)} | \varphi_j^{(\kappa)} \rangle$. Thus, if the constraint operators are Hermitian, initially orthonormal SPFs will remain orthonormal, a desirable feature. We hence restrict ourselves to the use of Hermitian constraint operators.

The MCTDH equations of motion in their full generality now read

$$i\dot{\mathbf{A}} = \boldsymbol{\mathcal{K}}_R \mathbf{A} \tag{3.18}$$

$$i\dot{\boldsymbol{\varphi}}^{(\kappa)} = [g^{(\kappa)}\mathbf{1}_{n_\kappa} + (1 - P^{(\kappa)})(\boldsymbol{\rho}^{(\kappa)})^{-1}\boldsymbol{\mathcal{H}}_R{}^{(\kappa)}]\boldsymbol{\varphi}^{(\kappa)} \tag{3.19}$$

where $\mathbf{1}_{n_\kappa}$ denotes the $n_\kappa \times n_\kappa$ unit matrix. The Hamiltonian and mean-field matrices, $\boldsymbol{\mathcal{K}}_R$ and $\boldsymbol{\mathcal{H}}_R^{(\kappa)}$, are formed using the operator

$$H_R = H - \sum_\kappa \hat{g}^{(\kappa)} \tag{3.20}$$

Whatever is chosen for the constraint operators, this will not affect the propagation of the total wavepacket, but will affect the propagation of the two parts of the equations of motion. Different choices lead to different sets of SPFs, connected by unitary transformations (the equations of motion for the SPFs differ by a Hermitian operator). The SPFs thus span the same space, irrespective of the constraint used. The previous equations of motion, Equations (3.4) and (3.5), are recovered when setting $\hat{g}^{(\kappa)} = 0$.

Finally, one may write the equations of motion in a form where only the matrix representations of the constraint operators appear:

$$i\dot{\mathbf{A}} = \left(\boldsymbol{\mathcal{K}} - \sum_\kappa \mathbf{g}^{(\kappa)}\right)\mathbf{A} \tag{3.21}$$

$$i\dot{\boldsymbol{\varphi}}^{(\kappa)} = [(\mathbf{g}^{(\kappa)})^{\mathrm{T}} + (1 - P^{(\kappa)})(\boldsymbol{\rho}^{(\kappa)})^{-1}\boldsymbol{\mathcal{H}}^{(\kappa)}]\boldsymbol{\varphi}^{(\kappa)} \tag{3.22}$$

with $\mathbf{g}^{\mathrm{T}}$ denoting the transpose of the matrix. This form makes it particularly transparent that the constraint determines the partitioning of the time derivative of the wavefunction into $\dot{A}$ and $\dot{\boldsymbol{\varphi}}$.

Various choices of the constraints have been tried. The most obvious is

$$\hat{g}^{(\kappa)} = 0 \tag{3.23}$$

which, as already noted, results in the equations of motion (3.4) and (3.5) discussed above. A second obvious choice is to partition the Hamiltonian into 'uncorrelated' parts, that act only on one particle, and the remaining 'correlated' part, that is,

$$H = \sum_{\kappa=1}^{p} h^{(\kappa)} + H_R \tag{3.24}$$

and choose

$$\hat{g}^{(\kappa)} = h^{(\kappa)} \tag{3.25}$$

This significantly changes the motion. The first choice minimizes the motion of the SPFs – only motion into space outside that spanned by the basis occurs. The latter choice treats the uncorrelated motion within the SPF space by rotation of these functions.

A further choice that has been tried is

$$g_{ij}^{(\kappa)} = \frac{\langle \Psi | H | \Psi_i^{(\kappa)} \varphi_j^{(\kappa)} \rangle - \langle \Psi_i^{(\kappa)} \varphi_j^{(\kappa)} | H | \Psi \rangle}{\rho_{ii}^{(\kappa)} - \rho_{jj}^{(\kappa)}} \quad \text{for } i \neq j \tag{3.26}$$

The resulting SPFs have the property that the density matrices defined in Equation (3.13) remain diagonal, that is, they are the 'natural orbitals'.

Experience has shown that the best choice depends on the integration scheme used for the propagation. Equation (3.23) is the best choice when using the efficient constant mean-field (CMF) integrator, while Equation (3.25) is the one to use otherwise. The use of natural orbitals was found to be numerically less efficient. However, the direct connection between the natural orbitals and populations of SPFs means that they can be used in a scheme to select and propagate only the most important configurations from the MCTDH wavefunction [42]. This attractive proposition, which aims to break the exponential scaling, has unfortunately failed to be efficient enough due to the book-keeping required when the full direct-product structure of the wavefunction is lost.

3.3 Efficiency and Memory Requirements

Standard wavepacket dynamics uses a wavefunction *Ansatz* like that of Equation (3.1), except with a set of time-independent basis functions for each DOF rather than a set of time-dependent functions for each particle. While the number of basis functions may vary for each DOF, if N is representative of this number, then the wavefunction is represented by N^f expansion coefficients. This is the basis of the exponential increase of computer resources with system size that plagues wavepacket dynamics. As $N \sim 50$ is reasonable, a four-dimensional system using double-precision complex arithmetic requires nearly 100 MB of memory just to store one wavefunction, while a five-dimensional system requires of the order of 4.8 GB. Clearly this scaling severely limits the size of systems treatable by these methods. We shall now analyse the requirements of the MCTDH method to show where its power, as well as its limitations, come from.

To solve the equations of motion for the A coefficients and SPFs, Equations (3.4) and (3.5), the elements of the Hamiltonian matrix, $\mathcal{K}$, need to be evaluated:

$$\langle \varphi_{j1}^{(1)} \cdots \varphi_{jp}^{(p)} | H | \varphi_{k_1}^{(1)} \cdots \varphi_{k_p}^{(p)} \rangle = \langle \varphi_{j1}^{(1)} \cdots \varphi_{jp}^{(p)} | T + V | \varphi_{k_1}^{(1)} \cdots \varphi_{k_p}^{(p)} \rangle \tag{3.27}$$

Elements of the mean-field matrices are also required, $\mathcal{H}^{(\kappa)}$, and the techniques described below can be used for these too.

If the basis functions are a *discrete variable representation* (DVR), this multidimensional integral would be straightforward. A set of DVR functions along a coordinate q_ν, $\{\chi^{(\nu)}(q_\nu)\}$, has the property that their matrix representation of the position operator, $\hat{q}_\nu$, is diagonal, that is,

$$\langle \chi_i^{(\nu)} | q_\nu | \chi_j^{(\nu)} \rangle = q_j^{(\nu)} \delta_{ij} \tag{3.28}$$

and the values q_ν provide a grid of points related to the DVR functions. As a result, if there are enough functions for the set to be effectively complete, the potential energy operator can be taken as diagonal in this basis,

$$\langle \chi_{i_1}^{(1)} \cdots \chi_{i_f}^{(f)} | V | \chi_{j1}^{(1)} \cdots \chi_{j_f}^{(f)} \rangle = V(q_{j_1}^{(1)}, \ldots, q_{j_f}^{(f)}) \, \delta_{i_1 j_1} \cdots \delta_{i_f j_f} \tag{3.29}$$

and the integral is obtained by evaluating the potential energy only at the grid point $q_{j_1}^{(1)}, \ldots, q_{j_f}^{(f)}$.

The kinetic energy operator usually only acts on a single coordinate, and matrix elements can be evaluated analytically in the related finite basis representation (FBR). The FBR–DVR transformation is then used to give $\{\phi^{(\nu)}\}$,

$$\langle \chi_i^{(\nu)} | T_\nu | \chi_j^{(\nu)} \rangle = \sum_{kl} U_{ik}^{(\nu)} \langle \phi_k^{(\nu)} | T_\nu | \phi_l^{(\nu)} \rangle (U_{lj}^{(\nu)})^\dagger \tag{3.30}$$

Thus the potential energy is obtained by evaluating the potential function at N^f points, and the kinetic energy by transforming N^2 matrices. At no time is it necessary to evaluate multidimensional integrals, and the full $N^f \times N^f$ Hamiltonian matrix does not need to be built.

DVRs are the time-independent bases used in standard wavepacket propagation calculations. A number of different DVRs have been developed, suitable for use for different types of coordinates. Examples are the harmonic oscillator DVR used for vibrational motion, Legendre DVR for rotations, and exponential and sine DVRs used for free motion with or without periodic boundary conditions. A related method is to use a collocation grid and fast Fourier transform (FFT) methods to evaluate the kinetic energy operator. An overview of the properties of different DVRs and FFT methods is given in Appendix B of Ref. [34].

Unfortunately, the evaluation of integrals in the SPF basis is not so straightforward. One way is to expand the SPFs using sets of DVR functions, which are then termed *primitive basis functions*,

$$\varphi_j^{(\kappa)}(Q_\kappa) = \sum_{k=1}^{N_\kappa} a_{kj}^{(\kappa)} \chi_k^{(\kappa)}(Q_\kappa) \tag{3.31}$$

In most cases DVRs are one-dimensional, and a multidimensional SPF is represented by a direct product of primitive basis functions, and the coefficients a and functions χ in Equation (3.31) must be interpreted in this way. For example, if the first particle, $\kappa = 1$, is a combination of DOFs q_1 and q_2, then an SPF is a two-dimensional function represented by the product

$$\varphi_j^{(1)}(Q_1) = \sum_{k_1=1}^{N_1} \sum_{k_2=1}^{N_2} a_{k_1 k_2 j}^{(1)} \chi_{k_1}^{(1)}(q_1) \chi_{k_2}^{(2)}(q_2) \tag{3.32}$$

The potential energy matrix elements can now be obtained by transforming from the SPF basis to the DVR. Using Equation (3.31) and the DVR potential energy (3.29), this is

$$\begin{aligned} &\langle \varphi_{i_1}^{(1)} \cdots \varphi_{i_p}^{(p)} | V | \varphi_{j_1}^{(1)} \cdots \varphi_{j_p}^{(p)} \rangle = \\ &\sum_{k_1 \ldots k_f} a_{k_1 \ldots k_{d_1}, i_1}^{(1)*} \cdots a_{k_{f-d_p+1} \ldots k_f, i_p}^{(p)*} a_{k_1 \ldots k_{d_1}, j_1}^{(1)} \cdots a_{k_{f-d_p+1} \ldots k_f, j_p}^{(p)} V(q_{k_1}^{(1)}, \ldots, q_{k_f}^{(f)}) \end{aligned} \tag{3.33}$$

where $q_{k_\nu}^{(\nu)}$ denotes the kth grid point of the νth primitive grid. The DVR is now being used to evaluate the multidimensional integral, which is equivalent to using a quadrature procedure. The kinetic energy can also be evaluated by an analogous transformation of the FBR representation on Equation (3.30).

While Equation (3.33) is completely general, it is unsuitable for our requirements, as it requires a transformation from the SPF basis to the full direct-product primitive grid. This is precisely what the MCTDH method sets out to avoid, as the full primitive grid for multidimensional systems has the dimensions of the standard wavepacket wavefunction discussed above. The advantages of the DVR can be used without the crippling scaling if the Hamiltonian is made up of products of functions with the same coordinates as the particles of the MCTDH wavefunction:

$$H(q_1, \ldots, q_f) = \sum_{r=1}^{n_s} c_r h_r^{(1)}(Q_1) \cdots h_r^{(p)}(Q_p) \tag{3.34}$$

The multidimensional integrals of Equation (3.33) are then reduced to products of low-dimensional integrals,

$$\langle \varphi_{j_1}^{(1)} \cdots \varphi_{j_p}^{(p)} | H | \varphi_{k_1}^{(1)} \cdots \varphi_{k_p}^{(p)} \rangle$$

$$= \sum_{r=1}^{n_s} c_r \langle \varphi_{j_1}^{(1)} | h^{(1)} | \varphi_{k_1}^{(1)} \rangle \cdots \langle \varphi_{j_p}^{(p)} | h^{(p)} | \varphi_{k_p}^{(p)} \rangle \qquad (3.35)$$

and these low-dimensional integrals can be easily evaluated using the particle primitive grids, which have the dimension N^d, where d is the dimensionality of the particle.

While the kinetic energy operator usually has the product form of Equation (3.34), often referred to as the MCTDH form, this is not true for potential energy functions. A notable exception is the vibronic coupling model Hamiltonian discussed in Chapter 18. For other functions, the potfit method has been developed to obtain the desired form. This is dealt with in Chapter 11. An alternative approach to using general potential functions is to use the SPFs to define a time-dependent DVR that follows the evolving wavepacket. This, the CDVR method, is detailed in Chapter 10.

We are now in a position to discuss the resources required for the MCTDH method. In comparison with the N^f numbers required to describe the standard wavepacket, the MCTDH wavefunction requires

$$\text{memory} \sim n^p + pnN^d \qquad (3.36)$$

where n is characteristic of the number of SPFs for the particles. The first term is due to the A coefficients and the second term is due to the representation of the SPFs through the primitive basis functions.

There are two limits to be examined. The first is when $p = f$ and $d = 1$, that is, all particles are one-dimensional. Here the first term dominates. Using reasonable values of $N = 50$ and $n = 10$, then for $f = 4$ the MCTDH wavefunction requires a tiny 0.18 MB and for $f = 5$ still only 1.56 MB. This is obviously much less than the memory required to store the full primitive grid. The method still hits the exponential wall , however, and 153 GB is needed for each wavefunction if $f = 10$. The other limit to be studied is when all DOFs are combined together so that only one particle is present. Thus $p = 1$ and $d = f$. In this limit, $n = 1$ and the first term is always 1. The second term then dominates and, of course, is simply the size of the full primitive grid, N^f, as in this limit the MCTDH method is identical to the standard wavepacket method. A single wavefunction now takes 1.5×10^9 GB. In between these two limits there is a trade-off between the memory required by the A coefficients and that required by the SPFs. Thus if two-dimensional particles are used in a 10-dimensional calculation, $f = 10$, $p = 5$, $d = 2$ and the memory required is 3.4 MB per wavefunction evenly distributed between the two parts.

The figures used above assume that $n = 10$ is a suitable figure regardless of how many DOFs are combined together into each particle. This is, of course, not the case. Imagine that n SPFs are required for each particle in a problem where $p = f$, that is, all particles are one-dimensional. If a second calculation is then made using two-dimensional particles, that is, $p = f/2$, then $\tilde{n}$, the number of SPFs required in the new calculation will be different from n, but $\tilde{n} < n^2$ ($\tilde{n} = dn$ is a reasonable rule of thumb). The upper limit is because correlations between these modes are now included at the SPF level. For large combinations, $\tilde{n} < n$ is possible, as in the limit that all DOFs are combined together $\tilde{n} = 1$: only a single SPF – the exact wavefunction – is required. When choosing which DOFs should be combined together, it is thus useful to put strongly correlated modes in one particle, as this significantly reduces the number of SPFs, and thus configurations, required. If the amount of correlation among the DOFs is not known, one should combine DOFs that are characterized by similar vibrational frequencies. One must be mindful, however, that the particle grid lengths do not get too long. For a balanced calculation, particles should be chosen with similar grid lengths.

The effort for the algorithm can be estimated by a sum of two terms:

$$\text{effort} \sim c_1 s p^2 n^{p+1} + c_2 s p n N^{2d} \tag{3.37}$$

where c_1 and c_2 are constants of proportionality. The first term is due to building the mean-field matrices and calculating the time derivative of the A coefficients. To build the mean fields, there are s terms in the Hamiltonian, and for each particle the A coefficient vector must be multiplied by the Hamiltonian matrices for all the other particles. The time derivative of the A coefficients is obtained at the end of these operations for virtually no cost. The second term is due to the operation of the Hamiltonian on the SPFs, that is, the operation of the s particle operators, represented in the particle primitive grids, on each SPF for each particle (for potential terms this becomes $spnN^d$, as the operator is diagonal in the primitive basis). The density matrices also need to be inverted, but this effort, which scales as n^3, is insignificant compared to these two terms.

Thus if p is large, the effort for the algorithm is dominated by the building of the mean-field matrices. If p is small and d is large, the second term, that for the propagation of the SPFs, dominates due to the high dimensionality of the functions. Again, we see the trade-off between the effort required for the coefficients and the SPFs, which can be altered by suitably combining DOFs together into particles, balancing the reduced effort due to low p with increased effort due to increasing N^d.

A final aspect of the MCTDH algorithm that affects its ability to efficiently solve the time-dependent Schrödinger equation is the ease of integration of the equations of motion. Wavepacket dynamics are an initial-value problem.

Starting from the wavepacket at $t = 0$ it is propagated forward in time by integrating the equations of motion, which are written above as derivatives of time. If the derivatives are smooth functions of time, then large time steps can be taken. Unfortunately, the MCTDH equations of motion are strongly coupled. All the sets of SPFs depend on each other, on the A coefficients through the mean fields, and the A coefficients depend on the SPFs through the Hamiltonian matrix, $\mathcal{K}$. The problem of integration will be addressed in Chapter 4, where the constant mean-field integrator that is tailored to the properties of the method will be described.

3.4 Multistate Calculations

Two different approaches have been used to treat systems in which more than one electronic state is involved. The first of these simply uses the equations of motion as written above, but with an extra DOF added to represent the electronic degree of freedom:

$$\Psi(q_1,\ldots,q_f,\alpha,t) = \sum_{j_1=1}^{n_1} \cdots \sum_{j_p=1}^{n_p} A_{j_1\ldots j_p}(t)\, \varphi_{j_1}^{(1)}(Q_1,t) \cdots \varphi_{j_{p-1}}^{(p-1)}(Q_{p-1},t)\varphi_{j_p}^{(p)}(\alpha,t) \quad (3.38)$$

where α labels the electronic state. As a complete set of electronic SPFs is used in general, that is, $n_p = \sigma$, where σ denotes the number of electronic states, the SPFs are time independent and chosen as $\varphi_{j_p}^{(p)}(\alpha,t) = \delta_{\alpha,j_p}$. Introducing electronic state functions $|\alpha\rangle$, one may rewrite the above equation as

$$\Psi = \sum_{j_1=1}^{n_1} \cdots \sum_{j_{p-1}=1}^{n_{p-1}} \sum_{\alpha=1}^{\sigma} A_{j_1\ldots j_{p-1},\alpha}\, \varphi_{j_1}^{(1)} \cdots \varphi_{j_{p-1}}^{(p-1)} |\alpha\rangle \quad (3.39)$$

This is called the *single-set* formulation, as one set of SPFs is used to treat the dynamics in all the electronic states.

In contrast, the *multi-set* formulation uses a different set of SPFs for each state. One writes the wavefunction as

$$\Psi = \sum_{\alpha=1}^{\sigma} \Psi^{(\alpha)} |\alpha\rangle \quad (3.40)$$

where each component function $\Psi^{(\alpha)}$ is expanded in MCTDH form as

$$\Psi^{(\alpha)}(q_1,\ldots,q_f,t) = \sum_{j_1^\alpha=1}^{n_1^\alpha} \cdots \sum_{j_p^\alpha=1}^{n_p^\alpha} A_{j_1^\alpha\ldots j_p^\alpha}^{(\alpha)}(t)\, \varphi_{j_1^\alpha}^{(1,\alpha)}(Q_1,t) \cdots \varphi_{j_p^\alpha}^{(p,\alpha)}(Q_p,t) \quad (3.41)$$

Note that different numbers of SPFs can be used for the different states, signified by the superscript α.

The equations of motion also now require state labels. The basic formulation selecting $\mathbf{g}^{(\kappa)} = 0$ for the constraint matrices is now

$$\mathrm{i}\dot{\mathbf{A}}^{(\alpha)} = \sum_{\beta=1}^{\sigma} \boldsymbol{\mathcal{K}}^{(\alpha\beta)}\mathbf{A}^{(\beta)} \tag{3.42}$$

$$\mathrm{i}\dot{\boldsymbol{\varphi}}^{(\kappa,\alpha)} = (1 - P^{(\kappa,\alpha)})(\boldsymbol{\rho}^{(\kappa,\alpha)})^{-1} \sum_{\beta=1}^{\sigma} \boldsymbol{\mathcal{H}}^{(\kappa,\alpha\beta)}\boldsymbol{\varphi}^{(\kappa,\beta)} \tag{3.43}$$

where the superscripts on the matrices denote that the matrix elements are with superscripted A coefficients and SPFs. Thus the particle Hamiltonian matrices used to build up the Hamiltonian matrix and mean-field operators are

$$\mathcal{K}_{JL}^{(\alpha\beta)} = \langle \Phi_J^{(\alpha)} | H^{(\alpha,\beta)} | \Phi_L^{(\beta)} \rangle \tag{3.44}$$

and

$$\mathcal{H}_{ab}^{(\kappa,\alpha\beta)} = \langle \Psi_a^{(\kappa,\alpha)} | H^{(\alpha,\beta)} | \Psi_b^{(\kappa,\beta)} \rangle \quad . \tag{3.45}$$

where $H^{(\alpha,\beta)} = \langle \alpha | H | \beta \rangle$ denotes the (α, β) electronic component of the Hamiltonian. If $\alpha \neq \beta$ the matrices $\boldsymbol{\mathcal{K}}$ and $\boldsymbol{\mathcal{H}}$ are in general not square and non-Hermitian. The single-set formulation requires fewer SPFs in total, and does not have to deal with the problem that the SPFs of different electronic states are not orthogonal to each other. In practice, however, the multi-set formulation has proved to be the more efficient one in most cases, as the SPFs adapt better to the different states and the total number of configurations required is less.

3.5 Parametrized Basis Functions: G-MCTDH

A final formulation of the MCTDH equations that will be outlined here is the G-MCTDH method introduced by Burghardt, Meyer and Cederbaum [43]. In this, the configurations for the wavefunction *Ansatz* (3.2) are written as

$$\Phi_J(Q_1, \ldots, Q_p, t) = \prod_{\kappa=1}^{m} \varphi_{j_\kappa}^{(\kappa)}(Q_\kappa, t) \prod_{\kappa=m+1}^{p} G_{j_\kappa}^{(\kappa)}(Q_\kappa, t) \tag{3.46}$$

where the first m particles are described by the flexible SPFs described above that are expressed using the primitive basis functions, and the remaining particles are described by SPFs that are defined using a small number of parameters. The idea is that, by propagating a limited set of parameters rather than the functions themselves, a huge saving of memory can be made. Part of the system can be treated using the usual grid-based wavepacket methods described above, and part using the parametrized functions. As the latter may introduce approximations into the dynamics, in this way a system can be described using a hierarchy of modes with a 'full quantum mechanical' part coupled to an 'approximate quantum mechanical' part.

One simple and suitable form for the parametrized functions is the Gaussian:

$$G_j^{(\kappa)}(\mathbf{Q}_\kappa, t) = \exp[\mathbf{Q}_\kappa \cdot \mathbf{a}_j^{(\kappa)}(t) \cdot \mathbf{Q}_\kappa + \xi_j^{(\kappa)}(t) \cdot \mathbf{Q}_\kappa + \eta_j^{(\kappa)}(t)] \tag{3.47}$$

The parametrized functions are thus referred to as Gaussian wavepackets (GWPs). The method is, however, completely general, and any function form could be used. In the limit that only GWPs are used to describe the wavefunction, the method is termed the *variational multiconfiguration Gaussian wavepacket* (vMCG) method.

Equations of motion can be set up using the variational principle as before. The main changes to those for the flexible SPFs and A coefficients are due to the non-orthonormality of the GWPs. Defining the particle GWP overlap and time-derivative overlap matrices:

$$S_{ij}^{(\kappa)} = \langle G_i^{(\kappa)} | G_j^{(\kappa)} \rangle \tag{3.48}$$

$$g_{ij}^{(\kappa)} = \mathrm{i}\langle G_i^{(\kappa)} | \dot{G}_j^{(\kappa)} \rangle \tag{3.49}$$

and using a configuration overlap matrix,

$$\mathcal{S}_{IJ} = \langle \Phi_I | \Phi_J \rangle \tag{3.50}$$

the equation of motion for the A coefficients can be written as

$$\mathrm{i}\dot{\mathbf{A}} = \boldsymbol{\mathcal{S}}^{-1}\left(\mathcal{K} - \sum_\kappa \mathbf{g}^{(\kappa)}\right)\mathbf{A} \tag{3.51}$$

The equations for the SPFs are unchanged, but it should be noted that the density-matrix elements (3.12) contain the overlap matrices, and Equation (3.13) must accordingly be rewritten. Finally, the equations of motion for the GWP parameters can be written as

$$\mathrm{i}\dot{\boldsymbol{\Lambda}}^{(\kappa)} = [\mathbf{C}^{(\kappa)}]^{-1}\mathbf{Y}^{(\kappa)} \tag{3.52}$$

where the parameters have been arranged in a vector, $\boldsymbol{\Lambda}$. The elements of $\mathbf{C}$ are complicated functions of the overlap and density matrices, and the elements of $\mathbf{Y}$ functions of the mean fields and Hamiltonian matrix elements. The reader is referred to Ref. [43] for the full definition.

A link can be made to the field of semiclassical methods, which, following the work of Heller [16], are based on GWPs that propagate along classical trajectories to model the true evolving wavefunction. In the G-MCTDH method, the GWPs do not follow classical trajectories, but are coupled. As a result, the method is able to treat phenomena such as tunnelling and non-adiabatic transitions in a straightforward manner. The convergence properties are also much better than classical trajectory-based methods.

4
Integration Schemes

Hans-Dieter Meyer, Fabien Gatti and Graham A. Worth

The efficiency of the MCTDH method strongly depends on the algorithm used for solving the equations of motion, Equations (3.18) and (3.19). This chapter therefore discusses different integration schemes of the MCTDH working equations.

4.1
The Variable Mean-Field (VMF) Integration Scheme

The MCTDH equations of motion (3.18) and (3.19) are a system of coupled nonlinear ordinary differential equations of first order. A straightforward way to solve the equations is to employ an all-purpose integration method [44,45]. Of all the integrators that have been tested, an Adams–Bashforth–Moulton predictor–corrector turned out to perform most efficiently in integrating the complete set of differential equations, but other methods, for example, Runge–Kutta, also perform well. To distinguish this approach from that described below, it is called the *variable mean-field* (VMF) scheme.

The VMF scheme is not the optimal method for solving the MCTDH equations, because the A-vector and the single-particle functions (SPFs) contain components that are highly oscillatory in time. This enforces small integration steps, at each of which the density and mean-field matrices have to be computed. The frequent calculation of the mean fields contributes dominantly to the computational effort. One therefore expects a significant speed-up of the MCTDH algorithm by employing an integration scheme that is specifically tailored to the solution of the MCTDH equations of motion. Such a method is discussed next.

Multidimensional Quantum Dynamics: MCTDH Theory and Applications.
Edited by Hans-Dieter Meyer, Fabien Gatti, and Graham A. Worth

ISBN: 978-3-527-32018-9

4.2
A Simple Constant Mean-Field (CMF) Integration Scheme

The constant mean-field (CMF) integration scheme takes advantage of the fact that the Hamiltonian matrix elements, $\mathcal{K}_{JL} = \langle \Phi_J | H | \Phi_L \rangle$, and the product of inverse density and mean-field matrices, $\boldsymbol{\rho}^{(\kappa)-1}\,\boldsymbol{\mathcal{H}}^{(\kappa)}$, generally change much more slowly in time than the MCTDH coefficients and the SPFs. For that reason, it is possible to use a wider meshed time discretization for the evaluation of the former quantities than for the propagation of the latter ones. In other words, during the integration of the equations of motion, one may hold the Hamiltonian matrix elements and the products of inverse density and mean-field matrices constant for some time τ.

This concept will now be discussed in more detail. For the sake of simplicity, we first consider a simplified variant that already demonstrates most of the properties and advantages of the integrator. The actual integration scheme is somewhat more subtle. An integration step in this simplified variant begins with the initial values $\mathbf{A}(t_0)$ and $\boldsymbol{\varphi}^{(\kappa)}(t_0)$ being employed to determine the Hamiltonian matrix $\boldsymbol{\mathcal{K}}(t_0)$, the density matrices $\boldsymbol{\rho}^{(\kappa)}(t_0)$, and the mean-field matrices $\boldsymbol{\mathcal{H}}^{(\kappa)}(t_0)$. With these matrices kept constant, the wavefunction is then propagated from t_0 to $t_1 = t_0 + \tau$. The propagated values $\mathbf{A}(t_1)$ and $\boldsymbol{\varphi}^{(\kappa)}(t_1)$ are used to compute $\boldsymbol{\mathcal{K}}(t_1)$, $\boldsymbol{\rho}^{(\kappa)}(t_1)$ and $\boldsymbol{\mathcal{H}}^{(\kappa)}(t_1)$. This procedure is reiterated until the desired final point of time is reached.

Using constraint operators $g^{(\kappa)} = 0$ the CMF equations of motion read:

$$\mathrm{i}\dot{\mathbf{A}} = \bar{\boldsymbol{\mathcal{K}}}\mathbf{A} \tag{4.1}$$

$$\begin{aligned} \mathrm{i}\dot{\boldsymbol{\varphi}}^{(1)} &= (1 - P^{(1)})(\bar{\boldsymbol{\rho}}^{(1)})^{-1}\bar{\boldsymbol{\mathcal{H}}}^{(1)}\boldsymbol{\varphi}^{(1)} \\ &\vdots \\ \mathrm{i}\dot{\boldsymbol{\varphi}}^{(p)} &= (1 - P^{(p)})(\bar{\boldsymbol{\rho}}^{(p)})^{-1}\bar{\boldsymbol{\mathcal{H}}}^{(p)}\boldsymbol{\varphi}^{(p)} \end{aligned} \tag{4.2}$$

Here the bar indicates that the corresponding term is evaluated at time $t_m = t_0 + m\tau$, with m being the number of steps made so far, and is then held constant over the CMF integration step $t_m \leqslant t \leqslant t_{m+1}$.

Supposing that a comparatively large update step τ can be chosen, the CMF integration scheme has several advantages. First of all and most importantly, the mean-field matrices have to be set up much less frequently.

Another important characteristic of the CMF scheme lies in the simplified structure of the system of differential equations. While the MCTDH equations of motion (3.4) and (3.5) are coupled, the CMF equations (4.1) and (4.2) separate into $p + 1$ disjoined subsystems. One of these subsystems is for the MCTDH coefficients; the others are for each of the p single-particle vectors $\boldsymbol{\varphi}^{(\kappa)}$. Splitting a large system of differential equations into smaller ones generally lessens the computational effort because the step size can then be adapted independently for each subsystem; if one subsystem requires small integration steps, the others remain unaffected.

Finally, we point out that the working equation (4.1) for the A-vector is linear with constant coefficients, which permits its solution by integrators explicitly designed for equations of that kind [46], such as the Chebyshev [47] or Lanczos [48] method. This further increases the efficiency of the CMF approach. The set of equations (4.2) for the SPFs is, however, still nonlinear due to the projection operator $P^{(\kappa)}$.

4.3 Why CMF Works

To understand why the CMF approach is successful, let us consider a very simple case, namely a separable Hamiltonian:

$$H = \sum_{\kappa=1}^{p} h^{(\kappa)} \tag{4.3}$$

For this Hamiltonian, the mean-field and Hamiltonian matrices take the simple forms

$$(\boldsymbol{\rho}^{(\kappa)})^{-1} \boldsymbol{\mathcal{H}}^{(\kappa)} = h^{(\kappa)} \, \mathbf{1}_{n_\kappa} \tag{4.4}$$

$$\mathcal{K}_{JL} = \sum_{\kappa=1}^{p} \langle \varphi_{j_\kappa}^{(\kappa)} | h^{(\kappa)} | \varphi_{l_\kappa}^{(\kappa)} \rangle \, \delta_{J^\kappa, L^\kappa} \tag{4.5}$$

The mean field, Equation (4.4), is obviously constant. Next we will show that the Hamiltonian matrix, Equation (4.5), is also constant when the Hamiltonian is Hermitian and the projected Hamiltonian terms commute with the projected constraint operators, $[Ph^{(\kappa)}P, Pg^{(\kappa)}P] = 0$.

The equations of motion for the SPFs read

$$\mathrm{i}\dot{\varphi}_j = g\varphi_j + (1 - P)(h - g)\varphi_j \tag{4.6}$$

where, for ease of notation, we have dropped the particle index κ. Taking the time derivative of the matrix elements of h, one easily finds, using Equation (4.6), that

$$\frac{\mathrm{d}}{\mathrm{d}t} \langle \varphi_j | h | \varphi_l \rangle = \mathrm{i} \langle \varphi_j | (h^\dagger - h)(1 - P) h | \varphi_l \rangle + \mathrm{i} \langle \varphi_j | gPh - hPg | \varphi_l \rangle \tag{4.7}$$

The right-hand side vanishes when $h^\dagger = h$ and $[PhP, PgP] = 0$. Note that the commutator vanishes for the two standard choices of the constraints, $g = 0$

and $g = h$ (see Equations (3.23) and (3.25)). Hence, for a separable Hermitian Hamiltonian, the CMF scheme is *exact*, and an arbitrarily large update step τ can be taken.

For a general, non-separable Hamiltonian, only its non-separable terms – which are in general smaller than the separable ones – make $\boldsymbol{\mathcal{K}}$ and $\boldsymbol{\rho}^{-1}\boldsymbol{\mathcal{H}}$ time dependent. Non-Hermitian terms do so as well. In particular, when a wavepacket enters a complex absorbing potential (CAP) [49], the CMF scheme must take smaller update times τ, which makes the method less efficient.

4.4 Second-Order CMF Scheme

The simple variant of the CMF integration scheme depicted above has to be improved further in order to be competitive. One reason is that in its current form the method defines – analogous to Euler's rule – a first-order integrator, which is known to perform quite poorly. Another reason is that the above scheme lacks a means of estimating the discretization error and adjusting the update step τ. The following modifications of the method eliminate these shortcomings.

The actual CMF integration scheme eliminates the first-order time changes of $\boldsymbol{\mathcal{K}}$ and $\boldsymbol{\rho}^{-1}\boldsymbol{\mathcal{H}}$ by essentially employing the trapezoidal rule when propagating the coefficients and using the midpoint rule when propagating the SPFs. To this end $\boldsymbol{\mathcal{K}}(0)$ and $\boldsymbol{\mathcal{K}}(\tau)$ are used when propagating $\mathbf{A}$ from $t = 0$ to $t = \tau$, while $\boldsymbol{\mathcal{H}}(\tau/2)$ is used when propagating $\boldsymbol{\varphi}$. For a comprehensive description of the method, see Refs. [34, 50].

To obtain an error estimate, the propagated wavefunction is compared to a predicted wavefunction, where the latter is computed with the Euler method, that is, essentially the simple CMF scheme of Equations (4.1) and (4.2). This error measure scales with τ^2. However, the error introduced in the wavefunction while propagating it over one update step τ scales with τ^3. This inconsistency is expected to be of little relevance for the automatic step size control as long as τ does not change by more than an order of magnitude during the propagation (which is usually the case). In a recent publication [51], Manthe devised a new CMF scheme, called CMF2, where this problem is eliminated by propagating the predicted function also with a second-order scheme but using a doubled update time. He reports speed-up factors between 1.3 and 1.8 when replacing CMF by CMF2. The CMF2 scheme requires on average fewer mean-field evaluations but more propagations than CMF. This speed-up may thus be related to the fact that the correlated discrete variable representation (CDVR) method (see Chapter 10) was used: the build-up of the mean fields dominates the total effort of an MCTDH propagation much more strongly when using CDVR than when using a product-form Hamiltonian, Equation (3.34).

A CMF integration is usually run with the constraint $g^{(\kappa)} = 0$. The other standard constraint, $g^{(\kappa)} = h^{(\kappa)}$, has also been tried, but was found to be less efficient in almost all cases. The CMF scheme conserves the norm of the wavefunction (when the Hamiltonian is Hermitian), but it does not (for time-independent Hamiltonians) strictly conserve the energy: only for $\tau \to 0$ is energy strictly conserved. As the CMF equations of motion for the coefficients are linear, a short iterative Lanczos (SIL) [46, 48] or – when the Hamiltonian is non-Hermitian – an Arnoldi–Lanczos [34, 52] integrator is used. The propagation of the SPFs is usually performed with a Bulirsch–Stoer (BS) or Runge–Kutta (RK) integrator [45].

In the majority of cases to date, CMF is faster than VMF, often by one order of magnitude. There are rare cases, however, where VMF is to be preferred over CMF.

5
Preparation of the Initial Wavepacket

Hans-Dieter Meyer, Fabien Gatti and Graham A. Worth

Solving the time-dependent Schrödinger equation requires the generation of an initial wavepacket. The definition of the initial state depends on the physical process under consideration. Two main strategies can be applied: (i) the initial wavefunction is written as a simple product of SPFs, or (ii) an eigenstate of an operator is taken as initial state. Finally, a wavepacket defined by either of the methods may be modified by applying an operator to it.

5.1
Initial Wavepacket as Hartree Product

Often the initial wavefunction, $\Psi(0)$, is a simple Hartree product, that is, a product of SPFs. Hence the A-vector is zero everywhere except for one entry where it is 1. What needs to be defined are the initial single-particle functions (SPFs). If the SPFs are not one-dimensional, their initial form is usually again a simple product of one-dimensional functions. These one-dimensional functions can be Gaussian functions:

$$\varphi(q) = (2\pi\Delta)^{-1/4}\, \mathrm{e}^{-1/4[(q-q_0)/\Delta]^2}\, \mathrm{e}^{\mathrm{i}p_0(q-q_0)} \tag{5.1}$$

Here q_0 and p_0 are the centre and initial momentum of the wavepacket and Δ is the width. (It is actually the width of $|\varphi(q)|^2$.) An orthonormal set of functions is constructed by multiplying the preceding functions with q, followed by Schmidt orthogonalization onto lower functions.

For angular coordinates, associated Legendre functions are an appropriate choice,

$$\tilde{P}_l^m(\cos\theta) = \sqrt{\frac{2l+1}{2}\frac{(l-m)!}{(l+m)!}}\, P_l^m(\cos\theta) \tag{5.2}$$

with $0 \leqslant m \leqslant l$. The parameter m denotes the magnetic quantum number and is treated as a fixed parameter. P_l^m is the (standard) associated Legendre function and $\tilde{P}_l^m$ is the L^2 normalized form of it. For a combined mode involving

Multidimensional Quantum Dynamics: MCTDH Theory and Applications.
Edited by Hans-Dieter Meyer, Fabien Gatti, and Graham A. Worth

ISBN: 978-3-527-32018-9

two angles, θ and ϕ, a normalized spherical harmonic, $Y_{lm}(\theta, \phi)$, may be an appropriate initial function for that mode.

Diagonalizing one-dimensional operators offers a general way of defining initial functions. The eigenfunctions of a suitably chosen operator serve as initial SPFs. When mode combination is used, one may even diagonalize a multidimensional operator of the same dimension as the combined mode.

To give an example, when atom–diatom scattering is studied in Jacobi coordinates, the initial wavepacket could be a Gaussian (with momentum) in the reaction coordinate, an associated Legendre function in the angle, and an eigenfunction of the diatom potential in the vibrational coordinate.

Finally, we note that it is sometimes necessary to take, as initial wavepacket, not a single Hartree product but a (small) sum of Hartree products. Such a necessity may occur when symmetry constraints have to be satisfied.

5.2 Eigenstates and Operated Wavefunctions

Instead of explicitly building an initial wavepacket, one may read a wavefunction that has been created in a previous calculation. Indeed, in many processes the initial state is the eigenfunction of another Hamiltonian. For example, in applications that involve an initial transition from the electronic ground state to an electronically excited state, the initial function is an eigenfunction (usually the vibrational ground state) of the electronic ground-state potential. This wavepacket is then placed on an excited electronic state energy surface (Condon approximation). The initial wavefunction is hence first generated by, for example, *improved relaxation* (see Chapter 8) using the electronic ground-state Hamiltonian and is then propagated on an electronically excited state (or on a manifold of states). Note that it is important that the initial wavefunction is generated within MCTDH, otherwise it would not have the required form, Equation (3.1).

If an infrared spectrum is to be calculated, either by simply Fourier transforming the autocorrelation function or by filter diagonalization [53–56], the initial state is the dipole surface operated ground state. The ground state is again easily computed by relaxation or improved relaxation. The application of a multidimensional correlated operator, however, is not simple, because it will change both the A-vector and the SPFs. This change should be such that the new wavefunction is optimally represented in the new SPFs. To this end, an iterative scheme was devised [38] that computes $\Psi = \hat{O}\tilde{\Psi}$ for any operator $\hat{O}$ that is given in MCTDH product form, Equation (3.34). Here $\tilde{\Psi}$ denotes a wavefunction of MCTDH form.

To arrive at an optimal MCTDH-representation we employ again a variational principle and require

$$\langle \delta\Psi \,|\, \Psi - \hat{O}\tilde{\Psi} \rangle = \delta \sum_{\kappa} \sum_{jl} \epsilon_{jl}^{(\kappa)} \left(\langle \varphi_j^{(\kappa)} \,|\, \varphi_l^{(\kappa)} \rangle - \delta_{jl} \right) . \tag{5.3}$$

The right hand side of Eq. (5.3) is included to ensure the ortho-normality of the SPFs of Ψ. The $\epsilon_{jl}^{(\kappa)}$ are Lagrange parameters.

Variation with respect to the coefficients yields

$$A_J = \sum_L \langle \Phi_J \,|\, \hat{O} \,|\, \tilde{\Phi}_L \rangle \, \tilde{A}_L \, , \tag{5.4}$$

where it is assumed that both sets of SPFs are orthonormal. Variation with respect to the SPFs yields

$$\sum_l \left(\rho_{jl}^{(\kappa)} - \epsilon_{jl}^{(\kappa)} \right) \varphi_l^{(\kappa)} = \sum_l \langle \Psi_j^{(\kappa)} \,|\, \hat{O} \,|\, \tilde{\Psi}_l^{(\kappa)} \rangle \, \tilde{\varphi}_l^{(\kappa)} \, . \tag{5.5}$$

Rather than to determine the values of the Lagrange parameters $\epsilon_{jl}^{(\kappa)}$, we drop the matrix $(\rho_{jl}^{(\kappa)} - \epsilon_{jl}^{(\kappa)})$, i. e. replace it by a unit matrix, and Gram-Schmidt orthogonalize the thus obtained functions. This is legitimate as only the space spanned by the SPFs matters. Orthogonal transformations among the SPFs are accounted for by the coefficients.

Eq. (5.5) is not an explicit equation, because the SPFs to be determined are already needed when evaluating the mean–fields $\langle \Psi_j^{(\kappa)} \,|\, \hat{O} \,|\, \tilde{\Psi}_l^{(\kappa)} \rangle$. The equations are thus to be solved iteratively. The iteration is started by setting

$$\varphi_j^{(\kappa)(0)} = \tilde{\varphi}_j^{(\kappa)} \tag{5.6}$$

$$A_J^{(0)} = \sum_L \langle \Phi_J^{(0)} \,|\, \hat{O} \,|\, \tilde{\Phi}_L \rangle \, \tilde{A}_L \, , \tag{5.7}$$

and then for $i = 0, 1, 2, \cdots$

$$a) \quad \varphi_j^{(\kappa)(i+1)} = \sum_l \langle \Psi_j^{(\kappa)(i)} \,|\, \hat{O} \,|\, \tilde{\Psi}_l^{(\kappa)} \rangle \, \tilde{\varphi}_l^{(\kappa)} \, , \tag{5.8}$$

$b)$ Gram–Schmidt orthogonalize the SPFs,

$$c) \quad A_J^{(i+1)} = \sum_L \langle \Phi_J^{(i+1)} \,|\, \hat{O} \,|\, \tilde{\Phi}_L \rangle \, \tilde{A}_L \, . \tag{5.9}$$

We still need a convergence criterion to stop the iteration. The coefficients and the SPFs do not necessarily converge, because, as already emphasized, only the space spanned by the SPFs is of relevance. Although the wavefunction Ψ converges, the coefficients and the SPFs may differ from the ones of a previous iteration by a unitary transformation. A convenient convergence criterion is given by the MCTDH projector

$$\delta = 1 - \mathrm{Tr}\{P^{(\kappa)(i)} P^{(\kappa)(i+1)} \hat{\rho}^{(\kappa)(i+1)}\} / \mathrm{Tr}\{\hat{\rho}^{(\kappa)(i+1)}\} . \tag{5.10}$$

This expression provides a measure of the similarity of the i-th and $(i+1)$-th orbital space. The MCTDH density operator $\hat{\rho}^{(\kappa)}$ is included to give the weakly occupied orbitals a small weight. δ vanishes when $P^{(\kappa)(i)}$ becomes equal to $P^{(\kappa)(i+1)}$, and the iteration is stopped for sufficiently small δ.

6
Analysis of the Propagated Wavepacket

Hans-Dieter Meyer, Fabien Gatti and Graham A. Worth

Wavepacket dynamics can be used to calculate a number of different properties. Scattering calculations can be used to calculate reaction cross-sections (Section 6.5) and spectra of various types can be calculated (Section 6.2). In addition, time-dependent expectation values can be calculated to extract information from the evolving wavefunction to help provide a mechanistic explanation of phenomena. Owing to the approximate nature of the method, it is also important that a runtime analysis is made to check the accuracy of a calculation (Section 6.1).

While all the techniques developed for grid-based methods can be used, evaluating the results from an MCTDH calculation needs to take account of the special structure of the wavefunction. Thus if the operator, $\hat{\Omega}$, has the MCTDH product form introduced for the Hamiltonian in Equation (3.34), then the calculation of an expectation value is a straightforward product of low-dimensional integrals in the space of the different modes:

$$\langle \Psi(t)|\hat{\Omega}|\Psi(t)\rangle = \sum_{r=1}^{n_s} c_r \sum_J \sum_L A_J^* A_L \prod_{\kappa=1}^{p} \langle \varphi_{j_\kappa}^{(\kappa)} | \hat{\omega}_r^{(\kappa)} | \varphi_{l_\kappa}^{(\kappa)} \rangle \tag{6.1}$$

Note that the two wavefunctions may be different as long as the primitive basis sets are the same.

If the operator does not have the product form, then a transformation must be made from the single-particle function (SPF) basis to the primitive basis to effect its evaluation. This will, however, be impossible if the primitive grid does not fit into memory, as will often be the case. In such cases, the accuracy afforded by Monte Carlo integration may be sufficient for analysis.

6.1
Runtime Analysis of Accuracy

The MCTDH method contains three areas that need to be checked for convergence with respect to the property of interest: the primitive basis set, the SPF basis set and the integrator. As explained in Chapter 3, when converged,

Multidimensional Quantum Dynamics: MCTDH Theory and Applications.
Edited by Hans-Dieter Meyer, Fabien Gatti, and Graham A. Worth

ISBN: 978-3-527-32018-9

the results are a numerically exact solution of the time-dependent Schrödinger equation.

The primitive grid must be large enough to describe the evolving wavepacket correctly, and prevent effects due to reflection from the edges. This can be monitored by checking the expectation values of the centre and width of the wavepacket to check that the grid is wide enough. A more detailed analysis checks that the population of the first and last grid points remains negligible. If fast Fourier transformation (FFT) is used to evaluate the kinetic energy, then the momentum spread and edges of the conjugate momentum grid must also be monitored, and the grid points must be close enough together to support the maximum momentum reached. For a discrete variable representation (DVR) basis, the grid points must be dense enough such that the underlying basis set consists of enough basis functions. The population of these basis functions – the so-called *second population* – must be monitored accordingly.

The SPF basis convergence is crucial for good results. While the only fail-safe method is to increase the basis set size until the property of interest converges, the MCTDH method contains a useful internal guide to the quality of a calculation through the *one-particle density matrices*. These are defined in Equation (3.13). The eigenfunctions of these matrices are the *natural orbitals*. They span the same space as the SPFs and an importance can be associated with each function through the eigenvalues of the density matrices, the *natural weights*. Thus if the weight of the least important natural orbital is small, the SPF basis is good. As a rule of thumb, when calculating averaged quantities such as spectra or expectation values, a calculation is converged if the smallest natural weight for each mode remains below 0.001. For more sensitive quantities, such as state-to-state cross-sections, the smallest natural weight must be much smaller than this for convergence. The natural weights should also be used to ensure that all modes are treated in a balanced way, that is, the space spanned is of a similar quality.

The convergence properties of the primitive and SPF bases are very different. If the primitive grids are sufficiently large and convergence sets in, this convergence is very fast – in fact, it is exponential. But a too small primitive basis will lead to grossly false results. The SPF convergence, on the other hand, is very steady. Even with a rather small set of SPFs, one often can reproduce the main features of, for example, a spectrum reasonably well. Increasing the numbers of SPFs will add fine details to the spectrum and improve the accuracy of transition frequencies and intensities.

A rigorous check of convergence is to compare the wavefunction from two calculations with different numbers of SPFs:

$$\Delta = ||\Psi_2 - \Psi_1|| \tag{6.2}$$

The structure of the MCTDH wavefunction means that the difference cannot be performed directly, but Δ may be obtained from

$$\|\Psi_2 - \Psi_1\|^2 = \|\Psi_1\|^2 + \|\Psi_2\|^2 + 2\,\mathrm{Re}\langle\Psi_1|\Psi_2\rangle \tag{6.3}$$

Note that an overlap between MCTDH wavefunctions (with identical primitive grids) is performed very efficiently. If Δ is negligible, the calculations are converged.

The final runtime errors can be due to the integration scheme used. During a long simulation, errors propagate and grow. For this reason, the Heidelberg MCTDH program uses robust integration schemes with adaptive time steps and good error control. It should also be noted that, despite the underlying approximations, the very efficient constant mean-field (CMF) scheme described in Chapter (4) converges on the true integration as the error parameter is tightened. The simplest monitors of integration are that the energy and norm are conserved, unless an absorbing potential is used. The energy is the expectation value of the Hamiltonian. Assuming that the SPFs remain orthonormal during the propagation, the norm of the wavefunction is given by the sum of the squares of the expansion coefficients:

$$\|\Psi\|^2 = \sum_J A_J^* A_J \tag{6.4}$$

A more detailed analysis would additionally check that the SPFs remain orthonormal.

6.2 Spectra

6.2.1 Photoabsorption Spectra

Perhaps the easiest experimental observable to obtain is a photoabsorption spectrum. The absorption spectrum, $\sigma(\omega)$, can be calculated from the Fourier transform of the autocorrelation function,

$$\sigma(\omega) \propto \omega \int_{-\infty}^{\infty} \mathrm{d}t\, C(t)\, \mathrm{e}^{\mathrm{i}(E_0+\omega)t} \tag{6.5}$$

where E_0 denotes the ground-state energy and where the autocorrelation function is defined as

$$C(t) = \langle\Psi(0)|\Psi(t)\rangle \tag{6.6}$$

A detailed derivation is given in Ref. [57]. See also Refs. [58,59] for a formula that provides absolute intensities.

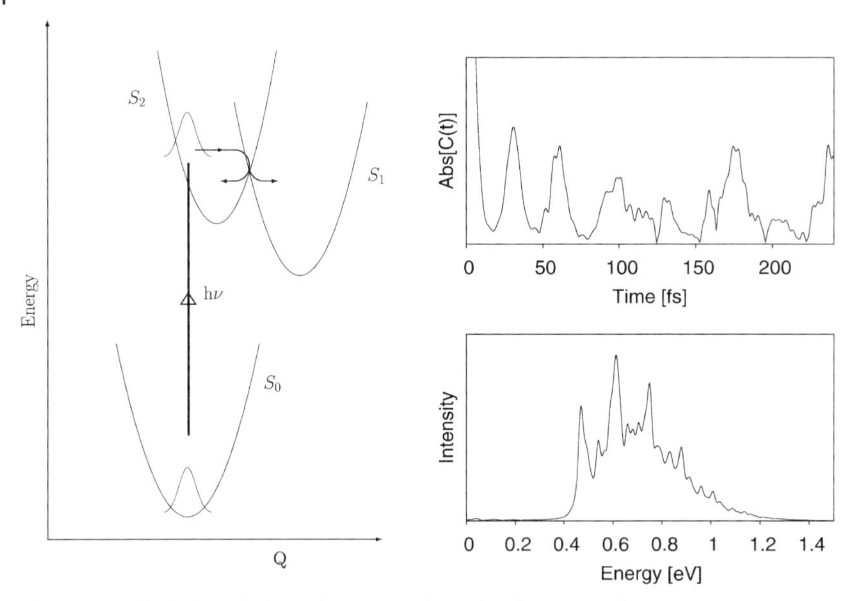

Figure 6.1 Vertical excitation of a system from the S_0 state to S_2.

To obtain the autocorrelation function, one must propagate an initial state, which is the transition operator operated ground state, $\Psi(0) = \hat{\mu}\Psi_{GS}$. For an electronic excitation in Condon approximation, the situation is particularly simple. The ground-state wavefunction is placed on the upper surface at the Franck–Condon point: 'vertical excitation'. This scheme is shown in Figure 6.1 for the excitation from S_0 to S_2 of a system in which the upper state is coupled to a dark S_1 state. It is then propagated on the excited state, that is, it evolves under the influence of the excited-state Hamiltonian, and $C(t)$ is the overlap of the evolving function with its initial form. Greater efficiency can be obtained if the initial wavepacket is real and the Hamiltonian symmetric ($H = H^{\mathrm{T}}$), as then [60, 61]

$$C(2t) = \langle \Psi^*(t) | \Psi(t) \rangle \tag{6.7}$$

that is, the overlap of the wavefunction at time t with itself (not its complex conjugate) gives the autocorrelation function at time $2t$. This means that $C(t)$ is obtained over twice the time of the propagation. This is a huge saving, not only directly due to the shorter propagation time, but also indirectly, as shorter propagations require fewer SPFs. The initial wavefunction also does not need to be stored.

Spectra from the autocorrelation function have been obtained in a number of different systems, such as photodissociation [60], photoabsorption [62] and photoelectron spectra [63]. Figure 6.1 shows the autocorrelation function and spectrum from a four-mode model used to investigate the photoinduced dynamics of pyrazine. Chapter 18 looks in more detail at the study of excited-state dynamics.

The main problem in obtaining spectra from the autocorrelation function is that the Fourier transform in Equation (6.5) goes to infinity, whereas the simulation time is finite. If $C(t) \to 0$ at large times, then there is no problem. This is the case in, for example, photodissociation, but not in bound-state problems such as the example in Figure 6.1. To remove errors due to the finite length (Gibbs phenomenon), $C(t)$ can be multiplied by a damping function such as

$$g_k(t) = \cos^k\left(\frac{\pi t}{2T}\right)\Theta\left(1 - \frac{|t|}{T}\right) \qquad \text{for } k = 0, 1, 2 \tag{6.8}$$

where $\Theta(1 - |t|/T)$ is the Heaviside function that switches from 1 to 0 at time T. This function smoothly forces $C(t)$ to be 0 at T. In practice, one hence solves the integral

$$\sigma(\omega) \propto \omega \,\mathrm{Re} \int_0^{\mathrm{T}} g_k(t) C(t)\, \mathrm{e}^{\mathrm{i}(E_0+\omega)t}\, \mathrm{d}t \tag{6.9}$$

where the integration over negative times is avoided by using $C(-t) = C^*(t)$, which holds for Hermitian Hamiltonians. The effect of the filter $g(t)$ is that the exact spectrum is convoluted with $\tilde{g}(\omega)$, the Fourier transform of $g(t)$:

$$\tilde{g}_0(\omega) = 2\sin(\omega T)/\omega \tag{6.10}$$

$$\tilde{g}_1(\omega) = \frac{4\pi T \cos(\omega T)}{(\pi - 2\omega T)(\pi + 2\omega T)} \tag{6.11}$$

$$\tilde{g}_2(\omega) = \frac{\pi^2 \sin(\omega T)}{\omega(\pi - \omega T)(\pi + \omega T)} \tag{6.12}$$

The three energy filters are displayed in Figure 6.2. Filter $\tilde{g}_0$ has the highest resolution, but it is useless in most cases because of its long oscillating tails and large negative contributions. The filter that is appropriate for most cases is $\tilde{g}_1$. Filter $\tilde{g}_2$ shows an even better suppression of negative parts, but at the cost of lower resolution. The full widths at half-maximum (FWHM) of the three filters are 2.5, 3.4 and 4.1 eV fs (or 20, 27 and 33 cm^{-1} ps) for $k = 0$, 1 and 2, respectively.

The autocorrelation function may additionally be multiplied by the factor

$$f(t) = \exp(-t/\tau) \tag{6.13}$$

This damping function is equivalent to convoluting the spectral lines with Lorentzian functions, the width of which can be related to a homogeneous broadening due to experimental resolution.

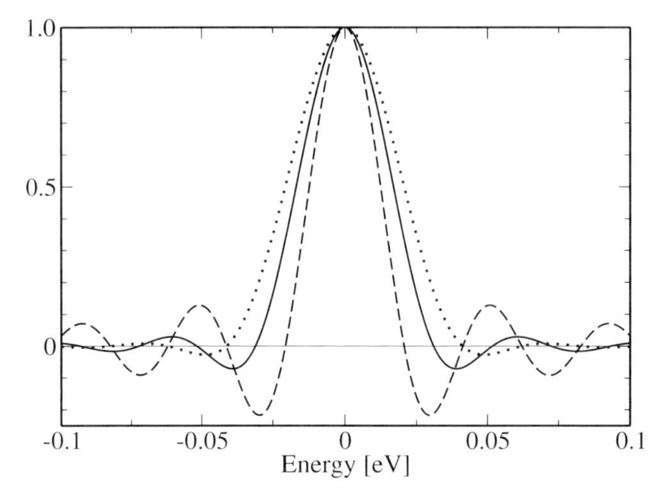

Figure 6.2 Filter functions $\tilde{g}_k$ for $k = 0$ (long-dashed line), $k = 1$ (full line) and $k = 2$ (dotted line). The length of the autocorrelation function is $T = 100$ fs.

6.2.2
Eigenvalues and Filter Diagonalization

The Fourier transformation of the autocorrelation function, as described above in Section 6.2.1, is not a suitable method if a detailed structure needs to be analysed. This is due to the very long propagation times required to provide narrow enough linewidths to observe the fine structure. This is especially a problem for an approximate method such as MCTDH, as the error grows as the propagation time gets longer.

Diagonalization of the Hamiltonian matrix in a suitable basis provides the better route to providing the eigenvalues and eigenvectors. Unfortunately, even using highly adapted DVR basis sets, this is really only possible for small systems with two or three degrees of freedom. Filter diagonalization (FD), first proposed by Neuhauser [64], is a hybrid approach that promises to combine the advantages of wavepacket propagation with obtaining accurate eigenvalues. It has been implemented together with MCTDH, and used to obtain the spectra of CO_2 [55] and HO_2 [56].

The idea is to build a suitable small basis to represent the Hamiltonian in a particular energy range. To do this, a number of wavepackets are propagated, $\Psi_\alpha(t)$, and Fourier-transformed on a set of, typically equidistant, energies $E_1, \ldots, E_L$ spanning the energy window. For one energy, E_k, this is

$$|\hat{\Psi}_{k\alpha}\rangle = \int_{\infty}^{\infty} \mathrm{d}t\, g(t)\, \mathrm{e}^{\mathrm{i}E_k t}\, |\Psi_\alpha(t)\rangle \tag{6.14}$$

Here $g(t)$ is a filtering function that goes to zero at T, the propagation time. Suitable functions are the g_k defined in Equation (6.8).

Introducing the matrix elements

$$H^{(n)}_{k\alpha l\beta} = \langle \hat{\Psi}_{k\alpha} | H^n | \hat{\Psi}_{l\beta} \rangle \tag{6.15}$$

the states in the chosen energy range are found by solving the generalized eigenvalue problem

$$\mathbf{H}^{(1)}\mathbf{B} = \mathbf{H}^{(0)}\mathbf{B}\hat{\epsilon} \tag{6.16}$$

where $\mathbf{H}^{(0)}$ is the overlap matrix.

Rather than calculating the time-independent wavefunctions explicitly, it is more efficient to reformulate the problem using the nth-order correlation functions:

$$c^{(n)}_{\alpha\beta}(t) = \langle \Psi_\alpha(0) | H^n | \Psi_\beta(t) \rangle = \langle \Psi^*_\alpha(\tfrac{1}{2}t) | H^n | \Psi_\beta(\tfrac{1}{2}t) \rangle \tag{6.17}$$

from which

$$H^{(n)}_{k\alpha l\beta} = \int_0^{2T} \mathrm{d}\tau \, G(\Delta E_{kl}, \tau) \, \mathrm{Re}[c^{(n)}_{\alpha\beta}(\tau) \, \mathrm{e}^{\mathrm{i}\bar{E}_{kl}\tau}] \tag{6.18}$$

where $\Delta E_{kl} = E_k - E_l$, $\bar{E}_{kl} = \frac{1}{2}(E_k + E_l)$ and G is a function that depends on the filter function $g(t)$. A discussion of different functions is given in Ref. [54]. As $c^{(n)}$ is related to the nth time derivative of $c^{(0)}$, $H^{(n)}$ can also be computed by an integral involving only the ordinary correlation function $c^{(0)}$. The spectral intensities can also be obtained [54,55].

A number of parameters need to be set. For example, the number of propagations, P, to be made to provide the $\Psi_\alpha(t)$ must be selected, as must suitable initial conditions for each run. If only one wavepacket is propagated, $P = 1$, the indices α andβ may be dropped from the equations above. The number of energy points, L, at which the Fourier transform is to be performed must also be selected. The matrix to be diagonalized will then have dimension LP. This number needs to be larger than the number of eigenfunctions contained in the window; a factor of at least 4 seems reasonable. These parameters must be set by trial and error until all eigenvalues are found.

One problem is that, like all numerical methods, FD can produce spurious eigenvalues. This can be a serious problem if there are errors in the propagation, something that cannot be avoided in MCTDH. A reliable way of removing the spurious eigenvalues is to solve the eigenvalue problem first using different filtering functions, and then using different variational principles. Eigenvalues that appear in all the calculations are real, the others are spurious [55].

This MCTDH/FD scheme works well for small systems like CO_2 [55] or HO_2 [56]. Figure 6.3 demonstrates the power of the FD scheme by comparing the FD spectrum with the Fourier transform of the autocorrelation function.

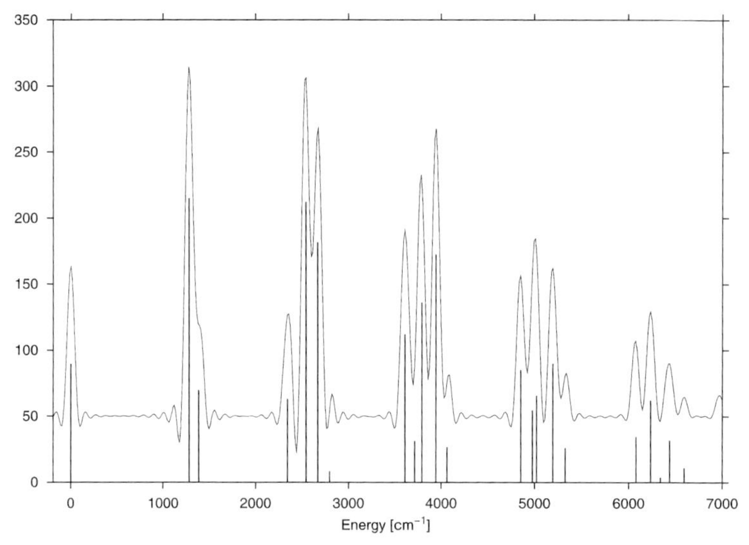

Figure 6.3 The vibrational spectrum of CO_2 as obtained by Fourier transform of the autocorrelation function and by FD using the same autocorrelation function. For better visibility, the Fourier spectrum is shifted upwards by 50 units.

Note that both spectra are obtained from the same data, namely the autocorrelation function of a single wavepacket propagation.

For large systems, however, one is plagued with a large number of spurious eigenvalues, and it is difficult to identify and remove them. For large systems one hence adopts a different strategy. One no longer tries to generate the full spectrum by one calculation, but concentrates on the computation of the eigenenergy of a selected state. An initial state is built that is as close to the sought state as possible. The FD spectrum of such a state will show one strong line, the sought eigenstate, and several small lines, real or spurious, of no further interest. This approach, which may be called *filtering by the initial state*, was successfully applied in Refs. [59,65].

6.2.3
Time-Resolved Spectra

The spectra produced from the autocorrelation function above are appropriate for steady-state conditions, or if the system is prepared by excitation by a delta pulse. In femtochemistry experiments, however, the system is prepared by a very short, but finite, laser pulse. The scenario is similar to that sketched in Figure 6.1(a), but the initial wavepacket must be prepared by explicit inclusion of the ground state and the laser pulse. The molecular Hamiltonian, $\hat{H}_{\mathrm{M}}$, must be augmented by the molecule–light interaction, $\hat{H}_{\mathrm{ML}}$,

$$\hat{H} = \hat{H}_{\mathrm{M}} + \hat{H}_{\mathrm{ML}} \tag{6.19}$$

which, with the molecule-radiation field treated in the dipole approximation, is

$$\hat{H}_{\mathrm{ML}} = \sum_{i \neq j} |\psi_i\rangle \, \boldsymbol{\mu}_{ij} \cdot \boldsymbol{\mathcal{E}}(t) \, \langle\psi_j| \tag{6.20}$$

How the relevant information can be obtained from wavepacket dynamics has been detailed in Ref. [66].

One technique that has shown its usefulness in providing dynamical data is time-resolved photoelectron spectroscopy [67]. After excitation, the excited-state dynamics are probed by ionization to a further state. This spectrum depends on the position of the excited-state wavepacket, and thus the dynamics can be extracted. It is straightforward to include the two laser pulses into an MCTDH simulation. The one new factor here is that bound states and ion states must be treated together.

If it is assumed that the ejected electron immediately leaves the molecule, that is, the molecular–ion wavefunction does not depend on the free-electron wavefunction, the wavefunction can be written as

$$|\Psi(R,t)\rangle = |S_0\rangle + \sum_{i>0} |S_i\rangle + \sum_j \int_0^\infty \mathrm{d}E \, |I_j(E)\rangle \tag{6.21}$$

where S_0 is the ground state, the sum over i is over the bound excited states, and the sum over j is over the ion states. The continuum of free-electron states then needs to be discretized for the simulation [68]. For standard wavepacket dynamics, Seel and Domcke [68] used a set of Legendre functions, effectively treating the electron continuum in a DVR in the same way as a coordinate. The photoelectron spectrum is then obtained from the population of the various continuum states,

$$I(E) \propto \sum_{i=1,2} | \langle\Psi(t \rightarrow \infty)|I_{0,i}(E)\rangle |^2 \tag{6.22}$$

Each state represents a free-electron energy.

Many points were required to cover the continuum. In MCTDH, the continuum can be treated very efficiently by including it as an extra degree of freedom in the wavefunction:

$$\Psi(\mathbf{Q},E,t) = \sum_s \psi^{(s)}(\mathbf{Q},E,t) \tag{6.23}$$

A standard DVR is then used to cover the desired energy range, and the wavefunction is treated with a set of SPFs. This has been used to calculate the photoelectron spectrum of benzene [69].

6.3 Optimal Control

The theoretical treatment that closely matches coherent experiments is known as optimal control theory [70–72]. The time evolution of a wavepacket, $\Psi(t)$, is guided towards a target function Ψ_{tar}. It is found that the maximum overlap at time T between the wavepacket and the target, $\langle \Psi_{\text{tar}} | \Psi(T) \rangle$, is obtained using the field

$$\mathbf{E}(t) = -\frac{2}{\hbar\lambda(t)} \, \text{Im}\langle \Theta(t) | \hat{\mu} | \Psi(t) \rangle \tag{6.24}$$

where $\hat{\mu}$ is the dipole moment and $\lambda(t)$ is a time-dependent function that keeps the electric field within physically realistic bounds and turns the field on and off smoothly.

The wavepacket is driven by the Schrödinger equation

$$i\hbar \frac{\partial}{\partial t} |\Psi(t)\rangle = H_{\text{mol}} + \hat{\boldsymbol{\mu}} \cdot \mathbf{E}(t) \, |\Psi(t)\rangle \tag{6.25}$$

where H_{mol} is the field-free molecular Hamiltonian, while $\Theta(t)$ evolves via a similar equation

$$i\hbar \frac{\partial}{\partial t} |\Theta(t)\rangle = H_{\text{mol}} + \hat{\boldsymbol{\mu}} \cdot \mathbf{E}(t) \, |\Theta(t)\rangle \tag{6.26}$$

but propagating back in time, starting at time $t = T$, with the function

$$\Theta(T) = |\Psi_{\text{tar}}\rangle \langle \Psi_{\text{tar}} | \Psi(T) \rangle \tag{6.27}$$

and ending at $t = 0$. The optimal field is then obtained using an iterative process: a guess field $\mathbf{E}_0(t)$ is used to obtain the zero-order $\Psi_0(t)$, which is then used to obtain a zero-order $\Theta_0(t)$ and a new guess field by solving Equations (6.24) and (6.26) simultaneously. This is repeated until convergence.

Optimal control has been implemented in the Heidelberg MCTDH package and applied to the control of the excitation of pyrazine using displaced ground-state vibrational states as a target [72]. Depending on the target state, up to 10 iterations were required for a converged result. In addition, the inclusion of the electric field makes the Hamiltonian time dependent, meaning that much smaller steps must be taken during the propagation.

6.4 State Populations

For the interpretation of photochemical processes, the state populations as a function of time are often required. The population is given by the expectation value of the state projection operator:

$$P_\alpha = \langle \Psi | \hat{P}_\alpha | \Psi \rangle = \langle \Psi | \alpha \rangle \langle \alpha | \Psi \rangle \tag{6.28}$$

The electronic states can be defined using two different pictures: the *diabatic* and *adiabatic*. Both contain useful information. The former can be related to the electronic configurations of a molecule, while the latter are energy ordered states [73].

Wavepacket dynamics are usually performed in the diabatic picture, in which inter-state couplings appear in the Hamiltonian as potential-like terms. The diabatic populations are then straightforward to obtain. In the multi-set formulation, the wavefunction has a component for each state, Equation (3.40). The population of state α is then the norm of this component,

$$P_\alpha^{(d)} = ||\Psi^{(\alpha)}||^2 \tag{6.29}$$

In the single-set formulation, Equation (3.39), the populations can be obtained from the density matrix for the electronic degree of freedom.

The adiabatic populations are not so easy to obtain. The diabatic and adiabatic wavefunctions are related by a position-dependent unitary transformation,

$$\mathbf{\Psi}^{(a)}(\mathbf{Q}) = \mathbf{U}(\mathbf{Q})\mathbf{\Psi}^{(d)}(\mathbf{Q}) \tag{6.30}$$

where the rotation matrix is given by the eigenvectors of the diabatic potential energy matrix at the point. This matrix transforms the diabatic potential matrix, $\mathbf{W}$, to the diagonal adiabatic potential matrix, $\mathbf{V}$,

$$\mathbf{U}^\dagger(\mathbf{Q})\mathbf{W}(\mathbf{Q})\mathbf{U}(\mathbf{Q}) = \mathbf{V}(\mathbf{Q}) \quad . \tag{6.31}$$

Thus the projection operator for the adiabatic state α is

$$\hat{P}_\alpha^{(a)} = \sum_{\beta,\gamma} |\beta\rangle U^\dagger_{\beta\alpha} U_{\alpha\gamma} \langle\gamma| \tag{6.32}$$

This operator unfortunately does not have the MCTDH form, and a straightforward way to obtain the adiabatic populations is to transform the wavefunction from the diabatic to the adiabatic representation on each point of the full primitive grid. This direct approach can only be used for small systems of three or at most four degrees of freedom. A much more efficient route to the adiabatic state populations is to build the matrix representation of the adiabatic projector, Equation (6.32), and then to potfit (see Chapter 11) the matrix elements of the projector. After a product form of the projector is generated, one may efficiently compute the adiabatic population for many different propagation times. This approach is discussed and applied in Ref. [74]. However, when building the projector, one has again to involve the full primitive grid. This limits this approach to systems with at most six or seven degrees of freedom.

For larger systems, one has to turn to more approximate methods. Fortunately the accuracy required for state populations is often not high, and so Monte Carlo integration can be used to solve the multidimensional integral in Equation (6.28). In Ref. [75] Monte Carlo integration was used to obtain adiabatic populations for a 14-dimensional system.

6.5 Reaction Probabilities

In a scattering calculation the system is divided into internal coordinates (vibrations and rotations) and the scattering coordinate. For example, in an atom–diatom system, A + BC, expressed in Jacobi coordinates (see Figure 6.4), the internal coordinate is the BC vibrational bond, r, and the scattering coordinate is the line joining the atom to the centre of mass of the diatom, R.

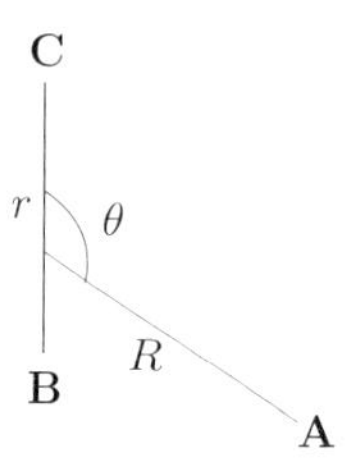

Figure 6.4 An atom–diatom scattering system defined in Jacobi coordinates.

The calculation starts with a wavepacket representing the system A + BC by a product of a particular internal eigenstate for the BC molecule and a (usually Gaussian) wavepacket in the scattering coordinate that locates A relative to BC:

$$\Psi(\mathbf{q}, R, t=0) = \psi_v(\mathbf{q})\,\chi(R) \tag{6.33}$$

with

$$H_{\text{int}}(\mathbf{q})\psi_v = \epsilon_v \psi_v \tag{6.34}$$

$$\chi(R) = \mathrm{e}^{-A(R-R_0)^2}\,\mathrm{e}^{\mathrm{i}p_0 R} \tag{6.35}$$

The scattering wavepacket is thus initially centred at R_0 and is given an initial (negative) momentum, p_0, so that the particle is moving towards the interaction region where the two species collide.

After the interaction, the wavepacket continues to evolve into the reactive product channels, B + AC and C + AB, as well as the non-reactive channel A + BC. The scattering process is then completely defined by knowledge of the S-matrix. This contains the probability of starting in one state and finishing in another, and has elements

$$S_{\beta\nu',\alpha\nu}(E) = \langle \Psi^{-}_{\beta\nu'}(E) | \Psi^{+}_{\alpha\nu}(E) \rangle \tag{6.36}$$

that is, the overlap between the exact time-independent scattering wavefunction, $\Psi^{+}_{\alpha\nu}$, with incoming (+) momentum (outgoing boundary condition) and the corresponding eigenfunction, $\Psi^{-}_{\beta\nu'}$, with outgoing (−) momentum (incoming boundary condition). The (+) state is incoming in reactant channel α, with internal state ν, and the (−) state is outgoing in the product channel β, with internal state ν'. Note that the initial and final states are at the same energy.

A number of ways of evaluating the S-matrix matrix elements have been developed (see Section 8.6 in Ref. [34] for a brief overview). For the MCTDH method, the most efficient way is to augment the system Hamiltonian, H, with a complex absorbing potential (CAP) [49,76,77] and to use this to analyse the flux through the possible channels. Thus

$$\tilde{H} = H - \mathrm{i}W \tag{6.37}$$

where

$$W = W_\alpha + W_\beta + \cdots \tag{6.38}$$

are real, positive-valued potentials in the channels α, β, and so on. They are only defined in the asymptotic regions, so that

$$W_\alpha = \Theta_\alpha W_\alpha \tag{6.39}$$

where Θ_α is the Heaviside function that provides the dividing surface defining the product α.

As the CAP is outside the interaction region, it will not affect the propagation of the wavepacket. The only place where it may interfere is with the initial wavepacket. This wavepacket is located in the asymptotic region of the reactant channel, which becomes the non-reactive product channel. Either a long grid is required here to make sure that the CAP is away from the initial wavepacket, or else an 'automatic' CAP can be used that is turned on once the wavepacket has left the asymptotic region and before it returns.

Using the CAP, the probability of going from an initial state $\alpha\nu$ to one product channel γ, irrespective of final internal state, can be obtained from [78]

$$\sum_{\nu'} |S_{\gamma\nu',\alpha\nu}(E)|^2 = \frac{2}{\pi|\Delta(E)|^2} \,\mathrm{Re} \int_0^{T} \mathrm{d}\tau \, g(\tau) \, \mathrm{e}^{\mathrm{i}E\tau} \tag{6.40}$$

with $g(\tau) = g_w(\tau) + g_\theta(\tau)$ and

$$g_w(\tau) = \int_0^{T-\tau} \mathrm{d}t \, \langle \Psi(t) | W_\gamma | \Psi(t+\tau) \rangle \tag{6.41}$$

$$g_\theta(\tau) = \tfrac{1}{2} \langle \Psi(T-\tau) | \Theta_\gamma | \Psi(T) \rangle \tag{6.42}$$

where T denotes the propagation time. The first term, g_w, accounts for the outgoing flux that is already absorbed by the CAP. The second term, g_θ, accounts for the probability density that is not yet absorbed by the CAP but has already entered the asymptotic region defined by Θ_γ. This contribution vanishes for $T \to \infty$.

The energy distribution contained in the initial wavepacket is formally given by $\Delta(E) = \langle \Psi^+_{\alpha\nu}(E) | \psi_\nu \chi \rangle$. If the initial wavepacket is located in the far asymptotic region where the potential is zero, the energy distribution is simply the coordinate–momentum Fourier transform of the translational wavepacket:

$$\Delta(E) = \sqrt{\frac{\mu_R}{2\pi p}} \int_0^\infty \mathrm{d}R \, \chi(R) \, \mathrm{e}^{\mathrm{i}pR} + E_\nu \tag{6.43}$$

where E_ν denotes the energy of the internal state ν. To be able to place the initial wavepacket closer to the scattering centre, we evaluate the energy distribution by

$$|\Delta_i(E)|^2 = \langle \chi | \delta(\hat{H}_\nu - E) | \chi \rangle \tag{6.44}$$

where $\hat{H}_\nu = \langle \psi_\nu | \hat{H} | \psi_\nu \rangle$ denotes a mean field. For more details, see Ref. [79]. Equation (6.44) is correct as long as the initial wavepacket is placed at a large enough distance from the scattering centre such that the potential cannot excite internal degrees of freedom. The potential itself (for example, its centrifugal part) does not need to be small over the width of χ.

By using projection operators onto a particular final state,

$$P_{\gamma\nu'} = |\psi_{\gamma\nu'}\rangle\langle\psi_{\gamma\nu'}| \tag{6.45}$$

the full state-to-state S-matrix can also be obtained from

$$|S_{\gamma\nu',\alpha\nu}(E)|^2 = \frac{2}{\pi |\Delta(E)|^2} \operatorname{Re} \int_0^{\mathrm{T}} \mathrm{d}\tau \, g_{\gamma\nu'}(\tau) \, \mathrm{e}^{\mathrm{i}E\tau} \tag{6.46}$$

with

$$g_{\gamma\nu'}(\tau) = \int_0^{T-\tau} \mathrm{d}t \, \langle \Psi(t) | P_{\gamma\nu'} W_\gamma P_{\gamma\nu'} | \Psi(t+\tau) \rangle + \tfrac{1}{2} \langle \Psi(T-\tau) | P_{\gamma\nu'} \Theta_\gamma P_{\gamma\nu'} | \Psi(T) \rangle \tag{6.47}$$

The CAP/flux method is very suitable for analysing reactive scattering events. Here one usually sums over all final internal states ν'. The use of projectors on particular final states is technically very difficult, because the construction of the projectors, Equation (6.45), is only simple in the product coordinates, whereas the propagated wavepacket is still in the educt coordinates. Turning to non-reactive inelastic scattering, the use of projectors to compute state-to-state cross-sections is straightforward [80,81].

However, when fully resolved state-to-state cross-sections are to be computed, we now prefer to use the method of Tannor and Weeks [82,83]. Within this approach the S-matrix element is given by a Fourier transform of a cross-correlation function,

$$|S_{\nu'\nu}(E)|^2 = \frac{1}{4\pi^2|\Delta'(E)|^2\,|\Delta(E)|^2}\left|\int_0^T \mathrm{e}^{\mathrm{i}Et} c_{\nu'\nu}(t)\,\mathrm{d}t\right|^2 \tag{6.48}$$

with

$$c_{\nu'\nu}(t) = \langle \psi_{\nu'}\,\chi' | \mathrm{e}^{-\mathrm{i}Ht} | \psi_\nu\,\chi \rangle \tag{6.49}$$

where ψ_ν and $\psi_{\nu'}$ are wavefunctions of internal motion as defined in Equation (6.34) and χ' denotes a wavefunction of the scattering coordinate similar to χ, Equation (6.35), but with outgoing (positive) momentum. $\Delta'(E)$ is defined similarly to $\Delta(E)$, but with χ replaced by χ'. To avoid long grids, the propagated wavepacket is absorbed by a CAP after it has passed the analyse wavepacket χ'. This approach was used in Refs. [84,85].

7
MCTDH for Density Operator

Hans-Dieter Meyer, Fabien Gatti and Graham A. Worth

7.1
Wavefunctions and Density Operators

A wavefunction describes a particular, well-defined state. A system at finite temperature, however, is an incoherent mixture of very many thermally excited states, $|\Psi_n\rangle$. To describe such a statistical mixture, a density operator is introduced,

$$\rho = \sum_n p_n |\Psi_n\rangle\langle\Psi_n| \tag{7.1}$$

where $0 \leqslant p_n \leqslant 1$ are the occupation probabilities, which add up to 1. (The density operator ρ should not be confused with the MCTDH one-particle reduced density.)

The time evolution of a density operator is given by the Liouville–von Neumann equation,

$$\mathrm{i}\dot{\rho} = \mathcal{L}(\rho) \tag{7.2}$$

where $\mathcal{L}$ is a linear super-operator, which reads

$$\mathcal{L}(\rho) = [H, \rho] \tag{7.3}$$

The most important advantage of the density-operator formalism is the possibility to include the effects of an environment on the system dynamics, thus allowing the description of *open* quantum systems and their non-equilibrium dynamics [86]. For open systems one divides the total system into two parts, system and bath (environment). The density operator for the system, ρ, is obtained by tracing out all bath degrees of freedom [86],

$$\rho = \mathrm{Tr}_{\mathrm{bath}}\{\rho_{\mathrm{tot}}\} \tag{7.4}$$

This approach is known as *reduced dynamics*. The reduced dynamic equations of motion for ρ are known but are very complicated because they contain a

Multidimensional Quantum Dynamics: MCTDH Theory and Applications.
Edited by Hans-Dieter Meyer, Fabien Gatti, and Graham A. Worth

ISBN: 978-3-527-32018-9

memory kernel, which makes them non-local in time. The Markov approximation [87,88] is often used to remove this non-locality. One of the prominent approaches in this context goes back to Lindblad [89–91],

$$\mathcal{L}(\rho) = [H,\rho] + \mathrm{i}\sum_j (V_j \rho V_j^\dagger - \tfrac{1}{2} V_j^\dagger V_j \rho - \tfrac{1}{2}\rho V_j^\dagger V_j) \tag{7.5}$$

where H is the Hamiltonian of the (reduced) system and where the operators V_j are to be determined – usually via perturbation theory – by analysing the Hamiltonian of the full system. The Lindblad form guarantees that the time evolution is completely positive. Other well-known expressions are due to Redfield [92] and Caldeira and Leggett [93], who proposed a perturbative treatment of the interaction and a bath at high temperature.

In this section we extend MCTDH to propagation of density operators. There are two different MCTDH expansions of density operators, called type I and type II. For type I, single-particle density operators are used to replace the single-particle functions of the wavepacket version. In the type II variant, the single-particle density operators are themselves represented by a product of single-particle functions. A more detailed discussion on MCTDH for density operators than the one given below can be found in Refs. [94,95], and applications of MCTDH for density operators are discussed in Refs. [96–98].

7.2 Type I Density Operators

We generalize the MCTDH *Ansatz*, Equation (3.1), to density operators,

$$\rho(Q_1,\ldots,Q_p,Q_1',\ldots,Q_p',t) = \sum_{\tau_1=1}^{n_1}\cdots\sum_{\tau_p=1}^{n_p} B_{\tau_1\ldots\tau_p}(t)\prod_{\kappa=1}^{p}\sigma_{\tau_\kappa}^{(\kappa)}(Q_\kappa,Q_\kappa',t) \tag{7.6}$$

where the $B_{\tau_1\ldots\tau_p}$ denote the expansion coefficients, which are now called B rather than A to avoid confusion with the wavefunction formalism. The $\sigma_{\tau_\kappa}^{(\kappa)}$ are the so-called *single-particle density operators* (SPDOs), analogous to the single-particle functions (SPFs) in the wavefunction scheme (see Section 3.1). A density operator has to be Hermitian. To ensure this property, we require that the coefficients are real and the SPDOs are Hermitian:

$$B_{\tau_1\ldots\tau_p} = B^*_{\tau_1\ldots\tau_p} \qquad \text{and} \qquad \sigma_{\tau_\kappa}^{(\kappa)} = \sigma_{\tau_\kappa}^{(\kappa)\dagger} \tag{7.7}$$

As shown in Ref. [95], the type I equations of motion conserve these properties.

To derive the equations of motion, one needs a Hilbert space structure and in particular a scalar product. For this purpose, we employ the Hilbert–Schmidt scalar product [99],

$$\langle\langle A|B\rangle\rangle = \mathrm{Tr}\{A^\dagger B\} \tag{7.8}$$

As in MCTDH for wavefunctions, the representation of the density operator (7.6) is not unique, and constraints are needed to ensure unique, singularity-free equations of motion. The constraints, which imply that the SPDOs are orthonormal, read

$$\langle\langle \sigma_\mu^{(\kappa)}(0)|\sigma_\nu^{(\kappa)}(0)\rangle\rangle = \delta_{\mu\nu} \tag{7.9}$$

$$\langle\langle \sigma_\mu^{(\kappa)}(t)|\dot{\sigma}_\nu^{(\kappa)}(t)\rangle\rangle = -\mathrm{i}\langle\langle \sigma_\mu^{(\kappa)}(t)|\mathcal{G}^{(\kappa)}\sigma_\nu^{(\kappa)}(t)\rangle\rangle \tag{7.10}$$

Here the *constraint super-operator* $\mathcal{G}^{(\kappa)}$ is a self-adjoint, but otherwise arbitrary, super-operator acting exclusively on the κth particle. In particular, one may set $\mathcal{G}^{(\kappa)} = 0$. For further reference, we introduce the total constraint super-operator

$$\mathcal{G} = \sum_{\kappa=1}^{p} \mathcal{G}^{(\kappa)} \tag{7.11}$$

which is just the sum of the individual constraint super-operators.

The derivation of the equations of motion is very similar to the wavefunction case. We need to define *single-hole density operators*

$$\Pi_\nu^{(\kappa)} = \langle\langle \sigma_\nu^{(\kappa)}|\rho\rangle\rangle = \sum_{\mathcal{T}^\kappa} B_{\mathcal{T}_\nu^\kappa} \prod_{\kappa'\neq\kappa} \sigma_{\tau_{\kappa'}}^{(\kappa')} \tag{7.12}$$

with which we define reduced density matrices, which are now called $\mathcal{D}$, as the symbol ρ is already used. For an explanation of the nomenclature, see Equation (3.10). The symbol $\mathrm{Tr}\{\cdot\}_\kappa$ indicates that the trace over the κth particle is not performed. Thus we have

$$\begin{aligned} \mathcal{D}_{\mu\nu}^{(\kappa)} &= \langle\langle \Pi_\mu^{(\kappa)}|\Pi_\nu^{(\kappa)}\rangle\rangle \\ &= \langle\langle \sigma_\nu|\mathrm{Tr}\{\rho^\dagger\rho\}_\kappa|\sigma_\mu\rangle\rangle \\ &= \sum_{\mathcal{T}^\kappa} B^*_{\mathcal{T}_\mu^\kappa} B_{\mathcal{T}_\nu^\kappa} \end{aligned} \tag{7.13}$$

and mean-field Liouvillian super-operators

$$\langle \mathcal{L} - \mathcal{G}\rangle_{\mu\nu}^{(\kappa)} = \langle\langle \Pi_\mu^{(\kappa)}|(\mathcal{L} - \mathcal{G})\Pi_\nu^{(\kappa)}\rangle\rangle \tag{7.14}$$

Finally, we define the type I projector

$$\mathcal{P}^{(\kappa)} = \sum_{\nu=1}^{n_\kappa} |\sigma_\nu^{(\kappa)}\rangle\rangle\langle\langle \sigma_\nu^{(\kappa)}| \tag{7.15}$$

and the Hartree product

$$\Omega_{\mathcal{T}} = \prod_{\kappa=1}^{p} \sigma_{\tau_\kappa}^{(\kappa)} \tag{7.16}$$

With the above definitions we can formulate the equations of motion:

$$\mathrm{i}\dot{B}_{\mathcal{T}} = \sum_{\mathcal{T}'} \langle\langle \Omega_{\mathcal{T}} | (\mathcal{L} - \mathcal{G}) \Omega_{\mathcal{T}'} \rangle\rangle B_{\mathcal{T}'} \tag{7.17}$$

and

$$\mathrm{i}\dot{\sigma}^{(\kappa)} = \boldsymbol{\mathcal{G}}^{(\kappa)} \sigma^{(\kappa)} + (1 - \mathcal{P}^{(\kappa)})(\boldsymbol{\mathcal{D}}^{(\kappa)})^{-1} \langle \boldsymbol{\mathcal{L}} - \boldsymbol{\mathcal{G}} \rangle^{(\kappa)} \sigma^{(\kappa)} \tag{7.18}$$

Note the similarity of these equations of motion with Equations (3.18) and (3.19).

7.3 Type II Density Operators

The SPDOs can also be written as 'ket–bra' products of wavefunctions. Doing so, the algorithm for the density operators becomes even more similar to that for the wavefunctions. One interprets the index τ_κ as composite index $\tau_\kappa = (j_\kappa, \ell_\kappa)$, and similarly $\mathcal{T} = (J, L)$. Setting

$$\sigma_{\tau_\kappa}^{(\kappa)}(Q_\kappa, Q'_\kappa, t) = |\varphi_{j_\kappa}^{(\kappa)}(Q_\kappa, t)\rangle \langle \varphi_{\ell_\kappa}^{(\kappa)}(Q'_\kappa, t)| \tag{7.19}$$

and

$$B_{\tau_1 \ldots \tau_p} = B_{j_1 \ldots j_p, \ell_1 \ldots \ell_p} = B^*_{\ell_1 \ldots \ell_p, j_1 \ldots j_p} \tag{7.20}$$

one arrives at the type II density-operator expansion:

$$\rho(Q_1, \ldots, Q_p, Q'_1, \ldots, Q'_p, t) = \sum_{j_1, \ell_1 = 1}^{n_1} \cdots \sum_{j_p, \ell_p = 1}^{n_p} B_{j_1 \ldots j_p, \ell_1 \ldots \ell_p}(t) \prod_{\kappa=1}^{p} |\varphi_{j_\kappa}^{(\kappa)}(Q_\kappa, t)\rangle \langle \varphi_{\ell_\kappa}^{(\kappa)}(Q'_\kappa, t)| \tag{7.21}$$

The hermiticity of B, Equation (7.20), ensures that ρ is Hermitian. The hermiticity is conserved during the propagation [95].

The constraints (7.10) must be translated into constraints for the SPFs, which requires the constraint super-operator to be given as

$$\mathcal{G}^{(\kappa)} = [g^{(\kappa)}, \cdot\,] \tag{7.22}$$

where the constraint operators are similar to those of Equation (3.15). In fact, using Equation (7.22) one arrives at Equation (3.15), that is, at the constraint equations for SPFs.

The type II equations of motion for propagating type II density operators now read [94, 95]:

$$\mathrm{i}\dot{B}_{J,L} = \langle \Phi_J | (\mathcal{L} - \mathcal{G})(\rho) | \Phi_L \rangle \tag{7.23}$$

and

$$\mathrm{i}\dot{\boldsymbol{\varphi}}^{(\kappa)} = g^{(\kappa)}\boldsymbol{\varphi}^{(\kappa)} + (1 - P^{(\kappa)})\,\mathrm{Tr}\{(\mathcal{L} - \mathcal{G})(\rho)\rho\}_\kappa(\boldsymbol{\mathcal{D}}^{(2),(\kappa)})^{-1}\boldsymbol{\varphi}^{(\kappa)} \tag{7.24}$$

where the reduced single-particle density matrix is given by

$$\mathcal{D}^{(2),(\kappa)}_{j\ell} = \langle\varphi^{(\kappa)}_\ell|\mathrm{Tr}\{\rho^2\}_\kappa|\varphi^{(\kappa)}_j\rangle = \sum_L\sum_{J^\kappa} B^*_{L,J^\kappa_\ell}\, B_{L,J^\kappa_j} \tag{7.25}$$

It is illustrative to exemplify these equations for the simplest case, $\mathcal{L} = [H,\,\cdot\,]$ and $\mathcal{G} = 0$. The equations of motion can then be written as

$$\mathrm{i}\dot{B}_{J,L} = \sum_K \langle\Phi_J|H|\Phi_K\rangle B_{K,L} - B_{J,K}\langle\Phi_K|H|\Phi_L\rangle \tag{7.26}$$

and

$$\mathrm{i}\dot{\varphi}^{(\kappa)}_m = (1 - P^{(\kappa)}) \sum_{J,L,K} (\boldsymbol{\mathcal{D}}^{(2),(\kappa)})^{-1}_{mk} B_{J,L}\, B_{L,K}\, \langle\Phi_{K^\kappa}|H|\Phi_{J^\kappa}\rangle\, |\varphi^{(\kappa)}_j\rangle \tag{7.27}$$

where Φ_{J^κ} denotes a Hartree product in which the SPF of the κth degree of freedom is missing. The indices j and k are the κth entry of the composite indices J and K, respectively. When deriving Equation (7.27) we have discarded the contribution of the term $\mathrm{Tr}\{\rho H\rho\}_\kappa$ because it will be annihilated by the projector $(1 - P^{(\kappa)})$. Only the term $\mathrm{Tr}\{H\rho^2\}_\kappa$ is kept.

When the density is a pure state, $\rho = |\Psi\rangle\langle\Psi|$, the coefficient matrix factorizes, $B_{J,L} = A_J A^*_L$, where A denotes the MCTDH coefficient vector of Ψ. Inserting this separation into Equations (7.25)–(7.27), one recovers the MCTDH wavefunction equations of motion (3.18) and (3.19). Hence, the closed system propagation of a pure-state density operator using type II expansion is equivalent to an MCTDH wavefunction propagation (that is, identical SPFs, B coefficients of ρ factorize into A coefficients of Ψ).

7.4 Properties of MCTDH Density Operator Propagation

MCTDH for wavefunctions conserves the total probability $\langle\Psi|\Psi\rangle$ and the total energy $\langle\Psi|H|\Psi\rangle$. This follows directly from the Dirac–Frenkel variational principle [34]. Similarly, the variational principle ensures that $\mathrm{Tr}\{\rho^2\}$ and $\mathrm{Tr}\{\rho^2 H\}$ are conserved when density operators are propagated using MCTDH for a closed system. Unfortunately, these are not the total probability, $\mathrm{Tr}\{\rho\}$, and total energy, $\mathrm{Tr}\{\rho H\}$. But, when the calculation converges, the latter quantities are rather well conserved. The lack of (exact) energy conservation made us think about using another variational principle [94,95], which does ensure exact energy conservation. However, this so-called *linear mean-field approach* was found to be less efficient and we do not discuss it here.

For closed systems and type II density operators one additionally can prove some interesting results [95]:

(i) $\mathrm{Tr}\{\rho^n\}$, $n = 1, 2, 3, \ldots$, is conserved.

(ii) A pure state remains pure.

(iii) If ρ is pure, type II propagation is equivalent to MCTDH propagation of wavefunctions.

Point (i) tells us that probability conservation is no problem (for type II densities and closed systems). Point (ii) follows from (i), but is worth mentioning explicitly. Point (iii) has already been discussed in the previous section.

We have introduced two types of MCTDH density-operator expansions, but have not yet discussed which expansion is superior under given conditions. Consider an uncorrelated system at high temperature. The type I density-operator propagation then becomes numerically exact with one single configuration. A type II propagation, on the other hand, would require many SPFs to correctly represent the thermal excitations. Going to the other extreme, a pure state and a strongly correlated system, one notices that type II now becomes much more efficient than type I. One needs more SPDOs for type I than SPFs for type II to account for the correlation, and it is more elaborate to propagate SPDOs than to propagate SPFs. To make the comparison more vivid, let us distinguish between correlation and mixing. In a coordinate representation, the density operator is a $2f$-dimensional function $\rho(\mathbf{q}, \mathbf{q}')$, where $\mathbf{q}$ are the coordinates of the f degrees of freedom of the system. If this function is non-separable with respect to the coordinates q_κ and $q_{\kappa'}$ of *different* degrees of freedom, $\kappa \neq \kappa'$, we speak of correlation between these degrees of freedom. On the other hand, if $\rho(\mathbf{q}, \mathbf{q}')$ is non-separable with respect to the coordinates q_κ and q'_κ of a *single* degree of freedom κ, we speak of mixing in this degree of freedom. In particular, a pure state is unmixed in all degrees of freedom. Using this terminology, one can express the different performance of type I and type II expansions as follows. The type I expansion is more efficient if there is more mixing than correlation, whereas the type II expansion is to be preferred if correlation is stronger than mixing.

8
Computing Eigenstates by Relaxation and Improved Relaxation

Hans-Dieter Meyer, Fabien Gatti and Graham A. Worth

8.1
Relaxation

The generation of a ground-state wavefunction is conveniently done by energy relaxation [100]. An initial wavepacket is propagated in negative imaginary time and then renormalized:

$$\Psi(t) = \mathrm{e}^{-Ht}\Psi(0)\|\mathrm{e}^{-Ht}\Psi(0)\|^{-1} \tag{8.1}$$

By expanding the wavefunction in the eigenfunctions of the Hamiltonian, it can be shown that $\Psi(t)$ converges to the ground state as $t \to \infty$.

Energy relaxation is not the most efficient way to produce a ground-state wavefunction, but here it is a convenient way. As the relaxation is performed by MCTDH, the computed ground-state wavefunction is automatically in MCTDH form and may serve as an initial state of a subsequent propagation with a different Hamiltonian. Energy relaxation can in principle also be used to produce excited states, by keeping $\Psi(t)$ orthogonal to already computed lower-lying states. But this is rather cumbersome to do.

Relaxation can be done in a direct manner, by simply propagating a wavepacket in imaginary time. A far better way is to use the *improved relaxation* algorithm described below. By combining diagonalization of a Hamiltonian together with the principle of relaxation, this method makes it possible to calculate any desired eigenfunction efficiently in MCTDH form. Owing to its performance, improved relaxation should now be used even when computing ground states.

8.2
Improved Relaxation

Improved relaxation [38,101] is essentially a multiconfiguration self-consistent field (MCSCF) approach. Initially, this method was implemented as a modi-

Multidimensional Quantum Dynamics: MCTDH Theory and Applications.
Edited by Hans-Dieter Meyer, Fabien Gatti, and Graham A. Worth

ISBN: 978-3-527-32018-9

fication of the usual relaxation method discussed above. Using the constant mean-field (CMF) propagation scheme (see Section 4.2), the propagation of the coefficients is decoupled from the propagation of the single-particle functions (SPFs). The modification consists in determining the A-vector no longer by relaxation but by diagonalization of the Hamiltonian matrix $\mathcal{K}$. Hence the name, 'improved relaxation'.

To derive the working equations of the improved relaxation algorithm in a rigorous way, we employ the variational principle of time-independent quantum mechanics,

$$\delta\left\{ \langle\Psi|H|\Psi\rangle - E\left(\sum_J A_J^* A_J - 1\right) - \sum_{\kappa=1}^{p} \sum_{j,l=1}^{n_\kappa} \epsilon_{jl}^{(\kappa)} \left(\langle\varphi_j^{(\kappa)}|\varphi_l^{(\kappa)}\rangle - \delta_{jl}\right) \right\} = 0 \tag{8.2}$$

where it is assumed that Ψ is of MCTDH form, Equation (3.1). The Lagrange multipliers E and $\epsilon_{jl}^{(\kappa)}$ are introduced to keep the A-vector normalized and the SPFs orthonormal, respectively.

Varying A_J^* yields

$$\sum_L \mathcal{K}_{JL} A_L = E A_J \tag{8.3}$$

Hence the coefficient vector is obtained as an eigenvector of the Hamiltonian represented in the basis of the configurations Φ_J.

From Equation (3.8) it follows that a variation of the total wavefunction with respect to an SPF yields a single-hole function. Using this and varying the expression (8.2) with respect to $\langle\varphi_k^{(\kappa)}|$ gives [102, 103]

$$\sum_{l=1}^{n_\kappa} \mathcal{H}_{jl}^{(\kappa)} \varphi_l^{(\kappa)} = \sum_{l=1}^{n_\kappa} \epsilon_{jl}^{(\kappa)} \varphi_l^{(\kappa)} \tag{8.4}$$

Projecting Equation (8.4) onto $\varphi_k^{(\kappa)}$, one arrives at

$$\epsilon_{jk}^{(\kappa)} = \sum_{l=1}^{n_\kappa} \langle\varphi_k^{(\kappa)}|\mathcal{H}_{jl}^{(\kappa)}|\varphi_l^{(\kappa)}\rangle \tag{8.5}$$

and from Equations (8.4) and (8.5) it then follows that

$$(1 - P^{(\kappa)}) \sum_{l=1}^{n_\kappa} \mathcal{H}_{jl}^{(\kappa)} \varphi_l^{(\kappa)} = 0 \tag{8.6}$$

A variationally optimal solution is found when Equations (8.3) and (8.6) are satisfied simultaneously.

Since Equation (8.6) holds for each j, it must hold for any linear combination of these equations as well. To arrive at a form similar to the equation of motion (3.5), we insert the inverse of the density matrix

$$\dot{\varphi}_j^{(\kappa)} := -(1 - P^{(\kappa)}) \sum_{k,l=1}^{n_\kappa} (\rho^{(\kappa)})^{-1}_{jk} \mathcal{H}^{(\kappa)}_{kl} \varphi_l^{(\kappa)} = 0 \tag{8.7}$$

where $\dot{\varphi}$ denotes the time derivative in negative imaginary time, $\dot{\varphi} = \partial\varphi/\partial\tau$ with $\tau = -it$. Equations (8.6) and (8.7) are equivalent as long as $\boldsymbol{\rho}^{-1}$ is non-singular. (Remember, one always uses a regularized $\boldsymbol{\rho}$, Equation (3.14), hence $\boldsymbol{\rho}^{-1}$ is always defined and regular.) Hence, the variational solution is found if and only if the time derivatives of the SPFs vanish. This suggests a convenient way to solve the variational equation (8.6). The SPFs are relaxed, that is, propagated in negative imaginary time, until their time derivative is sufficiently small. To show that relaxation indeed provides a solution of the variational equation, we inspect the time derivative of the energy during relaxation of the SPFs. A small calculation yields

$$\dot{E} = -2\,\mathrm{Re} \sum_{\kappa=1}^{p} \sum_{i=1}^{n_\kappa} \left\| \sum_{j=1}^{n_\kappa} (\rho^{(\kappa)})^{1/2}_{ij} \dot{\varphi}_j^{(\kappa)} \right\|^2 \leq 0 \tag{8.8}$$

This shows that $\dot{\varphi}_j^{(\kappa)} \to 0$ when $\tau \to \infty$, as the energy cannot decrease infinitely. This also shows that the energy always decreases when the SPFs are relaxed. Note that the SPFs are relaxed with respect to the mean fields, which depend on the A-vector. Orbital relaxation thus generates SPFs that are optimal for the sought state but not necessarily for the ground state.

The improved relaxation method proceeds as follows. First the user has to define an initial state, which should have some overlap with the eigenstate one wants to compute. The Hamiltonian is then diagonalized in the basis of the configurations of the initial state. After the diagonalization, the mean fields $\boldsymbol{\mathcal{H}}^{(\kappa)}$ are build and the SPFs are relaxed over a suitable time interval. The Hamilton matrix $\boldsymbol{\mathcal{K}}$ is then rebuilt in the new configurations (Hartree products) and diagonalized. The whole process is iterated until convergence is achieved. If the ground state is computed, the selection of the eigenvector of the Hamiltonian matrix is simple: one takes the eigenvector of lowest energy. When excited states are to be computed, that eigenvector is taken which corresponds to the wavefunction that has the largest overlap with the initial state. Note that the variational equation (8.2) and its consequences, Equations (8.3) and (8.6), are solved numerically exactly, provided the iteration scheme converges. Previous MCSCF approaches to solve the molecular vibrational problem are discussed in Refs. [103–105].

8.3
Technical Details

As the dimension of the space spanned by the configurations is rather large – typical values range from 3000 to 3 000 000 – the Davidson algorithm [106] is employed for the diagonalization of $\mathcal{K}$. The preconditioner for the Davidson step is, as usual, the diagonal of the matrix. However, the SPFs are first unitarily transformed to so-called energy orbitals to yield the most diagonally dominant representation possible. Energy orbitals diagonalize the trace of the matrix (of operators) $\mathcal{H}^{(\kappa)}$, that is,

$$\left\langle \varphi_i^{(\kappa)} \middle| \sum_l \mathcal{H}_{ll}^{(\kappa)} \middle| \varphi_j^{(\kappa)} \right\rangle = \delta_{ij} e_j^{(\kappa)} \tag{8.9}$$

where the eigenvalues $e_j^{(\kappa)}$ are called *orbital energies*. When excited states are computed, one may additionally improve the preconditioner by inverting, say, a 1000×1000 block around the energy of interest. This accelerates the convergence of the Davidson iterations.

In particular, when highly excited states are computed, it is of importance to provide an initial state that has a large overlap with the desired eigenstate. Several methods are used to generate suitable initial states. One may take a Hartree product – or a sum of a few Hartree products if some symmetry requirement is to be satisfied – as initial state. The SPFs of this Hartree product are conveniently chosen as eigenfunctions of particle Hamiltonians, that is, of one- or low-dimensional Hamiltonians that operate on the degrees of freedom of one MCTDH particle only. These Hamiltonians are usually derived from the full Hamiltonian by freezing all coordinates, except those of the considered particle.

Another method to generate initial states is first to compute the ground state (or some other eigenstate) and then to apply some operator to this state. This operator is usually a harmonic excitation operator, $a^\dagger$, or a variant thereof [107]. An initial state generated by either method may be 'purified', that is, orthogonalized against previously computed eigenstates.

An MCTDH propagation always works, whatever the numbers of SPFs. If there are too few configurations, the propagation will be less accurate, but usually it still describes the overall features rather well. This is in contrast to improved relaxation, which fails to converge when the configurational space is too small. There is no problem in computing the ground state – this will always converge – but converging to excited states becomes more difficult the higher the excitation energy, or – more precisely – the higher the density of states.

After each diagonalization of $\mathcal{K}$, one obtains an approximation to the sought exact eigenstate of the Hamiltonian. This approximate eigenstate contains, of course, contributions from many exact eigenstates. Problematic here are contributions from eigenstates with energies below the one sought. If these contributions are not small, the algorithm 'notices' that the energy can be lowered by optimizing the SPFs for the low-energy contributions to the approximate eigenstate. (Remember that the coefficients are kept constant during SPF relaxation.) The energy hence decreases in the following improved relaxation iterations until – based on the overlap criterion – another eigenvector of $\mathcal{K}$ is taken. In this case, the energy jumps up and is lowered again in the following iteration steps. The energy keeps on oscillating [101] and convergence cannot be achieved. Presently the only known solution to this problem is to increase the numbers of SPFs. A larger configuration space, however, quickly makes the calculation unfeasible, at least for larger systems. Note that, if convergence is achieved, the computed energies are variational, that is, they are upper bounds to the exact ones [108, 109].

Improved relaxation has been applied quite successfully to a number of problems (see, for example, Refs. [59, 101, 107]). For four-atom systems (six dimensions), it is in general possible to compute all eigenstates of interest. For a system as large as $H_5O_2{}^+$ (15 dimensions), it was, of course, only possible to converge low-lying states [59]. In this case, *filter diagonalization* (see Section 6.2.2) was used to study high-lying states.

The improved relaxation algorithm may be used in block form, that is, one may start with a block of initial vectors, which then converge collectively to a set of eigenstates. Formally, the different wavefunctions are treated as electronic states of one 'super-wavefunction'. As the single-set algorithm is used (see Section 3.4), there is one set of SPFs for all wavefunctions. The mean fields are hence state-averaged mean fields and the Davidson routine is replaced by a block Davidson one. The block form of improved relaxation is more efficient than the single-vector one when several eigenstates are to be computed. However, using the block form requires considerably more memory.

9
Iterative Diagonalization of Operators

Fermín Huarte-Larrañaga and Uwe Manthe

9.1
Operators Defined by Propagation

In this section, a scheme is presented that solves the eigenvalue problem by Lanczos iterative diagonalization for operators $\hat{O}$ that are implicitly defined by real or imaginary-time propagation. Typical examples are given by the Boltzmann operator $\hat{O} = e^{-\beta\hat{H}}$ or the thermal flux operator $\hat{O} = e^{-\beta\hat{H}/2}\,\hat{F}\,e^{-\beta\hat{H}/2}$ [110]. Here the application of $\hat{O}$ on the wavefunction is an imaginary-time propagation with $\beta = -it$. (The application of the flux operator $\hat{F}$ after half of the propagation time is simple, if $\hat{F}$ is chosen to operate on a single degree of freedom only.)

As a result of the action of such an operator $\hat{O}$, usually the single-particle functions (SPFs) employed in the resulting $\hat{O}\Psi$ will differ from those in Ψ. This results in problems when one tries to apply the usual schemes for iterative diagonalization. To understand the problem, let us consider the sum of two MCTDH wavefunctions, Ψ_1 and Ψ_2,

$$|\Psi\rangle = |\Psi_1\rangle + |\Psi_2\rangle \tag{9.1}$$

Unless both wavefunctions share the same set of SPFs (which is in general not the case), the combined set of SPFs of Ψ_1 and Ψ_2 is required in the representation of Ψ. Therefore, if n_j SPFs are employed to represent Ψ_1 in the jth degree of freedom and m_j in the case of Ψ_2, then a total of $n_j + m_j$ SPFs is needed to represent Ψ. Thus, for a general system with f degrees of freedom, the operation of summing two MCTDH wavefunctions would increase the number of configurations required by a factor of 2^f (assuming $n_j = m_j$). This increase makes the evaluation of time integrals or the application of standard iterative diagonalization schemes prohibitive.

Consider the standard Lanczos scheme for diagonalizing a Hermitian operator $\hat{O}$. Starting from an initial guess function Ψ_0, a sequence of functions Ψ_m

Multidimensional Quantum Dynamics: MCTDH Theory and Applications.
Edited by Hans-Dieter Meyer, Fabien Gatti, and Graham A. Worth

ISBN: 978-3-527-32018-9

is recursively generated,

$$\Psi_m = N\left(\hat{O}\Psi_{m-1} - \sum_{i=0}^{m-1} \Psi_i \langle \Psi_i | \hat{O} | \Psi_{m-1} \rangle \right) \tag{9.2}$$

where N is a normalization constant. After M consecutive applications of this recursion relation, $\hat{O}$ is represented and diagonalized in the $\{\Psi_1, \ldots, \Psi_M\}$ subspace. The number of iterations M can be increased until convergence. This scheme, as presented so far, is not suitable for the MCTDH approach, since $\hat{O}\Psi_{m-1}$ and the different Ψ_i are represented in different SPF bases and cannot therefore be summed without expanding the configuration space.

9.2
A Modified Lanczos Scheme

To overcome the problem, a Lanczos-type iterative diagonalization scheme adapted for the MCTDH approach has been developed [111]. The key point in the modification is noticing that the iteration sequence in Equation (9.2) can also be seen as deriving Ψ_m from the orthogonalization of $\hat{O}\Psi_{m-1}$ with respect to all Ψ_i (with $i < m$). Let us consider a system with f degrees of freedom and write the $\hat{O}\Psi_{m-1}$ term of the recursion relation in MCTDH form as

$$\hat{O}\Psi_{m-1} = \sum_J A_J \Phi_J \tag{9.3}$$

where $J \equiv (j_1, \ldots, j_f)$ is a collective index and Φ_J represents a configuration. One can then define a projection operator onto the space spanned by the SPFs of $\hat{O}\Psi_{m-1}$:

$$\hat{P} = \sum_J |\Phi_J\rangle\langle\Phi_J| \tag{9.4}$$

Applying this operator, any wavefunction Ψ_i can be split into two components: $\hat{P}\Psi_i$ is completely located within the space spanned by the SPFs of $\hat{O}\Psi_{m-1}$, and $(1 - \hat{P})\Psi_i$ contains all the remaining orthogonal components. Thus, $\hat{O}\Psi_{m-1}$ only needs to be orthogonalized with respect to $\hat{P}\Psi_i$. The iterative scheme then reads

$$\Psi_m = \tilde{N}\left(\hat{O}\Psi_{m-1} - \sum_{i=0}^{m-1} \hat{P}\Psi_i \alpha_i \right) \tag{9.5}$$

where the α are the solutions of the following set of linear equations:

$$\langle \hat{P}\Psi_j | \hat{O}\Psi_{m-1} \rangle = \sum_{i=0}^{m-1} \langle \hat{P}\Psi_j | \hat{P}\Psi_i \rangle \alpha_i \tag{9.6}$$

for $j = 0, \ldots, m-1$. Equation (9.5) is identical to Equation (9.2) if the number of SPFs is increased until $\hat{P}$ converges towards unity. Moreover, in contrast to Equation (9.2), now $\hat{O}\Psi_{m-1}$ and $\hat{P}\Psi_i$ are represented by the same set of SPFs, and the new Ψ_m wavefunction resulting from the recursion can be obtained within this configuration space. Applying Equation (9.5) recursively, a set of M wavefunctions is generated and the eigenvalues of $\hat{O}$ are obtained by diagonalizing the matrix $\langle\Psi_i|\hat{O}|\Psi_j\rangle$. It should be noted that the maximum order of this modified Lanczos scheme is limited by the number of configurations employed in the representation of the MCTDH wavefunction. If the number of iterations exceeds the number of configurations, the right-hand side of Equation 9.5 vanishes.

9.3 The State-Averaged MCTDH Approach

An alternative approach to facilitate iterative diagonalization of operators $\hat{O}$ employs the state-averaged MCTDH approach [112]. Here a set of n_{packet} f-dimensional wavefunctions $\Psi_m(q_1, q_2, \ldots, q_f, t)$ is represented within a common SPF basis:

$$\Psi_m(q_1, \ldots, q_f, t) = \sum_{j_1=1}^{n_1} \cdots \sum_{j_f=1}^{n_f} A_{j_1,\ldots,j_f,m}(t)\varphi_{j_1}^{(1)}(q_1, t), \ldots, \varphi_{j_f}^{(f)}(q_f, t) \tag{9.7}$$

for $m = 1, \ldots, n_{\text{packet}}$. Using such a set of n_{packet} wavefunctions Ψ_m, the n_{packet} eigenstates of an operator $\hat{O}$ with the largest absolute eigenvalues can be calculated by successive application of the operator $\hat{O}$ to this set of states and subsequent re-orthonormalization of the resulting set $\hat{O}\Psi_m$, $m = 1, \ldots, n_{\text{packet}}$. Since a common set of SPFs is employed by the $\hat{O}\Psi_m$, linear combinations of the wavefunctions can be computed straightforwardly and the re-orthonormalization step does not pose any problems.

10 Correlation Discrete Variable Representation (CDVR)

Fermín Huarte-Larrañaga and Uwe Manthe

10.1 Introduction

In this section we discuss a procedure to deal with Hamiltonian operators of the form

$$\hat{H} = \sum_{m} \prod_{\kappa=1}^{f} \hat{h}_{m,\kappa}(q_\kappa) + V(q_1, \ldots, q_f) \tag{10.1}$$

where $V(q_1, \ldots, q_f)$ is a potential energy operator in general form. As explained in the previous chapters, the efficiency of the MCTDH approach is crucially determined by the ability to evaluate the potential energy integrals, $\langle \varphi_{j_1}^{(1)} \cdots \varphi_{j_f}^{(f)} | \hat{V} | \varphi_{\ell_1}^{(1)} \cdots \varphi_{\ell_f}^{(f)} \rangle$, without performing multidimensional integrations on the primitive $N_1 \times N_2 \times \cdots \times N_f$ grid. The correlation discrete variable representation (CDVR) offers an efficient scheme to calculate the corresponding integrals for general potential energy surfaces.

The discrete variable representation (DVR) [18, 113, 114] has become a widely used numerical technique in the field of quantum dynamics, employing sets of optimally localized functions. A common approach to construct a DVR is based on the diagonalization of the position operator matrix representation [113,114],

$$X_{nm} = \langle \chi_n | \hat{q} | \chi_m \rangle = \sum_{i} \langle \chi_n | \xi_i \rangle g_i \langle \xi_i | \chi_m \rangle \tag{10.2}$$

where $\{\chi_i(q)\}$, $i = 1, \ldots, N$, is a set of orthonormal functions, $\hat{q}$ is the position operator, $\mathbf{X}$ is its matrix representation, and g_i are the eigenvalues of the position matrix, which are taken as grid points. The eigenvectors of $\mathbf{X}$ define the DVR functions. For a given basis set, Equation (10.2) defines a unique DVR provided that the matrix is non-degenerate (see below, Section 10.4, in the case of degeneracy). The DVR functions are orthonormal and, by construction, diagonalize the position operator, $\langle \xi_i | \hat{q} | \xi_j \rangle = g_i \delta_{ij}$.

Multidimensional Quantum Dynamics: MCTDH Theory and Applications.
Edited by Hans-Dieter Meyer, Fabien Gatti, and Graham A. Worth

ISBN: 978-3-527-32018-9

At this point, it is useful to demonstrate that the $\{\xi_i\}$ functions are optimally localized basis functions within the given space. This means that they minimize the functional

$$\sum_n \langle \chi_n | (\hat{q} - \langle \chi_n | \hat{q} | \chi_n \rangle)^2 | \chi_n \rangle = \sum_n \langle \chi_n | \hat{q}^2 | \chi_n \rangle - \langle \chi_n | \hat{q} | \chi_n \rangle^2 \tag{10.3}$$

where $\{|\chi_i\rangle\}$ can be any orthonormal basis spanning the underlying space. Since $\sum_n \langle \chi_n | \hat{q}^2 | \chi_n \rangle$ is invariant under unitary transformations, minimizing Equation (10.3) is equivalent to maximizing $\sum_n \langle \chi_n | \hat{q} | \chi_n \rangle^2$. Consider now a unitary transformation between the basis $|\chi_i\rangle$ and the optimally localized basis $|\tilde{\xi}_i\rangle$:

$$|\chi_i\rangle = \sum_j U_{ij} |\tilde{\xi}_j\rangle \tag{10.4}$$

Then, the functional to be maximized can be rewritten as

$$\sum_n \langle \chi_n | \hat{q} | \chi_n \rangle^2 = \sum_n (\mathbf{U}^\dagger \tilde{\mathbf{X}} \mathbf{U})^2_{nn} = \sum_n (\mathrm{e}^{\mathrm{i}\mathbf{A}}\, \tilde{\mathbf{X}}\, \mathrm{e}^{-\mathrm{i}\mathbf{A}})^2_{nn} \tag{10.5}$$

where $\tilde{\mathbf{X}}$ is the matrix representation of the position operator in the optimally localized basis, $\tilde{X}_{ij} = \langle \tilde{\xi}_i | \hat{q} | \tilde{\xi}_j \rangle$, and $\mathbf{U} = \mathrm{e}^{-\mathrm{i}\mathbf{A}}$, with $\mathbf{A}$ Hermitian. Here $(\cdot)^2_{nn}$ is to be read as $((\cdot)_{nn})^2$. Since the extremum condition is satisfied for $\mathbf{A} = 0$, one finds

$$\frac{\partial}{\partial A_{k\ell}} \sum_n (\mathrm{e}^{\mathrm{i}\mathbf{A}}\, \tilde{\mathbf{X}}\, \mathrm{e}^{-\mathrm{i}\mathbf{A}})^2_{nn} \Bigg|_{\mathbf{A}=0} = 2\mathrm{i} \tilde{X}_{\ell k} (\tilde{X}_{kk} - \tilde{X}_{\ell\ell}) = 0 \qquad \text{for all } k, \ell. \tag{10.6}$$

When the eigenvalues of the coordinate matrix are non-degenerate, Equation (10.6) yields the condition $\tilde{X}_{k\ell} = \tilde{X}_{kk}\delta_{k\ell}$. This can be found by diagonalizing the coordinate matrix, and the resulting eigenfunctions are identical to the DVR basis functions ($\xi \leftrightarrow \tilde{\xi}$).

In the DVR approach, the potential energy matrix is then approximated using a collocation scheme as

$$\langle \xi_k | V(q) | \xi_\ell \rangle = V(g_\ell) \delta_{k\ell}. \tag{10.7}$$

Since the wavefunction is typically more structured than the potential energy function, the numerical error resulting from the quadrature scheme is usually smaller than the error due to the finite size of the basis set [18].

10.2
Time-Dependent Discrete Variable Representation

A time-dependent discrete variable representation (TDDVR) approach could, in principle, be used for the evaluation of the matrix elements in the MCTDH equations of motion, evaluating the potential energy using a time-dependent grid representation [37]. Analogously to the original time-independent scheme, in the TDDVR formulation the time-dependent grid is introduced by diagonalizing the coordinate matrix in the time-dependent basis representation,

$$\langle \varphi_{\ell_\kappa}^{(\kappa)}(q_\kappa,t)|\hat{q}_\kappa|\varphi_{m_\kappa}^{(\kappa)}(q_\kappa,t)\rangle = \sum_{j=1}^{n_\kappa} \langle \varphi_{\ell_\kappa}^{(\kappa)}(q_\kappa,t)|\xi_j^{(\kappa)}\rangle g_j^{(\kappa)} \langle \xi_j^{(\kappa)}|\varphi_{m_\kappa}^{(\kappa)}(q_\kappa,t)\rangle \quad (10.8)$$

This expression is already written within an MCTDH context, κ labels a given coordinate of the multidimensional system and $\varphi_{\ell_\kappa}^{(\kappa)}(q_\kappa,t)$ denotes a single-particle function. The eigenvalues, $g_j^{(\kappa)}$, and the corresponding eigenvectors are, obviously, time dependent. These eigenvalues are the grid points that would be employed in the TDDVR scheme. In order to evaluate the potential energy elements, the potential energy matrix is transformed from the single-particle function (SPF) basis $\{\varphi\}$, to the grid-point basis $\{\xi\}$ by a unitary transformation. Then, the potential energy matrix is approximated in the time-dependent grid-point representation as

$$\langle \xi_{j_1}^{(1)} \cdots \xi_{j_f}^{(f)}|V(q_1,\ldots,q_f)|\xi_{\ell_1}^{(1)} \cdots \xi_{\ell_f}^{(f)}\rangle = V(g_{j_1}^{(1)},\ldots,g_{j_f}^{(f)})\,\delta_{j_1\ell_1}\cdots\delta_{j_f\ell_f} \quad (10.9)$$

This approximation could allow an efficient calculation of the Hamiltonian matrix that propagates the A coefficients of the MCTDH wavefunction. The one-dimensional mean-field potentials appearing in the equations of motion of the SPFs could correspondingly be approximated as

$$\begin{aligned}&\langle \xi_{j_1}^{(1)} \cdots \xi_{j_{\kappa-1}}^{(\kappa-1)} \xi_{j_{\kappa+1}}^{(\kappa+1)} \cdots \xi_{j_f}^{(f)}|V(q_1,\ldots,q_f)|\xi_{\ell_1}^{(1)} \cdots \xi_{\ell_{\kappa-1}}^{(\kappa-1)} \xi_{\ell_{\kappa+1}}^{(\kappa+1)} \cdots \xi_{\ell_f}^{(f)}\rangle \\ &= V(g_{j_1}^{(1)},\ldots,g_{j_{\kappa-1}}^{(\kappa-1)},q_\kappa,g_{j_{\kappa+1}}^{(\kappa-1)},\ldots,g_{j_f}^{(f)})\,\delta_{j_1\ell_1}\cdots\delta_{j_{\kappa-1}\ell_{\kappa-1}}\delta_{j_{\kappa+1}\ell_{\kappa+1}}\cdots\delta_{j_f\ell_f}\end{aligned} \quad (10.10)$$

However, a serious problem arises from the use of TDDVR as described above, and severe errors are introduced into the MCTDH scheme [115]. This can be understood as follows. The MCTDH expansion coefficients $A_{j_1,\ldots,j_f}(t)$ describe the correlation between the different degrees of freedom, while the separable dynamics is included in the time evolution of the SPFs $\varphi_{j_\kappa}^{(\kappa)}(q_\kappa,t)$. The size of the time-dependent SPF basis depends only on the amount of correlation and is thus independent of the separable dynamics. The SPF basis is in general small compared with the size of the underlying primitive grid, and a TDDVR based on this basis is hence adapted only to the description of the correlation. The number of time-dependent grid points is therefore too small to provide an accurate evaluation of the separable parts of the potential.

In order to illuminate this point, let us consider a completely uncorrelated system, which would be represented within the MCTDH frame by a single configuration with one time-dependent basis function for each degree of freedom. The corresponding separable potential energy function reads

$$V(q_1,\ldots,q_f) = V_1(q_1) + V_2(q_2) + \cdots + V_f(q_f) \tag{10.11}$$

The exact value of the potential energy matrix element in the time-dependent basis representation is

$$\langle \varphi^{(1)} \cdots \varphi^{(f)} | V | \varphi^{(1)} \cdots \varphi^{(f)} \rangle = \langle \varphi^{(1)} | V_1 | \varphi^{(1)} \rangle + \cdots + \langle \varphi^{(f)} | V_f | \varphi^{(f)} \rangle \tag{10.12}$$

However, the single potential energy matrix would be approximated within the TDDVR scheme as

$$\langle \varphi^{(1)} \cdots \varphi^{(f)} | V | \varphi^{(1)} \cdots \varphi^{(f)} \rangle \approx V_1(\langle \varphi^{(1)} | \hat{q}_1 | \varphi^{(1)} \rangle) + \cdots + V_f(\langle \varphi^{(f)} | \hat{q}_f | \varphi^{(f)} \rangle) \tag{10.13}$$

This expression does generally not yield a reasonable approximation to the potential energy value, because the $\varphi^{(\kappa)}$, the one-dimensional functions of the Hartree product, are not localized.

10.3 Correlation Discrete Variable Representation

The problems of the TDDVR scheme can be solved by a modification of the potential quadrature approximation in Equation (10.9). Instead of using the straightforward collocation approximation, the potential energy matrix elements are calculated as

$$\begin{aligned}
&\langle \xi_{j_1}^{(1)} \cdots \xi_{j_f}^{(f)} | V(q_1,\ldots,q_f) | \xi_{\ell_1}^{(1)} \cdots \xi_{\ell_f}^{(f)} \rangle \\
&\quad = V(g_{j_1}^{(1)},\ldots,g_{j_f}^{(f)})\, \delta_{j_1\ell_1} \cdots \delta_{j_f\ell_f} \\
&\qquad + \sum_{\kappa=1}^{f} \langle \xi_{j_\kappa}^{(\kappa)} | \Delta V(g_{j_1}^{(1)},\ldots,g_{j_{\kappa-1}}^{(\kappa-1)}, q_\kappa, g_{j_{\kappa+1}}^{(\kappa+1)},\ldots,g_{j_f}^{(f)}) | \xi_{\ell_\kappa}^{(\kappa)} \rangle \\
&\qquad \times \delta_{j_1\ell_1} \cdots \delta_{j_{\kappa-1}\ell_{\kappa-1}}\, \delta_{j_{\kappa+1}\ell_{\kappa+1}} \cdots \delta_{j_f\ell_f}
\end{aligned} \tag{10.14}$$

where the second term on the right-hand side is a correction,

$$\begin{aligned}
&\langle \xi_{j_\kappa}^{(\kappa)} | \Delta V(g_{j_1}^{(1)}, \dots, g_{j_{\kappa-1}}^{(\kappa-1)}, q_\kappa, g_{j_{\kappa+1}}^{(\kappa+1)}, \dots, g_{j_f}^{(f)}) | \xi_{\ell_\kappa}^{(\kappa)} \rangle \\
&\quad = \langle \xi_{j_\kappa}^{(\kappa)} | V(g_{j_1}^{(1)}, \dots, g_{j_{\kappa-1}}^{(\kappa-1)}, q_\kappa, g_{j_{\kappa-1}}^{(\kappa-1)}, \dots, g_{j_f}^{(f)}) | \xi_{\ell_\kappa}^{(\kappa)} \rangle \\
&\qquad\qquad - V(g_{j_1}^{(1)}, \dots, g_{j_f}^{(f)})\, \delta_{j_\kappa \ell_\kappa}
\end{aligned} \tag{10.15}$$

This modified time-dependent DVR is called correlation discrete variable representation (CDVR) [115]. In this scheme, the correction terms in Equation (10.15) are introduced to account implicitly for the separable parts of the potential energy. The CDVR thus gives a correct description of separable potentials

$$\begin{aligned}
&\langle \xi_{j_1}^{(1)} \cdots \xi_{j_f}^{(f)} | V_1(q_1) + \cdots + V_f(q_f) | \xi_{\ell_1}^{(1)} \cdots \xi_{\ell_f}^{(f)} \rangle \\
&\quad = \sum_{\kappa=1}^{f} \delta_{j_1 \ell_1} \cdots \delta_{j_{\kappa-1} \ell_{\kappa-1}} \langle \xi_{\ell_\kappa}^{(\kappa)} | V_\kappa(q_\kappa) | \xi_{j_\kappa}^{(\kappa)} \rangle \delta_{j_{\kappa+1} \ell_{\kappa+1}} \cdots \delta_{j_f \ell_f}
\end{aligned} \tag{10.16}$$

which is the exact expression of the potential matrix elements in Equation (10.12).

It should be noted that the potential matrix element appearing on the right-hand side of Equation (10.15) employs the same potential values that are already used in Equation (10.10). Hence the CDVR scheme does not significantly increase the computational effort compared to the simple TDDVR approach. In calculations employing the CDVR approach, the evaluation of the potential energy function on the time-dependent grid is typically the computationally expensive part: $n_1 \times n_2 \times \cdots \times n_f \times (1 + \sum_{\kappa=1}^{f} N_\kappa / n_\kappa)$ potential evaluations are required to compute all the matrix elements required by the MCTDH equations of motion.

The CDVR accuracy was originally tested for the photodissociation of NOCl in Ref. [115], with the error introduced found to be an order of magnitude smaller than the MCTDH truncation error. This initial finding has been confirmed in a large number of subsequent applications. Since the MCTDH equations of motion conserve the energy expectation value in the case of an accurate quadrature, the variation of the energy expectation value can be used to monitor the CDVR quadrature error [116]. In the limit of a complete basis set of single-particle functions, the CDVR approach becomes exact.

As commented in the previous chapters, the number of single-particle functions used in MCTDH simulations is typically much smaller than the number of basis functions used in standard approaches. The corresponding CDVR grid employed to calculate matrix elements is consequently rather small, often containing just a few points. If one attempts to overconverge the wavefunction by employing more SPFs than necessary, the almost redundant SPFs will have almost zero population. These low populated SPFs tend to reside in regions where the wavepacket is very small. The weights of the SPFs, however, are ignored when diagonalizing the position operator, and CDVR grid points are created in unwanted regions. This causes CDVR to lose accuracy. Thus, the accuracy of the CDVR quadrature cannot be arbitrarily improved in MCTDH calculations by adding basis functions at will, once the wavefunction is converged.

10.4 Symmetry-Adapted Correlation Discrete Variable Representation

Symmetry-related issues can become particularly important in MCTDH calculations that employ the CDVR scheme. If the Hamiltonian and the initial wavefunction are symmetric (or antisymmetric) with respect to $q \to -q$, then all single-particle functions will become even (odd) functions of q during the propagation. As has been commented above, whenever the position operator matrix representation has degenerate eigenvalues, a unitary transformation cannot be unequivocally determined, nor can a unique DVR be defined. Degeneracy affects the time-dependent DVR in the same way and $X_{ij}^{(\kappa)}(t)$ has only vanishing eigenvalues if all single-particle functions for the κth degree of freedom are symmetric (even or odd) functions of q_κ.

Such a degeneracy leads to a collapse of the time-dependent DVR grid and a complete breakdown of the CDVR approach, which was first observed [79, 117, 118] in calculations of the initial-state selected reaction probabilities for

$$H + H_2(v{=}0, j{=}0) \to H_2 + H$$

Simple ways to avoid this problem can be devised [118]: one can, for example, diagonalize the $\hat{q}^2$ operator instead of the $\hat{q}$ operator to obtain the (C)DVR grid points and the basis to grid transformation matrix.

10.5 Multidimensional Correlation Discrete Variable Representation

As described in the previous chapters, the efficiency of the MCTDH method can be further enhanced by using mode combination [34, 62] or even multilayer descriptions [119], grouping several physical coordinates together into a single logical coordinate and thereby introducing multidimensional single-particle functions. The CDVR corresponding to a set of multidimensional single-particle functions requires a multidimensional non-direct-product quadrature. Only recently has a practical scheme to construct multidimensional non-direct-product DVRs been presented [120] and utilized to introduce a corresponding CDVR approach [116].

In Section 10.1 it was shown that, for the one-dimensional case, the DVR basis constitutes an optimally localized basis set. Analogous criteria to Equation (10.3) can be established for basis functions that are optimally localized in multiple dimensions and therefore maximize

$$\sum_i \sum_n \langle \chi_n | \hat{q}_i | \chi_n \rangle^2 \tag{10.17}$$

where the index i runs over multiple coordinates. The Jacobi simultaneous diagonalization scheme [121] can be employed to find the extremum and thereby to obtain a non-direct-product multidimensional DVR. Once the multidimensional DVR is obtained, the potential energy matrix element can be approximated as

$$\langle \varphi_i | V(q_1, q_2, \ldots) | \varphi_j \rangle \approx \sum_k \langle \varphi_i | \xi_k \rangle V(g_k^{(1)}, g_k^{(2)}, \ldots) \langle \xi_k | \varphi_j \rangle \tag{10.18}$$

where $\langle \varphi_i | \xi_k \rangle$ is the matrix corresponding to the unitary transformation between the $\varphi_i(q_1, q_2, \ldots)$ basis and the localized basis $\xi_k(q_1, q_2, \ldots)$ and $g_k^{(i)} = \langle \xi_k | \hat{q}_i | \xi_k \rangle$ are the grid points. At this point, it should be mentioned that in multidimensional DVRs the grid points are generally not the eigenvalues of the coordinate matrices. However, the multidimensional basis functions are optimally localized and therefore the quadrature is still rather accurate. As in the one-dimensional case, the numerical inaccuracy resulting from the quadrature scheme is small compared to the basis set truncation error if the wavefunction is more structured than the potential energy function.

Using this multidimensional DVR approach, the CDVR approach was generalized to MCTDH calculations with mode combination [116]. Consider the wavefunction expressed in combined modes,

$$\Psi(Q_1, \ldots, Q_p, t) = \sum_{j_1=1}^{n_1} \cdots \sum_{j_p=1}^{n_p} A_{j_1,\ldots,j_p}(t) \prod_{\kappa}^{p} \varphi_{j_\kappa}^{(\kappa)}(Q_\kappa, t) \tag{10.19}$$

where Q_κ denotes a multidimensional particle coordinate and $\varphi_{j_\kappa}^{(\kappa)}(Q_\kappa, t)$ a multimode single-particle function. The transformation between the original time-dependent multidimensional basis $\varphi_i^{(\kappa)}(Q_\kappa, t)$ and the time-dependent multidimensional grid states $\xi_i^{(\kappa)}(Q_\kappa, t)$ is obtained by simultaneous diagonalization of the matrices

$$X_{\ell m}^{(\kappa,i)} = \langle \varphi_\ell^{(\kappa)}(Q_\kappa, t) | \hat{q}_i^{(\kappa)} | \varphi_m^{(\kappa)}(Q_\kappa, t) \rangle \tag{10.20}$$

for each κ and for $i = 1, 2, \ldots, d_\kappa$, where d_κ denotes the dimension of the κth combined mode. The matrix elements of the potential energy, which are now considered as a function of multidimensional coordinates, are approximated as

$$\begin{aligned}
&\langle \xi_{j_1}^{(1)}(Q_1,t)\cdots\xi_{j_p}^{(p)}(Q_p,t)|V(Q_1,\ldots,Q_p)|\xi_{\ell_1}^{(1)}(Q_1,t)\cdots\xi_{\ell_p}^{(p)}(Q_p,t)\rangle \\
&\quad = V(G_{l_1}^{(1)},\ldots,G_{l_f}^{(f)})\,\delta_{j_1\ell_1}\cdots\delta_{j_f\ell_f} \\
&\qquad + \sum_{\kappa=1}^{p}\langle \xi_{j_\kappa}^{(\kappa)}(Q_\kappa,t)|\Delta V(G_{j_1}^{(1)},\ldots,G_{j_{\kappa-1}}^{(\kappa-1)},Q_\kappa,G_{j_{\kappa+1}}^{(\kappa+1)},\ldots,G_{j_p}^{(p)})|\xi_{\ell_\kappa}^{(\kappa)}(Q_\kappa,t)\rangle \\
&\qquad\qquad \times\delta_{l_1\ell_1}\cdots\delta_{j_{\kappa-1}\ell_{\kappa-1}}\delta_{j_{\kappa+1}\ell_{\kappa+1}}\cdots\delta_{j_f\ell_f}
\end{aligned} \tag{10.21}$$

The former equation is formally identical to Equation (10.14) only that here the multidimensional coordinate $Q_\kappa = (q_1^{(\kappa)},\ldots,q_{d_\kappa}^{(\kappa)})$ replaces the one-dimensional coordinate q_κ of Equation (10.14), and $G_j^{(\kappa)}$ are the multi-dimensional time-dependent grid points that replace the one-dimensional grid points $g_j^{(\kappa)}$ in Equation (10.14),

$$G_j^{(\kappa)} = \left(\langle \xi_j^{(\kappa)}(Q_\kappa,t)|\hat{q}_1^{(\kappa)}|\xi_j^{(\kappa)}(Q_\kappa,t)\rangle,\ldots,\langle \xi_j^{(\kappa)}(Q_\kappa,t)|\hat{q}_{d_k}^{(\kappa)}|\xi_j^{(\kappa)}(Q_\kappa,t)\rangle\right) \tag{10.22}$$

The correction terms in Equation (10.21) are, analogously to Equation (10.15), given by

$$\begin{aligned}
&\langle \xi_{j_\kappa}^{(\kappa)}(Q_\kappa,t)|\Delta V(G_{j_1}^{(1)},\ldots,G_{j_{\kappa-1}}^{(\kappa-1)},Q_\kappa,G_{j_{\kappa+1}}^{(\kappa+1)},\ldots,Q_{j_p}^{(p)})|\xi_{\ell_\kappa}^{(\kappa)}(Q_\kappa,t)\rangle \\
&\quad = \langle \xi_{j_\kappa}^{(\kappa)}(Q_\kappa,t)|V(G_{j_1}^{(1)},\ldots,G_{j_{\kappa-1}}^{(\kappa-1)},Q_\kappa,G_{j_{\kappa-1}}^{(\kappa-1)},\ldots,Q_{j_p}^{(p)})|\xi_{\ell_\kappa}^{(\kappa)}(Q_\kappa,t)\rangle \\
&\qquad - V(G_{j_1}^{(1)},\ldots,G_{j_p}^{(p)})\,\delta_{j_\kappa\ell_\kappa}
\end{aligned} \tag{10.23}$$

11 Potential Representations (potfit)

Hans-Dieter Meyer, Fabien Gatti and Graham A. Worth

11.1 Expansion in Product Basis Sets

For an optimal performance of the MCTDH method, it is essential that the Hamiltonian operator is of product form, Equation (3.34). As already discussed (see also Section 12), the kinetic energy operator usually satisfies product form, and one is left with the problem of bringing the potential to product form.

Let us first note that the product form is not as artificial as it may look at first glance. A Taylor expansion of the potential satisfies product form, and many model potentials are of product form from the outset.

The most direct way to the product form is an expansion in a product basis. Hence we approximate some given potential V by

$$V^{\text{app}}(Q^{(1)}, \ldots, Q^{(p)}) = \sum_{j_1=1}^{m_1} \cdots \sum_{j_p=1}^{m_p} C_{j_1,\ldots,j_p} \, v_{j_1}^{(1)}(Q^{(1)}) \cdots v_{j_p}^{(p)}(Q^{(p)}) \qquad (11.1)$$

where p denotes, as before, the number of particles (or combined modes), and $Q^{(\kappa)}$ is the one- or multidimensional coordinate of the κth particle. The basis functions $v_{j_\kappa}^{(\kappa)}(Q^{(\kappa)})$ are called *single-particle potentials* (SPPs). The expansion orders, m_κ, must be chosen large enough to achieve an accurate expansion. On the other hand, they should be as small as possible, because the numerical effort of an MCTDH propagation grows (almost linearly) with the number of potential terms, that is, with the product of the expansion orders. Hence both the expansion coefficients and the SPPs should be optimized to provide the best approximative potential for a given set of expansion orders.

Before we turn to analyse this optimization problem, we simplify the problem somewhat. When DVRs are used to represent the wavefunctions, one needs to know the potential only at grid points. This allows us to work in finite-dimensional vector spaces. The full potential is now given by the set of

Multidimensional Quantum Dynamics: MCTDH Theory and Applications.
Edited by Hans-Dieter Meyer, Fabien Gatti, and Graham A. Worth

ISBN: 978-3-527-32018-9

numbers

$$V(Q_{i_1}^{(1)}, \ldots, Q_{i_p}^{(p)}) = V_{i_1,\ldots,i_p} \tag{11.2}$$

where $Q_i^{(\kappa)}$ denotes the coordinate of the ith grid point of the κth grid. Similarly we write

$$V_{i_1,\ldots,i_p}^{\text{app}} = \sum_{j_1=1}^{m_1} \cdots \sum_{j_p=1}^{m_p} C_{j_1,\ldots,j_p} \, v_{i_1 j_1}^{(1)} \cdots v_{i_p j_p}^{(p)} \tag{11.3}$$

with $v_{ij}^{(\kappa)} = v_j^{(\kappa)}(Q_i^{(\kappa)})$. The SPPs are assumed to be orthonormal on the grid, $\sum_i v_{ij}^{(\kappa)} v_{i\ell}^{(\kappa)} = \delta_{j\ell}$. Throughout this section we will use the letters i and k to label grid points and j and ℓ to label SPPs.

The task is now to determine optimal coefficients and SPPs. To this end we minimize

$$\Delta^2 = \sum_{i_1=1}^{N_1} \cdots \sum_{i_f=1}^{N_p} (V_{i_1,\ldots,i_p} - V_{i_1,\ldots,i_p}^{\text{app}})^2 \, w_{i_1,\ldots,i_p}^2 = \sum_I (V_I - V_I^{\text{app}})^2 \, w_I^2 \tag{11.4}$$

where I denotes a composite index that runs over all grid points and w_I is a weight function. The introduction of weights is important, because one does not want to minimize the error equally well for all grid points. The product representation should be accurate near the potential minimum and, for example, in the vicinity of a transition state, but may be rather inaccurate at regions where the potential is high, simply because the wavepacket does not visit regions with high potential values. The optimal choice for w^2 would be the time-averaged modulus square of the wavepacket, but this is impracticable. A simple and convenient choice is the Boltzmann weight

$$w_I = \mathrm{e}^{-\beta V_I} \tag{11.5}$$

with a suitable value for the parameter β.

11.2 Optimizing the Coefficients

Let us now determine the optimal coefficients $C_{j_1,\ldots,j_p} \equiv C_J$ for a fixed set of SPPs. To this end it is convenient to introduce a matrix–vector notation and to define configuration matrices $\mathbf{\Omega}$, that is, products of SPPs, and the diagonal weight matrix, $\mathbf{W}$,

$$\Omega_{I,J} = v_{i_1 j_1}^{(1)} \cdots v_{i_p j_p}^{(p)} \tag{11.6}$$

$$W_{I,K} = w_I^2 \, \delta_{I,K} \tag{11.7}$$

Because the SPPs are orthonormal, $\mathbf{\Omega}^{\mathrm{T}}\mathbf{\Omega} = \mathbf{1}$ holds. Equation (11.3) can be written compactly as

$$\mathbf{V}^{\mathrm{app}} = \mathbf{\Omega}\mathbf{C} \tag{11.8}$$

Minimizing Δ^2 of Equation (11.4) by varying only the coefficients yields

$$\mathbf{C} = (\mathbf{\Omega}^{\mathrm{T}}\mathbf{W}\mathbf{\Omega})^{-1}\mathbf{\Omega}^{\mathrm{T}}\mathbf{W}\mathbf{V} \tag{11.9}$$

and the approximate potential, Equation (11.3), may be written as

$$\mathbf{V}^{\mathrm{app}} = \mathbf{P}\mathbf{V} \tag{11.10}$$

with

$$\mathbf{P} = \mathbf{\Omega}(\mathbf{\Omega}^{\mathrm{T}}\mathbf{W}\mathbf{\Omega})^{-1}\mathbf{\Omega}^{\mathrm{T}}\mathbf{W} \tag{11.11}$$

where $\mathbf{P}$ is a projector, $\mathbf{P}^2 = \mathbf{P}$, but in general not a Hermitian one because $\mathbf{P}$ satisfies $\mathbf{P}^{\mathrm{T}}\mathbf{W} = \mathbf{W}\mathbf{P}$.

It should be clear that V^{app} reproduces the original potential *exactly* when the expansion orders approach the number of grid points, $m_\kappa = N_\kappa$. In this case $\mathbf{\Omega}$ becomes unitary and $\mathbf{P} = \mathbf{1}$.

11.3 Optimizing the Basis

The final task is to optimize the SPPs. Plugging Equation (11.10) into Equation (11.4) yields

$$\Delta^2 = \mathbf{V}^{\mathrm{T}}\mathbf{W}\mathbf{V} - \mathbf{V}^{\mathrm{T}}\mathbf{W}\mathbf{P}\mathbf{V} \tag{11.12}$$

that is, one has to optimize the SPPs such that $\mathbf{V}^{\mathrm{T}}\mathbf{W}\mathbf{P}\mathbf{V}$ becomes maximal. The resulting equations are unfortunately rather cumbersome and their solution is unclear. The problem originates from the fact that a matrix has to be inverted. However, the equations become much simpler if weights are ignored. (We will reintroduce weights later.) Setting $\mathbf{W} = \mathbf{1}$ yields

$$\mathbf{C} = \mathbf{\Omega}^{\mathrm{T}}\mathbf{V} \tag{11.13}$$

$$\mathbf{P} = \mathbf{\Omega}\mathbf{\Omega}^{\mathrm{T}} \tag{11.14}$$

The explicit form of Equation (11.13) reads

$$C_{j_1,\ldots,j_p} = \sum_{i_1=1}^{N_1} \cdots \sum_{i_p=1}^{N_p} v_{i_1 j_1}^{(1)} \cdots v_{i_p j_p}^{(p)} V_{i_1 \ldots i_p} \tag{11.15}$$

which shows that the coefficients are simply given by overlap of the SPPs with the potential.

We now turn to the problem of optimizing the SPPs, and first note that Equation (11.12) simplifies to $\Delta^2 = \|\mathbf{V}\|^2 - \|\mathbf{C}\|^2$. Hence the norm of the coefficient vector has to be maximized under the constraint that the SPPs remain orthonormal. This leads to the variational equation

$$\sum_J \left(\sum_I v_{i_1 j_1}^{(1)} \cdots v_{i_p j_p}^{(p)} V_{i_1 \ldots i_p} \right)^2 - \sum_\kappa \sum_{j,\ell} \epsilon_{j\ell}^{(\kappa)} \left(\sum_i v_{ij}^{(\kappa)} v_{i\ell}^{(\kappa)} - \delta_{j\ell} \right) = \max \quad (11.16)$$

where the $\epsilon_{j\ell}^{(\kappa)}$ are Lagrange parameters. Using the techniques developed in Section 8, one readily finds the solution of this variational principle,

$$\sum_i \tilde{\varrho}_{ki}^{(\kappa)} v_{ij}^{(\kappa)} = \sum_\ell \epsilon_{j\ell}^{(\kappa)} v_{k\ell}^{(\kappa)} \quad (11.17)$$

where the modified potential density matrices $\tilde{\varrho}_{ki}^{(\kappa)}$ are defined by [122]

$$\tilde{\varrho}_{ki}^{(\kappa)} = \sum_J^{\kappa} C_{j_1,\ldots,k,\ldots,j_p}^{(\kappa)} C_{j_1,\ldots,i,\ldots,j_p}^{(\kappa)} \quad (11.18)$$

and where $\sum_J^\kappa$ denotes a summation over indices of all particles except the κth one. The $C^{(\kappa)}$ are partially transformed coefficients, where the κth index is not transformed, but refers to grid points:

$$C_{j_1,\ldots,j_{\kappa-1},i_\kappa,j_{\kappa+1},\ldots,j_p}^{(\kappa)} = \sum_I^{\kappa} v_{i_1 j_1}^{(1)} \cdots v_{i_{\kappa-1} j_{\kappa-1}}^{(\kappa-1)} v_{i_{\kappa+1} j_{\kappa+1}}^{(\kappa+1)} \cdots v_{i_p j_p}^{(p)} V_{i_1,\ldots,i_p} \quad (11.19)$$

Similarly as in Chapter 8, one may remove the Lagrange parameters ϵ,

$$\sum_k \left(\delta_{k'k} - \sum_\ell v_{k'\ell}^{(\kappa)} v_{k\ell}^{(\kappa)} \right) \sum_i \tilde{\varrho}_{ki}^{(\kappa)} v_{ij}^{(\kappa)} = 0 \quad (11.20)$$

There is an obvious solution to this equation. As optimal SPPs one takes the m_κ eigenvectors of $\tilde{\varrho}^{(\kappa)}$ with largest eigenvalues. However, a solution can be obtained only iteratively. Starting with some guess for the SPPs, one builds the modified density matrices, diagonalizes these to obtain improved SPPs, and again builds the modified density matrices. The process is continued till self-consistency is reached. This algorithm is equivalent to the one discussed in the original potfit article [122]. Unfortunately it tends to converge to a local minimum, the absolute minimum of Δ^2 may not be found this way.

11.4
The potfit Algorithm

To avoid an iterative process, which is expensive, we simplify the algorithm further and replace the modified potential density matrices, $\tilde{\varrho}^{(\kappa)}$, by the potential density matrices $\varrho^{(\kappa)}$

$$\varrho^{(\kappa)}_{kk'} = \sum_I^\kappa V_{i_1,\ldots,i_{\kappa-1},k,i_{\kappa+1},\ldots,i_p} V_{i_1,\ldots,i_{\kappa-1},k',i_{\kappa+1},\ldots,i_p} = \sum_{I^\kappa} V_{I^\kappa_k} V_{I^\kappa_{k'}} \tag{11.21}$$

The eigenvectors of this positive definite matrix, ordered with respect to their eigenvalues, are taken as SPPs. This particular choice of SPPs is given a special name, *natural potentials*, and the resulting algorithm is called *potfit* [34,122,123]. The *potfit* algorithm was later rederived by mathematicians as a generalized low-rank approximation [124, 125].

The natural potentials are not the optimal SPPs – the latter are the solutions of Equation (11.20). However, for the special case of two particles, it is known [126] that the natural potentials provide an optimal approximation. For more than two particles, potfit is certainly not optimal, but we have observed that the algorithm usually provides fits that are close to optimal. Typically, the root-mean-square (rms) error, $\Delta_{\mathrm{rms}} = \sqrt{\Delta^2/N_{\mathrm{tot}}}$, drops by 10–20% when fully optimizing the SPPs. Only when too small values for the expansion orders were used, resulting in a rather inaccurate fit, do we observe a lowering of the rms error by about 40%. However, these investigations could only be done on small three-dimensional systems.

The N_{tot} that appears in the definition of the rms error is the total grid size, $N_{\mathrm{tot}} = \Pi_\kappa N_\kappa$. When weights are used, N_{tot} should be replaced by S, where $S = \sum_I w_I^2$ is the sum over all weights.

Consider the case where one uses a complete set of SPPs for all particles except the κth one. In this case, the rms error can be directly related to the eigenvalues of the neglected natural potentials [122],

$$\Delta_{\mathrm{rms}} = \sqrt{\sum_{j=m_\kappa+1}^{N_\kappa} \lambda_j^{(\kappa),\mathrm{red}}} \tag{11.22}$$

where $\lambda_j^{(\kappa),\mathrm{red}}$ denotes a reduced eigenvalue of the density matrix $\varrho^{(\kappa)}$, that is, $\lambda_j^{(\kappa),\mathrm{red}} = \lambda_j^{(\kappa)}/S$ (remember, $S = N_{\mathrm{tot}}$ if weights are not included). This leads to a bound on the rms error for the general case,

$$\Delta_{\mathrm{rms}} \leq \sqrt{\sum_{\kappa=1}^{p} \sum_{j=m_\kappa+1}^{N_\kappa} \lambda_j^{(\kappa),\mathrm{red}}} \tag{11.23}$$

This bound, which is also a reasonable estimate, is very useful for determining appropriate expansion orders.

11.5 Contraction Over One Particle

The number of expansion terms, $\prod_{\kappa=1}^{p} m_\kappa$, appearing in Equation (11.3) should be as small as possible, because this number determines the speed of an MCTDH calculation using this product-form potential. At virtually no cost, one can reduce the number of expansion terms by one expansion order m_κ. To this end we define *contracted expansion coefficients*

$$D^{(\kappa)}_{j_1,\ldots,j_{\kappa-1},i_\kappa,j_{\kappa+1},\ldots,j_f} = \sum_{j_\kappa=1}^{m_\kappa} C_{j_1,\ldots,j_f}\, v^{(\kappa)}_{i_\kappa j_\kappa} \tag{11.24}$$

As the expansion order m_κ will no longer appear in the working equation for $V^{\rm app}$, one may set $m_\kappa = N_\kappa$ and use the full set of natural potentials for this particular particle. In this case one performs a unitary transformation on the κth index of the potential to obtain the coefficient C, and then performs the inverse unitary transformation on the κth index of the coefficient. Hence, effectively, there is no transformation on the κth index and $D^{(\kappa)}$ becomes identical to $C^{(\kappa)}$, which is defined by Equation (11.19). Using contraction, the approximate potential is written as

$$\begin{aligned} V^{\rm app}_{i_1,\ldots,i_p} &= \sum_{j_1=1}^{m_1} \cdots \sum_{j_{\kappa-1}=1}^{m_{\kappa-1}} \sum_{j_{\kappa+1}=1}^{m_{\kappa+1}} \cdots \sum_{j_p=1}^{m_p} C^{(\kappa)}_{j_1,\ldots,j_{\kappa-1},i_\kappa,j_{\kappa+1},\ldots,j_p} \\ &\quad \times v^{(1)}_{i_1 j_1} \cdots v^{(\kappa-1)}_{i_{\kappa-1} j_{\kappa-1}}\, v^{(\kappa+1)}_{i_{\kappa+1} j_{\kappa+1}} \cdots v^{(p)}_{i_p j_p} \end{aligned} \tag{11.25}$$

This *contraction over the κth particle* is a very helpful trick, as it substantially reduces the numerical effort of the following MCTDH calculation without affecting the accuracy of the product expansion. One should contract over that particle which otherwise would require the largest expansion order.

11.6 Separable Weights

When deriving the potfit algorithm, we have ignored weights for the sake of simplicity of the method. But – as discussed above – the inclusion of appropriate weights is often very important, and in this and the following section we will discuss how weights can be reintroduced.

We want to minimize

$$\Delta^2 = \sum_I (V_I - V_I^{\rm app})^2 w_I^2 = \sum_I (w_I V_I - w_I V_I^{\rm app})^2 = \sum_I (\tilde{V}_I - \tilde{V}_I^{\rm app})^2 \tag{11.26}$$

where we have implicitly defined the weighted potentials $\tilde{V}$ and $\tilde{V}^{\rm app}$. This equation suggests the following procedure. Rather than fitting V, one fits the weighted potential $\tilde{V}$. The fit thus obtained, $\tilde{V}^{\text{app}}$, must then be divided by the weights to obtain the desired approximate potential, $V_I^{\text{app}} = \tilde{V}_I^{\text{app}}/w_I$. This last step, however, will destroy the product form of V^{app}, except when the weights themselves are separable,

$$w_{i_1,\ldots,i_p} = w_{i_1}^{(1)} \cdots w_{i_p}^{(p)} \tag{11.27}$$

In this case the approximate potential reads

$$V_{i_1\ldots i_p}^{\text{app}} = \sum_{j_1=1}^{m_1} \cdots \sum_{j_p=1}^{m_p} \tilde{C}_{j_1\ldots j_p} \frac{\tilde{v}_{i_1 j_1}^{(1)}}{w_{i_1}^{(1)}} \cdots \frac{\tilde{v}_{i_p j_p}^{(p)}}{w_{i_p}^{(p)}} \tag{11.28}$$

where $\tilde{v}^{(\kappa)}$ and $\tilde{C}_{j_1\ldots j_f}$ denote the natural potentials and expansion coefficients obtained by potfitting the weighted potential energy surface $\tilde{V}$. In practice, one would additionally contract the expansion over one particle (see Section 11.5).

Separable weights have been successfully applied; however, the required product form of the weights limits their usefulness.

11.7 Non-Separable Weights

To emulate non-separable weights, we introduce a reference potential V^{ref} such that the weighted difference between the potential and its product representation is identical to the difference between the reference potential and the product representation,

$$(V_I - V_I^{\text{app}})\, w_I^2 = V_I^{\text{ref}} - V_I^{\text{app}} \tag{11.29}$$

Then one can simply potfit the reference potential to obtain a product representation that is almost optimal with respect to the weighted sum of squared differences, Equation (11.4). (The fit is only 'almost optimal' because potfit is not strictly optimal.) Obviously, V^{ref} is given by

$$V_I^{\text{ref}} = w_I^2 V_I + (1 - w_I^2) V_I^{\text{app}} \tag{11.30}$$

The definition of the reference potential depends on V^{app}, which in turn depends on the reference potential. Hence, the equations must be solved iteratively. One first potfits V and evaluates the reference potential. Then the reference potential is potfitted and, with the new V^{app}, a new reference potential is built. The process is iterated until some break-off criterion is satisfied.

A question of great practical relevance is whether the iterations converge, and, if they do, whether the speed of convergence is sufficient. The only parameters that are available to steer the convergence are the weights. Multiplying all weights by a common factor does not change the solution to the variational problem (11.4). Scaling the weights to low values leads to a very slow convergence. Increasing the weights speeds up the convergence until a point is reached where convergence is turned into divergence.

When emulating non-separable weights, we have always used a special form of the weights. The weights are set to 1 within the so-called *relevant region* and are 0 otherwise. The relevant region is usually defined by an energy criterion, that is, it is the region where the potential is lower than some suitably chosen energy threshold. Restrictions on coordinates can be set as well when defining the relevant region. With such a definition of the weights, that is, 0 or 1, Equation (11.30) has a vivid interpretation. The reference potential is the original potential within the relevant region and the fitted potential otherwise. With this choice of the weights, we always observed convergence. Scaling the weights up improves the convergence speed, but when w_I^2 becomes larger than 2, we always observed divergence.

The relevant-region iteration procedure just discussed is very useful, as it allows one to emulate non-separable weights. The inclusion of such weights is often inevitable for obtaining an accurate product representation of the physically relevant part of the potential without going to high expansion orders. On the other hand, the iteration procedure is more expensive because potfit has to be performed several times.

11.8 Computational Effort and Memory Request

The potfit expansion was introduced to reduce the numerical labour when evaluating the mean fields. Consider the computation of the matrix element $\langle \Phi_J | V | \Phi_L \rangle$. Doing this integral on the primitive grid requires N^p multiplications. (Here we assume, for the sake of simplicity, that all particles have the same number of grid points, $N_\kappa = N$.) Doing the integral with a potfit expansion requires spN multiplications. The number of potential terms is, due to contraction, $s = m^{p-1}$, where, similar to above, $m_\kappa = m$ is assumed. The gain is hence

$$\text{gain}_{\text{CPU}} = \frac{1}{p}\left(\frac{N}{m}\right)^{p-1} \tag{11.31}$$

As an example let us consider a six-dimensional problem where each degree of freedom (DOF) is represented by 25 grid points. Combining always two DOFs into one particle, one arrives at $p = 3$ particles and $N = 625$ grid points per particle. Assuming $m = 35$, that is, 1225 potfit terms (after contraction), one has a gain of 106, which is a quite remarkable speed-up.

A potfit expansion does not only speed up the calculation, it also compacts the representation of the potential, leading to a much lower memory demand. The full potential consists of N^p data points, whereas a potfit expansion, Equation (11.25), takes $Nm^{p-1} + Nm(p-1)$ data points. For $p > 2$, the second term is negligible in comparison with the first one, and one arrives at a memory gain

$$\text{gain}_{\text{mem}} = \left(\frac{N}{m}\right)^{p-1} \tag{11.32}$$

which is even larger than the CPU gain.

The example potential (six dimensions, 25 points per DOF) consists of $N^p = 2.4 \times 10^8$ points and requires 1.8 GB of storage. The potfit consumes only 6 MB. Turning to larger systems, the reduction in memory is even more spectacular. Considering a 12-dimensional system with 10 points per DOF, one arrives at 10^{12} points or 8 TB of storage. This is not feasible. A potfit, however, assuming four three-dimensional particles and an expansion order of $m = 50$, takes only 1 GB of storage.

Unfortunately, one cannot potfit a 12-dimensional potential. The potfit algorithm – at least in its present implementation – requires the full potential to be kept in memory. Moreover, the number of multiplications to be performed is a multiple of the total number of grid points. This limits potfit to grid sizes of at most 10^9 points, or, in general, to potentials with at most six or seven DOFs. For larger problems, one has to turn to other strategies, for example, to a cluster expansion (also called n-mode representation [127, 128] or cut-HDMR [129, 130]). The clusters, which are of much lower dimension than the potential, are finally potfitted. For a brief discussion of this approach, see Chapter 23.

12
Kinetic Energy Operators

Hans-Dieter Meyer, Fabien Gatti and Graham A. Worth

12.1
Introduction

The convergence of the MCTDH method is determined by the correlation between MCTDH particles. A set of coordinates that minimizes the correlation will improve the convergence of MCTDH, while a coordinate set that introduces strong artificial correlations, that is, correlations that are entirely due to an unsuitable choice of coordinates, will slow down convergence. Let us consider an example. For small-amplitude motions around a well-defined equilibrium geometry, the vibrations are often rather harmonic. The well-known normal-mode rectilinear coordinates then make the Hamiltonian operator almost separable – it is exactly separable for infinitesimal delocalizations – and the use of these coordinates will be optimal in a low-energy domain. However, for more floppy systems exhibiting two or several minima, or at higher excitation energies, the vibrational amplitudes become larger and the rectilinear normal coordinates cease to describe the motion in a natural way. This introduces strong artificial correlations. In such situations, in particular when studying scattering, dissociation or isomerization problems, the use of appropriate internal coordinates becomes important. In general, curvilinear coordinates, involving angles, are the natural choice, as they usually lead to a more separable and hence less artificially correlated Hamiltonian operator.

Unfortunately, the use of curvilinear coordinates often leads to very involved expressions of the kinetic energy operator (KEO) [131,132], which need to be derived for a particular system [133–135] and are not easy to generalize. This is in contrast to the rectilinear coordinates, which simplify the mathematical formulation of the same operator in a systematic way. The problem is not primarily to derive a formula for the KEO. An algorithmic program such as MATHEMATICA [136] can be used to evaluate the operators analytically. A numerical computation of the action of the kinetic operator is also feasible, and several contributions have been made in this direction (see, for example,

Multidimensional Quantum Dynamics: MCTDH Theory and Applications.
Edited by Hans-Dieter Meyer, Fabien Gatti, and Graham A. Worth

ISBN: 978-3-527-32018-9

Refs. [137–140]). For efficient MCTDH calculations, the crucial point is to find a form that is (i) as compact as possible and (ii) of the required product form. In this context, one must note that the form of the KEO is not unique. For example,

$$\frac{\partial}{\partial R} f \frac{\partial}{\partial R}, \qquad f \frac{\partial^2}{\partial R^2} + \frac{\partial f}{\partial R} \frac{\partial}{\partial R} \quad \text{and} \quad \frac{1}{2}\left(f \frac{\partial^2}{\partial R^2} + \frac{\partial^2}{\partial R^2} f - \frac{\partial^2 f}{\partial R^2} \right)$$

are three different forms of the same operator.

The *polyspherical approach*, developed in a series of papers [141–151], accomplishes both of these requirements. It is a general formulation of the exact KEO of an N-atom system. The approach received its name, *polyspherical*, because the operators are often eventually expressed in terms of spherical coordinates. This approach, however, also possesses properties that can be (and have been) exploited for other kinds of coordinates than polyspherical ones.

The polyspherical approach is characterized by the following properties: (i) It explicitly provides rather compact and general expressions of the exact kinetic energy operator, including rotation and Coriolis coupling, and avoids the use of differential calculus when deriving these operators. (ii) Within this approach it is very easy to find a spectral basis set (for example, a basis set of spherical harmonics) that discards all singularities that may occur in the KEO. (iii) General expressions for the KEO are explicitly provided in two different forms. (iv) There is a large freedom in choosing the underlying set of vectors, which may be of Jacobi, Radau, valence or satellite type, or a combination of these. (v) When polyspherical coordinates are used, the KEO is always separable: that is, it can be written as a sum of products of monomodal operators. This property is, of course (see Section 3.3), very important for MCTDH, and the resulting operators have been used in several applications with MCTDH [65,74,81,101,152–156]. The main references for the polyspherical approach are listed in Table 12.1.

12.2 Vector Parametrization and Properties of Angular Momenta

Systems of N particles possess $3N$ degrees of freedom, of which three can be eliminated by translational invariance when there is no external field. The reduced system thus obtained has $3N - 3$ degrees of freedom, which implies that an N-body configuration can always be described by $N - 1$ vectors. Hereafter, G will denote the centre of mass of the molecular system, and the space-fixed (SF) frame will denote the inertial space-fixed frame whose origin coincides with G.

Table 12.1 Main references for the polyspherical approach.

Theory	References
Reviews	[150, 151]
Orthogonal vectors	[143]
Orthogonal vectors for tetraatomic molecules	[144]
Non-orthogonal vectors	[146]
Properties of projections of angular momenta	[149]
General expression for semi-rigid systems	[148]
Orthogonal vectors for semi-rigid systems	[167]
Separation into subsystems	[145]
Constrained operators	[142, 168]
KEO for molecules	**References**
NH_3	[145]
Toluene (9D)	[152]
$H_2 + H_2$	[81]
HCF_3	[169]
H_2CS	[101]
HFCO	[154]
HONO	[65]
$H_5O_2{}^+$	[156]
C_2H_4 (6D with corrections)	[74]
Ar_2HBr (4D)	[153]

12.2.1 Examples

To illustrate the parametrization of such vectors, let us consider two four-atom molecular systems. The first example is the molecular system H_2+H_2 described by three Jacobi vectors. The choice of the Jacobi vectors is not unique and several clustering schemes are possible. A natural choice for this system is given in Figure 12.1. Here, $\mathbf{R}_1$ and $\mathbf{R}_2$ are the vectors between the two pairs of H atoms, and $\mathbf{R}_3$ is the vector between the centres of mass of the two H_2 molecules. The corresponding kinetic energy is given by

$$2T^{\mathrm{SF}} = \sum_{i=1,2,3}^{\lambda=x,y,z} \mu_i (\dot{R}_{i\lambda^{\mathrm{SF}}})^2 \tag{12.1}$$

The reduced masses are given below in Equation (12.11). The kinetic energy is in diagonal form and this is why the Jacobi coordinates are said to be orthogonal coordinates [157].

The second example is the molecule HFCO described by three valence vectors, that is, vectors corresponding to bonds between atoms (see Figure 12.2). We use the following definitions: $\mathbf{R}_1 = \mathbf{r}_{\mathrm{H}} - \mathbf{r}_{\mathrm{C}}$, $\mathbf{R}_2 = \mathbf{r}_{\mathrm{F}} - \mathbf{r}_{\mathrm{C}}$ and

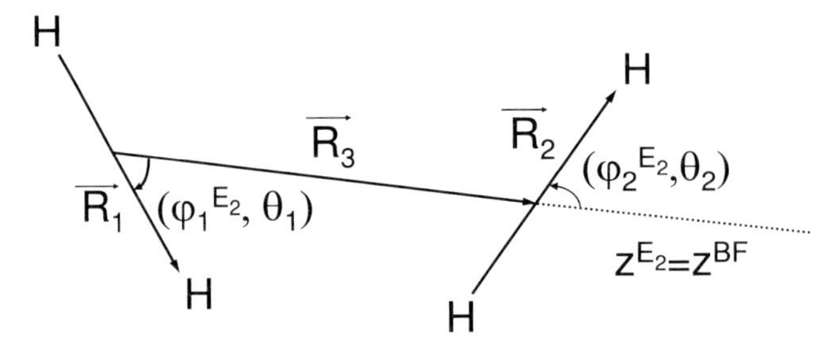

Figure 12.1 Definition of the three Jacobi vectors and the polyspherical coordinates for $H_2 + H_2$.

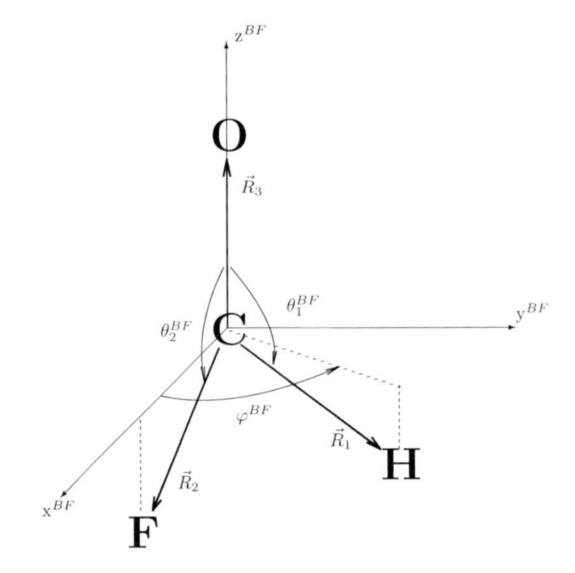

Figure 12.2 Definition of the three valence vectors and the six valence polyspherical coordinates for HFCO.

$\mathbf{R}_3 = \mathbf{r}_\text{O} - \mathbf{r}_\text{C}$. The corresponding kinetic energy is then non-diagonal,

$$2T^{\text{SF}} = \sum_{i,j=1,2,3}^{\lambda=x,y,z} \mu_{ij}(\dot{R}_{i\,\lambda^{\text{SF}}})(\dot{R}_{j\,\lambda^{\text{SF}}}) \tag{12.2}$$

and this is why the valence vectors are said to be non-orthogonal.

To each vector $\mathbf{R}_i$ ($i = 1, 2, 3$), one associates a conjugate momentum,

$$P_{i\,\lambda^{\text{SF}}} = \frac{\partial L}{\partial \dot{R}_{i\,\lambda^{\text{SF}}}} = \frac{\partial T^{\text{SF}}}{\partial \dot{R}_{i\,\lambda^{\text{SF}}}}$$

where L is the Lagrangian of the system. It is also possible to define an angular momentum for each vector $\mathbf{R}_i$, $\mathbf{L}_i = \mathbf{R}_i \times \mathbf{P}_i$, with

$$\mathbf{P}_i = P_{i\,x^{\mathrm{SF}}}\,\mathbf{e}_{x^{\mathrm{SF}}} + P_{i\,y^{\mathrm{SF}}}\,\mathbf{e}_{y^{\mathrm{SF}}} + P_{i\,z^{\mathrm{SF}}}\,\mathbf{e}_{z^{\mathrm{SF}}}$$

Note also the decomposition

$$\mathbf{P}_i = P_{r\,i}\,\mathbf{e}_{R_i} - \frac{\mathbf{e}_{R_i} \times \mathbf{L}_i}{R_i} \tag{12.3}$$

where $\mathbf{e}_{R_i} = \mathbf{R}_i / R_i$ denotes a unit vector, $R_i = \|\mathbf{R}_i\|$ a vector length, and $P_{r\,i} = \mathbf{e}_{R_i} \cdot \mathbf{P}_i$ a radial momentum. For the $H_2 + H_2$ example, it is straightforward to show that the KEO reads

$$2T^{\mathrm{SF}} = \sum_{i=1}^{3} \frac{(\mathbf{P}_i)^2_{\mathrm{SF}}}{\mu_i} = \sum_{i=1}^{3} \frac{P^2_{r\,i}}{\mu_i} + \frac{(\mathbf{L}_i)^2_{\mathrm{SF}}}{\mu_i R_i^2} \tag{12.4}$$

The notation $(\ )_{\mathrm{SF}}$ denotes that we use the projections of the SF momenta onto the SF axes to calculate the scalar product. The importance of this notation will appear below when we introduce the projections of the SF momenta onto the body-fixed axes. For HFCO, we obtain

$$2T^{\mathrm{SF}} = \sum_{i,j=1,2,3} (\mathbf{P}_i\,\mathbf{M}\,\mathbf{P}_j)_{\mathrm{SF}} \tag{12.5}$$

where the matrix $\mathbf{M}$ is given below in Equation (12.9).

12.2.2 General Formulation

12.2.2.1 Defining a Set of $N-1$ Vectors and the Corresponding Classical Kinetic Energy

Let $\mathbf{R}_1,\ \mathbf{R}_2,\ \ldots,\ \mathbf{R}_{N-1}$ be any set of vectors chosen for the description of a molecular system. The set $\mathbf{R}_1,\ \mathbf{R}_2,\ \ldots,\ \mathbf{R}_{N-1}$ can be orthogonal vectors such as Radau or Jacobi vectors or non-orthogonal vectors such as valence vectors joining two atoms. The classical SF kinetic energy can be written as

$$2T^{\mathrm{SF}} = \sum_{i,j=1,\ldots,N-1}^{\lambda=x,y,z} P_{i\,\lambda^{\mathrm{SF}}}\,M_{ij}\,P_{j\,\lambda^{\mathrm{SF}}} \tag{12.6}$$

where M_{ij} are the elements of the mass matrix $\mathbf{M}$. To determine $\mathbf{M}$, we introduce a (redundant) set of position vectors, $\mathbf{r}_1,\ \mathbf{r}_2,\ \ldots,\ \mathbf{r}_N$, where $\mathbf{r}_i$ points from the centre of mass G to the position of the ith atom. The kinetic energy, relative to the centre of mass, is simply given by

$$2T^{\mathrm{SF}} = \sum_{i=1,\ldots,N}^{\lambda=x,y,z} \frac{(p_{i\,\lambda^{\mathrm{SF}}})^2}{m_i} \tag{12.7}$$

where $\mathbf{p}_i$ is the conjugate momentum associated with $\mathbf{r}_i$, and m_i the mass of the ith atom. Let us now return to a set of $(N-1)$ internal vectors, $\mathbf{R}_1, \mathbf{R}_2, \ldots, \mathbf{R}_{N-1}$, chosen to describe the same N-atom system. Then there exists an $N \times (N-1)$ matrix $\mathbf{A}$ that connects the two sets of coordinates:

$$\begin{pmatrix} \mathbf{R}_1 \\ \mathbf{R}_2 \\ \vdots \\ \mathbf{R}_{N-1} \end{pmatrix} = \mathbf{A} \begin{pmatrix} \mathbf{r}_1 \\ \mathbf{r}_2 \\ \vdots \\ \mathbf{r}_N \end{pmatrix} \tag{12.8}$$

The symmetric mass-dependent constant matrix $\mathbf{M}$ for the internal set of $N-1$ vectors appearing in the KEO is given by $\mathbf{M} = \mathbf{A}\mathbf{m}^{-1}\mathbf{A}^{\mathrm{T}}$, where $\mathbf{m}$ denotes the diagonal matrix of particle masses. In this formulation, it is thus very easy to switch from one set of vectors to another, just by changing the matrix $\mathbf{M}$. For the KEO of HFCO, using the particular choice of valence vectors displayed in Figure 12.2, one obtains

$$\mathbf{M} = \begin{bmatrix} \frac{1}{m_{\mathrm{H}}} + \frac{1}{m_{\mathrm{C}}} & \frac{1}{m_{\mathrm{C}}} & \frac{1}{m_{\mathrm{C}}} \\ \frac{1}{m_{\mathrm{C}}} & \frac{1}{m_{\mathrm{C}}} + \frac{1}{m_{\mathrm{F}}} & \frac{1}{m_{\mathrm{C}}} \\ \frac{1}{m_{\mathrm{C}}} & \frac{1}{m_{\mathrm{C}}} & \frac{1}{m_{\mathrm{C}}} + \frac{1}{m_{\mathrm{O}}} \end{bmatrix} \tag{12.9}$$

If the vectors are orthogonal, $\mathbf{M}$ is diagonal and $M_{ii} = 1/\mu_i$, with μ_i being the reduced mass associated with the vector $\mathbf{R}_i$. In other words, the kinetic energy can be described as the kinetic energy of $N-1$ fictitious particles of masses μ_i,

$$2T^{\mathrm{SF}} = \sum_{i=1,\ldots,N-1}^{\lambda=x,y,z} \frac{\left(P_{i\,\lambda^{\mathrm{SF}}}\right)^2}{\mu_i} \tag{12.10}$$

For the $H_2 + H_2$ Jacobi coordinate system of Figure 12.1, one obtains

$$\mu_1 = \frac{m_{\mathrm{H}}}{2}, \quad \mu_2 = \frac{m_{\mathrm{H}}}{2} \quad \text{and} \quad \mu_3 = m_{\mathrm{H}} \tag{12.11}$$

12.2.2.2 **Introduction of the Body-Fixed Frame and Quantization**

In order to separate the overall rotation from the internal deformation, let us now introduce a body-fixed (BF) frame, $\{G; \mathbf{e}_{x^{\mathrm{BF}}}, \mathbf{e}_{y^{\mathrm{BF}}}, \mathbf{e}_{z^{\mathrm{BF}}}\}$. The BF frame is linked to the system, and its axes $\mathbf{e}_{x^{\mathrm{BF}}}$, $\mathbf{e}_{y^{\mathrm{BF}}}$ and $\mathbf{e}_{z^{\mathrm{BF}}}$ rotate when the particles move. The orientation of the BF frame with respect to the SF frame is determined by three Euler angles: α, β, γ. After definition of the three Euler angles, the deformations of the molecules are described by $3N - 6$ internal BF coordinates. This separation greatly facilitates the construction of the irreducible representations of the rotation group symmetry and hence reduces the size of the calculations needed to solve the Schrödinger equation. At this level, the exact definition of the BF frame is not yet specified, but the classical kinetic energy can already be recast as follows:

$$2T^{\mathrm{SF}} = \sum_{i,j=1,\ldots,N-1}^{\lambda=x,y,z} P_{i\,\lambda^{\mathrm{BF}}}\, M_{ij}\, P_{j\,\lambda^{\mathrm{BF}}} \tag{12.12}$$

where $P_{i\,\lambda^{\mathrm{BF}}}$ are the Cartesian components of the SF (and not BF! – see below) conjugate momenta in the BF frame [149], or, in other words, the projections of the SF conjugate momenta onto the BF axes:

$$\mathbf{P}_i = P_{i\,x^{\mathrm{BF}}}\,\mathbf{e}_{x^{\mathrm{BF}}} + P_{i\,y^{\mathrm{BF}}}\,\mathbf{e}_{y^{\mathrm{BF}}} + P_{i\,z^{\mathrm{BF}}}\,\mathbf{e}_{z^{\mathrm{BF}}} \tag{12.13}$$

According to Refs. [132,149], the exact quantal counterpart of Equation (12.12) is given by

$$2\hat{\mathbf{T}} = \sum_{i,j=1,\ldots,N-1}^{\lambda=x,y,z} \hat{P}^{\dagger}_{i\,\lambda^{\mathrm{BF}}}\, M_{ij}\, \hat{P}_{j\,\lambda^{\mathrm{BF}}} \tag{12.14}$$

12.2.2.3 Introduction of the Body-Fixed Projections of the Angular Momenta Associated With the $N - 1$ Vectors

It should be clear that the kinetic energy operator will always be the space-fixed one, and not the body-fixed kinetic energy, even after introduction of the BF frame. Moreover, the quantum mechanical counterpart of Equation (12.3) is given by

$$\hat{\mathbf{P}}_i = \hat{P}_{r\,i}\,\mathbf{e}_i - \frac{\mathbf{e}_i \times \hat{\mathbf{L}}_i}{R_i} \tag{12.15}$$

In order to simplify the notation, we will drop the caret ($\hat{}$) for the conjugate momenta operators in the following. After the introduction of a BF frame, projections of the angular momenta onto BF axes will be used. For instance, in the special case of orthogonal vectors, Equation (12.14) yields

$$2\hat{\mathbf{T}} = \sum_{i=1}^{N-1} \frac{P^{\dagger}_{r\,i}\, P_{r\,i}}{\mu_i} + \frac{(\mathbf{L}^{\dagger}_i \cdot \mathbf{L}_i)_{\mathrm{BF}}}{\mu_i R_i^2} \tag{12.16}$$

Let us explain why it is necessary to introduce BF projections of the momenta when making a change of coordinates to separate the overall rotation from the internal vibrations. On the one hand, the action of SF components of the angular momenta on the primitive basis functions expressed in BF coordinates becomes very complicated. On the other hand, the action of the BF components onto the primitive basis functions expressed in BF coordinates is rather simple – see, for example, the action of the KEO onto a basis of spherical harmonics of BF spherical coordinates discussed in Section 12.3 below. As BF coordinates are needed to separate the overall rotation from the internal vibrations, it becomes necessary to express the KEO in terms of BF components of momenta and angular momenta as in Equations (12.14) and (12.16), rather than in terms of the SF components.

If the set of vectors is non-orthogonal, the structure of the kinetic energy operator is more complex. The KEO, irrespective of the set of vectors, can be written as a function of $N-1$ radial and angular momenta,

$$\hat{T} = f(P_{r\,i}(i=1,\ldots,N-1);\ \mathbf{L}_i(i=1,\ldots,N-1)) \tag{12.17}$$

It is emphasized again that all angular momenta that will appear in the kinetic energy operators (such as in Equation (12.16)) are all *computed* in the SF frame but *projected* onto the axes of several frames (for example, the BF frame in Equation (12.16)). The introduction of these projections is necessary when using the BF coordinate, but raises a new technical problem. The projections of the angular momenta onto the SF axes satisfy the usual commutation relations, and their action onto a basis set of spherical harmonics in terms of the SF spherical coordinates is well known (see, for instance, Ref. [158]). But the projections of the same angular momenta onto the axes of a moving frame may satisfy unusual, non-definite commutation relations [149]. Luckily, this problem does not occur for all angular momenta. If a vector is not involved in the definition of a frame F, the expression for the projection of the corresponding angular momentum onto the F axes expressed in the coordinates in this frame is identical to the usual one in a frame SF. For instance, if one vector $\mathbf{R}_j$ is not involved in the definition of the BF frame, the projections of $\mathbf{L}_j$ onto the BF axes and expressed in terms of the BF Cartesian coordinates are given by

$$\begin{bmatrix} L_{j\,x^{\mathrm{BF}}} \\ L_{j\,y^{\mathrm{BF}}} \\ L_{j\,z^{\mathrm{BF}}} \end{bmatrix} = \begin{bmatrix} R_{j\,x^{\mathrm{BF}}} \\ R_{j\,y^{\mathrm{BF}}} \\ R_{j\,z^{\mathrm{BF}}} \end{bmatrix} \times \begin{bmatrix} \frac{1}{\mathrm{i}}\frac{\partial}{\partial R_{j\,x^{\mathrm{BF}}}} \\ \frac{1}{\mathrm{i}}\frac{\partial}{\partial R_{j\,y^{\mathrm{BF}}}} \\ \frac{1}{\mathrm{i}}\frac{\partial}{\partial R_{j\,z^{\mathrm{BF}}}} \end{bmatrix} \tag{12.18}$$

It is easy to verify that the usual commutation relations,

$$[L_{j\,\lambda^{\mathrm{BF}}}, L_{j\,\nu^{\mathrm{BF}}}] = \mathrm{i}\sum_{\rho} \epsilon_{\lambda\nu\rho} L_{j\,\rho^{\mathrm{BF}}} \tag{12.19}$$

are satisfied, where ϵ denotes the well-known totally antisymmetric tensor, that is, $\epsilon_{\lambda\nu\rho} = 1$ ($\epsilon_{\lambda\nu\rho} = -1$) if $\{\lambda\nu\rho\}$ is an even (odd) permutation of {xyz}, and 0 otherwise. This property is very helpful, since combined with the previous vector parametrization of N-atom systems, it will allow us to derive kinetic energy operators in a compact and general form without making use of differential calculus.

12.3
A General Expression of the KEO in Standard Polyspherical Coordinates

12.3.1
General Expression

Let us now consider our particular definition of the BF frame and the parametrization of the $N-1$ vectors.

12.3.1.1 **Definition of the BF frame: Figure 12.3**

(i) Let $\{G; \mathbf{e}_{x^{E_1}}, \mathbf{e}_{y^{E_1}}, \mathbf{e}_{z^{E_1}}\}$ be the E_1 frame resulting from the first Euler rotation, that is, whose origin coincides with G, where $\mathbf{e}_{z^{E_1}} = \mathbf{e}_{z^{\mathrm{SF}}}$ and where $\mathbf{e}_{x^{E_1}}$ and $\mathbf{e}_{y^{E_1}}$ are obtained by a rotation through the angle α about $\mathbf{e}_{z^{E_1}}$.

(ii) Let $\{G; \mathbf{e}_{x^{E_2}}, \mathbf{e}_{y^{E_2}}, \mathbf{e}_{z^{E_2}}\}$ be the E_2 frame resulting from the first two Euler rotations, where $\mathbf{e}_{y^{E_2}} = \mathbf{e}_{y^{E_1}}$ and where $\mathbf{e}_{x^{E_2}}$ and $\mathbf{e}_{z^{E_2}}$ are obtained by a rotation through the angle β about $\mathbf{e}_{y^{E_2}}$.

(iii) Let $\{G; \mathbf{e}_{x^{E_3}}, \mathbf{e}_{y^{E_3}}, \mathbf{e}_{z^{E_3}}\} = \{G; \mathbf{e}_{x^{\mathrm{BF}}}, \mathbf{e}_{y^{\mathrm{BF}}}, \mathbf{e}_{z^{\mathrm{BF}}}\}$ be the E_3 or BF frame, that is, the frame resulting from the three Euler rotations, where $\mathbf{e}_{z^{E_3}} = \mathbf{e}_{z^{E_2}}$ and where $\mathbf{e}_{x^{E_3}}$ and $\mathbf{e}_{y^{E_3}}$ are obtained by a rotation through the angle γ about $\mathbf{e}_{z^{E_2}}$. The BF or E_3 frame is oriented such that $\mathbf{e}_{z^{\mathrm{BF}}} = \mathbf{e}_{z^{E_2}}$ is parallel to the vector $\mathbf{R}_{N-1}$ and that $\mathbf{R}_{N-2}$, $\mathbf{e}_{x^{\mathrm{BF}}}$ and $\mathbf{e}_{z^{\mathrm{BF}}}$ lie in the same half-plane (see Figure 12.3). It is important to note that only two vectors are involved in this definition of the BF frame. This property will greatly simplify the quantum mechanical expression of the kinetic energy operator. Indeed, as outlined in Section 12.2, the expressions for the angular momenta associated with the remaining $N-3$ vectors are the normal ones when projected onto the BF axes and expressed in terms of BF coordinates. It should be pointed out that this is very different from the Eckart frame [159], which minimizes the Coriolis coupling. The definition of this frame, however, depends on all BF vectors rather than on two.

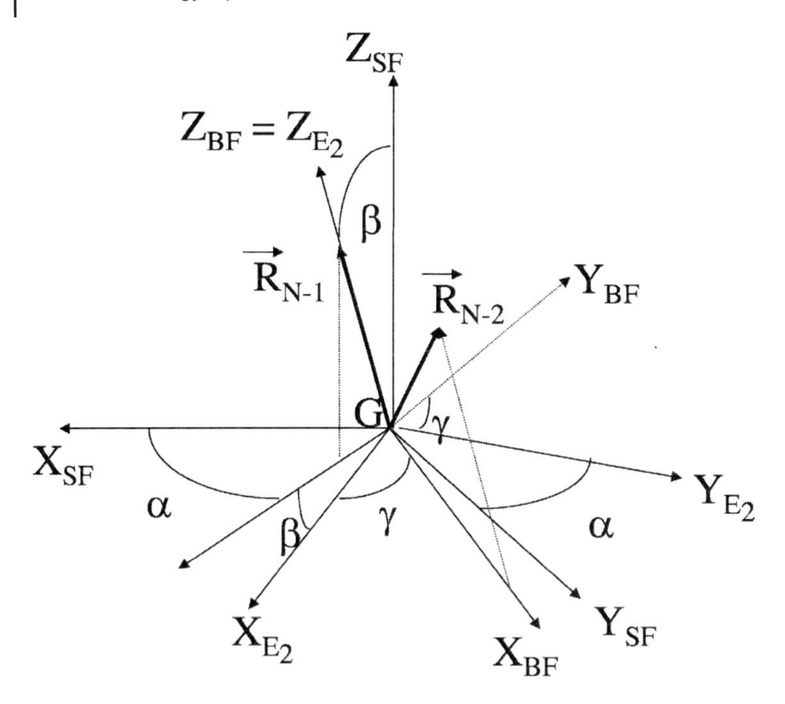

Figure 12.3 Definition of the BF Frame and of the three Euler angles. The E_2 frame is the frame obtained after the first two Euler rotations.

12.3.1.2 Polyspherical Parametrization

Let us now parametrize the vectors by spherical coordinates called the 'polyspherical coordinates'. The spherical coordinates in a given frame F are defined as

$$R_{i\,x^{\mathrm{F}}} = R_i \sin\theta_i^{\mathrm{F}} \cos\varphi_i^{\mathrm{F}}, \quad R_{i\,y^{\mathrm{F}}} = R_i \sin\theta_i^{\mathrm{F}} \sin\varphi_i^{\mathrm{F}} \quad \text{and} \quad R_{i\,z^{\mathrm{F}}} = R_i \cos\theta_i^{\mathrm{F}}$$

with $0 \leqslant \theta_i^{\mathrm{SF}} \leqslant \pi$ and $0 \leqslant \varphi_i^{\mathrm{SF}} < 2\pi$. The parametrization of the standard polyspherical type consists of three Euler angles for the overall rotation of the BF frame and $3N-6$ internal coordinates. With our definition of the BF, it is clear that the first two Euler angles are simply the two spherical angles of $\mathbf{R}_{N-1}$ in the SF frame, $\alpha = \varphi_{N-1}^{\mathrm{SF}}$ and $\beta = \theta_{N-1}^{\mathrm{SF}}$. The third Euler angle is given by $\gamma = \varphi_{N-2}^{E_2}$. The other $3N-6$ coordinates are the BF spherical coordinates, that is, the $N-1$ vector lengths $R_i \in [0, \infty)$, $N-2$ (BF) planar angles $\theta_i^{\mathrm{BF}} \in [0, \pi]$ between the vectors $\mathbf{R}_{N-1}$ and $\mathbf{R}_i$, and $N-3$ (BF) dihedral angles $\varphi_i^{\mathrm{BF}} \in [0, 2\pi[$ between the two vectors $\mathbf{R}_i$, $\mathbf{R}_{N-2}$ around the vector $\mathbf{R}_{N-1}$. They are depicted in Figure 12.4.

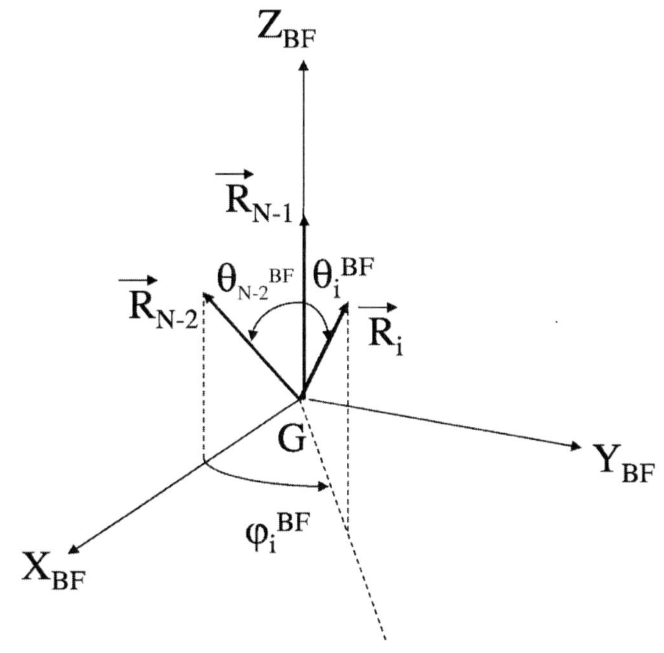

Figure 12.4 Definition of the standard polyspherical coordinates: the $N-1$ vector lengths $R_i \in [0,\infty)$, $N-2$ (BF) planar angles $\theta_i^{\mathrm{BF}} \in [0,\pi]$, and $N-3$ (BF) dihedral angles $\varphi_i^{\mathrm{BF}} \in [0,2\pi)$.

It is worth mentioning that a Euclidean normalization is used [132], that is, the elementary volume reads

$$\mathrm{d}\tau = R_{N-1}^2 \sin\beta \, \mathrm{d}R_{N-1} \, \mathrm{d}\alpha \, \mathrm{d}\beta \, R_{N-2}^2 \sin\theta_{N-2}^{\mathrm{BF}} \, \mathrm{d}R_{N-2} \, \mathrm{d}\gamma \, \mathrm{d}\theta_{N-2}^{\mathrm{BF}} \times \prod_{i=1}^{N-3} R_i^2 \sin\theta_i^{\mathrm{BF}} \, \mathrm{d}R_i \, \mathrm{d}\varphi_i^{\mathrm{BF}} \, \mathrm{d}\theta_i^{\mathrm{BF}} \quad (12.20)$$

12.3.1.3 **Properties of the BF Projections of the Angular Momenta**

In the expression of the kinetic energy operator in Equation (12.17), $N-1$ angular momenta appear. Since only $\mathbf{R}_{N-1}$ and $\mathbf{R}_{N-2}$ are involved in the definition on the BF frame, the projections of all the $N-3$ angular momenta $\mathbf{L}_1, \ldots, \mathbf{L}_{N-3}$ onto the BF axes are regular. But what about the last two angular momenta, $\mathbf{L}_{N-1}$ and $\mathbf{L}_{N-2}$? First, it is trivial to eliminate $\mathbf{L}_{N-1}$ by substituting $\mathbf{L}_{N-1}$ with $\mathbf{J} - \sum_{i=1}^{N-2} \mathbf{L}_i$, where $\mathbf{J}$ is the total angular momentum, which is a constant of the motion when no external field is present. Equation (12.17) can then be recast as

$$\hat{T} = f(P_{ir}(i=1,\ldots,N-1);\ \mathbf{L}_i(i=1,\ldots,N-2),\mathbf{J}) \quad (12.21)$$

and in the particular case of orthogonal vectors, Equation (12.16) yields

$$2\hat{\mathbf{T}} = \sum_{i=1}^{N-1} \frac{P_{r\,i}^{\dagger} P_{r\,i}}{\mu_i} + \sum_{i=1}^{N-2} \frac{(\mathbf{L}_i^{\dagger} \cdot \mathbf{L}_i)_{\mathrm{BF}}}{\mu_i R_i^2} + \frac{\left((\mathbf{J}^{\dagger} - \sum_{i=1}^{N-2} \mathbf{L}_i^{\dagger}) \cdot (\mathbf{J} - \sum_{i=1}^{N-2} \mathbf{L}_i)\right)_{\mathrm{BF}}}{\mu_{N-1} R_{N-1}^2} \tag{12.22}$$

Here, we utilize the BF projections of the (SF) angular momenta. However, because of their different properties or expressions, it is crucial to distinguish three types of angular momenta.

(i) *The total angular momentum* $\mathbf{J}$. According to Ref. [160], we obtain $[J_{\lambda\mathrm{BF}}, J_{\nu\mathrm{BF}}] = -\mathrm{i}\sum_{\rho} \epsilon_{\lambda\nu\rho} J_{\rho\mathrm{BF}}$, which are the well-known anomalous commutation relations [161] with self-adjoint components

$$\begin{bmatrix} J_{x\mathrm{BF}} \\ J_{y\mathrm{BF}} \\ J_{z\mathrm{BF}} \end{bmatrix} = \begin{bmatrix} -\dfrac{\cos\gamma}{\sin\beta} & \sin\gamma & \cot\beta\,\cot\gamma \\ \dfrac{\sin\gamma}{\sin\beta} & \cos\gamma & -\cot\beta\,\sin\gamma \\ 0 & 0 & 1 \end{bmatrix} \begin{bmatrix} \dfrac{1}{\mathrm{i}}\dfrac{\partial}{\partial\alpha} \\ \dfrac{1}{\mathrm{i}}\dfrac{\partial}{\partial\beta} \\ \dfrac{1}{\mathrm{i}}\dfrac{\partial}{\partial\gamma} \end{bmatrix} \tag{12.23}$$

(ii) *The angular momenta* $\mathbf{L}_i$ *(*$i = 1, \ldots, N-3$*) associated with the rotating vectors* $\mathbf{R}_i$ *(*$i = 1, \ldots, N-3$*) not involved in the definition of the BF frame.* These are characterized by the usual formula

$$\begin{bmatrix} L_{i\,x\mathrm{BF}} \\ L_{i\,y\mathrm{BF}} \\ L_{i\,z\mathrm{BF}} \end{bmatrix} = \begin{bmatrix} -\sin\varphi_i^{\mathrm{BF}} & -\cos\varphi_i^{\mathrm{BF}}\cot\theta_i^{\mathrm{BF}} \\ \cos\varphi_i^{\mathrm{BF}} & -\sin\varphi_i^{\mathrm{BF}}\cot\theta_i^{\mathrm{BF}} \\ 0 & 1 \end{bmatrix} \begin{bmatrix} \dfrac{1}{\mathrm{i}}\dfrac{\partial}{\partial\theta_i^{\mathrm{BF}}} \\ \dfrac{1}{\mathrm{i}}\dfrac{\partial}{\partial\varphi_i^{\mathrm{BF}}} \end{bmatrix} \tag{12.24}$$

Hence they obey the normal commutation relations and are self-adjoint.

(iii) *The angular momentum* $\mathbf{L}_{N-2}$ *associated with* $\mathbf{R}_{N-2}$. Since $\mathbf{R}_{N-2}$ is involved in the definition of the BF frame, nothing can be said *a priori* about the properties of $\mathbf{L}_{N-2}$ onto the BF axes. When such a case occurs, one proceeds as follows. One starts with the projections of the same angular momentum but onto *another* frame. Here, it is judicious to start with the projections of $\mathbf{L}_{N-2}$ onto the E_2-axes, because $\mathbf{R}_{N-2}$ is not used when defining the E_2 frame. (We recall that the E_2 frame is the frame resulting from the first two Euler rotations only, see above.) Indeed, $\mathbf{R}_{N-2}$ is needed only for defining the third Euler angle. Consequently, the E_2

components of $\mathbf{L}_{N-2}$ have a regular form

$$\begin{bmatrix} L_{N-2x^{E_2}} \\ L_{N-2y^{E_2}} \\ L_{N-2z^{E_2}} \end{bmatrix} = \begin{bmatrix} -\sin\varphi_{N-2}^{E_2} & -\cos\varphi_{N-2}^{E_2}\cot\theta_{N-2}^{E_2} \\ \cos\varphi_{N-2}^{E_2} & -\sin\varphi_{N-2}^{E_2}\cot\theta_{N-2}^{E_2} \\ 0 & 1 \end{bmatrix} \begin{bmatrix} \dfrac{1}{\mathrm{i}}\dfrac{\partial}{\partial\theta_{N-2}^{\mathrm{BF}}} \\ \dfrac{1}{\mathrm{i}}\dfrac{\partial}{\partial\varphi_{N-2}^{E_2}} \end{bmatrix} \tag{12.25}$$

These components obey the normal commutation relations and are self-adjoint. The BF projections of $\mathbf{L}_{N-2}$ are obtained by rotating the E_2 projections of $\mathbf{L}_{N-2}$ through the angle $\varphi_{N-2}^{E_2} = \gamma$ around the z_{E_2} axis onto the E_2 projections of $\mathbf{L}_{N-2}$,

$$\begin{bmatrix} L_{N-2\,x^{\mathrm{BF}}} \\ L_{N-2\,y^{\mathrm{BF}}} \\ L_{N-2\,z^{\mathrm{BF}}} \end{bmatrix} = \begin{bmatrix} 0 & \cos\varphi_{N-2}^{E_2} & \sin\varphi_{N-2}^{E_2} \\ 0 & -\sin\varphi_{N-2}^{E_2} & \cos\varphi_{N-2}^{E_2} \\ 0 & 0 & 1 \end{bmatrix} \begin{bmatrix} L_{N-2\,x^{E_2}} \\ L_{N-2\,y^{E_2}} \\ L_{N-2\,z^{E_2}} \end{bmatrix} \tag{12.26}$$

A final change of coordinates is then performed:

$$\begin{aligned} \gamma &= \varphi_{N-2}^{E_2} \\ \theta_{N-2}^{\mathrm{BF}} &= \theta_2^{E_2} \\ \theta_i^{\mathrm{BF}} &= \theta_i^{E_2} \quad \text{for } i = 1, \ldots, N-3 \\ \varphi_i^{\mathrm{BF}} &= \varphi_i^{E_2} - \varphi_{N-2}^{E_2} \quad \text{for } i = 1, \ldots, N-3 \end{aligned} \tag{12.27}$$

Applying the chain rule to $\partial/\partial\theta_{N-2}^{E_2}$ and $\partial/\partial\varphi_{N-2}^{E_2}$ gives respectively

$$\begin{aligned} \frac{\partial}{\partial\theta_{N-2}^{E_2}} &= \frac{\partial}{\partial\theta_{N-2}^{\mathrm{BF}}} \\ \frac{\partial}{\partial\varphi_{N-2}^{E_2}} &= \frac{\partial}{\partial\gamma} - \sum_{i=1}^{N-3} \frac{\partial}{\partial\varphi_i^{\mathrm{BF}}} \end{aligned} \tag{12.28}$$

Inserting these relations into Equation (12.26), one obtains

$$\begin{bmatrix} L_{N-2\,x^{\mathrm{BF}}} \\ L_{N-2\,y^{\mathrm{BF}}} \\ L_{N-2\,z^{\mathrm{BF}}} \end{bmatrix} = \begin{bmatrix} -\cot\theta_{N-2}^{\mathrm{BF}} & 0 & \cot\theta_{N-2}^{\mathrm{BF}} \\ 0 & 1 & 0 \\ 1 & 0 & -1 \end{bmatrix} \begin{bmatrix} \dfrac{1}{\mathrm{i}}\dfrac{\partial}{\partial\gamma} \\ \dfrac{1}{\mathrm{i}}\dfrac{\partial}{\partial\theta_{N-2}^{\mathrm{BF}}} \\ \displaystyle\sum_{j=1}^{N-3} \frac{1}{i}\frac{\partial}{\partial\varphi_j^{\mathrm{BF}}} \end{bmatrix} \tag{12.29}$$

that is,

$$\mathbf{L}_{N-2} = \begin{pmatrix} -\cot\theta_{N-2}^{\mathrm{BF}}\left(J_{z^{\mathrm{BF}}} - \sum_{i=1}^{N-3} L_{iz^{\mathrm{BF}}}\right) \\ -\mathrm{i}\dfrac{\partial}{\partial\theta_{N-2}^{\mathrm{BF}}} \\ J_{z^{\mathrm{BF}}} - \sum_{i=1}^{N-3} L_{iz^{\mathrm{BF}}} \end{pmatrix} \tag{12.30}$$

which satisfy unusual commutation relations,

$$\begin{aligned} \left[L_{N-2\,x^{\mathrm{BF}}}, \hat{L}_{N-2\,y^{\mathrm{BF}}}\right] &= \frac{1}{(\sin\theta_{N-2}^{\mathrm{BF}})^2}\left(\frac{\partial}{\partial\gamma} - \sum_{i=1}^{N-3}\frac{\partial}{\partial\varphi_i^{\mathrm{BF}}}\right) \\ \left[L_{N-2\,y^{\mathrm{BF}}}, L_{N-2\,z^{\mathrm{BF}}}\right] &= 0 \\ \left[L_{N-2\,z^{\mathrm{BF}}}, L_{N-2\,x^{\mathrm{BF}}}\right] &= 0 \end{aligned} \tag{12.31}$$

The y BF component of $\mathbf{L}_{N-2}$ is not Hermitian and $\hat{\mathbf{L}}_{N-2}^{\dagger}$ is given by the following equation (we use Equation (12.20)):

$$(\mathbf{L}_{N-2})^{\dagger} = \mathbf{L}_{N-2} + \begin{pmatrix} 0 \\ \mathrm{i}\cot\theta_{N-2}^{\mathrm{BF}} \\ 0 \end{pmatrix} \tag{12.32}$$

Two points must be addressed here: (i) The BF component of $\mathbf{L}_{N-2}$ does not commute with those of the other angular momenta (that is, $J_{z^{\mathrm{BF}}}$ and $L_{iz^{\mathrm{BF}}}$, $i = 1, \ldots, N-3$, see Equation (12.32)). (ii) The fact that we have used an intermediate frame (the E_2 frame here) to determine the projections of $\mathbf{L}_{N-2}$ onto the BF axes is of general character and can be used to determine the projections of other angular momenta with a non-regular behaviour.

12.3.1.4 General Expression of the KEO in Polyspherical Coordinates

Since we have the expression of all components of angular momenta appearing in the KEO (Equations (12.21) and (12.22)) and since the expression for all angular momenta $\mathbf{L}_i$ ($i = 1, \ldots, N-3$) is very simple, we can now provide a general expression of the KEO for the family of polyspherical coordinates, irrespective of the number of atoms and the set of vectors. One just has to remember that $\mathbf{L}_{N-2}$ has a particular behaviour. For instance, a KEO in orthogonal coordinates, Equations (12.16) and (12.22), can be written as

$$2\hat{\mathbf{T}} = \sum_{i=1}^{N-1} \frac{P_{ri}^{\dagger} P_{ri}}{\mu_i} + \sum_{i=1}^{N-2} \frac{(\mathbf{L}_i^{\dagger}\mathbf{L}_i)_{\mathrm{BF}}}{\mu_i R_i^2} + \frac{(\mathbf{J}^{\dagger}\mathbf{J} - \sum_{i=1}^{N-2} \mathbf{L}_i^{\dagger}\mathbf{J} - \mathbf{J}^{\dagger} \sum_{i=1}^{N-2} \mathbf{L}_i + \sum_{i,j=1,j\neq i}^{N-2} \mathbf{L}_i^{\dagger}\mathbf{L}_j)_{\mathrm{BF}}}{\mu_{N-1} R_{N-1}^2} \tag{12.33}$$

All the explicit expressions of the projections onto the BF axes are given by Equations (12.23), (12.24) and (12.29). The explicit expression in terms of angular momenta (similar to Equation (12.33) but with some rearrangements) for orthogonal coordinates is given in Ref. [143]; the expression in non-orthogonal coordinates is given in Ref. [146]. Finally, the general expression explicitly in terms of the polyspherical coordinates and their conjugate momenta (and not in terms of the angular momenta) is provided in Ref. [148].

12.3.1.5 Introduction of a Primitive Basis Set of Spherical Harmonics

Let us now introduce an appropriate angular basis set. In the absence of an external field, SF is isotropic, that is, the orientation of z^{SF} is arbitrary, and any observable must be α independent. The overall rotation of the molecule can thus be described by the following basis set [158, 162]:

$$\langle(\alpha), \beta, \gamma | J, 0, \Omega\rangle = Y_J^{\Omega}(\beta, \gamma)(-1)^{\Omega} \tag{12.34}$$

where $\Omega = \sum_{i=1}^{N-2} \Omega_i$ is the projection of the total angular momentum onto z^{BF}, and Ω_i is the projection of $\mathbf{L}_i$ onto the same axis (note that $\Omega_{N-1} = 0$, since $\mathbf{R}_{N-1}$ is parallel to z^{BF}). The current element of the working angular function basis for the BF spherical angles of vectors $\mathbf{R}_i$ ($i = 1, \ldots, N-2$) reads

$$\begin{aligned}
&\langle(\alpha), \beta, \gamma, \theta_{N-2}^{\mathrm{BF}}, \varphi_1^{\mathrm{BF}}, \theta_1^{\mathrm{BF}}, \ldots, \varphi_{N-3}^{\mathrm{BF}}, \theta_{N-3}^{\mathrm{BF}} \mid \\
&\qquad J, \ell_1, \Omega_1, \ldots, \ell_{N-3}, \Omega_{N-3}, \ell_{N-2}, \Omega_{N-2}\rangle \\
&= Y_J^{\Omega}(\beta, \gamma)(-1)^{\Omega} \tilde{P}_{\ell_{N-2}}^{\Omega_{N-2}}(\cos(\theta_{N-2}^{\mathrm{BF}})) Y_{\ell_1}^{\Omega_1}(\theta_1^{\mathrm{BF}}, \varphi_1^{\mathrm{BF}}) \ldots Y_{\ell_{N-3}}^{\Omega_{N-3}}(\theta_{N-3}^{\mathrm{BF}}, \varphi_{N-3}^{\mathrm{BF}})
\end{aligned} \tag{12.35}$$

where $\tilde{P}_{\ell}^{\Omega}(cos(\theta))$ is a normalized associated Legendre function times $(-1)^{\Omega}$ and $Y_{\ell}^{\Omega}(\theta, \varphi)$ is a spherical harmonic,

$$Y_{\ell}^{\Omega}(\theta, \varphi) = \tilde{P}_{\ell}^{\Omega}(\cos\theta)(2\pi)^{-1/2} \exp(\mathrm{i}\Omega\varphi)$$

We introduce this basis set for two reasons: (i) The action of the general kinetic energy operator on this angular basis set is very simple (especially for orthogonal vectors) and, as shown below, can be determined analytically. (ii) This spectral representation has the advantage of analytically treating all angular singularities that may appear in the kinetic energy operator. Indeed, if some angles θ_i^{BF} ($i = 1, \ldots, N-2$) are equal to 0 or π, several terms in the KEO can diverge, for example, all terms including $1/\sin\theta_i^{BF}$ (see, for example, Equations (12.23), (12.24) and (12.29)). Within the framework of the angular primitive basis set, the treatment of all singularities is straightforward and analytical.

Concerning the action of the projections of the angular momenta onto the BF axes on the previous primitive basis functions, two cases must be distinguished: (i) The BF component of the total angular momentum and the operators $\mathbf{L}_i$ ($i = 1, \ldots, N-3$) have the usual expressions. Their action on the spherical harmonics in BF coordinates, Equation (12.35), is the usual one and hardly needs mentioning. (ii) For the angular momentum $\mathbf{L}_{N-2}$, this action is less straightforward. However, it was proved that, if care is taken to account for the unusual features of the BF components of $\mathbf{L}_{N-2}$, this action is eventually a *normal one*. For instance, one obtains

$$\begin{aligned}
&\hat{L}_{(N-2)\pm^{BF}}\langle \text{angles}|\ldots\rangle_J \\
&\quad = c_\pm(\ell_{N-2},\Omega_{N-2})\langle \text{angles}|J,\ell_1,\Omega_1,\ldots,\ell_{N-3},\Omega_{N-3},\ell_{N-2},\Omega_{N-2}\pm 1\rangle \\
&\hat{L}_{(N-2)z^{BF}}\langle \text{angles}|\ldots\rangle_J = (\Omega_{N-2})\langle \text{angles}|\ldots\rangle_J \qquad (12.36) \\
&\hat{J}_{\mp^{BF}}\hat{L}_{(N-2)\pm^{BF}}\langle \text{angles}|\ldots\rangle_J \\
&\quad = c_\pm(J,\Omega)c_\pm(\ell_{N-2},\Omega_{N-2}) \\
&\qquad \times \langle \text{angles}|J,\ell_1,\Omega_1,\ldots,\ell_{N-3},\Omega_{N-3},\ell_{N-2},\Omega_{N-2}\pm 1\rangle
\end{aligned}$$

with $c_\pm(J,\Omega) = \sqrt{J(J+1)-\Omega(\Omega\pm 1)}$ and $\Omega = \sum_{i=1}^{N-2}\Omega_i$, that is, $\Omega_{N-2} = \Omega - \sum_i^{N-3}\Omega_i$. The symbol $\langle \text{angles}|\ldots\rangle_J$ denotes the current element of the working angular basis set. Even though these results are not obvious, they are not surprising, as all angular momenta are computed in the SF frame and several changes of coordinates and projections onto adequate axes is all that has been performed.

12.4 Examples

For concrete applications with MCTDH, two cases must be distinguished.

(1) Geometries corresponding to angular singularities in the KEO are accessible during the physical process. This is usually true for scattering systems.

In this case, one should express the KEO in angular momenta together with a basis set of spherical harmonics. The action of the KEO is rather simple in this basis set and the singularities are analytically removed. In a grid representation, however, one has to deal with complicated multi-dimensional discrete variable representations (DVRs).

(2) The potential prevents the system from reaching singular geometries. Such a system is called *semi-rigid*, although it may exhibit large-amplitude motions. For semi-rigid systems one should express the KEO explicitly in derivative operators $\partial/\partial q_i$, where q_i denotes the $3N-6$ polyspherical coordinates. The number of terms in the KEO increases, but only simple one-dimensional DVRs are required.

12.4.1 Scattering Systems: $H_2 + H_2$

Let us now consider an example that has been treated using MCTDH: the $H_2 + H_2$ collision [81]. The molecular system is parametrized by three Jacobi vectors as depicted in Figure 12.1, and the reduced masses are given by Equation (12.11). The KEO appears as a particular case of Equation (12.33) for $N = 3$:

$$2\hat{\mathbf{T}} = \frac{(\mathbf{J}^\dagger\mathbf{J} - (\mathbf{L}_1^\dagger + \mathbf{L}_2^\dagger)\mathbf{J} - \mathbf{J}^\dagger(\mathbf{L}_1 + \mathbf{L}_2) + \mathbf{L}_1^\dagger\mathbf{L}_1 + \mathbf{L}_2^\dagger\mathbf{L}_2 + \mathbf{L}_2^\dagger\mathbf{L}_1 + \mathbf{L}_1^\dagger\mathbf{L}_2)_{\mathrm{BF}}}{\mu_3 R_3^2} + \sum_{i=1}^{3}\frac{P_{ri}^\dagger P_{ri}}{\mu_i} + \frac{(\mathbf{L}_1^\dagger\mathbf{L}_1)_{\mathrm{BF}}}{\mu_1 R_1^2} + \frac{(\mathbf{L}_2^\dagger\mathbf{L}_2)_{\mathrm{BF}}}{\mu_2 R_2^2} \quad (12.37)$$

Now, it can be proved that $(\mathbf{L}_1^\dagger + \mathbf{L}_2^\dagger)\mathbf{J} + \mathbf{J}^\dagger(\mathbf{L}_1 + \mathbf{L}_2) = 2\mathbf{J}(\mathbf{L}_1 + \mathbf{L}_2)$ and that $\mathbf{L}_2^\dagger\mathbf{L}_1 + \mathbf{L}_1^\dagger\mathbf{L}_2 = 2\mathbf{L}_1\mathbf{L}_2$. The order in which the operators appear is important, as they do not commute! Moreover, since the BF components of $\mathbf{J}$ are Hermitian, Equation (12.37) yields

$$2\hat{\mathbf{T}} = -\sum_{i=1}^{3}\frac{1}{\mu_i R_i}\frac{\partial^2}{\partial R_i^2}R_i + \left(\frac{1}{\mu_1 R_1^2} + \frac{1}{\mu_3 R_3^2}\right)(\mathbf{L}_1^\dagger\mathbf{L}_1)_{\mathrm{BF}} + \left(\frac{1}{\mu_2 R_2^2} + \frac{1}{\mu_3 R_3^2}\right)(\mathbf{L}_2^\dagger\mathbf{L}_2)_{\mathrm{BF}} + \frac{(\mathbf{J}^2 - 2\mathbf{J}(\mathbf{L}_1 + \mathbf{L}_2) + 2\mathbf{L}_1\mathbf{L}_2)_{\mathrm{BF}}}{\mu_3 R_3^2} \quad (12.38)$$

The angular basis, Equation (12.35), becomes

$$\langle \mathrm{angles}|J,\Omega,\ell_1,\Omega_1,\ell_2,\Omega_2\rangle = Y_J^{\Omega}(\beta,\gamma)(-1)^{\Omega}Y_{\ell_2}^{\Omega_2}(\theta_2^{\mathrm{BF}},0)Y_{\ell_1}^{\Omega_1}(\theta_1^{\mathrm{BF}},\varphi_1^{\mathrm{BF}}) \quad (12.39)$$

with $\Omega = \Omega_1 + \Omega_2$. The actions of all projections of angular momenta appearing in Equation (12.38) onto the angular basis functions of Equation (12.39) are the usual ones, which is not obvious because $\mathbf{R}_2$ is involved in the definition of the BF frame. For instance, one obtains

$$\begin{aligned}
&(\mathbf{L}_i^\dagger \mathbf{L}_i)_{\mathrm{BF}} \langle \text{angles} | J, \Omega, \ell_1, \Omega_1, \ell_2, \Omega_2 \rangle \\
&\quad = \ell_i(\ell_i + 1) \langle \text{angles} | J, \Omega, \ell_1, \Omega_1, \ell_2, \Omega_2 \rangle \qquad (i = 1, 2) \\
&J_{\mp \mathrm{BF}} (L_{1\,\pm\mathrm{BF}} + L_{2\,\pm\mathrm{BF}}) \langle \text{angles} | J, \Omega, \ell_1, \Omega_1, \ell_2, \Omega_2 \rangle \\
&\quad = c_\pm(J, \Omega) c_\pm(\ell_1, \Omega_1 \pm 1) \langle \text{angles} | J, \Omega \pm 1, \ell_1, \Omega_1 \pm 1, \ell_2, \Omega_2 \rangle \\
&\qquad + c_\pm(J, \Omega) c_\pm(\ell_2, \Omega_2) \langle \text{angles} | J, \Omega \pm 1, \ell_1, \Omega_1, \ell_2, \Omega_2 \pm 1 \rangle \\
&L_{1\,\pm\mathrm{BF}} L_{2\,\mp\mathrm{BF}} \langle \text{angles} | J, \Omega, \ell_1, \Omega_1, \ell_2, \Omega_2 \rangle \\
&\quad = c_\pm(\ell_1, \Omega_1) c_\mp(\ell_2, \Omega_2) \langle \text{angles} | J, \Omega, \ell_1, \Omega_1 \pm 1, \ell_2, \Omega_2 \mp 1 \rangle
\end{aligned} \tag{12.40}$$

with $L_+ = L_x + \mathrm{i}L_y$ and $L_- = L_x - \mathrm{i}L_y$. It should be noted that the basis set of Equation (12.39) along with an appropriate pseudospectral approach [163–165] eliminates all the singularities appearing in the KEO. Such a pseudospectral approach, called 'extended Legendre DVR' [165], is implemented in the Heidelberg MCTDH package. The explicit expression of the KEO is given in Ref. [81].

12.4.2
Semi-Rigid Molecules: HFCO

Let us consider now a second example, the HFCO [154] and DFCO [155] molecules. The system is parametrized by three valence vectors as depicted in Figure 12.2. In Ref. [155], highly excited eigenstates of the system as well as the intramolecular vibrational energy redistribution (IVR), initiated by an excitation of an out-of-plane bending mode, have been studied. In the corresponding energy domain, the singular geometries, $\theta_i = 0$ or π with $i = 1, 2$, are physically not accessible. The KEO appears as a particular case of Equation (12.21) for $N = 3$. As no singularity appears in the kinetic energy operator, we use the expression of the KEO not in angular momenta but explicitly in terms of the six polyspherical coordinates. As explained in Section 12.3.1.4, one can give a general and compact matrix expression of the KEO of Equation (12.21) in which only polyspherical coordinates and their associated derivative operators appear. The notation 'matrix' is used because the matrix G_{ij} serves to define the KEO (see Equation (12.41) below). This general expression, which holds for any number of atoms and all sets of vectors, and includes overall rotation and Coriolis coupling, is given in Ref. [148]. For the particular case of HFCO and vanishing total angular momentum, $J = 0$, this operator can be written in the following form:

$$\hat{T} = \frac{1}{2} \sum_{i,j=1}^{6} \hat{p}_i^\dagger \, G_{ij} \, \hat{p}_j \tag{12.41}$$

where $\hat{p}_j = -\mathrm{i}\partial/\partial q_j$ denote the momenta conjugate to the polyspherical coordinates q_j. The matrix $\mathbf{G}$ depends on the polyspherical coordinates and on the mass matrix $\mathbf{M}$, which in the present case is given by Equation (12.9). The general expression in Ref. [148] provides the matrix elements G_{ij} appearing in Equation (12.41). The final expression for HFCO is given in Equation (15) of Ref. [154]. Let us consider just one term appearing in this operator,

$$-\sum_{i=1}^{2} \left(\frac{M_{ii}}{2R_i^2 \sin^2 \theta_i^{\mathrm{BF}}} + \frac{M_{33} \cos^2 \theta_i^{\mathrm{BF}}}{2R_3^2 \sin^2 \theta_i^{\mathrm{BF}}} - \frac{M_{i3} \cos \theta_i^{\mathrm{BF}}}{R_3 R_i \sin^2 \theta_i^{\mathrm{BF}}} \right) \frac{\partial^2}{\partial \varphi_{\mathrm{BF}}^2} \tag{12.42}$$

Since $\theta_i^{\mathrm{BF}} = 0$ and π are not physically accessible, the singularities originating from the terms $\sin^{-2} \theta_i^{\mathrm{BF}}$ do not appear. The MCTDH calculations can thus be performed with a simple one-dimensional DVR for each degree of freedom. Equation (12.42) clearly shows that all terms are products of one-dimensional functions, which is the form perfectly adapted to MCTDH. In addition, since the Heidelberg MCTDH package is capable of parsing all functions and derivative operators appearing in the general expression for the KEO given in Ref. [148], the implementation of the operator of HFCO is straightforward, that is, without coding a new routine.

12.5 Extensions

12.5.1 Separation Into Subsystems

The general approach described in Section 12.2 can be applied to derive kinetic energy operators in types of coordinates other than the standard polyspherical coordinates. The standard polyspherical coordinates are defined with respect to two vectors only: $\mathbf{R}_{N-1}$ and $\mathbf{R}_{N-2}$. For large systems, this can introduce artificial couplings between motions that are physically decoupled, for example, between atoms that are located in very different parts of the molecule. A simple solution is to integrate atoms into subsystems and to define for each subsystem an intermediate BF frame. In each subsystem, the angles are defined with respect to the intermediate frame, which decouples the motions between atoms belonging to different subsystems. Examples can be found for $H_5O_2{}^+$ [156] and NH_3 [145].

12.5.2
Constrained Operators

For large systems, the sheer number of internal coordinates makes a reduction of dimensionality unavoidable. A widely used approach is the rigid constraint one, which freezes selected bond lengths, angles or entire atomic groups, that is, $q_i'' = q_i''|_0$, where the double prime denotes coordinates that are to be frozen and where $q_i''|_0$ denotes a fixed position, usually the equilibrium one. Deriving the corresponding constrained KEO is a difficult task [142, 166], except when the matrix **G** happens to be diagonal. In this case one may simply erase the frozen degrees of freedom from the KEO. However, the general case is more complicated as $q_i'' = q_i''|_0$ corresponds to $\dot{q}_i'' = 0$, which is *not* equivalent to $p_{q''} = 0$ when **G** is non-diagonal. Corrections must be added that generally do not have the direct-product form needed for MCTDH. In practice, one starts with the usual expression in polyspherical coordinates (as Equation (12.41) for HFCO), and reduces it by removing terms in $\partial/\partial q_i''$. One makes additional approximations for the corrections in order to assure a direct-product form and then adds these simplified corrections to the reduced operator. See Ref. [74] for an application of this approach to C_2H_2. There one finds also a discussion on the importance of the correction terms.

It is convenient to calculate the corrections numerically by employing the program TNUM [139]. TNUM is a program that numerically calculates a KEO in any set of coordinates. However, TNUM is not directly useful for MCTDH as it conceals the product form of the KEO, but it was very useful when checking that an analytic KEO in polyspherical coordinates is correctly implemented in MCTDH [156].

Part 2 Extension to New Areas

Multidimensional Quantum Dynamics: MCTDH Theory and Applications.
Edited by Hans-Dieter Meyer, Fabien Gatti, and Graham A. Worth

ISBN: 978-3-527-32018-9

13
Direct Dynamics With Quantum Nuclei

Benjamin Lasorne and Graham A. Worth

13.1
Introduction

A serious bottleneck to quantum dynamics studies is the production of accurate potential energy surfaces (PESs). These functions, which must be represented on the full multidimensional grid used in a simulation, often need months of tedious work that must be repeated for each new system. Furthermore, to fit the PES of an f-dimensional system, N^f energy points are required (with N points per degree of freedom), and this number quickly becomes prohibitive. This situation is even worse when several electronic states are coupled, as intersecting PESs and non-adiabatic couplings need to be fitted using data from non-routine quantum chemistry methods that may be extremely expensive if high accuracy is required. Most of the grid is never visited by the system and so this represents much wasted effort.

We are particularly interested in molecular photophysics and photochemistry that involve radiationless transitions between electronic states (internal conversion or intersystem crossing), which are due to non-adiabatic coupling between electronic states and nuclear motion (vibronic couplings). These quantum effects become intense for nuclear geometries where two adiabatic electronic states approach degeneracy, that is, in regions where two adiabatic potential energy surfaces come into contact. Of particular importance are regions close to conical intersections (see, for example, Ref. [4] and references therein), a seam of crossing between two PESs where the non-adiabatic coupling is singular.

The signature of conical intersections is found in many electronic spectra [4, 73], and they are also recognized as a mechanistic feature in photochemistry [170, 171], as important as transitions structures in thermal chemistry. Owing to the quantum mechanical nature of non-adiabatic processes, simulations of photochemical mechanisms and calculations of absorption spectra

Multidimensional Quantum Dynamics: MCTDH Theory and Applications.
Edited by Hans-Dieter Meyer, Fabien Gatti, and Graham A. Worth

ISBN: 978-3-527-32018-9

where these are important must be carried out using quantum dynamics, or at least with a semiclassical approach [172].

In order to treat problems involving several coupled electronic states, standard MCTDH calculations can be used jointly with an analytical model such as the vibronic coupling Hamiltonian developed by Köppel *et al.* [173] (see Chapter 18). The validity of the vibronic models currently in use is only very local, which limits their field of application to the calculation of spectra of photoabsorption dominated by ultrafast non-adiabatic effects ($\approx$ 100–200 fs) and accompanied by quite small geometrical deformations (for example, [62, 73, 174]). Thus, the standard MCTDH approach has not yet been extensively applied to the description of photochemical processes involving large-amplitude motions, such as photoisomerizations, where a ground-state reactant is ultimately converted into a ground-state product via a non-adiabatic mechanism induced by light absorption (a recent example is given in Ref. [74]).

Direct dynamics aims to free calculations from the *a priori* need of an analytic PES by calculating the potential energy locally at points along trajectories, which can be done by a quantum chemistry program. By focusing only on where the system actually goes, direct dynamics breaks the exponential scaling of the resources with system size. As the surfaces are only known locally, direct dynamics methods are typically based on classical trajectories. A variety of semiclassical methods use classical trajectories to simulate the wavepacket evolution in terms of a swarm of trajectories. These include trajectory surface hopping [175] and Ehrenfest dynamics [176–178], which incorporate non-adiabatic effects. Owing to their simplicity and ease of implementation, they can produce good estimates for the time-scale and yield of a radiationless process. Unfortunately, they are expensive methods to converge, needing many trajectories, something to be avoided in direct dynamics as each trajectory requires many evaluations of the (expensive) electronic wavefunction. Furthermore, it is not possible to know the error introduced by the approximations in these methods, which can occasionally fail badly (for a comparison of methods, see Refs. [179, 180]). Both the trajectory surface hopping and Ehrenfest methods have been used in direct dynamics calculations [181–185].

Full quantum dynamical direct dynamics has more recently become viable using on-the-fly Gaussian wavepacket (GWP) dynamics. The key is that GWPs are localized in space so that a local harmonic approximation (LHA) of the PES around their centre is a reasonable approximation [16]. In recent reviews [186, 187], we provide an overview of the main non-adiabatic direct dynamics methods (both semiclassical trajectory and GWP methods), aiming to show their different properties, and using examples to show the types of systems being treated. Direct dynamics implementations of GWP methods have been reported in the form of the *ab initio* multiple spawning

(AIMS) method [188–191] and the direct dynamics variational multiconfiguration Gaussian wavepacket (DD-vMCG) method presented here. It is also worth mentioning a recent application using a GWP with on-the-fly transformation of curvilinear coordinates (for quantum dynamics) into Cartesian coordinates of the nuclei (for quantum chemistry) [192].

The vMCG equations of motion come from the MCTDH evolution equations and have been implemented in a development version of the Heidelberg MCTDH package. In the following, we review the theoretical background of the vMCG method and its direct dynamics implementation, DD-vMCG, and mention the applications achieved up to now. The main advantages of the method over other direct dynamics formulations are that the GWP basis functions are coupled directly and do not follow classical trajectories. As a result, the convergence properties are superior. This has been demonstrated qualitatively in a comparison with surface hopping [193] and in a full test of the convergence on an autocorrelation function [194].

13.2 Variational Multiconfiguration Gaussian Wavepackets

In the G-MCTDH method (see Section 3.5) some or all of the MCTDH single-particle functions (SPFs) are replaced by time-dependent parametrized GWPs [43,195,196]. The vMCG method is simply the G-MCTDH method in the limit that only multidimensional GWPs are used for representing the nuclear wavepacket. There is, thus, no mean-field and all nuclear coordinates are combined in a single particle. The use of Gaussian functions introduces the complication of a non-orthonormal basis set. On the other hand, their localized character frees the method from the requirement of a global PES fitted on a grid if the potential energy function is expanded locally around the centre of each Gaussian function in the form of an LHA representation, as is traditionally done in GWP methods [16].

At present, only rectilinear coordinates have been implemented. They can be given as normal coordinates at a given reference point (all $3N-6$ internal degrees of freedom, N being the number of atoms in the molecule or a selected subset in a reduced-dimensionality approach) or as Cartesian coordinates of the nuclei in a reference frame that can be free or constrained to separate overall rotations and translations from internal deformations.

13.2.1 Gaussian Wavepacket *Ansatz*

The vMCG *Ansatz* for a molecular wavepacket spread over a set of coupled electronic states $\{|s\rangle\}$ and describing f nuclear coordinates, collectively denoted $\mathbf{x} = \{x_j\}$, reads

$$|\Psi(\mathbf{x},t)\rangle = \sum_s \sum_j A_j^{(s)}(t) g_j^{(s)}(\mathbf{x},t)|s\rangle \tag{13.1}$$

In contrast with the standard MCTDH approach, where SPFs are expanded onto a time-independent basis set, each time-dependent function in Equation (13.1) is a GWP given as the exponential of a multidimensional quadratic form

$$g_j^{(s)}(\mathbf{x},t) = \exp[\mathbf{x} \cdot \zeta_j^{(s)}(t) \cdot \mathbf{x} + \boldsymbol{\xi}_j^{(s)}(t) \cdot \mathbf{x} + \eta_j^{(s)}(t)] \tag{13.2}$$

parametrized by time-dependent complex quantities: an $(f \times f)$-dimensional symmetrical matrix, $\zeta_j^{(s)}$, an f-dimensional vector, $\boldsymbol{\xi}_j^{(s)}$, and a scalar, $\eta_j^{(s)}$. (Note that a possible non-dimensionless term related to the norm of the GWP is formally included within the, otherwise dimensionless, real part of the scalar term.) For each GWP in the expansion, the $f(f+1)/2 + f + 1$ complex parameters, and the complex expansion coefficient, $A_j^{(s)}$, evolve according to a variational optimization of the solution of the time-dependent Schrödinger equation as shown below in Section 13.2.2.

In the multidimensional form, Equation (13.2), the time-evolving correlation between the different degrees of freedom is taken into account by the off-diagonal terms of the width matrix, $\zeta_j^{(s)}$. This allows the Gaussian wavepacket to change shape. However, it is also possible to keep this matrix diagonal, or even fixed, without violating the variational character of the method. Such 'separable' or 'frozen' Gaussian functions, as these two forms are respectively known, have far fewer parameters than 'thawed' Gaussian functions, and are numerically easier to treat. More of them may be required for convergence, but tests seem to indicate that they are preferred.

In the following, only separable or frozen GWPs are considered. The diagonal elements of the width matrix are real parameters. Using the relationships

$$\zeta_{j\alpha\alpha}^{(s)} = -\frac{1}{4\sigma_{j\alpha}^{(s)2}} \tag{13.3}$$

$$\xi_{j\alpha}^{(s)} = \frac{x_{j\alpha}^{(s)}}{2\sigma_{j\alpha}^{(s)2}} + \mathrm{i}\frac{p_{j\alpha}^{(s)}}{\hbar} \tag{13.4}$$

$$\eta_j^{(s)} = \mathrm{i}\gamma_j^{(s)} - \sum_\alpha \frac{x_{j\alpha}^{(s)2}}{4\sigma_{j\alpha}^{(s)2}} + \mathrm{i}\frac{p_{j\alpha}^{(s)} x_{j\alpha}^{(s)}}{\hbar} \tag{13.5}$$

where the index α runs over the different degrees of freedom, Equation (13.2) can be rewritten as follows to conform with the more familiar Heller expression [197] of a multidimensional separable GWP: a real-valued Gaussian function (spatial amplitude envelope) multiplied by a complex factor (norm and global phase) and a Fourier function (plane wave giving a group velocity),

$$g_j^{(s)}(\mathbf{x},t) = \exp[\mathrm{i}\gamma_j^{(s)}(t)] \prod_\alpha \exp\left(-\frac{1}{4\sigma_{j\alpha}^{(s)2}}[x_\alpha - x_{j\alpha}^{(s)}(t)]^2\right) \times \exp\left(\mathrm{i}\frac{p_{j\alpha}^{(s)}(t)}{\hbar}[x_\alpha - x_{j\alpha}^{(s)}(t)]\right) \tag{13.6}$$

where, along each degree of freedom α (nuclear coordinate x_α): $\sigma_{j\alpha}^{(s)}$ is the width (spatial standard deviation); $x_{j\alpha}^{(s)}$ and $p_{j\alpha}^{(s)}$ are the mean position and mean momentum, respectively, defining the trajectory followed by the centre of the function in the phase space; and $\gamma_j^{(s)}$ is the global complex phase of the function (real part defining an arbitrary gauge angle and imaginary part defining the norm of the GWP).

As in any multiconfigurational formalism, there is always an undetermined phase between expansion coefficients and functions, Equation (13.1), which imposes the use of a constraint that must be chosen conveniently to make the equations easier to solve numerically. The usual choice, further discussed below in Section 13.2.2, consists in using the imaginary part of $\gamma_j^{(s)}$ to keep the GWPs normalized and to set the real part of $\eta_j^{(s)}$ to zero in Equation (13.6). This results in the following expression for a separable GWP:

$$g_j^{(s)}(\mathbf{x},t) = \prod_\alpha (2\pi\sigma_{j\alpha}^{(s)2})^{-1/4} \exp\left(-\frac{1}{4\sigma_{j\alpha}^{(s)2}}[x_\alpha - x_{j\alpha}^{(s)}(t)]^2\right) \times \exp\left(\mathrm{i}\frac{p_{j\alpha}^{(s)}(t)}{\hbar}x_\alpha\right) \tag{13.7}$$

13.2.2 Equations of Motion

Using the standard MCTDH approach, a variational solution of the time-dependent Schrödinger equation for the expansion coefficients leads to [43, 187, 195]

$$\mathrm{i}\hbar \dot{A}_j^{(s)} = \sum_{l,m} [\mathbf{S}^{-1(s)}]_{jl} (H_{lm}^{(ss)} - \mathrm{i}\hbar\tau_{lm}^{(s)}) A_m^{(s)} + \sum_{s' \neq s} H_{lm}^{(ss')} A_m^{(s')} \tag{13.8}$$

where

$$S_{jl}^{(s)} = \langle g_j^{(s)} | g_l^{(s)} \rangle \tag{13.9}$$

$$H_{jl}^{(ss')} = \langle g_j^{(s)} | \hat{H}^{(ss')} | g_l^{(s')} \rangle \tag{13.10}$$

$$\tau_{jl}^{(s)} = \langle g_j^{(s)} | \dot{g}_l^{(s)} \rangle \tag{13.11}$$

are the elements of the overlap matrix, Hamiltonian matrix (for the operator connecting electronic states s and s') and time-derivative matrix, respectively. Equation (13.8) differs from a standard MCTDH equation only by the term including the time-derivative matrix, which simply accounts for the non-orthogonality of the Gaussian basis set.

Using the vector $\mathbf{\Lambda} = [\boldsymbol{\lambda}_j^{(s)}]$, where $\boldsymbol{\lambda}_j^{(s)} = [\lambda_{ja}^{(s)}]$ collects together all the time-dependent parameters defining each GWP ($a = \alpha$ for the f linear parameters, $\xi_{j\alpha}^{(s)}$, in the frozen case and, in addition, $a = f + \alpha$ for the f quadratic parameters, $\zeta_{j\alpha}^{(s)}$, in the separable case) , one can write the equations of motion (time evolution of the width of the GWP and trajectory followed by its centre in the phase space) [43, 187, 195] as

$$i\hbar\dot{\mathbf{\Lambda}} = \mathbf{C}^{-1}\mathbf{Y} \tag{13.12}$$

where

$$C_{j\alpha,l\beta}^{(s)} = \rho_{jl}^{(ss)} (S_{jl}^{(s,\alpha\beta)} - [\mathbf{S}^{(s,\alpha 0)} \mathbf{S}^{-1(s)} \mathbf{S}^{(s,0\beta)}]_{jl}) \tag{13.13}$$

$$Y_{j\alpha}^{(s)} = \sum_l \sum_{s'} \rho_{jl}^{(ss')} (H_{jl}^{(ss',\alpha 0)} - [\mathbf{S}^{(s,\alpha 0)} \mathbf{S}^{-1(s)} \mathbf{H}^{(ss')}]_{jl}) \tag{13.14}$$

where the density matrix, $\rho_{jl}^{(ss')} = A_j^{(s)*} A_l^{(s')}$, is defined as usual, and, in addition to the definitions of Equations (13.9)–(13.11),

$$S_{jl}^{(s,\alpha 0)} = \left\langle \frac{\partial g_j^{(s)}}{\partial \lambda_{j\alpha}^{(s)}} \middle| g_l^{(s)} \right\rangle \tag{13.15}$$

$$S_{jl}^{(s,\alpha\beta)} = \left\langle \frac{\partial g_j^{(s)}}{\partial \lambda_{j\alpha}^{(s)}} \middle| \frac{\partial g_l^{(s)}}{\partial \lambda_{l\beta}^{(s)}} \right\rangle \tag{13.16}$$

$$H_{jl}^{(ss',\alpha 0)} = \left\langle \frac{\partial g_j^{(s)}}{\partial \lambda_{j\alpha}^{(s)}} \middle| \hat{H}^{(ss')} \middle| g_l^{(s')} \right\rangle \tag{13.17}$$

The equation of motion for the GWP parameters, Equation (13.12), shows that they are coupled both directly and through the expansion coefficients. This introduces a quantum character to the basis set. As a result, unlike classical-based methods, vMCG is able to describe quantum events such as tunnelling [195] and curve crossing [194, 196, 199] without recourse to additional parameters. The structure of the equations of motion in terms of a classical part and a purely quantum part is further analysed in Refs. [194] and [187].

Although the equations look complicated, they are straightforward to evaluate, as all matrix elements have analytical expressions when using an LHA representation for the diabatic PESs and the off-diagonal potential-like interstate-coupling function, and rectilinear coordinates (Cartesian coordinates of the nuclei or linear combinations such as normal coordinates defined at a reference point) to express the kinetic energy operator (KEO). In this context, all matrix elements can be expressed in terms of Gaussian moments [43, 187, 195, 198].

The scalar parameters, $\eta_j^{(s)}$ in Equation (13.2) or $\gamma_j^{(s)}$ in Equation (13.6), are somewhat arbitrary, as they can be combined with the expansion coefficients, $A_j^{(s)}$, with no loss of information, as discussed above in Section 13.2.2. Thus, in contrast to the quadratic and linear parameters, $\zeta_j^{(s)}$ and $\xi_j^{(s)}$, respectively, the time evolution of the scalar parameters, $\eta_j^{(s)}$, is not determined by the variational principle, as this is made redundant by the variational optimization of the coefficients, $A_j^{(s)}$. They may, however, be used to change the characteristics of the basis set. As expressed in Equation (13.7), the most useful choice uses real scalar parameters that also keep the GWPs normalized. This results in a more stable propagation.

Explicit time derivatives of all the parameters, including the scalar one, are required to compute the time-derivative matrix, $\boldsymbol{\tau}^{(s)}$, Equation (13.11), by using the chain rule on the time derivative of the GWPs,

$$\tau_{jl}^{(s)} = S_{jl}^{(s)}\dot{\eta}_l^{(s)} + \sum_a S_{jl}^{(s,0a)}\dot{\lambda}_{la}^{(s)} \tag{13.18}$$

Since the GWPs are normalized, the corresponding diagonal elements, $\tau_{jj}^{(s)} = \langle g_j^{(s)}|\dot{g}_j^{(s)}\rangle$, are purely imaginary. Thus Equation (13.18) can be rearranged to provide an equation of motion for the real part of the scalar parameters,

$$\begin{aligned}
\dot{\eta}_j^{(s)} &= -\Re \sum_\alpha [S_{jj}^{(s,0\alpha)} \dot{\xi}_{j\alpha}^{(s)} + S_{jj}^{(s,0\alpha+f)} \dot{\zeta}_{j\alpha}^{(s)}] \\
&= -\frac{1}{2\sigma_{j\alpha}^{(s)}} \sum_\alpha \left[\frac{x_{j\alpha}^{(s)} \dot{x}_{j\alpha}^{(s)}}{\sigma_{j\alpha}^{(s)}} - \frac{x_{j\alpha}^{(s)2} \dot{\sigma}_{j\alpha}^{(s)}}{\sigma_{j\alpha}^{(s)2}} + \dot{\sigma}_{j\alpha}^{(s)} \right] \\
&= \frac{1}{4\zeta_{j\alpha}^{(s)}} \sum_\alpha \left[2\Re\xi_{j\alpha}^{(s)} \Re\dot{\xi}_{j\alpha}^{(s)} - \frac{(\Re\xi_{j\alpha}^{(s)})^2 \dot{\zeta}_{j\alpha}^{(s)}}{\zeta_{j\alpha}^{(s)}} + \dot{\zeta}_{j\alpha}^{(s)} \right]
\end{aligned} \tag{13.19}$$

where use was made of $S_{jj}^{(s,0\alpha)} = x_{j\alpha}^{(s)}$ and $S_{jj}^{(s,0\alpha+f)} = x_{j\alpha}^{(s)2} + \sigma_{j\alpha}^{(s)2}$. This can also be derived directly from Equations (13.3)–(13.5) and (13.7). The imaginary part of $\dot{\eta}_j^{(s)}$ is simply set to zero so that the GWP 'phase' is incorporated into the expansion coefficient evolution.

13.2.3 Integration Scheme

Using a simple predictor–corrector integration scheme, the coupling between GWPs and the expansion coefficients in the vMCG method leads to very small integration step sizes. The numerical instabilities due to singularities in the C-matrix and overlap matrices when GWPs are either unpopulated or overlap extensively also cause problems. The recent development of an efficient integration scheme based on the constant mean-field integrator developed for standard MCTDH calculations (Chapter 4) [50], and tailored to the vMCG method, has cured many of the problems [194]. In this approach, an auxiliary orthogonal basis set is used. In addition to the vMCG *Ansatz*, Equation (13.1), an MCTDH-like *Ansatz* is also taken,

$$|\Psi(\mathbf{x}, t)\rangle = \sum_s \sum_r B_r^{(s)}(t) \varphi_r^{(s)}(\mathbf{x}, t) |s\rangle \tag{13.20}$$

where both basis sets are assumed to span the same Hilbert space, $\{|g_j^{(s)}\rangle\} = \{|\varphi_r^{(s)}\rangle\}$.

Since the GWP basis set is non-orthogonal, the transformation matrix between the two basis sets is not simply the overlap matrix. In terms of the overlap,

$$D_{jr}^{(s)} = \langle g_j^{(s)} | \varphi_r^{(s)} \rangle \tag{13.21}$$

the transformation from GWPs to SPFs is

$$|\varphi_r^{(s)}\rangle = \sum_i \sum_j [\mathbf{S}^{-1(s)}]_{ij} D_{jr}^{(s)} |g_i^{(s)}\rangle \tag{13.22}$$

and the transformation matrix, $\mathbf{D}'^{(s)}$, is defined by $\mathbf{D}^{(s)} = \mathbf{S}^{(s)}\mathbf{D}'^{(s)}$ in order to get

$$|\varphi_r^{(s)}\rangle = \sum_i D_{ir}'^{(s)} |g_i^{(s)}\rangle \tag{13.23}$$

Since the auxiliary basis set is orthonormal, the reciprocal transformation matrix is the adjoint matrix of $\mathbf{D}^{(s)}$,

$$\begin{aligned} |g_i^{(s)}\rangle &= \sum_r |\varphi_r^{(s)}\rangle\langle\varphi_r^{(s)}|g_i^{(s)}\rangle \\ &= \sum_r D_{ri}^{\dagger(s)} |\varphi_r^{(s)}\rangle \end{aligned} \tag{13.24}$$

Note that $\mathbf{D}^{(s)}$ is not Hermitian, but $\mathbf{D}^{\dagger(s)}$ and $\mathbf{D}'^{(s)}$ are the inverse of each other. Further, they satisfy $\mathbf{D}^{(s)}\mathbf{D}^{\dagger(s)} = \mathbf{S}^{(s)}$ and $\mathbf{D}'^{(s)}\mathbf{D}^{\dagger(s)} = \mathbf{1}$; but $\mathbf{D}^{\dagger(s)}\mathbf{D}^{(s)} \neq \mathbf{S}^{(s)}$.

Using $\mathbf{H}'^{(ss')} = \mathbf{D}^{\dagger(s)}\mathbf{S}^{-1(s)}\mathbf{H}^{(ss')}\mathbf{D}'^{(s)}$ and $\mathbf{B}^{(s)} = \mathbf{D}^{\dagger(s)}\mathbf{A}^{(s)}$, the equation of motion for the A-coefficients, Equation (13.8), is rearranged to

$$\mathrm{i}\hbar\dot{\mathbf{B}}^{(s)} = \sum_{s'} \mathbf{H}'^{(ss')}\mathbf{B}^{(s')} - \mathrm{i}\hbar\mathbf{D}^{\dagger(s)}[\dot{\mathbf{D}}'^{(s)} + \mathbf{S}^{-1(s)}\boldsymbol{\tau}^{(s)}\mathbf{D}'^{(s)}]\mathbf{B}^{(s)} \tag{13.25}$$

This expression can be simplified by using

$$\dot{\mathbf{D}}'^{(s)} = \mathbf{S}^{-1(s)}\dot{\mathbf{D}}^{(s)} - \mathbf{S}^{-1(s)}\boldsymbol{\tau}^{\dagger(s)}\mathbf{D}'^{(s)} - \mathbf{S}^{-1(s)}\boldsymbol{\tau}^{(s)}\mathbf{D}'^{(s)} \tag{13.26}$$

and

$$\dot{\mathbf{D}}^{(s)} = \boldsymbol{\tau}^{\dagger(s)}\mathbf{D}'^{(s)} + \mathbf{D}^{(s)}\boldsymbol{\Theta}^{(s)} \tag{13.27}$$

where $\Theta_{qr}^{(s)} = \langle\varphi_q^{(s)}|\dot{\varphi}_r^{(s)}\rangle$ is set to zero (usual MCTDH phase constraint). One thus gets an MCTDH-like equation of motion for the B-coefficients:

$$\mathrm{i}\hbar\dot{\mathbf{B}}^{(s)} = \sum_{s'} \mathbf{H}'^{(ss')}\mathbf{B}^{(s')} \tag{13.28}$$

The auxiliary basis set is built using a Gram–Schmidt orthonormalization at the start of a propagation and the B-coefficients are propagated in place of the A-coefficients. The D-matrix must also be propagated using Equation (13.27), but the efficiency and stability compensates for the extra parameters in the propagation. The basis functions still evolve according to the equation of motion, Equation (13.12), governing the propagation of the Λ-vector of Gaussian parameters.

13.2.4 Initial Wavepacket

For simulating a Franck–Condon transition, the initial wavepacket, taken as a separable Gaussian product state representing the ground state of vibration in the electronic ground state, is used as a generic Gaussian function to define the rest of the variational Gaussian basis set when the conditions involve more than one function. In contrast to other GWP methods, vMCG is insensitive to the choice of initially non-populated basis functions. The variational nature means that the basis functions will follow the evolving wavepacket as well as possible and the same result is obtained irrespective of where they all start. The numerics of the propagation are, however, affected. It is found that it is best if all the functions are initially arranged closely packed around the initially populated GWP. This makes the basis set effectively complete for the first few integration steps so that the initial motion is mostly in the expansion coefficients allowing a smoother start.

In practice, the set of Gaussian functions are generated around the original function with the same widths, but positioned so that the overlap between any two functions is less than a fixed value (which can be chosen between 0.5 and 0.8 for numerical stability). Any new functions in the expansion are added in a shell-filling way: they correspond to a positive displacement along each successive coordinate until a half shell is reached, then a negative displacement until a full shell is completed ($2f + 1$ functions if f is the number of coordinates) and so on.

The wavepacket, placed at $t = 0$ on the excited electronic state, is then a superposition of all these functions, with the expansion coefficients set to 0 for all functions except the original, which has a weight of 1. A new option has been implemented and used, which allows one to build explicitly the initial wavepacket by choosing the initial values of the coefficients, $\mathbf{A}$, and parameters, $\boldsymbol{\Lambda}$, for all the GWPs. In this way, it is possible to fit the initial wavepacket to functions that are not Gaussian. This can be required when dealing, for instance, with forbidden electronic transitions, where the initial wavepacket needs to account for Herzberg–Teller selection rules (Franck–Condon wavepacket multiplied by a coordinate-dependent transition dipole moment).

13.2.5 Direct Dynamics Implementation

Using an LHA approximation for the coupled PESs at the centre of each GWP is a reasonable approximation because Gaussian functions are localized in space. In addition, using many frozen GWPs rather than a few thawed GWPs reduces the numerical errors, as the LHA will have a better range of applicability. This representation can be obtained directly from electronic structure calculations. The corresponding direct dynamics implementation of vMCG is termed DD-vMCG [198–200].

For two coupled electronic states, DD-vMCG propagations are carried out in a diabatic representation, which makes the LHA easier to define, because non-adiabatic couplings enter the Hamiltonian as a local function acting as a potential-like operator. In contrast, the adiabatic representation is most easily connected to on-the-fly computations, because the adiabatic surfaces are those obtained directly from a quantum chemistry code. At the moment, a simple form of the regularized diabatic states method of Köppel [201, 202] has been implemented (see also Ref. [196] for more details about the implementation) to transform the adiabatic information from the quantum chemistry calculations into the LHA expansion of both diabatic PESs and the coupling function. In this method, the gradient difference and non-adiabatic coupling at a point on the conical intersection seam [4] are used within a linear vibronic coupling Hamiltonian to estimate the transformation matrix.

The DD-vMCG propagation code has been implemented in a development version of the Heidelberg MCTDH package interfaced with a development version of the GAUSSIAN program [203]. Around the centre of each Gaussian function of the vMCG expansion, the LHA representation of the pair of diabatic PESs and the corresponding coupling function has to be regularly updated for each electronic state, which implies a calculation of the local adiabatic Hessian, gradient and energy for both electronic states and a subsequent diabatic regularization. The update must be carried out at the centre of each GWP in the expansion. This step is the most expensive of the whole direct dynamics process and can be run in parallel. In particular, analytical Hessian calculations are a bottleneck. Effort could be saved here by the use of Hessian update procedures, for example based on gradients. This is something to be implemented in future versions to apply the method to larger systems.

Owing to the small time steps of the nuclear dynamics, it is unnecessary to recalculate the surfaces at each step, as the surfaces do not change that fast. Instead, the LHA surfaces can be used as a model over which the system is propagated for a short while before being updated. An update time of 0.5 or 1 fs led to no significant loss of accuracy in all applications.

A further saving of effort has been implemented in the form of a database to store quantum chemistry calculations. The potential energy, gradient and Hessian of each electronic state are stored for each geometry where a calculation was made. If a GWP describes an oscillatory motion and passes through the space already sampled, the information for the LHA at the current point is not recalculated but rather is taken from the database by extrapolating the existing data through a harmonic expansion. The database is used if it contains at least one sufficiently close geometry, defined as when the relative standard deviation from the current geometry does not exceed a tunable criterion (typically 1% in most applications). When two database geometries are close, two-point-averaged extrapolations are used, with weights depending on the proximity to the current point.

A further advantage of using a database is that information from previous calculations can be reused, increasing thus the speed of successive calculations on a given system. The database thus grows, which means that global PES representations are gradually built in a similar way to the production of a PES in the GROW method [204].

The geometry and the first and second derivatives of the potential energy are stored in the database in Cartesian coordinates, which implies the use of a particular frame. However, the program should not ignore a database record differing from the current structure by only a rigid rotation. To deal with this, the code compares both geometries in a common standard orientation, uniquely defined in a conventional Eckart frame. Details are given in Ref. [198].

Another issue comes from the fact that identical nuclei in a molecule will lead to symmetry degeneracy in the complete nuclear permutation group. All permutation versions are thus considered as a unique species when comparing structures so that efficient use may be made of the stored information. This is further discussed in Ref. [199].

13.3 Applications

The first application of the vMCG method was to the four-dimensional Henon–Heiles potential, a set of coupled oscillators with linear–bilinear couplings [195]. The spectrum was accurately obtained with only a few GWP functions. Furthermore, the potential is unbound and it was noted that it was possible for the GWPs to go over the barrier and escape, thereby simulating tunnelling. The capabilities of the vMCG method were more recently demonstrated in benchmark calculations on the pyrazine absorption spectrum [187, 194] Calculations using GWPs that follow classical trajectories exhibited a much slower convergence and required far more effort despite the fact that they are much easier to integrate.

The convergence characteristic is of great importance for direct dynamics, whereby saving basis functions saves trajectories, and thus fewer points need to be calculated. The first application of DD-vMCG was to a classic problem in non-adiabatic dynamics: the butatriene cation [196], in which two states are strongly coupled with a conical intersection. This non-adiabatic coupling plays a crucial role in the photoelectron spectrum, being responsible for a range of unexpected states, which is seen as a 'mystery band' [205, 206].

The PESs of the cation were calculated using a complete active space self-consistent field (CASSCF) wavefunction with five electrons spread over the six π orbitals of the planar molecule and a 3-21G* basis set. To compare to standard MCTDH calculations, an analytic vibronic coupling model Hamiltonian was also calculated at the same level of theory [196].

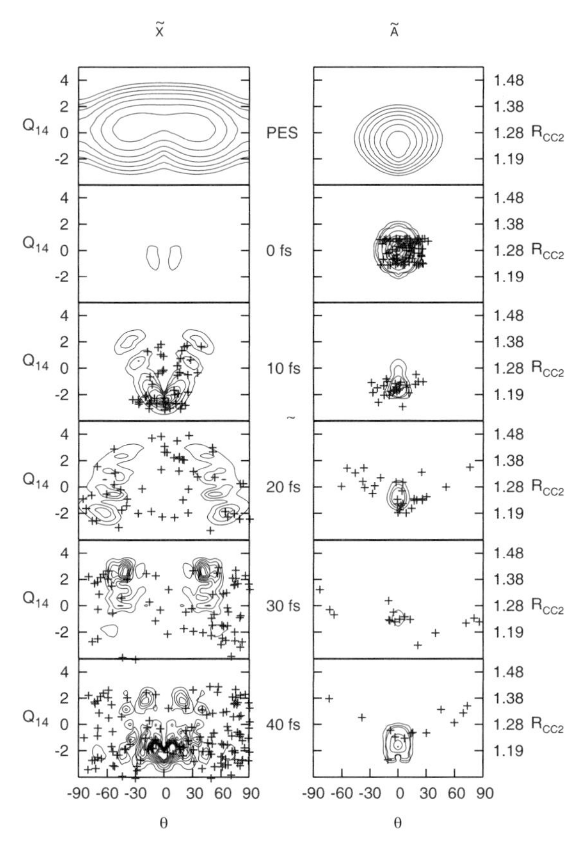

Figure 13.1 The time evolution of the adiabatic $\tilde{X}$ and $\tilde{A}$ reduced densities (contour plots) in the space of the two major nuclear modes of the butatriene cation $\tilde{X}/\tilde{A}$ states calculated using the model Hamiltonian and standard MCTDH calculations; 80 direct dynamics surface-hopping trajectories are shown for comparison (crosses). The uppermost frame contains a cut through the adiabatic PESs in this space with all other coordinates set to zero. Taken from Ref. [193].

At the time it was not possible to converge the results. It was, however, shown that, with 16 GWPs per state, vMCG was able to describe qualitatively the passage through the intersection, including the bifurcation along the coupling mode after crossing to the adiabatic ground state. With just eight GWPs per state, DD-vMCG calculations were performed and shown to provide reasonable state populations at a lower cost than comparable direct dynamics surface-hopping calculations [175, 207]. Figure 13.1 shows a comparison between direct dynamics semiclassical trajectories and standard MCTDH calculations [193], and Figure 13.2 shows a comparison between standard MCTDH

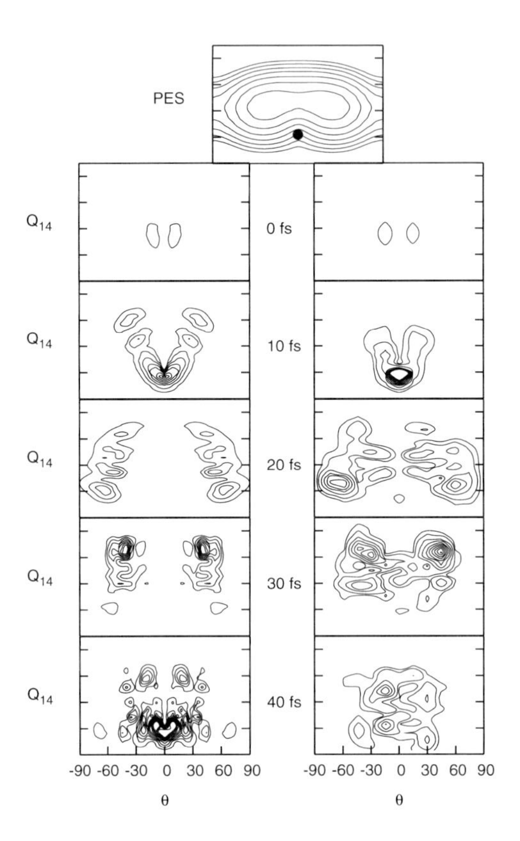

Figure 13.2 The time evolution of the adiabatic $\tilde{X}$ reduced density (contour plots) in the space of the two major nuclear modes of the butatriene cation $\tilde{X}/\tilde{A}$ states calculated using the model Hamiltonian: full quantum dynamics (left-hand column), and vMCG dynamics using 16 GWPs per state (right-hand column). The uppermost frame contains a cut through the $\tilde{X}$ adiabatic potential energy surface in this space with all other coordinates set to zero. Taken from Ref. [196].

calculations and vMCG calculations [196] (both MCTDH and vMCG calculations used the model Hamiltonian). This shows that the 80 semiclassical trajectories fail to keep with the evolving wavepacket unlike the fewer (32) 'quantum trajectories'.

The next application of DD-vMCG analysed its performance in greater detail, looking at the photodissociation of NOCl as a small test system [198, 200]. Again, this is a classic well-studied system in quantum dynamics [37, 208]. NOCl dissociates after excitation to the S_1 state directly on the adiabatic surface with excitation of the NO bond. Of particular interest here was the choice of coordinates. Direct dynamics calculations are easier to perform in rectilinear coordinates because the corresponding KEO is straightforward to derive and apply, as it does not depend on the position of the point. They are also to be preferred for GWP-based simulations, as the rectilinear KEO matrix elements are simple to evaluate analytically using Gaussian moments. Cartesian coordinates are also used to describe the PES in the quantum chemistry calculations, so the two parts of the calculation are easy to match up. For quantum dynamics calculations, however, curvilinear coordinates are often the more natural choice (see Chapter 12). They usually enable a better separation of rotational and internal motion, and eigenfunctions have a much simpler form, as these coordinates can be better related to quantum numbers. Implementing the use of curvilinear coordinates here would imply the on-the-fly calculation of the local expression of the curvilinear KEO. This could be achieved by interfacing the vMCG code to the TNUM code of Lauvergnat [139, 192].

For NOCl, while the ground-state wavefunction is nearly separable in Jacobi coordinates, using rectilinear coordinates introduces spurious coupling between modes that is seen in the calculated spectrum as extra high-energy structure. From the spectrum, 13 GWPs gave fairly converged DD-vMCG calculations, while a single GWP gave very poor results. The best approach was found to be the use of Cartesian coordinates of the Jacobi vectors. Including the centre-of-mass motion in a naive nine-dimensional Cartesian calculation was not bad, but did allow rotational contamination of the initial wavepacket, thus inducing a broader spectrum.

The use of a database proved essential in the most recent application. In this, DD-vMCG calculations were used to explore the S_0/S_1 conical intersection seam of benzene [199, 209]. A CASSCF wavefunction was used, with six electrons spread over the six π orbitals of the planar molecule and a 6-31G* basis set. A database with approximately 20 000 entries was generated during a series of calculations, which maps much of the space that is relevant to describe the access to the extended seam of conical intersection. By using different initial conditions, it was possible to see how exciting certain vibrational modes by specific amounts (by giving an initial momentum) made possible an increase of the population undergoing non-adiabatic transitions. It was also analysed how targeting specific regions of the crossing seam could be either in favour of the reactant regeneration or likely to help the formation of products. Typical trajectories are shown in Figure 13.3, with different initial conditions leading to different outcomes.

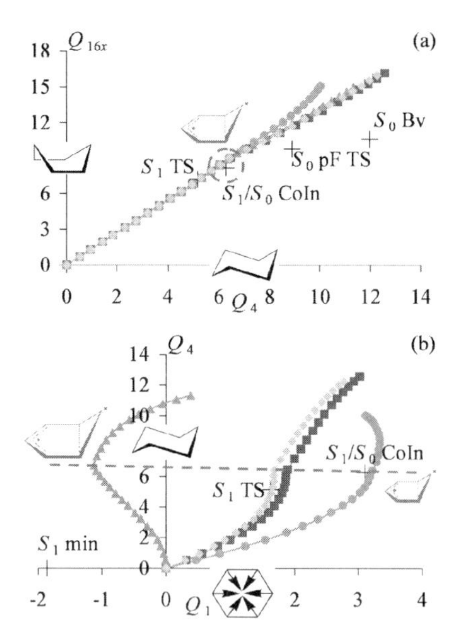

Figure 13.3 Targeting the S_1/S_0 conical intersection seam in benzene. The four trajectories are expectation values from DD-vMCG calculations with different initial momenta. (a) Expectation values in the ν_4/ν_{16a} (boat/chair) plane showing that the trajectories all target the intersection (marked S_1/S_0 CoIn). (b) Changing the momentum in the ν_1 breathing mode gives different behaviour – that desired is the middle two trajectories that cross the seam and carry on towards the photochemical products, whereas the other two trajectories curve back towards the reactant (benzene). Taken from [199].

13.4 Conclusions

The effect of non-adiabatic vibronic coupling is often important in electronic spectroscopy and photochemistry. Direct quantum dynamics simulations based on the expansion of the non-adiabatic wavepacket on a basis set of time-dependent Gaussian functions are an attractive variant of standard MCTDH calculations, promising to simulate general molecules in a straightforward manner. Modern computational techniques and computer resources are now able to cope with the huge computational effort required in these calculations, and results have been obtained on a range of systems.

Further developments in algorithms and computer power will mean that direct quantum dynamics will become ever more feasible, opening the door to quantum dynamics calculations being performed in the same straightforward way as electronic structure calculations. The result will enable us not only to treat the photochemistry of polyatomic molecules, but also to test and develop theory to meet the challenge these systems provide.

One must accept that quantum dynamics simulations are limited by the accuracy of the quantum chemistry used. An analytic function can be improved by the inclusion of all available information, such as experimental data. This is not possible in direct quantum dynamics. For excited electronic states, the CASSCF method allows calculation of analytic gradients and Hessians and provides thus the best-quality *ab initio* data that can be used in direct quantum dynamics calculations to date. Therefore, future implementations of analytic second derivatives in methods able to add dynamical correlation to CASSCF wavefunctions, for example, multireference configuration interaction (MRCI), restricted active space self-consistent field (RASSCF), or complete active space with second-order perturbation theory (CASPT2), are expected to improve the now limited accuracy of direct quantum dynamics simulations.

As a final note it must be remarked that, while this chapter has focused on running direct dynamics calculations as calculations in their own right, they are also used in the development of standard potential surfaces. For example, Bowman and co-workers used direct dynamics trajectories as a staring point in the development of multidimensional surfaces such as those for $H_5O_3{}^+$ [210] and $CH_5{}^+$ [211]. A second example is the work of the Collins group [212]. Direct dynamics trajectories provide sample points where the energy, first and second derivatives of the potential are calculated and stored. Powerful interpolation procedures are then used to provide the potential at any point [204, 213]. The resulting computer code, GROW, is able to provide high-quality surfaces for adiabatic reactions [214]. The DD-vMCG method, with its superior convergence properties, may play a useful rolc, building on the database of points already generated by calculations.

14
Multilayer Formulation of the Multiconfiguration Time-Dependent Hartree Theory

Haobin Wang and Michael Thoss

14.1
Introduction

In this chapter we discuss the multilayer (ML) formulation of the MCTDH theory. The ML-MCTDH method [119] represents an extension of the standard MCTDH approach [30, 34, 37, 38] that allows accurate quantum dynamical simulations for significantly larger and more complex systems. The key factor for this powerful extension is the more flexible and efficient way of representing the overall wavefunction in the ML-MCTDH method. This can be rationalized by the analogous comparison between the standard MCTDH approach with conventional wavepacket propagation methods. In the latter methods, the wavefunction is expressed rigidly as a linear combination of time-independent configurations (orbitals) multiplied by time-dependent coefficients. Thus, only the expansion coefficients vary during the time evolution. In contrast to that, in the MCTDH scheme both the expansion coefficients and the orbitals are time dependent according to the variation principle. This greatly reduces the redundant parameters involved in conventional wavepacket propagation methods and enables one to study dynamical processes involving a few tens of degrees of freedom. However, the scalability to larger systems of the MCTDH method is limited due to the fact that the configurations (orbitals) are expanded in terms of time-independent basis functions. It is thus natural to exploit the possibility of applying the basic MCTDH strategy also at the level of the configurations (orbitals), which is the ML-MCTDH theory to be described in this chapter.

Before describing in detail the theory and derivation of the ML-MCTDH method, two issues deserve to be mentioned. First, the ML-MCTDH theory is a method for wavepacket propagation that is based rigorously on the variational principle, just like the MCTDH method or any conventional wavepacket approaches. The fact that the ML-MCTDH method is applicable to more complex dynamical problems with many degrees of freedom (for exam-

Multidimensional Quantum Dynamics: MCTDH Theory and Applications.
Edited by Hans-Dieter Meyer, Fabien Gatti, and Graham A. Worth

ISBN: 978-3-527-32018-9

ple, a few hundred) is not based on any sacrifice of theoretical rigorousness but rather on a better, more flexible way of expressing the variational functional, which is known to be a key component for an effective variational calculation. Second, the multilayer construction of the wavefunction does not impose any additional limitations compared with the original MCTDH method. Any Hamiltonian that can be treated via the MCTDH approach can equally be handled with the ML-MCTDH theory. The only potential drawback of the ML-MCTDH method is its complexity with respect to implementation. This is, however, offset by the tremendous gain in computational efficiency the method can offer.

In the following section the theoretical formulation of the ML-MCTDH method will be presented. To facilitate a better understanding of the basis of its efficiency, we first give a brief outline of the conventional basis set approach to wavepacket propagations from a variational point of view. We then summarize briefly the derivation of the MCTDH method, discuss why it is a significant improvement over the conventional approach, and analyse its current limitations for treating even larger systems. This only serves as a basis for the subsequent derivation of the ML-MCTDH method. For a more complete derivation and analysis of the MCTDH theory, the reader should consult Chapter 3. Finally, we present a detailed derivation of the ML-MCTDH theory and analyse the essential features that make this method more effective.

14.2 From Conventional Wavepacket Propagation to ML-MCTDH Theory: A Variational Perspective

14.2.1 Conventional Approach Based on Time-Independent Configurations

The equations of motion for a variational basis set approach to quantum dynamics can be obtained from the Dirac–Frenkel variational principle [32, 33]

$$\left\langle \delta\Psi(t) \middle| \mathrm{i}\frac{\partial}{\partial t} - \hat{H} \middle| \Psi(t) \right\rangle = 0 \tag{14.1}$$

In most conventional wavepacket propagation methods, the wavefunction is expressed as a linear combination of *time-independent* states (configurations) $|\Phi_J\rangle$,

$$|\Psi(t)\rangle = \sum_J A_J(t)|\Phi_J\rangle \tag{14.2}$$

where J is a multidimensional index that runs through all combinations of basis functions in f degrees of freedom, that is,

$$\sum_J A_J(t)|\Phi_J\rangle \equiv \sum_{j_1}\sum_{j_2}\cdots\sum_{j_f} A_{j_1 j_2 \cdots j_f}(t) \prod_{\nu=1}^{f} |\phi_{j_\nu}^{(\nu)}\rangle \tag{14.3}$$

For convenience, the basis functions for each degree of freedom are usually chosen to be orthonormal

$$\langle \phi_n^{(\nu)} | \phi_m^{(\nu)} \rangle = \delta_{nm} \tag{14.4a}$$

which yields orthonormal time-independent configurations

$$\langle \Phi_L | \Phi_J \rangle = \delta_{LJ} \tag{14.4b}$$

Since the configurations $|\Phi_J\rangle$ are considered to be fixed and time independent, the variation and the time derivative of the wavefunction are given by

$$\delta|\Psi(t)\rangle = \sum_J \delta A_J(t)\, |\Phi_J\rangle \tag{14.5a}$$

$$\frac{\partial}{\partial t}|\Psi(t)\rangle = \sum_J \dot{A}_J(t)\, |\Phi_J\rangle \tag{14.5b}$$

After substitution into Equation (14.1) and application of the orthonormality condition of Equation (14.4), this results in the usual expression for the time evolution of the expansion coefficients,

$$\mathrm{i}\dot{A}_J(t) = \langle \Phi_J | \hat{H} | \Psi(t) \rangle = \sum_L \langle \Phi_J | \hat{H} | \Phi_L \rangle A_L(t) \tag{14.6}$$

Equation (14.6) is the scheme used in most of the current time-dependent wavepacket propagation methods. Considerable effort has been devoted to optimizing various aspects of the calculation, including the choice of basis functions and/or discrete variable representations (DVRs), coordinate systems, and methods of propagation. However, the fact that in this scheme the configurations are fixed and only the expansion coefficients are varied – which corresponds to a full configuration-interaction (FCI) approach in electronic structure calculations – severely restricts its practical applicability. As is evident from Equation (14.3), the total number of coefficients $A_J(t) = A_{j_1 j_2 \cdots j_f}(t)$ or combinations of basis functions (configurations)

$$|\Phi_J\rangle = \prod_{\nu=1}^{f} |\phi_{j_\nu}^{(\nu)}\rangle$$

scales exponentially versus the number of degrees of freedom. As a result, a dynamical calculation for more than about 10 degrees of freedom, which corresponds roughly to 10^{10} states, is rather unrealistic with the current computer hardware. Hence, the application of wavepacket propagation schemes based on Equation (14.6) is limited to systems of four or five atoms.

In many physical situations, however, only a relatively small number of *time-dependent* combinations of the fixed configurations are important for describing the quantum dynamics accurately. Therefore, it appears to be more appropriate to describe the wavefunction in the functional of the variational procedure in terms of *time-dependent* configurations. This idea is the conceptual basis of the MCTDH approach [30, 34, 37, 38].

14.2.2
The Multiconfiguration Time-Dependent Hartree Method

As described in Chapter 3, within the MCTDH method [30,34,37,38] the wavefunction is expanded in *time-dependent* configurations

$$|\Psi(t)\rangle = \sum_J A_J(t)|\Phi_J(t)\rangle \tag{14.7}$$

The major difference from the conventional approach, Equation (14.2), is thus that the configurations (orbitals) are also optimized in the variational procedure, a common and popular practice in electronic structure theory calculations for time-independent variations. In this way, the Dirac–Frenkel variational procedure can be separated explicitly into two parts:

$$\left\langle \delta\Psi(t) \middle| \mathrm{i}\frac{\partial}{\partial t} - \hat{H} \middle| \Psi(t) \right\rangle_{\text{coefficients}} = 0 \tag{14.8a}$$

and

$$\left\langle \delta\Psi(t) \middle| \mathrm{i}\frac{\partial}{\partial t} - \hat{H} \middle| \Psi(t) \right\rangle_{\text{orbitals}} = 0 \tag{14.8b}$$

where in Equation (14.8a) only the expansion coefficients are varied and in Equation (14.8b) only the orbitals. For situations where quantum statistics have been built into the Hamiltonian operator and is not explicitly expressed in the wavefunction, each configuration is expressed in the form of a Hartree product,

$$|\Phi_J(t)\rangle = \prod_{\kappa=1}^{p} |\varphi_J^{(\kappa)}(t)\rangle \tag{14.9}$$

where $|\varphi_J^{(\kappa)}(t)\rangle$ is the single-particle function (SPF) for the κth single particle (SP) group and p is the total number of SP groups. The expansion scheme for the MCTDH wavefunction thus reads

$$|\Psi(t)\rangle = \sum_{j_1}\sum_{j_2}\cdots\sum_{j_p} A_{j_1 j_2 \ldots j_p}(t) \prod_{\kappa=1}^{p} |\varphi_{j_\kappa}^{(\kappa)}(t)\rangle \tag{14.10}$$

For a fixed number of configurations, the MCTDH method can be considered as the time-dependent analogue of complete active space self-consistent field (CASSCF) methods in electronic structure theory. Thus, similar procedures can be used to separate the orbitals into active and inactive spaces, although the latter is usually eliminated in most MCTDH applications.

Without loss of generality, we consider the situation where SPFs are orthonormal,

$$\langle \varphi_n^{(\kappa)}(t) | \varphi_m^{(\kappa)}(t) \rangle = \delta_{nm} \tag{14.11}$$

In differential form, this is equivalent to

$$\langle \varphi_n^{(\kappa)}(0) | \varphi_m^{(\kappa)}(0) \rangle = \delta_{nm} \tag{14.12a}$$

$$\left\langle \varphi_n^{(\kappa)}(t) \middle| \mathrm{i}\frac{\partial}{\partial t} \varphi_m^{(\kappa)}(t) \right\rangle = \langle \varphi_n^{(\kappa)}(t) | \hat{g}_\kappa | \varphi_m^{(\kappa)}(t) \rangle \tag{14.12b}$$

Here, $\hat{g}_\kappa$ is an arbitrary Hermitian operator that does not affect the quality of the multiconfigurational expansions of the wavefunction, but determines the explicit form of the equations of motion (see Section 3.2). For simplicity we have chosen $\hat{g}_k = 0$ in this chapter, which yields the differential orthonormality condition

$$\langle \varphi_n^{(\kappa)}(0) | \varphi_m^{(\kappa)}(0) \rangle = \delta_{nm} \tag{14.13a}$$

$$\left\langle \varphi_n^{(\kappa)}(t) \middle| \frac{\partial}{\partial t} \varphi_m^{(\kappa)}(t) \right\rangle \equiv \langle \varphi_n^{(\kappa)}(t) | \dot{\varphi}_m^{(\kappa)}(t) \rangle = 0 \tag{14.13b}$$

As described in Chapter 3, the variations in Equation (14.8), with the orthonormal constraints above, lead to the MCTDH equations of motion:

$$\mathrm{i}\dot{A}_J(t) = \langle \Phi_J(t) | \hat{H} | \Psi(t) \rangle = \sum_L \langle \Phi_J(t) | \hat{H} | \Phi_L(t) \rangle A_L(t) \tag{14.14a}$$

$$\mathrm{i} | \underline{\dot{\varphi}}^{(\kappa)}(t) \rangle = [1 - \hat{P}^{(\kappa)}(t)][\hat{\rho}^{(\kappa)}(t)]^{-1} \langle \hat{H} \rangle^{(\kappa)}(t) | \underline{\varphi}^{(\kappa)}(t) \rangle \tag{14.14b}$$

Equation (14.14a) is similar to Equation (14.6), except that the Hamiltonian matrix $\mathcal{K}_{JL}(t) \equiv \langle \Phi_J(t) | \hat{H} | \Phi_L(t) \rangle$ now becomes time dependent. Equation (14.14b) describes the time-dependent change of the SPFs, which does not appear in the conventional method. Various quantities are defined in Chapter 3 of this book:

$$| \underline{\varphi}^{(\kappa)}(t) \rangle = \{ | \varphi_1^{(\kappa)}(t) \rangle, | \varphi_2^{(\kappa)}(t) \rangle, \ldots \}^{\mathrm{T}}$$

denotes the symbolic column vector of the SPFs for the κth SP group, $\langle \hat{H} \rangle^{(\kappa)}(t)$ denotes the mean-field operator, $\hat{\rho}^{(\kappa)}(t)$ is the reduced density matrix,

$$\langle \hat{H} \rangle_{nm}^{(\kappa)}(t) = \langle \Psi_n^{(\kappa)}(t) | \hat{H} | \Psi_m^{(\kappa)}(t) \rangle \tag{14.15a}$$

$$\rho_{nm}^{(\kappa)}(t) = \langle \Psi_n^{(\kappa)}(t) | \Psi_m^{(\kappa)}(t) \rangle \tag{14.15b}$$

and $P^{(\kappa)}(t)$ is the SP space projector for the κth SP group,

$$P^{(\kappa)}(t) = \sum_m |\varphi_m^{(\kappa)}(t)\rangle\langle\varphi_m^{(\kappa)}(t)| \tag{14.16}$$

In the expressions above, the single-hole function $|\Psi_n^{(\kappa)}(t)\rangle$ for the κth SP group is defined as [30, 34, 37, 38] (see Equation (3.10))

$$|\Psi_n^{(\kappa)}(t)\rangle = \sum_{j_1}\cdots\sum_{j_{\kappa-1}}\sum_{j_{\kappa+1}}\cdots\sum_{j_p} A_{j_1\cdots j_{k-1} n j_{k+1}\cdots j_p}(t) \prod_{\lambda\neq\kappa}^{p} |\varphi_{j_\lambda}^{(\lambda)}(t)\rangle \tag{14.17a}$$

so that

$$|\Psi(t)\rangle = \sum_n |\varphi_n^{(\kappa)}(t)\rangle|\Psi_n^{(\kappa)}(t)\rangle \tag{14.17b}$$

In practice, one solves the above coupled differential equations (14.14a) and (14.14b) simultaneously using an ordinary differential equation (ODE) integrator or a more elaborate integrator as described in Chapter 4. Three more technical details that affect the efficiency and reliability of the method are worth mentioning. First, it is advantageous to subtract the average energy from the Hamiltonian. This reduces the phase oscillation in the time evolution and does not change any physical quantities to be evaluated. Second, the time-dependent projection operator $1 - \hat{P}^{(\kappa)}(t)$ keeps the SPFs orthonormal within the same SP group. A naive implementation, however, does not fulfil this requirement due to numerical round-off errors. To remedy the numerical deficiency, the mathematically equivalent form

$$[1 - \hat{P}^{(\kappa)}(t)]^2 = [1 - \hat{P}^{(\kappa)}(t)] \tag{14.18}$$

can be used in Equation (14.14b). Similar tricks are often used in carrying out numerical Gram–Schmidt orthogonalizations. Third, it should be noted that $[\hat{\rho}^{(\kappa)}(t)]^{-1}$ denotes the pseudo-inverse. It casts the equations for the SPFs,

$$\mathrm{i}\sum_m \rho_{nm}^{(\kappa)}(t)|\dot{\varphi}_m^{(\kappa)}(t)\rangle = [1 - \hat{P}^{(\kappa)}(t)]\sum_m \langle\hat{H}\rangle_{nm}^{(\kappa)}(t)|\varphi_m^{(\kappa)}(t)\rangle \tag{14.19}$$

into the form of Equation (14.14b), which is based on the explicit variation of the SPFs,

$$\mathrm{i}\left\langle \delta\varphi_n^{(\kappa)}(t)\middle| \sum_m \rho_{nm}^{(\kappa)}(t)\middle|\dot{\varphi}_m^{(\kappa)}(t)\right\rangle = \left\langle \delta\varphi_n^{(\kappa)}(t)\middle|[1 - \hat{P}^{(\kappa)}(t)]\sum_m \langle\hat{H}\rangle_{nm}^{(\kappa)}(t)\middle|\varphi_m^{(\kappa)}(t)\right\rangle \tag{14.20}$$

On the other hand, Equation (14.19) can also be solved using the singular value decomposition (SVD) technique. Our experience shows that the pseudo-inverse solver is more efficient.

At first glance, the expansion in the MCTDH approach, Equation (14.10), resembles that in the FCI-type expression of Equation (14.3), that is, the total number of time-dependent configurations still scales exponentially versus the total number of SP groups p. However, the MCTDH method is applicable to more complex systems for two reasons: (i) The base of the exponential increase in labour of the MCTDH approach is usually much smaller than in a conventional FCI method, because the number of physically important SPFs is usually much smaller than the number of time-independent basis functions in the FCI approach. (ii) Each SP group can contain several physical degrees of freedom so that the number of SP groups (p) is usually much less than the number of physical degrees of freedom (f). As a result, the overall computational effort in the MCTDH approach scales more slowly with respect to the number of degrees of freedom, which makes the MCTDH method capable of handling rather large molecular systems in a numerically converged way [62,63,75,119,215–226].

The main limitation of the MCTDH approach outlined above lies in its way of constructing the SPFs, which is based on a multidimensional expansion employing *time-independent Hartree* products,

$$|\varphi_n^{(\kappa)}(t)\rangle = \sum_I B_I^{\kappa,n}(t)|u_I^\kappa\rangle \equiv \sum_{i_1}\sum_{i_2}\cdots\sum_{i_{F(\kappa)}} B_{i_1 i_2 \ldots i_{F(\kappa)}}^{\kappa,n}(t)\prod_{q=1}^{F(\kappa)}|\phi_{i_q}^{\kappa,q}\rangle \tag{14.21}$$

Here $F(\kappa)$ is the number of degrees of freedom contained in the κth SP group, and $|\phi_{i_q}^{\kappa,q}\rangle$ denotes the corresponding time-independent primitive basis functions for the qth degree of freedom within this SP group. The FCI-type expansion of the SPFs in Equation (14.21) is usually limited to a few ($\sim$10) degrees of freedom, and the multiconfigurational expansion of the wavefunction in Equation (14.10) is typically limited to about 10 SP groups. As a result, a routine MCTDH calculation is limited to systems with a few tens of quantum degrees of freedom (which is already many more than can be treated via conventional approaches). A further improvement can be achieved by employing static basis set contraction techniques [119,216,218,219,223,224] (for example, adiabatic reduction [227]), but due to the limitations discussed above the quantum dynamical treatment of a system with more than 100 degrees of freedom is currently not feasible.

As described in the next section, the ML-MCTDH theory [119] successfully circumvents these limitations of the MCTDH method. The basic idea of the ML-MCTDH formulation is to use *dynamic* contraction of the basis functions that constitute the original SPFs by building further layers in the

MCTDH functional. This strategy can also be viewed as 'cascading' the original MCTDH method to the SPFs [38], that is, using the MCTDH strategy to treat each SP group.

14.2.3 The Multilayer Formulation of the MCTDH Theory

The motivation of formulating the ML-MCTDH method is to extend the applicability of the original MCTDH method to substantially larger systems. For this purpose, the number of physical degrees of freedom that each SP group can accommodate needs to be increased significantly. To this end, the FCI-type construction of the SPFs in Equation (14.21) is replaced by a *time-dependent* multiconfigurational expansion

$$|\varphi_n^{(\kappa)}(t)\rangle = \sum_I B_I^{\kappa,n}(t)|u_I^\kappa(t)\rangle \tag{14.22a}$$

that is, the basic strategy of the MCTDH method, Equation (14.7), is adopted to treat each SPF.

We start our discussion with the two-layer formulation and then generalize it to an arbitrary number of layers. For clarity, we refer in the following to the SP introduced in the previous MCTDH section as the *level-one* (L1) SP, which in turn contains several *level-two* (L2) SPs,

$$|u_I^\kappa(t)\rangle = \prod_{q=1}^{Q(\kappa)} |v_{i_q}^{(\kappa,q)}(t)\rangle \tag{14.22b}$$

Similar to Equation (14.10), the L1 SP function $|\varphi_n^{(\kappa)}(t)\rangle$ is expanded in the time-dependent L2 SPFs as

$$\begin{aligned} |\varphi_n^{(\kappa)}(t)\rangle &= \sum_I B_I^{\kappa,n}(t)|u_I^\kappa(t)\rangle \\ &\equiv \sum_{i_1}\sum_{i_2}\cdots\sum_{i_{Q(\kappa)}} B_{i_1 i_2 \ldots i_{Q(\kappa)}}^{\kappa,n}(t) \prod_{q=1}^{Q(\kappa)} |v_{i_q}^{(\kappa,q)}(t)\rangle \end{aligned} \tag{14.23}$$

Here, $Q(\kappa)$ denotes the number of L2 SP groups in the κth L1 SP and $|v_{i_q}^{(\kappa,q)}(t)\rangle$ is the L2 SPF for the qth L2 SP group. Both are contained in the κth L1 SP group. The expansion of the overall wavefunction can thus be written in the form

$$\begin{aligned} &|\Psi(t)\rangle \\ &= \sum_{j_1}\sum_{j_2}\cdots\sum_{j_p} A_{j_1 j_2 \ldots j_p}(t) \prod_{\kappa=1}^{p} \left[\sum_{i_1}\sum_{i_2}\cdots\sum_{i_{Q(\kappa)}} B_{i_1 i_2 \ldots i_{Q(\kappa)}}^{\kappa,j_\kappa}(t) \prod_{q=1}^{Q(\kappa)} |v_{i_q}^{(\kappa,q)}(t)\rangle \right] \end{aligned} \tag{14.24}$$

As a result, the Dirac–Frenkel variational principle can be written explicitly as

$$\left\langle \delta\Psi(t) \middle| \mathrm{i}\frac{\partial}{\partial t} - \hat{H} \middle| \Psi(t) \right\rangle_{\text{top coefficients}} = 0 \tag{14.25a}$$

$$\left\langle \delta\Psi(t) \middle| \mathrm{i}\frac{\partial}{\partial t} - \hat{H} \middle| \Psi(t) \right\rangle_{\text{L1 SPFs}} = 0 \tag{14.25b}$$

$$\left\langle \delta\Psi(t) \middle| \mathrm{i}\frac{\partial}{\partial t} - \hat{H} \middle| \Psi(t) \right\rangle_{\text{L2 SPFs}} = 0 \tag{14.25c}$$

Thus, the equations of motion within the ML-MCTDH approach are obtained from variation of the overall wavefunction $|\Psi(t)\rangle$ with respect to three parts:

(i) the top (L1) expansion coefficients $A_J(t)$ in Equation (14.7);

(ii) the L1 SPFs, that is, the L2 expansion coefficients $B_I^{\kappa,n}(t)$ in Equation (14.22a); and

(iii) the L2 SPFs, $|v_{i_q}^{(\kappa,q)}(t)\rangle$ in Equation (14.23), which are yet to be specified.

The first two parts give equations that are in the same form as Equations (14.14a) and (14.14b):

$$\mathrm{i}\dot{A}_J(t) = \sum_L \langle \Phi_J(t)|\hat{H}|\Phi_L(t)\rangle A_L(t) \tag{14.26a}$$

$$\mathrm{i}|\dot{\underline{\varphi}}^{(\kappa)}(t)\rangle_{\text{L2 coefficients}} = [1 - \hat{P}^{(\kappa)}(t)][\hat{\rho}^{(\kappa)}(t)]^{-1}\langle \hat{H}\rangle^{(\kappa)}(t)|\underline{\varphi}^{(\kappa)}(t)\rangle \tag{14.26b}$$

Here, the symbolic notation on the left-hand side of Equation (14.26b) means that the time derivative of the L1 SPFs is only taken with respect to the L2 expansion coefficients $B_I^{\kappa,n}$ and does not act on the L2 configuration $|u_I^{\kappa}(t)\rangle$ (see Equation (14.23)).

It is important to point out that, in the ML-MCTDH formulation, various quantities, such as the configurations $|\Phi_J(t)\rangle$, the Hamiltonian matrix $\mathcal{K}_{JL}(t) = \langle \Phi_J(t)|\hat{H}|\Phi_L(t)\rangle$, the L1 mean-field operator $\langle \hat{H}\rangle^k(t)$ and the L1 SPF $|\underline{\varphi}^{(\kappa)}(t)\rangle$, depend on the L2 SPFs $|v_{i_q}^{(\kappa,q)}(t)\rangle$. Therefore, these quantities need to be built explicitly from the bottom-layer SPFs and basis functions and are thus more complicated to evaluate than in the original MCTDH method. For this reason, the ML-MCTDH approach is not expected to be more efficient than the standard MCTDH method if applied to systems with only a few degrees of freedom. Rather, it is designed to treat more complex problems with many degrees of freedom.

In order to carry out the third task (iii) above, the variation with respect to the L2 SPFs, it is useful to define the L2 single-hole function $|g_{n,r}^{(\kappa,q)}(t)\rangle$ in the same spirit as in Equation (14.17),

$$|\varphi_n^k(t)\rangle = \sum_r |v_r^{(\kappa,q)}(t)\rangle\, |g_{n,r}^{(\kappa,q)}(t)\rangle \tag{14.27}$$

It should be noted that the L2 single-hole function $|g_{n,r}^{(\kappa,q)}(t)\rangle$ depends on the L1 SPF index n, but the L2 SPF $|v_r^{(\kappa,q)}(t)\rangle$ does not. For simplicity, the L2 SPFs are again chosen to be orthonormal similar to Equation (14.13),

$$\langle v_r^{(\kappa,q)}(0)|v_s^{(\kappa,q)}(0)\rangle = \delta_{rs} \tag{14.28a}$$

$$\left\langle v_r^{(\kappa,q)}(t)\left|\frac{\partial}{\partial t}v_s^{(\kappa,q)}(t)\right.\right\rangle \equiv \langle v_r^{(\kappa,q)}(t)|\dot{v}_s^{(\kappa,q)}(t)\rangle = 0 \tag{14.28b}$$

which also leads to the orthonormality condition for the L2 configurations,

$$\langle u_l^\kappa(0)|u_j^\kappa(0)\rangle = \delta_{lj} \tag{14.28c}$$

$$\langle u_l^\kappa(t)|\dot{u}_j^\kappa(t)\rangle = 0 \tag{14.28d}$$

The variation of the L2 SPFs is thus given by

$$\begin{aligned} |\delta\Psi(t)\rangle_{\text{L2 SPF}} &= \sum_{\kappa=1}^{p}\sum_n |\delta\varphi_n^{(\kappa)}(t)\rangle_{\text{L2 SPF}}\,|\Psi_n^{(\kappa)}(t)\rangle \\ &= \sum_{\kappa=1}^{p}\sum_n\sum_{q=1}^{Q(\kappa)}\sum_r |\delta v_r^{(\kappa,q)}(t)\rangle|g_{n,r}^{(\kappa,q)}(t)\rangle|\Psi_n^{(\kappa)}(t)\rangle \end{aligned} \tag{14.29}$$

As the variation can be carried out independently for each L2 SPF $|\delta v_r^{(\kappa,q)}(t)\rangle$, the summation over κ, q and r can be dropped. On the other hand, it is important to note that, due to the dependence of the single-hole functions $|g_{n,r}^{(\kappa,q)}(t)\rangle$ on the L1 SPF index n, the summation over n has to be retained for variations of L2 SPFs. Application of the Dirac–Frenkel variational principle leads to a modification of Equation (14.20),

$$\begin{aligned} &\mathrm{i}\sum_n\left\langle \delta\varphi_n^{(\kappa)}(t)_{\text{L2 SPF}}\left|\sum_m \rho_{nm}^{(\kappa)}(t)\right|\dot{\varphi}_m^{(\kappa)}(t)\right\rangle \\ &\quad = \sum_n\left\langle \delta\varphi_n^{(\kappa)}(t)_{\text{L2 SPF}}\left|[1-\hat{P}^{(\kappa)}(t)]\sum_m\langle\hat{H}\rangle_{nm}^{(\kappa)}(t)\right|\varphi_m^{(\kappa)}(t)\right\rangle \end{aligned} \tag{14.30a}$$

It should be noted that, unlike in Equation (14.26b), $|\dot{\varphi}_m^{(\kappa)}(t)\rangle$ represents the full time derivative of the L1 SPF $|\varphi_m^{(\kappa)}(t)\rangle$. Upon substitution of Equation (14.29), Equation (14.30a) gives

$$\begin{aligned} \mathrm{i}\sum_n\sum_m \rho_{nm}^{(\kappa)}(t)\langle\delta v_r^{(\kappa,q)}(t)|\langle g_{n,r}^{(\kappa,q)}(t)|\dot{\varphi}_m^{(\kappa)}(t)\rangle \\ = \sum_n\sum_m\langle\delta v_r^{(\kappa,q)}(t)|\langle g_{n,r}^{(\kappa,q)}(t)|[1-\hat{P}^{(\kappa)}(t)]\langle\hat{H}\rangle_{nm}^{(\kappa)}(t)|\varphi_m^{(\kappa)}(t)\rangle \end{aligned} \tag{14.30b}$$

Furthermore, based on Equations (14.23) and (14.27), the full time derivative of the L1 SPF can be expressed as

$$\begin{aligned} \mathrm{i}\sum_m \rho_{nm}^{(\kappa)}(t)|\dot{\varphi}_m^{(\kappa)}(t)\rangle &= \mathrm{i}\sum_m \rho_{nm}^{(\kappa)}(t)\sum_I \dot{B}_I^{\kappa,m}(t)|u_I^{\kappa}(t)\rangle \\ &+\mathrm{i}\sum_m \rho_{nm}^{(\kappa)}(t)\sum_{\eta=1}^{Q(\kappa)}\sum_s|\dot{v}_s^{(\kappa,\eta)}(t)\rangle|g_{m,s}^{(\kappa,\eta)}(t)\rangle \end{aligned} \tag{14.31}$$

According to Equation (14.26b) the first term on the right-hand side can be further reduced to (see Equation (14.19) for an equivalent but more analogous expression)

$$\begin{aligned} \mathrm{i}\sum_I\sum_m \rho_{nm}^{(\kappa)}(t)|u_I^{\kappa}(t)\rangle\dot{B}_I^{\kappa,m}(t) \\ \equiv \mathrm{i}\sum_I\sum_m \rho_{nm}^{(\kappa)}(t)|u_I^{\kappa}(t)\rangle\langle u_I^{\kappa}(t)|\dot{\varphi}_m^{(\kappa)}(t)\rangle_{\text{L2 coefficients}} \\ = \sum_I |u_I^{\kappa}(t)\rangle\left\langle u_I^{\kappa}(t)\middle|[1-\hat{P}^{(\kappa)}(t)]\sum_m\langle\hat{H}\rangle_{nm}^{(\kappa)}(t)\middle|\varphi_m^{(\kappa)}(t)\right\rangle \\ \equiv \hat{P}_{\text{L2}}^{(\kappa)}(t)[1-\hat{P}^{(\kappa)}(t)]\sum_m\langle\hat{H}\rangle_{nm}^{(\kappa)}(t)\sum_s|v_s^{(\kappa,q)}(t)\rangle|g_{m,s}^{(\kappa,q)}(t)\rangle \end{aligned} \tag{14.32}$$

where we have substituted Equation (14.27) for $|\varphi_m^{(\kappa)}(t)\rangle$ and defined the L2 configurational space projection operator

$$\hat{P}_{\text{L2}}^{(\kappa)}(t) \equiv \sum_I |u_I^{\kappa}(t)\rangle\langle u_I^{\kappa}(t)| \tag{14.33}$$

With Equation (14.32), we can simplify Equation (14.31) as

$$\begin{aligned} \mathrm{i}\sum_m \rho_{nm}^{(\kappa)}(t)|\dot{\varphi}_m^{(\kappa)}(t)\rangle &= \hat{P}_{\text{L2}}^{(\kappa)}(t)[1-\hat{P}^{(\kappa)}(t)]\sum_m\langle\hat{H}\rangle_{nm}^{(\kappa)}(t)\sum_s|v_s^{(\kappa,q)}(t)\rangle|g_{m,s}^{(\kappa,q)}(t)\rangle \\ &+\mathrm{i}\sum_m \rho_{nm}^{(\kappa)}(t)\sum_{\eta=1}^{Q(\kappa)}\sum_s|\dot{v}_s^{k,\eta}(t)\rangle|g_{m,s}^{(\kappa,\eta)}(t)\rangle \end{aligned} \tag{14.34}$$

Substituting Equation (14.34) into Equation (14.30b) and using the orthonormality condition in Equation (14.28), the only term retained in the summation over η is $\eta = q$. This gives

$$\left\langle \delta v_r^{(\kappa,q)}(t) \middle| \mathrm{i} \sum_n \sum_m \rho_{nm}^{(\kappa)}(t) \sum_s \langle g_{n,r}^{(\kappa,q)}(t) | g_{m,s}^{(\kappa,q)}(t) \rangle \middle| \dot{v}_s^{(\kappa,q)}(t) \right\rangle$$

$$= \left\langle \delta v_r^{(\kappa,q)}(t) \middle| \sum_n \sum_m \langle g_{n,r}^{(\kappa,q)}(t) | [1 - \hat{P}_{\mathrm{L2}}^{(\kappa)}(t)][1 - \hat{P}^{(\kappa)}(t)] \right.$$

$$\left. \times \langle \hat{H} \rangle_{nm}^{(\kappa)}(t) \sum_s | g_{m,s}^{(\kappa,q)}(t) \rangle \middle| v_s^{(\kappa,q)}(t) \right\rangle \tag{14.35}$$

This expression can be simplified by using the following identities:

$$[1 - \hat{P}_{\mathrm{L2}}^{(\kappa)}(t)] \hat{P}^{(\kappa)}(t) = 0 \tag{14.36a}$$

$$\hat{P}_{\mathrm{L2}}^{(\kappa)}(t) | g_{n,r}^{(\kappa,q)}(t) \rangle = | g_{n,r}^{(\kappa,q)}(t) \rangle \sum_l | v_l^{(\kappa,q)}(t) \rangle \langle v_l^{(\kappa,q)}(t) |$$

$$\equiv | g_{n,r}^{(\kappa,q)}(t) \rangle \hat{P}_{\mathrm{L2}}^{(\kappa,q)}(t) \tag{14.36b}$$

where we have introduced the projection operator in the L2 SP space,

$$\hat{P}_{\mathrm{L2}}^{(\kappa,q)}(t) \equiv \sum_l | v_l^{(\kappa,q)}(t) \rangle \langle v_l^{(\kappa,q)}(t) | \tag{14.36c}$$

Thus, Equation (14.35) becomes

$$\left\langle \delta v_r^{(\kappa,q)}(t) \middle| \mathrm{i} \sum_n \sum_m \rho_{nm}^{(\kappa)}(t) \sum_s \langle g_{n,r}^{(\kappa,q)}(t) | g_{m,s}^{(\kappa,q)}(t) \rangle \middle| \dot{v}_s^{(\kappa,q)}(t) \right\rangle$$

$$= \left\langle \delta v_r^{(\kappa,q)}(t) \middle| [1 - \hat{P}_{\mathrm{L2}}^{(\kappa,q)}(t)] \sum_n \sum_m \langle g_{n,r}^{(\kappa,q)}(t) | \langle \hat{H} \rangle_{nm}^{(\kappa)}(t) \right.$$

$$\left. \times \sum_s | g_{m,s}^{(\kappa,q)}(t) \rangle \middle| v_s^{(\kappa,q)}(t) \right\rangle \tag{14.37}$$

We now introduce compact notations for the reduced density matrix and mean-field operator in the second layer:

$$\varrho_{rs}^{(\kappa,q)}(t) = \sum_n \sum_m \rho_{nm}^{(\kappa)}(t) \langle g_{n,r}^{(\kappa,q)}(t) | g_{m,s}^{(\kappa,q)}(t) \rangle \tag{14.38a}$$

$$\langle \hat{\mathcal{H}} \rangle_{rs}^{(\kappa,q)}(t) = \sum_n \sum_m \langle g_{n,r}^{(\kappa,q)}(t) | \langle \hat{H} \rangle_{nm}^{(\kappa)}(t) | g_{m,s}^{(\kappa,q)}(t) \rangle \tag{14.38b}$$

With this notation, Equation (14.37) simplifies to

$$\left\langle \delta v_r^{(\kappa,q)}(t) \middle| \mathrm{i} \sum_s \varrho_{rs}^{(\kappa,q)}(t) \middle| \dot{v}_s^{(\kappa,q)}(t) \right\rangle$$
$$= \left\langle \delta v_r^{(\kappa,q)}(t) \middle| [1 - \hat{P}_{\mathrm{L2}}^{(\kappa,q)}(t)] \sum_s \langle \hat{\mathcal{H}} \rangle_{rs}^{(\kappa,q)}(t) \middle| v_s^{(\kappa,q)}(t) \right\rangle \tag{14.39}$$

If we terminate the ML-MCTDH hierarchy at the second layer, the L2 SPFs are expanded in time-independent orthonormal configurations,

$$\begin{aligned} |v_r^{(\kappa,q)}(t)\rangle &= \sum_\alpha C_\alpha^{\kappa,q,r}(t) |w_\alpha^{\kappa,q}\rangle \\ &\equiv \sum_{\alpha_1} \sum_{\alpha_2} \cdots \sum_{\alpha_{M(\kappa,q)}} C_{\alpha_1 \alpha_2 \ldots \alpha_{M(\kappa,q)}}^{\kappa,q,r}(t) \prod_{\tau=1}^{M(\kappa,q)} |\phi_{\alpha_\tau}^{\kappa,q,\tau}\rangle \end{aligned} \tag{14.40}$$

where $M(\kappa, q)$ denotes the number of Cartesian degrees of freedom for the qth L2 SP group of the κth L1 SP group, and $|\phi_{\alpha_\tau}^{\kappa,q,\tau}\rangle$ denotes the α_τth primitive basis function for the τth degree of freedom within this particular L2 SP group. In this way the variation of $|\delta v_r^{(\kappa,q)}(t)\rangle$ can be turned into a regular derivative with respect to each of the expansion coefficients $C_\alpha^{\kappa,q,r}(t)$. Thus, the working equations for the L2 SPFs are given by

$$\mathrm{i} \sum_s \varrho_{rs}^{k,q}(t) |\dot{v}_s^{k,q}(t)\rangle = [1 - \hat{P}_{\mathrm{L2}}^{k,q}(t)] \sum_s \langle \hat{\mathcal{H}}(t) \rangle_{rs}^{k,q} |v_s^{k,q}(t)\rangle \tag{14.41}$$

where it is understood that the time derivatives are taken with respect to $C_\alpha^{\kappa,q,s}(t)$. The explicit solution is similar to that given by Equation (14.26b),

$$\mathrm{i}|\underline{\dot{v}}^{(\kappa,q)}(t)\rangle = [1 - \hat{P}_{\mathrm{L2}}^{(\kappa,q)}(t)][\hat{\varrho}^{(\kappa,q)}(t)]^{-1} \langle \hat{\mathcal{H}} \rangle^{(\kappa,q)}(t) |\underline{v}^{(\kappa,q)}(t)\rangle \tag{14.42}$$

Again, $|\underline{v}^{(\kappa,q)}(t)\rangle = \{|v_1^{(\kappa,q)}(t)\rangle, |v_2^{(\kappa,q)}(t)\rangle, \ldots\}^{\mathrm{T}}$ denotes the symbolic column vector of (the coefficients of) the L2 SPFs.

To summarize, the working equations for the two-layer ML-MCTDH are given by (see Equations (14.26a), (14.26b) and (14.41)):

$$\mathrm{i}\dot{A}_J(t) = \sum_L \langle \Phi_J(t) | \hat{H} | \Phi_L(t) \rangle A_L(t) \tag{14.43a}$$

$$\mathrm{i} \sum_m \rho_{nm}^{(\kappa)}(t) \dot{B}_{mj}^{\kappa}(t) = \left\langle u_j^{\kappa}(t) \middle| [1 - \hat{P}^{(\kappa)}(t)] \sum_m \langle \hat{H} \rangle_{nm}^{(\kappa)}(t) \middle| \varphi_m^{(\kappa)}(t) \right\rangle \tag{14.43b}$$

$$\mathrm{i} \sum_s \varrho_{rs}^{(\kappa,q)}(t) \dot{C}_\alpha^{\kappa,q,s}(t) = \left\langle w_\alpha^{(\kappa,q)} \middle| [1 - \hat{P}_{\mathrm{L2}}^{(\kappa,q)}(t)] \sum_s \langle \hat{\mathcal{H}} \rangle_{rs}^{(\kappa,q)}(t) \middle| v_s^{(\kappa,q)}(t) \right\rangle \tag{14.43c}$$

One immediately realizes that the equations for the expansion coefficients of the first-layer SPFs, that is, for the L2 coefficients B^{κ}_{mj} in Equation (14.43b), have exactly the same form as those for the second layer, that is, the coefficients $C^{k,q,s}_{\alpha}(t)$ in Equation (14.43c), though the specific ways of evaluating various operators are different. To be consistent in the notation and to highlight the possibility of extending the formulation to more layers, we denote $C^{k,q,s}_{\alpha}(t)$ as the level-three (L3) coefficient.

Up to this point, we have terminated the ML-MCTDH expansion at the second layer, that is, we have expressed the L2 SPFs as linear expansions of *time-independent* configurations

$$|w^{\kappa,q}_{\alpha}\rangle \equiv \prod_{\tau=1}^{M(\kappa,q)} |\phi^{\kappa,q,r}_{\alpha_\tau}\rangle$$

The expansions can be recursively generalized to more than two layers. This is evident by replacing $|w^{\kappa,q}_{\alpha}\rangle$ in Equation (14.43c) with $|w^{\kappa,q}_{\alpha}(t)\rangle$, that is, *time-dependent configurations*, which introduce another layer (the third layer in this case). However, none of the previous derivations is affected by this replacement. As a result, Equations (14.43a)–(14.43c) remain the same. An additional set of equations for the newly introduced layer can be derived using exactly the same procedure as that for the second layer, which also has the same form as Equation (14.43c). By mathematical induction, the equations of motion for the ML-MCTDH theory with an arbitrary number of layers can be formally written as:

$$\mathrm{i}|\dot{\Psi}(t)\rangle_{\text{L1 coefficients}} = \hat{H}(t)|\Psi(t)\rangle \tag{14.44a}$$

$$\mathrm{i}|\underline{\dot{\varphi}}^{k}(t)\rangle_{\text{L2 coefficients}} = [1 - \hat{P}^{(\kappa)}(t)][\hat{\rho}^{(\kappa)}(t)]^{-1}\langle\hat{H}\rangle^{(\kappa)}(t)|\underline{\varphi}^{(\kappa)}(t)\rangle \tag{14.44b}$$

$$\mathrm{i}|\underline{\dot{v}}^{(\kappa,q)}(t)\rangle_{\text{L3 coefficients}} = [1 - \hat{P}^{(\kappa,q)}_{\text{L2}}(t)][\hat{\varrho}^{(\kappa,q)}_{\text{L2}}(t)]^{-1}\langle\hat{\mathcal{H}}\rangle^{(\kappa,q)}_{\text{L2}}(t)|\underline{v}^{(\kappa,q)}(t)\rangle \tag{14.44c}$$

$$\mathrm{i}|\underline{\dot{\xi}}^{(\kappa,q,\gamma)}(t)\rangle_{\text{L4 coefficients}} = [1 - \hat{P}^{(\kappa,q,\gamma)}_{\text{L3}}(t)][\hat{\varrho}^{(\kappa,q,\gamma)}_{\text{L3}}(t)]^{-1}\langle\hat{\mathcal{H}}\rangle^{(\kappa,q,\gamma)}_{\text{L3}}(t)|\underline{\xi}^{(\kappa,q,\gamma)}(t)\rangle \tag{14.44d}$$

and so on, where

$$\varrho^{(\kappa,q)}_{\text{L2};rs}(t) = \sum_{n}\sum_{m}\rho^{(\kappa)}_{nm}(t)\langle g^{(\kappa,q)}_{\text{L2};n,r}(t)|g^{(\kappa,q)}_{\text{L2};m,s}(t)\rangle \tag{14.45a}$$

$$\varrho^{(\kappa,q,\gamma)}_{\text{L3};rs}(t) = \sum_{n}\sum_{m}\varrho^{(\kappa,q)}_{\text{L2};nm}(t)\langle g^{(\kappa,q,\gamma)}_{\text{L3};n,r}(t)|g^{(\kappa,q,\gamma)}_{\text{L3};m,s}(t)\rangle \tag{14.45b}$$

and so on, and where

$$\langle\hat{\mathcal{H}}\rangle_{\mathrm{L2};rs}^{(\kappa,q)}(t) = \sum_{n}\sum_{m}\langle g_{\mathrm{L2};n,r}^{(\kappa,q)}(t)|\langle\hat{H}\rangle_{nm}^{(\kappa)}(t)|g_{\mathrm{L2};m,s}^{(\kappa,q)}(t)\rangle \tag{14.46a}$$

$$\langle\hat{\mathcal{H}}\rangle_{\mathrm{L3};rs}^{(\kappa,q,\gamma)}(t) = \sum_{n}\sum_{m}\langle g_{\mathrm{L3};n,r}^{(\kappa,q,\gamma)}(t)|\langle\hat{\mathcal{H}}\rangle_{\mathrm{L2};nm}^{(\kappa,q)}(t)|g_{\mathrm{L3};m,s}^{(\kappa,q,\gamma)}(t)\rangle \tag{14.46b}$$

and so on. For clarity, we have inserted the subscripts 'L2', 'L3', ... to indicate the second layer, the third layer, and so on. The multilayer hierarchy is terminated at a particular level by expanding the SPFs in the deepest layer in terms of time-independent configurations.

Finally, we comment briefly on the evaluation of the various quantities in the ML-MCTDH equations of motion. In the ML-MCTDH approach, different parts of the Hamiltonian operator have to be built from the deepest layer L, that is, from the time-independent configurations (basis functions). At a particular step during time evolution, these pieces of operators are gathered to build parts of the Hamiltonian operator for the next upper layer $L-1$ using the relation between the basis functions and the SPFs in the $(L-1)$th layer. Using these parts, one obtains the operators for the $(L-2)$th layer, and so on, until all layers are reached. We refer to this procedure as 'bottom-up'. On the other hand, from Equations (14.45) one can see that the reduced density matrices need to be built from the first layer, and recursively generated to the second layer, the third layer, ..., until reaching the deepest layer. This procedure is thus 'top-down'.

For the mean-field operators the procedure is a mixture of the above two. First, starting from the deepest layer, one builds operator parts for each layer via the 'bottom-up' procedure as described above. Then, according to Equation (14.46), the matrix part of the mean-field operators are built from the first layer, and recursively generated to deeper layers via the 'top-down' procedure. The overall procedure is thus more complicated.

14.3 Concluding Remarks

In this chapter, we have presented a detailed derivation of the ML-MCTDH theory. This method is based rigorously on the Dirac–Frenkel variational principle [32, 33] and is thus a formally exact approach. Compared to the original ('single-layer') MCTDH method, the inclusion of several dynamically optimized layers builds in more flexibility in the variational functional, which is known to be important in a practical variational calculation. As a result, the ML-MCTDH method has the capability of treating substantially more physical degrees of freedom than the original MCTDH method without sacrifice of the theoretical rigorousness. Furthermore, the ML-MCTDH theory can be applied to any form of the Hamiltonian that is allowed by the original MCTDH approach.

We close this chapter with a few historical remarks and a brief overview of recent applications of the ML-MCTDH method. The concept of the ML-MCTDH theory had been around for several years before its actual implementation and application to real physical problems. This is largely due to the complexity in turning the sophisticated mathematical expressions into practical computer programs. The first report of a detailed derivation and successful implementation of the ML-MCTDH method appeared in 2003, where up to 1000 degrees of freedom were treated rigorously, employing two upper dynamical layers and one deeper static layer [119]. The system considered in that application was significantly larger than any other systems that had been treated before by quantum wavepacket methods, and demonstrated the performance of the method. Since then more dynamical layers have been implemented and better algorithms have been designed to increase the size of the systems that can be studied and to improve the efficiency of the calculations. Our current implementation is capable of handling four dynamical layers and one static layer, and has been applied to physical problems ranging from a few tens to more than 10 000 degrees of freedom.

The ML-MCTDH method has been applied to simulate the quantum dynamics of a variety of different complex systems [119,228–236]. A particularly important area for the application of the ML-MCTDH theory is condensed-phase physics and chemistry. Many interesting physical models exist in this field, but in most cases there are no effective methods available that can provide numerically exact results over a broad range of physical parameters. For some of these models, the ML-MCTDH method has already proven to be a very powerful method that can provide benchmark results for a broad range of parameters. One example is the so-called 'spin-boson model', where two discrete states are linearly coupled to a bath of harmonic oscillators. Generalizations of this model may include more electronic states, such as in heterogeneous charge transfer, or anharmonic intramolecular or bath modes. These types of models form the basis for nearly all modern electron transfer theories. A wealth of publications have been devoted to the approximate treatment of such models. Examples of methods that aim at a numerically exact method are approaches based on the influence-functional path integral approach [237–240]. However, these methods invariably encounter difficulties in situations with stronger coupling, slow characteristic time-scales of the bath, and/or anharmonic interactions. On the contrary, the ML-MCTDH theory has been applied successfully to study realistic electron transfer reactions in solution and at surfaces [119,228,229,232,233,235], where perturbation theories and path integral approaches are often not applicable. These studies have not only provided a better understanding of the fundamental mechanisms of electron transfer processes in a condensed-phase environment, but also served as benchmarks for the development of more approximate theories. From the numerical perspective, it is fair to say that the spin-boson model, which has been a challenge for theoretical physicists and chemists for several decades, has finally been solved within the ML-MCTDH framework.

Besides applications to simulate electron transfer reactions, the ML-MCTDH method has been used to study proton transfer processes in the condensed phase [230, 234] and photoisomerization reactions at conical intersections [231]. In addition to the simulation of photoinduced electronic and nuclear dynamics, the ML-MCTDH method has also been applied to the simulation of stationary and time-resolved spectra [229,236] and to the calculation of reaction rates employing the flux correlation function formalism [230, 234].

The ML-MCTDH program package we have developed is a mature combination of basic theories and a variety of efficient numerical techniques. These include the iterative definition of numerical basis functions, adiabatic basis set contraction techniques, an automated procedure to select the number of basis functions for all degrees of freedom, importance sampling of the initial wavefunction and Monte Carlo sampling of the thermal ensemble for simulations at finite temperature, mixed quantum–classical and quantum–semiclassical approaches, as well as various models using time correlation functions, for example, the flux correlation function for evaluating rate constants and the time-dependent polarization for simulating nonlinear spectroscopy. Furthermore, a serious effort has been made to implement the ML-MCTDH algorithm recursively, which is possible for certain forms of the Hamiltonian.

The major challenge for a broader applicability of the ML-MCTDH method is the complexity in designing efficient algorithms, writing compact codes, and building user-friendly interfaces. This is partly due to the fact that methodology development is often less rewarding than creative applications of existing theories. Still, the success of the ML-MCTDH theory to date suggests that it holds great promise for accurate quantum dynamical simulations of many other interesting problems. It can be anticipated that more joint effort will be devoted to developing this theory into a general tool for simulating dynamical processes in complex systems.

15 Shared Memory Parallelization of the Multiconfiguration Time-Dependent Hartree Method

Michael Brill and Hans-Dieter Meyer

15.1 Motivation

Over the last decade multiprocessor hardware has become more and more common, and research groups can afford distributed memory clusters or have access to supercomputing facilities. In addition, even desktop computers are often now equipped with multi-core processors for shared memory use. However, the Heidelberg MCTDH package is implemented for single-processor use and cannot profit from the current development of computer hardware. Both classes of parallel machines, those with distributed and those with shared memory, must be made accessible to MCTDH software. There have been attempts to develop software for parallel hardware in the Bielefeld code of Manthe [115] using the correlation discrete variable representation (CDVR; Chapter 10) or in the fermionic version of the MCTDH code of Scrinzi [241] (MCTDHF; Chapter 16). These codes are comparable to the Heidelberg one but differ decisively at certain points, thus prohibiting the use of the same parallelization strategies. Therefore, a parallelization scheme for the Heidelberg code has been developed. A distributed memory parallelization of the Heidelberg MCTDH code requires a reformulation of many of the implemented algorithms and will hence result in a major change of the code. As a first step, we therefore concentrated on parallelization for shared memory hardware, a type of hardware that is becoming more and more accessible.

15.2 Shared Memory Parallelization of MCTDH

As already mentioned, the MCTDH algorithm is a complex intricate scheme for the propagation of wavepackets. The complexity is rooted in the structure of the equations of motion (15.1) and (15.2). All data of a propagation

Multidimensional Quantum Dynamics: MCTDH Theory and Applications.
Edited by Hans-Dieter Meyer, Fabien Gatti, and Graham A. Worth

ISBN: 978-3-527-32018-9

step of the MCTDH coefficients (A-vector) are needed to propagate all single-particle functions (SPFs) and vice versa. This dependence makes it very difficult to partition data and calculations to distribute them to different processors. Hence a shared memory parallelization scheme was developed, avoiding the need to distribute the data to different processors and allowing much of the existing program structure to be kept.

For the implementation of shared memory parallelization, several programming standards are available. The two most common schemes are the Open MP standard and parallelization with POSIX threads. It was decided to use the second variant because not every FORTRAN compiler supports Open MP, but virtually all C compilers support the POSIX standard. POSIX provides a huge family of C-functions for the creation and manipulation of threads.

15.2.1 Equations of Motion and Runtime Distribution

In Section 3.1 the equations of motion (3.4) and (3.5) are introduced:

$$\mathrm{i}\dot{\mathbf{A}} = \mathcal{K}\mathbf{A} \tag{15.1}$$

$$\mathrm{i}\dot{\boldsymbol{\varphi}}^{(\kappa)} = (1 - P^{(\kappa)})(\boldsymbol{\rho}^{(\kappa)})^{-1}\boldsymbol{\mathcal{H}}^{(\kappa)}\boldsymbol{\varphi}^{(\kappa)} \tag{15.2}$$

For an efficient integration of these equations, usually the constant mean-field (CMF) approach, discussed in Section 4.2, is used. Within this integration scheme, the equations of motion, (15.1) and (15.2), are decoupled. Hence three main tasks of MCTDH can be distinguished, which can be parallelized separately:

1. Calculate the Hamiltonian matrix and the mean fields.
2. Propagate the MCTDH expansion coefficients.
3. Propagate the single-particle functions.

The sum of the computation time for the three tasks is the major part of the real time of an MCTDH run (usually $\geq 99\%$). The sharing of the workload between these parts depends strongly on the physical system that is treated. Three main distribution schemes for the computation time can be found:

(i) The computation time is distributed equally to the three tasks. The contribution of each part varies between 20% and 50% depending on the investigated system. Such calculations are called 'balanced'.

(ii) The propagation of the SPFs contributes only a little to the computation time ($\leqslant 5\%$). The other two parts contribute in a range of 30% to 70%, where the A-vector propagation usually takes the larger part. These calculations are called 'A-vector dominated'.

(iii) The propagation of the SPFs takes more than 60% of the computation time. Such an 'SPF-dominated' situation often occurs for small systems where parallelization is not needed. It happens also for over-combined calculations, that is, calculations in which too many degrees of freedom (DOFs) are combined to build an MCTDH particle. Over-combinations are inefficient and should be avoided.

Hence all three parts of MCTDH must be parallelized to achieve an adequate speed-up. In most cases over 65% of the computation time is taken by the mean-field calculation and the propagation of the expansion coefficients. The parallelization of these two tasks can be done efficiently and, in fact, is similar for both cases. The propagation of the SPFs is more difficult to parallelize, and this parallelization is less efficient compared to the other two cases.

15.2.2
Parallelization of the MCTDH Coefficients Propagation

For the propagation of the MCTDH coefficient vector, an efficient evaluation of the right-hand side of the differential equation (15.1) is important. This evaluation is done in two steps by the MCTDH code. In the first step, the expectation values of the Hamiltonian terms are computed:

$$h_{rjl}^{(\kappa)} = \langle \varphi_j^{(\kappa)} | \hat{h}_r^{(\kappa)} | \varphi_l^{(\kappa)} \rangle \tag{15.3}$$

Here $\hat{h}_r^{(\kappa)}$ denotes the operator on the κth particle in the rth Hamiltonian term as introduced in Equation (3.34). But the major part of the work is done in the second step. There, the results $h_{rjl}^{(\kappa)}$ are used to calculate the right-hand side of the equation of motion for the MCTDH coefficients:

$$\mathrm{i}\dot{A}_J(t) = \sum_L \langle \Phi_J | H | \Phi_L \rangle A_L(t) = \sum_{r=1}^{s} c_r \sum_{l_1=1}^{n_1} \cdots \sum_{l_p=1}^{n_p} h_{rj_1l_1}^{(1)} \cdots h_{rj_pl_p}^{(p)} A_{l_1 \ldots l_p} \tag{15.4}$$

Here one again makes use of the product structure of the Hamiltonian. The matrix multiplications are done sequentially:

$$A_{j_1 \cdots j_p}^{(\kappa,r)} = \sum_{l_\kappa=1}^{n_\kappa} h_{rj_\kappa l_\kappa}^{(\kappa)} A_{j_1 \ldots l_\kappa \ldots j_p}^{(\kappa-1,r)} \tag{15.5}$$

for $\kappa = 1, \ldots, p$ and $A^{(0,r)} = A$. Finally, the $A^{(p,r)}$ are summed. It is this sum over r that is parallelized, that is, matrix products with different label r are distributed over the available processors. The first step, the computation of the matrix elements, Equation (15.3), is parallelized in a similar way; different labels r are distributed to the processors.

In the case of the A-vector propagation, the efficiency of the parallelization depends on the size of the Hamiltonian. Naturally, the parallelization works better for Hamiltonians with many terms, for example, when the Hamiltonian contains a potfit.

15.2.3
Parallelization of the Mean-Field Computation

Although the parallelization of MCTDH has been done for the more efficient CMF integration scheme, which reduces the number of mean-field evaluations (Section 4.2), a major part of the computation time (about 20% to 40%) is still usually spent in mean-field computations. Hence an efficient parallelization is needed here.

The computation of the mean fields is split into two tasks. First, the mean-field tensor must be calculated,

$$\mathcal{H}_{rjl}^{(\kappa)} = c_r \sum_J{}^{\kappa} A_{J_j^\kappa}^* \prod_{\lambda=1,\lambda\neq\kappa}^{p} \sum_{l_\lambda} \langle \varphi_{j_\lambda}^{(\lambda)} | \hat{h}_r^{(\lambda)} | \varphi_{l_\lambda}^{(\lambda)} \rangle A_{L_l^\kappa} \tag{15.6}$$

which determines the mean-field operator

$$\hat{\mathcal{H}}_{jl}^{(\kappa)} = \sum_r^{s} \mathcal{H}_{rjl}^{(\kappa)} \hat{h}_r^{(\kappa)} \tag{15.7}$$

Second, another routine is used to sum up the action of the mean fields for the diagonal Hamiltonian terms (potential part),

$$\mathcal{M}_{kl\alpha}^{(\kappa)} = \sum_r^{s} \mathcal{H}_{rkl}^{(\kappa)} \langle \chi_\alpha^{(\kappa)} | \hat{h}_r^{(\kappa)} | \chi_\alpha^{(\kappa)} \rangle \tag{15.8}$$

where $\chi_\alpha^{(\kappa)}$ denotes a discrete variable representation (DVR) function. The idea behind this concept, which is described in Section 5.2.5 of Ref. [34], is to speed up the SPF propagation.

As in the case of the A-vector propagation, the two routines mainly consist of a loop over the Hamiltonian terms. The mean-field tensor $\mathcal{H}_{rjl}^{(\kappa)}$ has to be evaluated for each Hamiltonian term, and for the computation of $\mathcal{M}_{kl\alpha}^{(\kappa)}$ the summation runs over the potential part of the Hamiltonian. Hence the parallelization is implemented as in the case of the A-vector propagation, the loops over the Hamiltonian terms being distributed over the available processors.

The parallelization of the $\mathcal{M}_{kl\alpha}^{(\kappa)}$ computation becomes inefficient if a particle is present that is much larger than the others. The different threads then spend most of the computation time of $\mathcal{M}_{kl\alpha}^{(\kappa)}$ for this particle. During this computation, one thread locks the memory to prevent other POSIX threads from writing to the same part of the memory. Hence the other threads may stay idle. This problem is diminished by an alternative parallelization strategy: if a very large particle is present, the loop over l is parallelized rather than the sum over r.

15.2.4 Parallelization of the SPFs Propagation

For the propagation of the SPFs, the evaluation of the right-hand side of the equations of motion (15.2) is the time-determining step. By splitting the Hamiltonian into uncorrelated and correlated terms (see Section 3.23),

$$H = \sum_{\kappa=1}^{p} \hat{h}^{(\kappa)} + \sum_{r=1}^{s} c_r \prod_{\kappa=1}^{p} \hat{h}_r^{(\kappa)} \tag{15.9}$$

the equations of motion for the SPFs can be rewritten as

$$\mathrm{i}\dot{\varphi}_j^{(\kappa)} = (1 - P^{(\kappa)}) \left(h^{(\kappa)} \varphi_j^{(\kappa)}(t) + \sum_{k,l=1}^{n_p} (\rho^{(\kappa)^{-1}})_{jk} \sum_{r=1}^{s} \mathcal{H}_{rkl}^{(\kappa)} \hat{h}_r^{(\kappa)} \varphi_l^{(\kappa)}(t) \right) \tag{15.10}$$

The computation of the right-hand side of this equation is done in several steps. First, the sum over the Hamiltonian terms is evaluated. Afterwards, the inverse density matrix is applied. Then the action of the uncorrelated Hamiltonian terms on the SPFs is added. Finally, the projection is carried out.

The parallelization has to be done for each of the steps separately. The summation over the Hamiltonian terms is parallelized as in the cases above, that is, the A-vector propagation and the mean-field computation. But here the loop runs only over the kinetic Hamiltonian terms, which represent only a small part of the Hamiltonian. Owing to the summation, Equation (15.8), there is only one potential term.

Each of the steps left mainly consists of a loop over the SPFs for the particle considered. Thus the loops over the SPFs are parallelized. For small calculations, this parallelization scheme is disadvantageous because the number of processors (for example, eight) can be larger than the SPF basis for a certain particle. Hence this scheme is better suited for larger calculations with large numbers of SPFs.

15.2.5 Parallelization Scheme

The parallelization of the different parts of MCTDH is included in a scheduler–worker scheme. The main program has the role of a scheduler that does all the calculations that are not time consuming and controls the other threads. These threads are the worker threads performing the time-consuming computations on different processors. There are, usually, as many threads as processors available. Figure 15.1 shows the scheme in detail. Each time the main program comes to a point where a computation can be made in parallel, the barriers 1 to P are opened. The main program halts at the following barriers $P+1$ to $2P$. The barriers 1 to P are inside the threads; when these barriers are open, the threads can do the work. As soon as the computations are finished, the barriers $P+1$ to $2P$ are opened and the threads again wait at the barriers 1 to P. By opening barriers $P+1$ to $2P$, the main program takes control again and proceeds with further calculations. All barriers are closed immediately after a thread has passed a barrier.

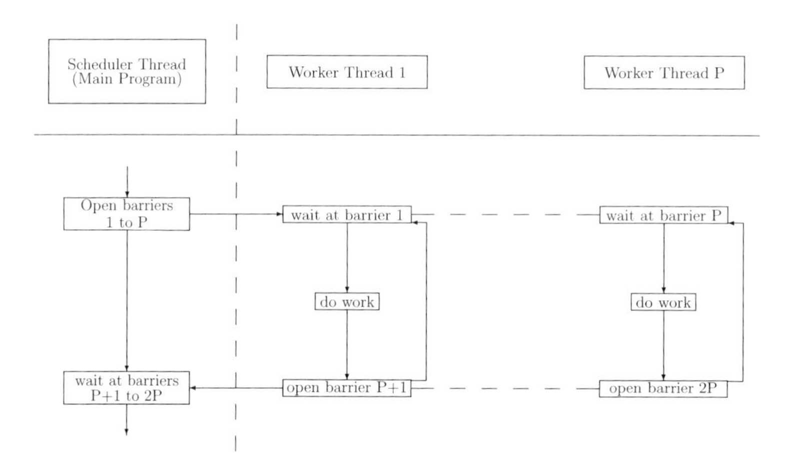

Figure 15.1 Organization of the threads in the scheduler–worker scheme.

15.2.6 Load Balancing and Memory Requirements

Two further aspects must be considered regarding the parallelization of MCTDH, that is, the memory requirements and the load balancing. For an efficient parallelization, it must be guaranteed that all available processors are used for the computations made. This means that the code must take care that the work is distributed equally to the processors. In the current version of MCTDH, this is ensured using a global counter variable combined with a mutual exclusion. As already mentioned, the parallelization is used to distribute the work of certain loops to the processors. The counter variable of the loops is seen by each thread. If a thread has finished its work in one step of the loop, the mutual exclusion is locked; then the counter variable is increased; afterwards the mutual exclusion is unlocked again, and the thread starts working in the step determined by the counter variable. The mutual exclusion ensures that no other thread can get the same value for the counter variable, thus ensuring that the work is done only once. All threads have the same work to do, and only at the end of the loop can it happen that some threads stay idle till the loop is finished.

The second problem that must be considered are the memory requirements. MCTDH needs memory to store different objects that are used during a propagation, for example, the wavefunction (A-vector, SPFs). Further memory is used for the computation itself, the so-called *work arrays*, where intermediate results are stored. Considering several threads working on different parts of the same problem (for example, mean-field tensor computation), they all need their own work array to store data, sum results, build matrices and so on. Because these threads are calculating different parts of the problem, they need the same amount of work memory exclusively. Hence, for each thread, extra memory must be allocated. The memory requirements for an MCTDH run with P processors are

$$M(P) = M_{\text{objects}} + P \times M_{\text{work}} \tag{15.11}$$

where M_{objects} denotes the memory needed for the storage of all objects (for example, the wavefunction) and M_{work} is the memory needed for the work arrays. Thus the required memory increases moderately with the number of processors. Examples are given in Table 15.1, where the memory requirements are given for a serial calculation (M_0) and a parallel calculation with eight processors (M_8).

Table 15.1 Runtime distribution and memory requirements for the benchmark systems. Columns 2–5 show the division of time between parts of the program. T_5 is the time for a 5 fs propagation. M_0/M_8 is the ratio of memory for a serial calculation to that for a parallel calculation with eight processors. (a and b refer to two different mode combination schemes, see text.)

System	A-vector	Mean field	SPF	Rest	T_5	M_0/M_8
$H_2 + H_2$	56.8%	42.1%	0.7%	0.4%	32m 51s	56 MB/66 MB
C_2H_4	42.0%	19.5%	34.0%	4.5%	19m 58s	108 MB/182 MB
$H_5O_2{}^+$-a	31.9%	36.4%	31.4%	0.3%	10h 26m 49s	361 MB/663 MB
$H_5O_2{}^+$-b	60.0%	39.7%	0.2%	0.1%	7h 47m 19s	258 MB/707 MB

15.3 Results and Conclusion

15.3.1 Benchmark Systems

For the test of the parallelization of the MCTDH code, four different benchmark systems were used. These systems are:

(i) the $H_2 + H_2$ inelastic scattering [84,85];

(ii) the quantum molecular non-adiabatic dynamics of C_2H_4 [74]; and

(iii) the quantum molecular dynamics of $H_5O_2{}^+$ using two different mode combinations [59,156].

Table 15.1 compares the computation time spent in the three main tasks. The runtime needed for a 5 fs propagation is also given. Clearly, the computation time of the molecular dynamics of C_2H_4 and $H_5O_2{}^+$-a is equally distributed over the three parallelized tasks. Hence the parallelization must work efficiently for all tasks to obtain good results. On the other hand, the $H_2 + H_2$ scattering and the molecular dynamics of $H_5O_2{}^+$-b are A-vector-dominated cases and less than 1% of the computation time is spent in the propagation of the SPFs. In Tables 15.2 and 15.3 the main parameters of the three calculations are compared. In the case of C_2H_4, only six of the 12 internal degrees of freedom are used, the others being frozen at their equilibrium values (for details, see [74]). The $H_2 + H_2$ system is characterized by seven DOFs because the KEO is seven dimensional for total $J > 0$ [84,85]. The three other examples are for $J = 0$.

Table 15.2 Major parameters for the benchmark systems: number of DOFs, number of electronic states, number of particles, grid size and the size of the Hamiltonian.

System	DOFs	Electronic states	Particles	Grid	Hamiltonian terms
$H_2 + H_2$	7	1	4	7.29×10^7	3439
C_2H_4	6	3	4	3.14×10^9	126
$H_5O_2{}^+$-a	15	1	5	1.29×10^{15}	1511
$H_5O_2{}^+$-b	15	1	6	4.82×10^{16}	2529

The sizes of the systems differ strongly. This is essential to give an overview of the efficiency of the parallelization. The $H_2 + H_2$ scattering and the molecular dynamics of C_2H_4 are the smaller systems, and a parallel code is not needed to perform the calculations. These two examples are only given as a reference. The dynamics of $H_5O_2{}^+$ is a problem that cannot be solved with a serial code within reasonable real times.

Table 15.3 Mode sizes (number of SPFs times grid size) and number of configurations for the benchmark systems. In the case of C_2H_4, the particles for the third electronic DOF are larger than those of the first and second states. The two numbers displayed refer to the different electronic states.

System	Particle 1	Particle 2	Particle 3	Particle 4	Particle 5	Particle 6	Configurations
$H_2 + H_2$	1536	256	1386	1386			10 368
C_2H_4	600/720	5120	7680/8960	7680/8960			209 280
$H_5O_2{}^+$-a	354 375	15 680	4416	3402	3402		216 000
$H_5O_2{}^+$-b	8100	15 435	5000	2312	4851	4851	705 600

15.3.2 Amdahl's Law

To measure the acceleration of parallel computations, one defines the speed-up:

$$S(P) = \frac{t_1}{t_P} \tag{15.12}$$

This is the time needed by a single-processor calculation divided by the time that is needed by a P-processor calculation. In an ideal parallel program, S is equal to P.

One can divide the computation time into two parts, a serial part q and a parallel part p, with $q + p = 1$. The part q is not affected by the number of processors used, but the part p is perfectly parallel. The speed-up can be calculated from the knowledge of p (with $q = 1 - p$):

$$S(P) = \frac{t_0(q+p)}{t_0(q+p/P)} = \frac{1}{q+p/P} = \frac{P}{P-p(P-1)} = \frac{P}{1+q(P-1)} \tag{15.13}$$

This law is called Amdahl's law and describes the theoretical speed-up of a parallel program. In the following, this formula is used to measure the parallel part of MCTDH for the benchmark systems. The parameter p is determined via a least-squares fit to the measured speed-up.

15.3.3 Results

For the four benchmark systems, MCTDH runs have been performed with an increasing number of processors (one, two, four and eight) on the XC1 system in Karlsruhe that is part of the High Performance Computing Centre Stuttgart (HLRS). These are Itanium processors with 1600 MHz and 6 MB level 3 cache. The Intel *ifort* compiler was used, as this compiler showed a better performance than the *gfortran* compiler also tested.

In Figure 15.2 the results for the quantum molecular dynamics of C_2H_4 and $H_5O_2{}^+$-a are displayed. The dotted and dashed lines are the fits based

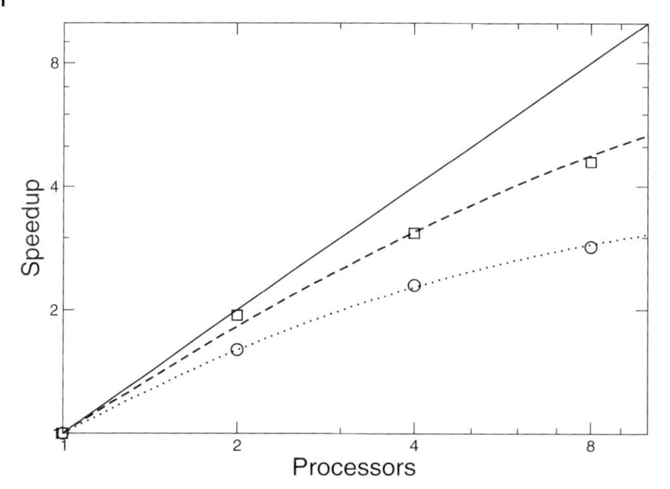

Figure 15.2 Speed-up behaviour of the C_2H_4 (circles, dotted line, $p = 0.75$) and the $H_5O_2{}^+$-a (squares, dashed line, $p = 0.90$) calculations. The circles and squares are the measured values; the dotted and dashed lines are the fits to Amdahl's law; and the solid line is the ideal (linear) speed-up.

on Amdahl's law to the data (circles and squares); the solid line shows the ideal linear speed-up. The corresponding parallel parts are $p_{C_2H_4} = 0.75$ and $p_{H_5O_2{}^+\text{-a}} = 0.90$, respectively. In the C_2H_4 case the parallel algorithm performs quite poorly. This is caused by the small size of the problem. The number of Hamiltonian terms is rather small. This downgrades the parallelization of the A-vector propagation and of the mean-field computation. Additionally, the particles are small, and thus the parallelization of the SPF propagation is not very efficient. This is because the algorithm has to synchronize the scheduler and worker threads many times for only a small amount of work. The parallelization works much better in the case of the $H_5O_2{}^+$-a propagation. In this case the Hamiltonian and the particles are bigger, causing a better overall parallelization. But the parallelization of the $\mathcal{M}^{(\kappa)}_{kl\alpha}$ computation (see Equation (15.8)) does not work well for the $H_5O_2{}^+$-a propagation, because there is one particle that is much larger than all the others (see Table 15.3). This constellation always prohibits a good parallelization, which can be compensated only partially (see Section 15.2.3).

In Figure 15.3 the corresponding results are displayed for the $H_2 + H_2$ scattering and the molecular dynamics of $H_5O_2{}^+$-b. Again, the circles and squares show the measured data, and the dotted and dashed lines are the fits. The fitted values for the parallel part are $p_{H_2+H_2} = 0.98$ and $p_{H_5O_2{}^+\text{-b}} = 0.94$. These are the above-mentioned cases where the computation time of the SPF propagation is negligible. For these examples MCTDH is very well parallelized. This is because the propagation of the A-vector and the computation of the

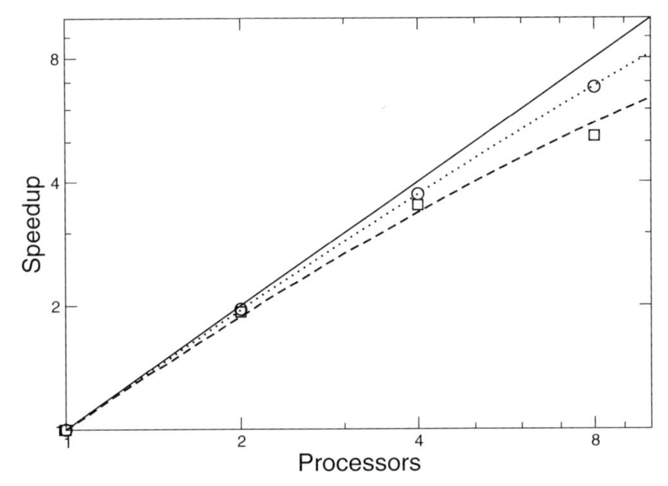

Figure 15.3 Speed-up behaviour of the $H_2 + H_2$ (circles, dotted line, $p = 0.98$) and the $H_5O_2{}^+$-b (squares, dashed line, $p = 0.94$) calculations. The circles and squares are the measured values; the dotted and dashed lines are the fits to Amdahl's law; and the solid line is the ideal (linear) speed-up.

mean fields generate the major part of the computation time. The large number of Hamiltonian terms ensures an efficient parallelization.

15.3.4
Conclusion and Outlook

The existing serial Heidelberg MCTDH code was extended to make use of parallel shared memory hardware. A scheduler–worker parallelization scheme was developed and applied. The three main tasks of the MCTDH propagation algorithm, that is, the propagation of the A-vector, the propagation of the SPFs and the computation of the mean fields, are parallelized. The resulting code was tested by means of benchmarks systems to determine the efficiency of the parallelization. It turned out that, for special cases where the propagation of the SPF is negligible, the code is well parallelized. In the general case, the problem arises that the parallelized routines for the SPF propagation contain – usually short – loops over the SPFs, which makes the parallelization less efficient. In both cases, that is, loops over SPFs and over Hamiltonian terms, the parallelization works better the larger the size of the loops. The degree of parallelization lies between $p = 0.75$ (worst case tested) and $p = 0.98$ (best case tested), leading to speed-ups between 3 and 7 when running on eight processors. Hence the parallelization of the wavepacket propagation of MCTDH can be seen as a success.

The described development improved the MCTDH code essentially. However, the present implementation is restricted to shared memory hardware. The number of processors combined in parallel shared memory machines is still growing, but at a lower rate than the number of processors in the distributed memory computing clusters. Hence it is an important task to use the experience of the shared memory parallelization of MCTDH to develop a parallel code for distributed memory machines. This, however, requires a severe restructuring of the existing code.

Finally, there are more programs in the Heidelberg MCTDH package than just *MCTDH* that are worth parallelizing. Most notably there is *potfit*. Owing to its different structure, it seems to be suitable for distributed memory parallelization.

16
Strongly Driven Few-Fermion Systems – MCTDHF

Gerald Jordan and Armin Scrinzi

The equations of motion (3.4) and (3.5) are invariant under the exchange of particle coordinates, if the underlying Hamiltonian has such a symmetry. Therefore, starting from a properly antisymmetrized initial state, antisymmetry of the solution will be maintained and MCTDH correctly describes the dynamics of fermionic systems without imposing any constraints. However, such a description would be highly redundant and computationally exceedingly expensive. In this chapter we describe the modifications needed to reduce redundancy in few-fermion systems and to obtain an efficient, parallelizable implementation. The application to few-electron molecules in strong laser fields requires a careful treatment of spin and Coulomb interaction. Different from most MCTDH applications, the computational effort in our typical MCTDHF calculation is dominated by operations related to the single-particle functions (SPFs). The procedures reviewed here were implemented in the MCTDHF code, which is described in greater detail and illustrated with examples in Refs. [242–244].

16.1
Equations of Motion for Indistinguishable Particles

For indistinguishable particles, a fully general *Ansatz* for a p-particle wavefunction can be made using a single set of SPFs $\boldsymbol{\varphi}$ for all particles,

$$\Psi(Q_1,\ldots,Q_p) = \sum_{j_1=1}^{n} \cdots \sum_{j_p=1}^{n} A_{j_1\ldots j_p}\varphi_{j_1}(Q_1)\cdots\varphi_{j_p}(Q_p) \tag{16.1}$$

where the coordinates $Q_\kappa = (\mathbf{r}_\kappa, s_\kappa)$ comprise spatial and spin coordinates. Antisymmetry of the fermionic wavefunction is reflected in the antisymmetry of the expansion coefficients under exchange of any two of their indices,

$$X_{\kappa\lambda}A_{j_1\ldots j_p} = -A_{j_1\ldots j_p} \tag{16.2}$$

where by $X_{\kappa\lambda}$ we denote the interchange of the κth with the λth index.

Multidimensional Quantum Dynamics: MCTDH Theory and Applications.
Edited by Hans-Dieter Meyer, Fabien Gatti, and Graham A. Worth

ISBN: 978-3-527-32018-9

As the Hamiltonian of a system of indistinguishable particles is invariant under exchange of any two of the particles,

$$H(\ldots, Q_\kappa, \ldots, Q_\lambda, \ldots) = H(\ldots, Q_\lambda, \ldots, Q_\kappa, \ldots) \tag{16.3}$$

the matrix $\mathcal{K}$ of Equation (3.4) computed with identical sets of SPFs $\boldsymbol{\varphi}^{(\kappa)} \equiv \boldsymbol{\varphi}$ is also exchange symmetric,

$$X_{\kappa\lambda} \mathcal{K} X_{\kappa\lambda} = \mathcal{K} \tag{16.4}$$

and the MCTDH equations of motion (3.5) all collapse to a single equation:

$$\dot{\boldsymbol{\varphi}} = (1 - P)(\boldsymbol{\rho})^{-1} \mathcal{H} \boldsymbol{\varphi} \tag{16.5}$$

This shows that, for exchange-symmetric Hamiltonians, MCTDH rigorously conserves exchange symmetry, if one starts from a symmetric state.

After this observation, any difference between MCTDH and MCTDH-Fock can be considered purely technical. Indeed, we found this a useful guideline for the practical implementation of MCTDHF, as it allowed us to focus on computational efficiency and, in addition, it keeps the code open for extensions to more exotic systems, such as boson–fermion mixtures. The differences with the greatest impact in practice are the following.

(i) There are usually only two-particle interactions, which, however, can become locally very strong or even singular, as in the case of the Coulomb interaction.

(ii) Owing to antisymmetry, for n SPFs and p particles, the number of independent expansion coefficients $A_{j_1 \ldots j_p}$ is only $\binom{n}{p}$ for MCTDHF rather than n^p for MCTDH.

(iii) There is a natural grouping of the degrees of freedom into those belonging to each single particle, that is, coordinates and spin, which restricts the choice of coordinate systems and the ways in which spin or rotational symmetries can be incorporated. The degrees of freedom of a single particle can be strongly correlated, for example, near the singularities of the single-particle potential.

(iv) For our typical applications, the single-particle Hamiltonian explicitly and strongly depends on time.

16.1.1
Model System: Laser-Driven Few-Electron Systems

Our model MCTDHF applications are strong-field phenomena of atoms and molecules in ultrashort infrared (IR) laser and extreme ultraviolet (XUV) pulses. The Hamiltonian of these systems has the general form:

$$H = \sum_{\kappa=1}^{p} \left(\tfrac{1}{2}[\mathbf{p}_\kappa + \mathbf{A}(t)]^2 + V(\mathbf{r}_\kappa) + \sum_{\lambda<\kappa}^{p} V_{12}(\mathbf{r}_\kappa, \mathbf{r}_\lambda) \right) \tag{16.6}$$

Here $\mathbf{p}_\kappa$ is the canonical momentum of the κth electron, $\mathbf{A}(t)$ is the electric vector potential, V denotes the single-particle potential, and V_{12} is the electron–electron interaction, for which we use the Coulomb repulsion $|\mathbf{r}_\kappa - \mathbf{r}_\lambda|^{-1}$ in most cases.

This specific class of systems favours particular choices for the implementation of MCTDHF. A large number of experiments involves linearly polarized lasers with near-IR wavelength, short pulse duration and significant intensity. At typical parameters, ionization ranges from a few per cent to total depletion. The detached electrons undergo a violent quiver motion in the polarization direction of the laser with excursion amplitudes of several tens of atomic units and kinetic energies of a few atomic units.

On a technical level this requires large simulation volumes with a significant density of discretization points for a faithful representation of momenta. Linear polarization in the z-direction suggests the use of cylindrical coordinates (ϕ, ρ, z) with large z-range and smaller ρ-range. To represent bound states accurately near the nuclei, a finer discretization is required than for the free motion in the laser field. On the other hand, hard electron–electron collisions are unimportant for continuum electrons and the short-range behaviour of the Coulomb repulsion is not as critical as near the nuclei. Therefore, for both the electron orbitals and the electron–electron interaction, we use a finite-element discretization, where the size of the finite elements is tunable in a position-dependent way.

16.1.2 Spin

When we consider systems without spin forces, spin is strictly conserved. Rather than explicitly enforcing a given spin symmetry, we take a similar attitude as for the implementation of exchange symmetry, that is, all spin states are propagated and redundant operations are bypassed in the code. For the SPFs we choose single spin eigenfunctions and set equal to 0 all one-particle integrals for SPFs with different spins and all mean-field operators that would connect different spins. For simplicity, some redundancy is admitted in the evaluation of the right-hand side of Equation (3.4).

With this approach, we do not need to distinguish 'unrestricted' Hartree–Fock from the mostly employed 'restricted' scheme, where SPFs are subjected to the constraint that their *spatial* parts must be either identical or orthogonal. Our only fundamental requirement is orthogonality of the SPFs including the spin coordinate. In multiconfiguration Hartree–Fock, this distinction loses much of its relevance, as any unrestricted scheme can be accommodated within a larger restricted scheme with comparable computational effort, by simply splitting the spatial orbitals into mutually orthogonal parts. In practice, we mostly use the restricted version, employing SPFs with pairwise identical spatial parts and, again, avoiding redundant operations.

To obtain an initial state with spin quantum numbers S and S_z by imaginary-time propagation (see Chapter 8), it is sufficient to start from a guess state with quantum numbers S and S_z. We first construct a set of SPFs $\boldsymbol{\varphi}$, usually with the eigenfunctions of some single-particle Hamiltonian $h(Q)$ for the spatial orbitals. A suitable guess coefficient vector $\boldsymbol{A}_{SS_z}$ is obtained by projection onto the subspace with the desired spin:

$$\boldsymbol{A}_{SS_z} = \mathcal{P}_S \mathcal{Q}_{S_z} \boldsymbol{A} \tag{16.7}$$

A straightforward way of constructing the projection operators $\mathcal{P}_S$ and $\mathcal{Q}_{S_z}$ for small numbers of SPFs is by diagonalizing the $\binom{n}{p} \times \binom{n}{p}$ spin matrices,

$$\boldsymbol{S}_{JL} = \langle \Phi_J | S^2 | \Phi_L \rangle \quad \text{and} \quad \boldsymbol{M}_{JL} = \langle \Phi_J | S_z | \Phi_L \rangle \tag{16.8}$$

where Φ_J denotes the Slater determinant for the SPFs $\{\varphi_{j_\kappa}, \ \kappa = 1, \ldots, p\}$. Here one must make sure that the matrices $\boldsymbol{S}$ and $\boldsymbol{M}$ commute, that is, that the determinants Φ_J cover complete spin subspaces, for example, by starting from a maximal set of all doubly occupied spatial orbitals.

16.2 Computation of Operators

16.2.1 $\mathcal{K}$ and Mean-Field Operators

If we assume for our Hamiltonian the general form

$$H(t) = \sum_{\kappa=1}^{p} h(Q_\kappa, t) + \sum_{\lambda<\kappa} H(Q_\kappa, Q_\lambda) := H_1 + H_2 \tag{16.9}$$

we can split the computation of $\boldsymbol{\mathcal{K}}$ into one- and two-particle operators,

$$\boldsymbol{\mathcal{K}} = \boldsymbol{\mathcal{K}}_1 + \boldsymbol{\mathcal{K}}_2 \tag{16.10}$$

with

$$(\boldsymbol{\mathcal{K}}_1 \boldsymbol{A})_{i_1 \ldots i_p} = \sum_{k=1}^{n} \sum_{\kappa=1}^{p} h_{i_\kappa k} A_{i_1 \ldots i_\kappa \to k \ldots i_p} \tag{16.11}$$

where we have introduced the single-particle matrix elements $h_{i_\kappa k} = \langle\varphi_{i_\kappa}|h|\varphi_k\rangle$. The left-hand side is calculated for the $\binom{n}{p}$ ordered p-tuples $i_1 < \ldots < i_p$. The notation $i_1 \ldots i_\kappa \rightarrow k \ldots i_p$ means that the index i_κ is replaced with k. Owing to antisymmetry of the A, there are at most $n-p+1$ non-vanishing terms in the right-hand side sum over k, which gives a count of $\binom{n}{p} \times \binom{n-p+1}{1} \times \binom{p}{1} \ll n^p$ multiplications. Similarly we have

$$(\mathcal{K}_2 A)_{i_1 \ldots i_p} = \sum_{\kappa=1}^{p} \sum_{\kappa<\lambda} \sum_{k<l} (H_{i_\kappa i_\lambda kl} - H_{i_\kappa j_\lambda lk}) A_{i_1 \ldots i_\kappa \rightarrow k \ldots i_\lambda \rightarrow l \ldots i_p} \tag{16.12}$$

with a corresponding multiplication count of $\binom{n}{p} \times \binom{n-p+2}{2} \times \binom{p}{2}$. For systems of only a few fermions and moderate correlation – which holds for a large class of chemical systems – all these numbers remain small and constitute only a small fraction of the computational effort.

For internal reference it is convenient to sort the indices $j_1 < \ldots < j_p$ lexicographically, which allows for easy incrementation of a given multi-index. This ordering is obtained by the assignment

$$(j_1 \ldots j_p) \rightarrow J = 1 + \sum_{\kappa=1}^{p} \sum_{k=1}^{j_i - j_{i-1} - 1} \binom{n - j_{i-1} - k}{p - \kappa} \tag{16.13}$$

and $j_0 = 0$.

16.2.2 Spatial Discretization

The cost of computing and applying the one-particle operators is problem dependent. For our model application of laser-driven molecules, there is no single coordinate system that allows an efficient separation of the SPFs. Near the nuclei, spherical symmetry dominates; but for detached electrons in a linearly polarized field, cylindrical coordinates are most appropriate. The multilayer approach should be efficient in this situation (see Chapter 14), but it has not yet been implemented for MCTDHF. Rather, in the present code, we represent the SPFs on a three-dimensional product basis of one-dimensional finite-element (FE) functions.

We use FEs of arbitrary order as described in Ref. [245]. A given coordinate q is split into intervals $[q_{n-1}, q_n]$ and the finite elements are completely defined by type and number of approximation functions, $\chi_{nk}^{(q)}(q)$, $k = 0, \ldots, K$, on that interval. The only requirement for the $\chi_{nk}^{(q)}(q)$ is that they be differentiable and that they can be linearly combined such that they all vanish at the endpoints of the FE, except for two of them: these are non-zero at the lower and upper boundary, respectively,

$$\chi_{nK}^{(q)}(q_n) = \chi_{n0}^{(q)}(q_{n-1}) = 1 \tag{16.14}$$

$$\chi_{nk}^{(q)}(q_{n-1}) = \chi_{nk}^{(q)}(q_n) = 0 \quad \text{else} \tag{16.15}$$

A spatial orbital in three dimensions is written as

$$\chi_c(r,s,t) = \sum_{l,m,n} \sum_{i,j,k} c_{ijk}^{(lmn)} \chi_{li}^{(r)}(r) \chi_{mj}^{(s)}(s) \chi_{nk}^{(t)}(t) \tag{16.16}$$

The sums extend over all 'voxels', that is, triplets of one-dimensional elements (l,m,n), and over all functions on each voxel.

Neighbouring voxels are connected by the requirement of continuity at q_n, which translates into a linear constraint on the expansion coefficients of the SPF, for example, for the t coordinate,

$$c_{ijK}^{lm(n-1)} = c_{ij0}^{lmn} \tag{16.17}$$

Note that, for the discretization of differential operators up to the second order, continuity of the derivative is not required: formally, a δ-like second derivative appears, but is always integrated over with a continuous function [245]. Dirichlet boundary conditions can be implemented by setting the desired values for the first coefficient of the first element and the last coefficient of the last element, respectively.

Any set of linearly independent functions with suitable continuity properties and boundary values can be used to represent the solution within one element. General global bases are included as limiting cases in this scheme, when a single element covers the whole coordinate range. To obtain a (global) discrete variable representation (DVR) of the basis, one transforms it such that it diagonalizes the coordinate operator (see Chapter 3). Our FE scheme also includes the case of the so-called FE-DVR method [246], where polynomial finite-element functions are represented by their values at the quadrature points of a Lobatto scheme, which includes the points at the element boundaries. This allows an easy implementation of continuity conditions. The same Lobatto quadrature can be employed for most integrations, if one makes minor compromises on the quadrature error.

For the implementation, in particular for parallelization, it is convenient to formulate the continuity condition for coordinate q at boundary q_n with the help of a projection operator $\mathcal{Q}_n^{(q)}$,

$$\mathcal{Q}_n^{(q)} = 1 - \mathcal{P}_n^{(q)} = 1 - |d_n\rangle\langle d_n| \tag{16.18}$$

which acts on the coefficients $\boldsymbol{c}$. The 'discontinuity vector'

$$\langle d_n| = \frac{1}{\sqrt{2}}(\ldots,0,0,-1,0,\ldots,0,1,0,0,\ldots) \tag{16.19}$$

has the values -1 and $+1$ at the position of the pairs of coefficients c that need to be identical for continuity at q_n. The projector ensuring continuity at all element boundaries on coordinate q is given by

$$\mathcal{Q}^{(q)} = \prod_n (1 - \mathcal{P}_n^{(q)}) = 1 - \sum_n \mathcal{P}_n^{(q)} =: 1 - \mathcal{P}^{(q)} \tag{16.20}$$

The second equality holds because of the mutual orthogonality of the projectors $\mathcal{P}_n^{(q)} \mathcal{P}_m^{(q)} = 0$ for $n \neq m$. A coefficient vector for a continuous function in three dimensions is obtained from an arbitrary coefficient vector $\boldsymbol{c}$ by applying the projectors for all three directions,

$$\boldsymbol{c}^{(\text{cont})} = \mathcal{Q}^{(r)} \mathcal{Q}^{(s)} \mathcal{Q}^{(t)} \boldsymbol{c} := \mathcal{Q} \boldsymbol{c} \tag{16.21}$$

There are two main reasons to use the more general FE framework rather than a DVR or global basis set formulation: (i) accurate integrations near the singularities of the potentials require transformations to denser quadrature grids, which can be easily administered with local functions; and (ii) local representations of all operators, including the differential operators, are needed for efficient parallelization.

In our model application, we need very large grids: the oscillation amplitude of the free electrons in typical fields may reach several tens of atomic units. Some of the most prominent strong-field phenomena such as high-harmonic generation or non-sequential double ionization occur only when electrons return to the molecule. During the excursion, electron kinetic energies can reach several atomic units. For a proper representation of momentum and excursion of the electrons, we need a typical number of 500 discretization points in the polarization direction of the laser, and at least several tens of points in the perpendicular direction. Even when we only admit cylindrically symmetric SPFs, we obtain on the order of 10^4 discretization points per SPF. In spite of the large extension of the simulation box, large parts of the wavefunction may reach the box boundaries and need to be absorbed. For this, we employ standard complex absorbing potentials (CAPs) – see Section 6.5.

The distribution of the FE boundaries can be adjusted to the specific physical system. At large distances, electron momenta are predictably lower, while the highest momenta appear in the surroundings of the original bound system. Unnecessary stiffness of the discretized equations results when the sum of the potential and kinetic energies locally present in the discretization significantly exceeds the energies in the actual solution. A dense grid near a negative potential singularity does not significantly impact on the stiffness of the system, but the same density in regions of zero potential may severely slow down time propagation. Variable grids therefore have the two-fold advantage of a significantly lower number of grid points and fewer stiffness-related problems.

The use of the FE discretization is motivated by its generality and our specific model application with a large fraction of the system in continuum states. In an alternative approach, Gaussian functions were successfully employed in cases where the dynamics remains mostly within the bound-state regime [247,248].

16.2.3
One-Particle Operators

With an FE product basis, the total single-particle Hamiltonian can be written as a sum over 'voxel' Hamiltonians,

$$hc = \mathcal{Q}\sum_v h^v \mathcal{Q}c \tag{16.22}$$

for the volumes $[r_l, r_{l+1}] \times [s_m, s_{m+1}] \times [t_n, t_{n+1}]$ with $v = (l, m, n)$. Assuming locality of h, the voxel discretization is diagonal with respect to the voxel index v, except for overlaps between neighbouring voxels introduced by the continuity constraint $\mathcal{Q}$. The time-independent parts of h^v are computed during an initial set-up stage. The h^v are in general full matrices, but their tensor-product structure is exploited wherever possible. Potentials are applied by transformations to quadrature grids, which again have tensor-product form. Only near the singularities of non-separable potentials, where one would need large product grids for accurate quadrature and correspondingly expensive transformations, are all single-electron terms united in a full three-dimensional voxel matrix h^v. The computationally optimal crossover point between the methods is determined during set-up by measuring the CPU time needed for application of h^v in either form.

When solving the single-particle equations in FE discretization, one obtains equations of the form

$$\mathcal{S}\mathrm{i}\frac{\mathrm{d}}{\mathrm{d}t}c = hc \tag{16.23}$$

where

$$\mathcal{S} := \mathcal{Q}\mathcal{S}_0\mathcal{Q} = \mathcal{Q}\sum_v \mathcal{S}^v \mathcal{Q} \tag{16.24}$$

with the voxel overlap matrices $\mathcal{S}^v = \mathcal{S}^{(r,l)} \otimes \mathcal{S}^{(s,m)} \otimes \mathcal{S}^{(t,n)}$ and

$$\mathcal{S}^{(q,i)}_{kk'} = \langle \chi^{(q)}_{i,k} | \chi^{(q)}_{i,k'} \rangle, \qquad (q,i) = (r,l), (s,m), (t,n) \tag{16.25}$$

Bringing $\mathcal{S}$ to diagonal form would be tantamount to making the transition to a global basis. For large polynomial FE orders K, an approximate diagonal form can be obtained if one admits quadrature errors in the order $2K + 2$. This is how explicit application of $\mathcal{S}^{-1}$ is avoided in the FE-DVR method [246]. Alternatively, a parallelizable method for computing $\mathcal{S}^{-1}hc$ exactly is described below.

16.2.4 Two-Particle Operators

Conceptually, two-particle interactions are much simpler than the many-particle interactions arising in general MCTDH applications. For example, with the Schmidt decomposition [249], we can uniquely identify approximations of optimal rank for a given L^2 accuracy (these are the optimal two-particle 'natural potentials', Section 11.4). However, the fact that the interaction is in general singular at the point of coalescence, that it may be long range, and that we are dealing with very extended orbitals, pose great computational challenges.

With the single-particle grid size on the scale of a few thousand, the exact discretization of the two-particle operators would lead to matrix sizes of several millions by several millions. For polynomial FE functions, the size of the exact discretization for two-particle operators can be strongly reduced, but always remains significantly larger than the dimension of the single-particle space. This is usually much too large even to calculate the discretization of H, let alone apply it in each time step. We therefore approximate the interaction potential in three stages by (i) representing it on a coarse grid that is adjusted to the physical problem in question, (ii) factorizing it by the H-matrix technique, and (iii) making a locally weighted low-rank approximation of the H-matrix.

16.2.4.1 Representation of *H* on a Coarse Grid

As electron–electron interactions near the nuclei are more important, we choose an FE discretization of H with high accuracies only in the vicinity of the molecule and much coarser grids outside. In addition, the order of the FE representation can be chosen mostly lower than the order of the FE on the orbitals. For efficient quadrature it is advantageous if the FE boundaries for H are a subset of those for the orbitals.

For a given FE basis, the interaction potential is

$$H(Q_1, Q_2) \approx \sum_{v,v'} G_v(Q_1)(\mathcal{S}^{-1}\boldsymbol{H}\mathcal{S}^{-1})_{vv'}G_{v'}(Q_2) =: \boldsymbol{GWG}^{\mathrm{T}} \tag{16.26}$$

where $v = (l, m, n, i, j, k)$ and $v' = (l', m', n', i', j', k')$ label the FE product functions $G_v(Q) = \chi_{li}^{(r)}(q)\chi_{mj}^{(s)}(s)\chi_{nk}^{(t)}(t)$. The matrix

$$\boldsymbol{H}_{vv'} = \int \mathrm{d}Q_1\, \mathrm{d}Q_2\, G_v(Q_1)H(Q_1, Q_2)G_{v'}(Q_2) \tag{16.27}$$

is the discretized potential matrix and the overlap matrix $\mathcal{S}$ is defined as in Equation (16.25). As we are approximating a multiplication operator, we do not need to enforce continuity across element boundaries. This has the technical advantage of a strictly local overlap matrix $\mathcal{S}$ that is easily inverted or diagonalized. It has the additional benefit of maintaining voxel-wise locality also after application of $\mathcal{S}^{-1}$, which is important for the H-matrix technique discussed in the following section.

16.2.4.2 *H*-Matrix Representation

For an efficient computation of the two-particle integrals, it is advantageous to bring H to the form

$$H(Q_1, Q_2) \approx \sum_{m=1}^{M} U_m(Q_1) w_m V_m(Q_2) \tag{16.28}$$

where the functions $U_m(Q_1)$ and $V_m(Q_2)$ and the numbers w_m can be assumed to be real. From Equation (16.26) one readily obtains one such representation with the help of the eigenvectors $W\boldsymbol{u}_m = \boldsymbol{u}_m w_m$,

$$H(Q_1, Q_2) \approx \sum_{m=1}^{M} (\boldsymbol{u}_m \cdot \boldsymbol{G}(Q_1)) w_m (\boldsymbol{u}_m \cdot \boldsymbol{G}(Q_2)) \tag{16.29}$$

In this particular case, the left and right functions are identical, $U_m = V_m = \boldsymbol{u}_m \cdot \boldsymbol{G}$. For eigenvalues sorted by decreasing modulus $|w_1| > |w_2| > \ldots$, this so-called Schmidt decomposition constitutes the L^2-optimal approximation to Equation (16.26) for a given rank M.[1] Another well-known example of such a decomposition is the multipole expansion of the Coulomb potential.

Both the Schmidt decomposition and the multipole expansion have global functions U_m and V_m, which makes each term in the sum (16.28) a non-local operator. A systematic way to generate *localized* factor functions and to take advantage of local smoothness of H is by hierarchical matrices (H-matrices) [250]. This technique relies on the fact that almost any two-particle interaction will be diagonally dominated and increasingly smooth far from the diagonal. This is best illustrated by the Coulomb potential, where for $|\boldsymbol{r}_1| \ll |\boldsymbol{r}_2|$ a few terms of the multipole expansion suffice for an accurate description of the interaction. If we choose a discretization (like finite elements) that to some extent preserves locality properties, this factorization into a few products carries over to the discrete matrix. The strategy is first to approximate large off-diagonal blocks of the interaction matrix by a few products and then to repeat the procedure for smaller blocks that are closer to the diagonal. One thus obtains a hierarchy of matrix blocks, each of which can be approximated by

1) A generalization of the Schmidt decomposition to larger dimensions is given by the *potfit* algorithm – see Section 11.4.

a low-rank matrix, until one reaches the diagonal elements (see Figure 16.1). The sum of the ranks of all blocks of this decomposition, that is, the number of terms in Equation (16.28), will in general be larger than the rank of a Schmidt decomposition of comparable accuracy. However, owing to the small size of the majority of the individual blocks, the operations count for applying the H-matrix approximation can be much smaller.

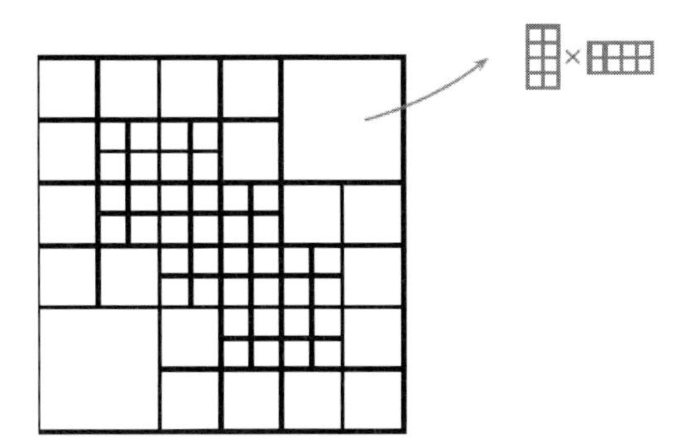

Figure 16.1 Schematic representation of a partitioning of the two-particle potential. Partitioning is denser near the centre of coordinate space by an *a priori* choice of the discretization. The discrete matrix is divided into blocks that are smaller near the diagonal. The off-diagonal blocks allow efficient low-rank approximation.

We use this decomposition to bring the two-particle interaction into the form (16.28). As in our case the rank of the approximation, that is, the number of terms in the sum, affects the operations count for the computation of the matrix $\mathcal{K}_2$, we set a lower limit to the block size of the H-matrix decomposition.

As a last fine tuning of the approximation, we choose different approximation accuracies for H-matrix blocks depending on their importance for the physical process. This is in spirit similar to the initial choice of a coarser FE grid. However, it provides additional flexibility, as it can be made on the two-particle space rather than on the single-particle coordinates: in the terminology of Chapter 11 it is a non-separable weight. In addition, it is more systematic, as the neglected terms can be directly related to the L^2 error of the approximation.

16.3 Parallelization

On distributed memory machines, the key to efficient large-scale parallelization are locality and synchronization. Our model application has the main compute load on the SPFs, for which locality of differential and multiplication operators translates into data locality, if one distributes the spatial part of each SPF over the compute nodes. Application of the projector $\mathcal{Q}$ for wavefunction continuity then involves only nearest-neighbour communication. Somewhat more involved is the application of a non-diagonal FE inverse overlap matrix, $\mathcal{S}^{-1}$, which we discuss below. The most challenging part is the parallel computation of the mean-field potentials, as the two-particle interactions are fundamentally non-local. However, strong interactions usually occur over a short range, that is, locally, while the long-range part of the interaction is well described by low-rank approximations. With the H-matrix decomposition, we have already taken advantage of such a structure. Below, we show how the H-matrix structure can be used for parallelization.

For simplicity, we assume that only the most extended coordinate is distributed over compute nodes. For our model systems this is usually the z-coordinate. The z-axis is split at some finite-element boundaries and each chunk is assigned to a different compute node. In the MCTDHF applications so far, the coefficient vector A_J is comparatively small and does not require parallelization.

Figure 16.2 summarizes the work flow of our parallelized MCTDHF implementation. Starting from a set of old orbitals $\boldsymbol{\varphi}$ and coefficients $\boldsymbol{A}$, we calculate the mean fields $\mathcal{H}$, which requires communication across all nodes. Applying the mean fields and the single-particle operators in the equations of motion, we compute new coefficients $\boldsymbol{A}$ and provisional new orbitals $\boldsymbol{\varphi}$. The latter are then subjected to the continuity condition, which requires mostly nearest-node communication. This procedure is repeated at each time step.

16.3.1
Application of the Inverse Overlap Matrix $\mathcal{S}^{-1}$

With high-order FE methods, the inverse overlap matrix can be approximately transformed to the unit matrix by local transformations (see discussion in Section 16.2.2). However, the exact overlap matrix can also be applied with only little extra inter-node communication. One reason why one may want to make this small effort is to maintain the variational property of the FE discretization.

To solve an equation of the form (16.23), we must invert the matrix $\mathcal{S}$ in the subspace of $\mathcal{Q}$. For that purpose we use the identity

$$\mathcal{S}^{-1} = (\mathcal{Q}\mathcal{S}_0\mathcal{Q})^{-1} = \mathcal{Q}\mathcal{S}_0^{-1}\mathcal{Q} - \mathcal{Q}\mathcal{S}_0^{-1}\mathcal{P}(\mathcal{P}\mathcal{S}_0^{-1}\mathcal{P})^{-1}\mathcal{P}\mathcal{S}_0^{-1}\mathcal{Q} \tag{16.30}$$

which holds for any pair of mutually orthogonal projectors $\mathcal{P} + \mathcal{Q} = 1$. The inversion is to be understood in the sense

$$(\mathcal{Q}\mathcal{S}_0\mathcal{Q})^{-1}\mathcal{Q}\mathcal{S}_0\mathcal{Q} = \mathcal{Q} \tag{16.31}$$

and analogously for $(\mathcal{P}\mathcal{S}_0^{-1}\mathcal{P})^{-1}$. The advantage of replacing the inversion of $\mathcal{Q}\mathcal{S}_0\mathcal{Q}$ with the inversion of $\mathcal{P}\mathcal{S}_0^{-1}\mathcal{P}$ lies in the fact that the number of $\mathcal{P}_n$

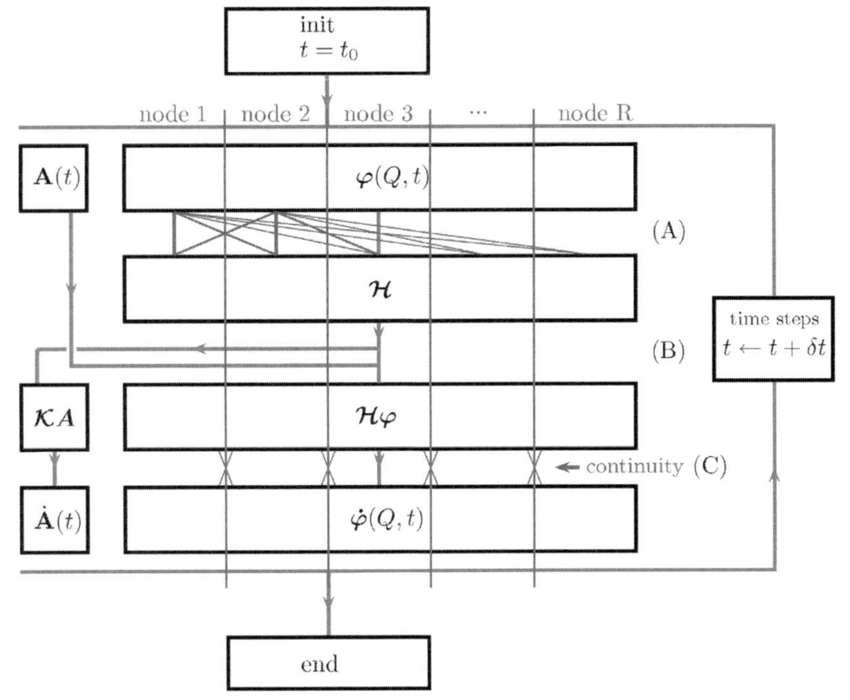

Figure 16.2 Parallelization of an MCTDHF calculation. Node boundaries are indicated by vertical lines. (A) The calculation of the mean-field potentials involves all-to-all communication. (B) Their application is strictly local. (C) Continuity conditions require mostly nearest-neighbour communication. Node boundaries are adjusted during computation.

that connect coefficients on different compute nodes is much smaller than the dimension of $\mathcal{Q}$, which equals the number of independent coefficients in the z-direction. Only this small number of coefficients needs to be exchanged in an all-to-all type of communication.

16.3.2 Parallel Computation of Mean Fields

With a factorization of the form (16.26) the computation of the integrals can be split into three steps. First, the single-particle factor integrals are calculated locally,

$$U_{mik,\nu} = \langle \varphi_i | U_m | \varphi_k \rangle_\nu \tag{16.32}$$

where the subscript ν indicates that integrations are restricted to the νth compute node. This integration needs to be done only for U_m, as, due to the exchange symmetry of the two-particle operator, for each m there is an m' such that $W_{mik} = U_{m'ik}$. The integrals are then added up over all nodes ν where $U_m(Q)$ has non-vanishing support,

$$U_{mik} = \sum_{\nu} U_{mik,\nu} \tag{16.33}$$

With an H-matrix partitioning as sketched in Figure 16.1, many U_m have their support only on a single node and require no summation and no communication in this step, and a large number of m involve only two or a few nodes, causing little communication. After summation of the integrals, they are distributed to those nodes that compute separate parts of the $\mathcal{K}$-matrix and the mean-field potentials. Finally, the application of the mean-field potentials is strictly local and multiplication of $\boldsymbol{A}$ by the distributed $\mathcal{K}$-matrix requires all-to-all communication of the coefficient vector $\boldsymbol{A}$, which has moderate size.

16.3.3 Dynamic Load Balancing

Synchronization is deteriorated by imbalance of the compute load introduced by the strong spatial variation of the potentials and the spatial weighting of the H-matrix decomposition. The exact load distribution is difficult to predict *a priori*. For typical applications, we have observed load imbalances of 50% or more with a corresponding loss in scalability. We have therefore implemented a scheme where self-timing of the code is used to redistribute the load over the processors.

Dynamic load balancing causes little overhead, as imbalances originate in the time-independent part of the operators and only few adjustments occur after the initial establishment of a balanced distribution.

16.4 Observables and Transformations

The computation of observables is in general not time critical in MCTDHF applications. However, antisymmetry can be exploited to obtain compact and easy-to-handle expressions for the observables. Here we illustrate this on a few examples.

16.4.1 Orbital Transformations

The MCTDHF wavefunction is invariant under the transformation

$$\Psi = \sum_{j_1 \dots j_p} \sum_{k_1 \dots k_p} \sum_{l_1 \dots l_p} \left(\prod_{\kappa=1}^{p} U_{l_\kappa j_\kappa} \right) A_{j_1 \dots j_p} \left(\prod_{\lambda=1}^{p} U^*_{l_\lambda k_\lambda} \varphi_{k_\lambda} \right) \tag{16.34}$$

for unitary matrices U. For example, natural orbitals $\varphi_l^{(nat)} = \sum_k U_{kl}^* \varphi_k$ are obtained if U diagonalizes the single-particle density matrix (3.13). In the corresponding transformation of the A-coefficients, antisymmetry can be exploited,

$$A_{l_1...l_p}^{(nat)} = \sum_{j_1}^{n} \cdots \sum_{j_p}^{n} \left(\prod_{\kappa=1}^{p} U_{l_\kappa j_\kappa} \right) A_{j_1...j_p} = \sum_{j_1<...<j_p} \det(U_{l_1...l_p}^{j_1...j_p}) A_{j_1...j_p} \tag{16.35}$$

where

$$U_{l_1...l_p}^{j_1...j_p} := \begin{pmatrix} U_{l_1 j_1} & \cdots & U_{l_1 j_p} \\ \vdots & & \vdots \\ U_{l_p j_1} & \cdots & U_{l_p j_p} \end{pmatrix} \tag{16.36}$$

denotes the submatrix of the total transformation matrix for the respective left and right index sets.

16.4.2
Projections Onto Multiparticle States

The overlap between two multiparticle wavefunctions constructed from the same orthonormal set of SPFs is obtained from the inner product of the two coefficient vectors. For the calculation of overlaps with different sets of SPFs, for example, the autocorrelation function $\langle \Psi(0) | \Psi(t) \rangle$, one can take advantage of antisymmetry by observing that

$$\begin{aligned} \langle \Psi_a | \Psi_b \rangle &= \sum_{j_1...j_p} \sum_{l_1...l_p} A_{j_1...j_p}^* \left(\prod_{\kappa=1}^{p} \mathcal{O}_{j_\kappa l_\kappa} \right) B_{l_1...l_p} \\ &= p! \sum_{j_1<...<j_p} \sum_{l_1<...<l_p} A_{j_1...j_p}^* B_{l_1...l_p} \det(\mathcal{O}_{j_1...j_p}^{l_1...l_p}) \end{aligned} \tag{16.37}$$

where $\mathcal{O}_{jl} = \langle \varphi_j^a | \varphi_l^b \rangle$ and the matrix $\mathcal{O}_{j_1...j_p}^{l_1...l_p}$ is constructed analogously to (16.36).

16.4.3
One- and Two-Particle Expectation Values

For a system of indistinguishable particles, the general form of a one-particle observable is the sum of one operator in the single-particle space over all single-particle coordinates,

$$X(Q_1 \dots Q_p) = \sum_{\kappa=1}^{p} x(Q_\kappa) \tag{16.38}$$

The expectation value of X with an MCTDHF wavefunction Ψ is therefore just p times the expectation value for the first particle,

$$\langle\Psi|X|\Psi\rangle = p\sum_{k=1}^{n}\sum_{l=1}^{n}\mathcal{X}_{kl}\sum_{j_2\dots j_p} A^*_{kj_2\dots j_p}A_{lj_2\dots j_p} =: p\,\mathrm{Tr}(\mathcal{X}^*\rho) \tag{16.39}$$

where $\mathcal{X}$ denotes the matrix $(\mathcal{X})_{jl} = \langle\varphi_j|x|\varphi_l\rangle$ and ρ is the single-particle density matrix (3.13).

Two-particle observables have the general form

$$Y = \sum_{\kappa,\lambda=1}^{p} y(Q_\kappa, Q_\lambda) \tag{16.40}$$

Analogous to single-particle expectation values, general two-particle expectation values can be written as

$$\langle\Psi|Y|\Psi\rangle = p^2\,\mathrm{Tr}(\mathcal{Y}^*\sigma) \tag{16.41}$$

where we have introduced the two-particle density matrix

$$\sigma_{kl,mn} = \sum_{j_3\dots j_p} A^*_{klj_3\dots j_p}A_{mnj_3\dots j_p} \tag{16.42}$$

and the SPF matrix $\mathcal{Y}_{kl,mn} = \langle\langle\varphi_k\varphi_l|y|\varphi_m\varphi_n\rangle\rangle$, with the double brackets $\langle\langle\cdots\rangle\rangle$ indicating integration over the two-particle coordinates.

16.4.4 All-Particle Observables

The scheme set up above for one- and two-particle observables can be generalized to many-particle observables. One such observable that unexpectedly involves all particles is the probability of single or multiple ionization, which we define as the probability of finding – at large times – any m particles *outside* some binding area B, while the remaining particles are *inside* this area. The corresponding operator is

$$Z_m = \sum_{T_m}\prod_{\kappa\in T_m} O_B(Q_\kappa)\prod_{\lambda\notin T_m} P_B(Q_\lambda) \tag{16.43}$$

where T_m extends over all m-element subsets of $\{1,\dots,p\}$ and

$$O_B(Q) = \begin{cases} 0 & \text{for } Q\in B \\ 1 & \text{else}\end{cases} \tag{16.44}$$

and $P_B = 1 - O_B$. The expectation value for this p-particle operator is

$$\langle\Psi|Z_m|\Psi\rangle = p!\sum_{j_1<\dots<j_p}\sum_{l_1<\dots<l_p} A^*_{j_1\dots j_p}Z^{l_1\dots l_p}_{j_1\dots j_p}A_{l_1\dots l_p} \tag{16.45}$$

where, for observables of the form (16.43), the matrix elements $Z^{l_1 \dots l_p}_{j_1 \dots j_p}$ can be evaluated as

$$Z^{l_1 \dots l_p}_{j_1 \dots j_p} = \sum_{R,S} \sigma(S)\sigma(R) \det(O^S_R) \det(P^{C(S)}_{C(R)}) \tag{16.46}$$

and the sum extends over all ordered m-element subsets $R = (r_1, \dots, r_m)$ and $S = (s_1, \dots, s_m)$ of $\{j_1 \dots j_p\}$ and $\{l_1 \dots l_p\}$, respectively. $C(R)$ denotes the complement of R, that is, those indices of $(j_1 \dots j_p)$ that do not occur in R, and similarly for $C(S)$. The single-particle submatrices O^S_R and $P^{C(S)}_{C(R)}$ of the matrices $O_{jl} = \langle \varphi_j | O_B | \varphi_l \rangle$ and $P_{jl} = \langle \varphi_j | P_B | \varphi_l \rangle$ are defined as in (16.36). The sign $\sigma(X)$ is the sign of the permutation $(X, C(X))$ relative to the original ordering of the indices.

By replacing in Equation (16.43) the product $\prod^p_{\lambda=m+1} P_B(Q_\lambda)$ with a projector $P_i = |i\rangle\langle i|$ onto ground or excited states $|i\rangle$ of the m-fold ionized system, one obtains, without further changes, ionization into the ionic ground state and shake-up. In that case, one must ensure (approximate) orthogonality of the projectors $O_B P_i \approx 0$, that is, the binding area B must be chosen large enough to accommodate the state $|i\rangle$.

16.4.5 Spectra

One very important observable for our model application are electron spectra. The calculation of spectra for time-independent Hamiltonians with absorbing boundary conditions is described in Section 6.5. The method relies on the assumption that particles beyond a certain distance follow a free time evolution. For Hamiltonians with a long-range time-dependent term – like the dipole interaction in a laser field – this method is not applicable, as the spectral distribution keeps changing for all particles, no matter how far from the system they are. However, one can assume to good approximation that the interaction with the binding system and between particles becomes negligible at large distances from the interaction centre. Based on this assumption, in Ref. [243] a scheme for calculating spectra is derived. The momentum spectrum of detached electrons is the sum over contributions from all possible ionization channels c,

$$\sigma(\mathbf{p}) = \lim_{T \to \infty} \sum_c |b_c(\mathbf{p}, T)|^2 \tag{16.47}$$

It can be shown [243] that the channel amplitudes $b_c(\mathbf{p}, T)$ at large time T can be written as

$$b_c(\mathbf{p}, T) = \mathrm{i} \int_{-\infty}^{\mathrm{T}} \mathrm{d}t \, \langle c, \mathbf{p}; t | W | \Psi(t) \rangle \tag{16.48}$$

if the time evolution of $\Psi(t)$ is calculated with the CAP $-\mathrm{i}W$. The time-dependent channel wavefunction has the product form

$$|c, \mathbf{p}; t\rangle = \Phi_c(t)\chi_{\mathbf{p}}(Q_p, t) \tag{16.49}$$

where, for large t, the function $\Phi_c(t)$ evolves into the ionic bound-state function of the given channel and $\chi_{\mathbf{p}}(Q_p, t)$ tends to the free single-electron state for electron momentum $\mathbf{p}$. The two factors of the channel wavefunction do not need to be explicitly antisymmetrized, as only its antisymmetric content contributes to the matrix element in Equation (16.48).

This approach assumes that, during the time evolution, Φ_c remains located inside the simulation box and is not affected by absorption at the boundaries. Otherwise, it is an arbitrary function. However, usually one associates a channel with the ion remaining in an energy eigenstate Φ_α with quantum numbers α. In this case Φ_c is determined by the *final* (rather than initial) condition

$$\lim_{t\to\infty} \Phi_c(t) = \Phi_\alpha \tag{16.50}$$

For excited ionic channels Φ_α, one must exclude artefacts in the time propagation due to the nonlinearity of the MCTDHF equations. This can be achieved by the simultaneous propagation of several excited states [243].

16.5 Applications

16.5.1 Ionization of Linear Molecules

The ionization of three-dimensional (3D) linear model molecules with two to six nuclei at a constant internuclear separation was investigated in [251]. It was assumed that there is one active electron per nucleus, while the remaining electrons are taken into account by a screening of the nuclear potentials $V(\mathbf{r})$ in Equation (16.6). The screening was adjusted to obtain a constant ionization potential of 0.3 a.u. ($\sim$8 eV), independent of the number of nuclei. Two different internuclear separations, $R = 1.4$ and $R = 3$ a.u., were used in the calculations. As initial states we used the singlet neutral ground states of the molecules. The model molecules are exposed to a laser pulse with linear polarization along the molecular axis (z-direction) and a width of one optical cycle. We use the near-IR laser wavelength $\lambda = 800$ nm and intensity $I = 2.5 \times 10^{13}$ W cm^{-2}, which leads to nearly 90% depletion for the largest molecule at $R = 3$ a.u. For $R = 1.4$ a.u., a slightly higher intensity of $I = 3.5 \times 10^{13}$ W cm^{-2} was used.

Figure 16.3 shows the depletion of the ground state after passage of the laser pulse for molecules with an increasing number of nuclei. Contrary to the trend suggested by experiments, we found that for both internuclear separations, $R = 1.4$ and $R = 3$ a.u., depletion increases with the number of nuclei. The impact of correlation on these findings is only small: TDHF agrees qualitatively with MCTDHF where TDHF yields are typically only below MCTDHF by about 10%.

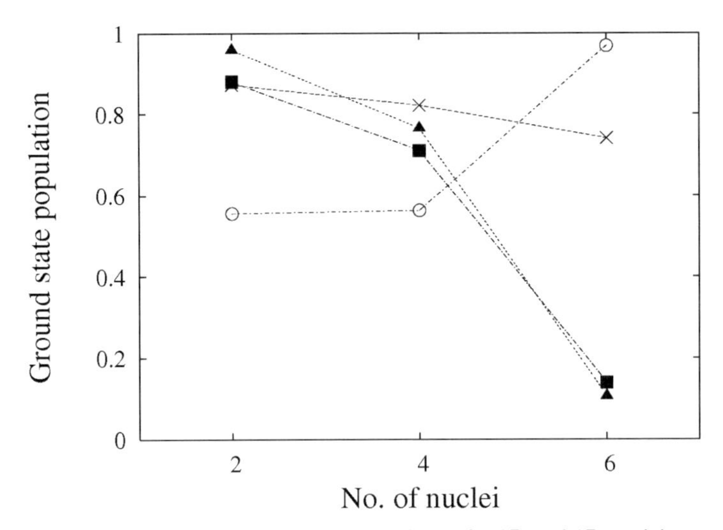

Figure 16.3 Residual ground-state population for 3D and 1D model molecules as a function of molecule size. In 3D, population decreases with molecule size for internuclear separations $R = 1.4$ a.u. (crosses) and $R = 3$ a.u. (full squares), and the Hartree–Fock result is similar to the correlated result (triangles, $R = 3$ a.u.). In contrast, 1D models show an increase of residual population with size (circles).

Interestingly, we observe a strong influence of dimensionality on our results: the same calculations in one dimension (1D) yield the opposite trend for ionization as a function of molecule size, and also show a much larger deviation between TDHF and MCTDHF yields.

16.5.1.1 High-Harmonic Spectra of Molecules

The radiative response of molecules to non-perturbative laser fields is calculated from the Fourier transform of the time-dependent dipole,

$$\mathbf{d}(t) = \left\langle \Psi(t) \middle| \sum_{\kappa=1}^{p} \mathbf{r}_\kappa \middle| \Psi(t) \right\rangle \tag{16.51}$$

The standard assumption in the literature is that the high-frequency part of this response can be well described by single-electron models [252], possibly with simple corrections for multi-electron effects [253, 254]. MCTDHF calculations show that this assumption is not in general correct.

As a model, we use a diatomic homonuclear molecule with internuclear separation $R = 2.8$ a.u. and four active electrons [255]. Core electrons are accounted for by nuclear screening, with the parameters adjusted to give an ionization potential close to the ionization potential of the N_2 molecule, which is frequently used for high-harmonic generation. A linearly polarized laser pulse with the standard wavelength of 800 nm, a full width at half-maximum (FWHM) of one optical cycle and an intensity of 3×10^{14} W cm^{-2} was used.

The influence of correlation on the high-harmonic spectrum is not very pronounced. In Figure 16.4 we show the convergence of the spectrum for the four-electron system with $n = 4, 6, 8$ orbitals. The shape of the spectrum is already obtained to good approximation by the single-configuration TDHF calculation. As the dipole response (16.51) depends only on the *single*-electron density $\sum_{ij} \varphi_i(x)\varphi_j^*(x)\rho_{ij}$, this only tells us that TDHF produces a sufficiently accurate single-electron density.

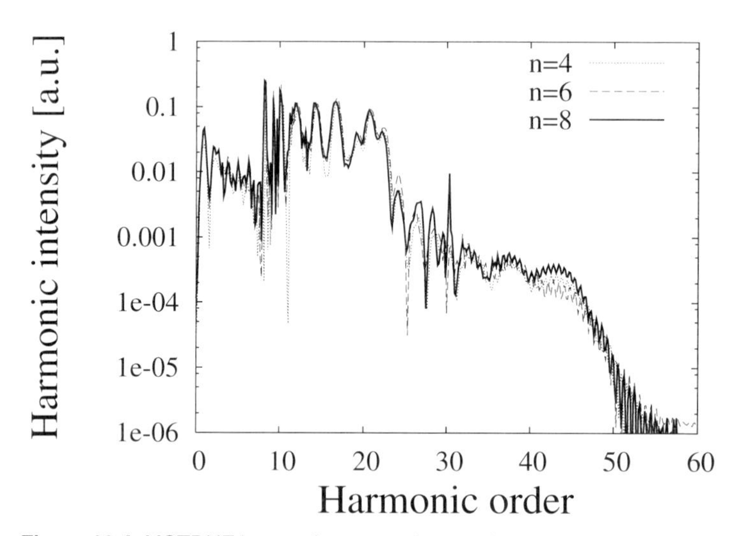

Figure 16.4 MCTDHF harmonic spectra for a molecule with four active electrons using an increasing number of orbitals n, with $n = 4$ corresponding to TDHF. While a full quantitative convergence is not reached, the main qualitative features are already present in TDHF.

Taking into account correlation becomes important when one wants to analyse the underlying mechanisms in terms of separate electrons. To assess the influence of multi-electron effects on high-harmonic generation, we make a comparison of the MCTDHF result with the popular one-electron model proposed in Ref. [252], its multi-electron extension [253,254], and a single-electron Schrödinger equation with an effective Hamiltonian [255]. Figure 16.5 shows that none of these models can even qualitatively reproduce the spectral features of the MCTDHF multi-electron solution.

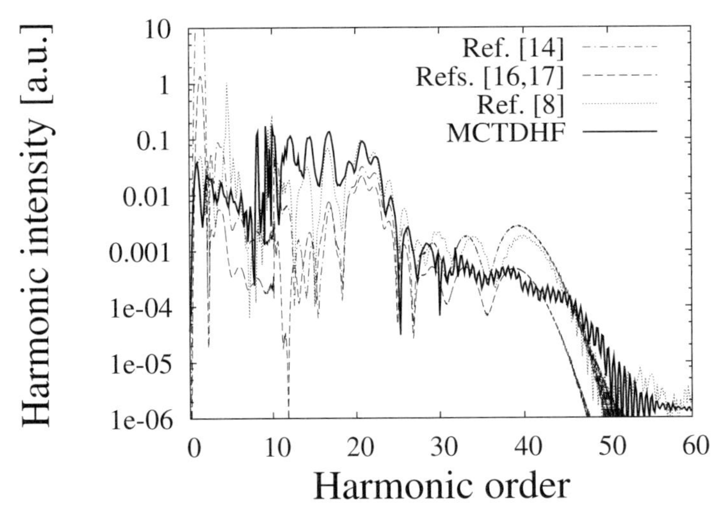

Figure 16.5 Comparison of various simplifying models for molecular high-harmonic generation with the correlated MCTDHF harmonic spectra. None of the models can reproduce the spectral features of the multi-electron calculation.

We attribute this failure to the fact that the one-electron models do not account for the polarization of the ionic core. To demonstrate that core polarization is crucial, we factorize the total MCTDHF wavefunction into a product of an ionic core wavefunction Φ and a single-electron orbital ϕ:

$$\Psi' = \frac{1}{N(t)} \mathcal{A}[\Phi(Q_1, \ldots, Q_{p-1}; t)\phi(Q_p; t)] \tag{16.52}$$

Here $N(t)$ is a time-dependent normalization factor and

$$\mathcal{A} = \frac{1}{\sqrt{p}} \left(1 - \sum_{\kappa=1}^{p} P_{\kappa p}\right)$$

is the antisymmetrization operator. $P_{\kappa p}$ transposes electron κ and p. The idea of this adiabatic factorization is that the (active) ionized electron moves on larger length and faster velocity scales than the rest of the electrons. For a given Φ, the single-electron orbital ϕ is determined such that the overlap of Ψ' with the MCTHDF solution Ψ is maximized. For Φ we choose either (i) the 'static', field-free ionic ground state, $\Phi = \Phi_{\text{stat}} := \Psi_{\text{ion}}(t{=}0)$, or (ii) the 'dynamic' ionic wavefunction evolving in the laser field, $\Phi(t) = \Phi_{\text{dyn}} := \Psi_{\text{ion}}(t)$.

We find that only for Φ_{dyn} does the factorization (16.52) lead to qualitatively correct high-harmonic spectra. This tells us that – at least in our reasonably realistic few-electron models – the polarization of the core contributes crucially to the high-harmonic spectra, although the harmonic response of the core itself is rather weak. We ascribe the great importance of core polarization to heterodyning effects that are also responsible for the multi-electron corrections described in Refs. [253, 254].

Technically, it is interesting to remark that the factorization (16.52) reproduces the harmonic spectra only when the full MCTDHF wavefunction is used. For the (single-configuration) TDHF functions, the procedure fails: although TDHF produces approximately correct electron densities, the TDHF wavefunction is too crude an approximation to be interpreted in terms of individual electrons.

16.5.2
Cold Fermionic Atoms

Cold fermionic atoms in a two-dimensional harmonic trap are described by the Hamiltonian

$$H = \sum_{i=1}^{N} \left(-\frac{\Delta_i}{2m} + V_{\text{trap}}(\mathbf{r}_i) + \sum_{j<i} V_{\text{int}}(\mathbf{r}_i, \mathbf{r}_j) \right) \tag{16.53}$$

where N is the total number of particles and m is the mass of the atom, which in our calculations is that of ^{40}K. The trapping potential $V_{\text{trap}}(\mathbf{r})$ consists of an elliptic harmonic oscillator with a central Gaussian barrier of width $a = 0.5\,\mu\text{m}$,

$$V_{\text{trap}}(\mathbf{r}) = \frac{1}{2} m\omega_x^2 x^2 + \frac{1}{2} m\omega_y^2 y^2 + c\frac{1}{\pi a} \exp\left(-\frac{\mathbf{r}^2}{2a^2} \right) \tag{16.54}$$

The parameter c controlling the central barrier is chosen to be 0.023 eV, which allows for two well-separated wavepackets. The frequencies ω_x and ω_y are $\omega_x = 4.9\,\text{kHz}$ and $\omega_y = 6.9\,\text{kHz}$, respectively. These extensions of the trap are comparable to the experimental conditions used, for example, in [256]. The repulsive short-range interaction with a $2d$-scattering length a_{2d} has the form

$$V_{\text{int}}(\mathbf{r}_i, \mathbf{r}_j) = \frac{4\pi a_{2d}}{m} \frac{1}{2\pi\sigma^2} \exp\left(-\frac{(\mathbf{r}_i - \mathbf{r}_j)^2}{2\sigma^2} \right) \tag{16.55}$$

Here we approximate the delta function typically used as the particle–particle interaction for a Bose–Einstein condensate by a Gaussian, which is more realistic and easier to treat numerically.

We used MCTDHF to simulate the formation of interference patterns after releasing the atoms from the trap ground state. Without atomic repulsion the interference patterns are in accordance with a simple shell model for the trap. The *gerade* and *ungerade* orbitals of the double well are pairwise near-degenerate. For atomic numbers $N = 2$, 6 and 10, we have an open-shell situation where two atoms occupy a single spatial orbital that extends over both

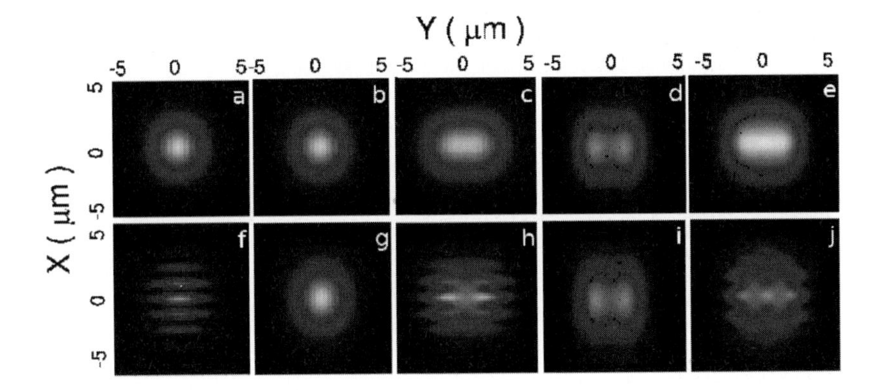

Figure 16.6 Interference patterns observed 320 μs after the harmonic potential was switched off. The results without particle–particle interaction are presented in the lower row (f–j). In the upper row (a–e) we chose the interaction strengths large enough so that the interference pattern disappears because the fermions localize on each side of the well. The number of particles increases from $N = 2$ in the first column to $N = 10$ in the last column.

wells. When released, these extended states produce the interference patterns seen in the lower row of Figure 16.6. For $N = 4$ and 8 the *gerade* and *ungerade* orbitals are both occupied, which fills in the interference patterns. Interferences are destroyed when the interaction energy exceeds the *gerade–ungerade* splitting: analogous to the 'fermionization' of bosonic systems, atoms become localized in either the left or the right well and lose the ability to produce interference patterns (Figure 16.6, upper row).

17
The Multiconfigurational Time-Dependent Hartree Method for Identical Particles and Mixtures Thereof

Ofir E. Alon, Alexej I. Streltsov and Lorenz S. Cederbaum

17.1
Preliminary Remarks

The quantum world is full of identical particles. Some familiar examples from nuclear, atomic, molecular and condensed-matter physics include: protons and neutrons in nuclei; electrons in atoms, molecules and quantum dots; and atoms themselves in quantum fluids and degenerate quantum gases. Mixtures of identical particles occur as well in many natural and man-made systems. Examples of these include: protons and neutrons in nuclei; electrons and protons in certain molecules; bosonic ^{4}He and fermionic ^{3}He as interacting quantum fluids; electrons and positrons in matter–antimatter systems; and various combinations of bosonic and fermionic atoms as degenerate quantum gases. What is the quantum dynamics of systems comprising identical particles or mixtures of identical particles? This is a challenging, fundamental and important problem in contemporary physics [257–260].

The MCTDH approach discussed in the previous chapters is formulated as a distinguishable-particle propagation theory [30, 34, 37]. MCTDH has in principle no restrictions on the treatment of the dynamics of identical particles. However, because of the large amount of redundancies in the distinguishable-particle multiconfigurational expansion of the MCTDH wavefunction, it becomes computationally difficult to treat more than a handful of identical particles with MCTDH.

In the present chapter we apply the MCTDH approach to systems of identical particles – bosons or fermions – and mixtures thereof. For this, it is helpful to resort to second quantization representation of the quantum many-body problem, which is naturally designed to treat identical particles. In this way, the indistinguishability of particles and permutational symmetries of the many-body wavefunction are directly taken into account.

There is another important difference between systems of distinguishable degrees of freedom, say vibrations in molecules, and systems of identical par-

Multidimensional Quantum Dynamics: MCTDH Theory and Applications.
Edited by Hans-Dieter Meyer, Fabien Gatti, and Graham A. Worth

ISBN: 978-3-527-32018-9

ticles. The latter interact, in many cases, via two-body interactions. This, as we shall see below, offers an appealing representation of the equations of motion in terms of *reduced one- and two-body density matrices*, which opens avenue for new types of approximations. In the present context, reduced one- and two-body density matrices were first used to derive the stationary many-body states within the general variational theory with complete self-consistency for trapped bosonic systems – the multiconfigurational Hartree for bosons (MCHB) [261]. It is instructive to mention that reduced density matrices and their use in electronic structure theory have long been an active research field [262–264].

The sections presented below develop the theory and discuss some applications to many-boson systems.

17.2 Bosons or Fermions? – Unifying MCTDHB and MCTDHF

The bosonic version of MCTDH (MCTDHB) was developed recently [265,266]. The fermionic version of MCTDH (MCTDHF) was developed not long before that by three independent groups [242, 247, 267] (see Chapter 16). In this section, we unify the MCTDHB and MCTDHF approaches within a single formulation based on second quantization and reduced density matrices [268]. Note that the form of MCTDHF in the literature [242, 247, 267] did not exploit *reduced two-body density matrices*.

17.2.1 Basic Ingredients

Consider N identical particles, either bosons or fermions. Let us recruit the field operator $\hat{\mathbf{\Psi}}(\mathbf{x})$, which annihilates a particle at coordinate $\mathbf{x}$ and satisfies the common commutation (anticommutation) relations for bosons (fermions):

$$\hat{\mathbf{\Psi}}(\mathbf{x})\{\hat{\mathbf{\Psi}}(\mathbf{x}')\}^{\dagger} \mp \{\hat{\mathbf{\Psi}}(\mathbf{x}')\}^{\dagger}\hat{\mathbf{\Psi}}(\mathbf{x}) = \delta(\mathbf{x}-\mathbf{x}') \tag{17.1}$$

Here and throughout this chapter the upper sign refers to bosons and the lower one to fermions. The coordinate $\mathbf{x} \equiv \{\mathbf{r}, \sigma\}$ groups spatial degrees of freedom and (possible) spin. Correspondingly, the following shorthand notation is used throughout this chapter:

$$\delta(\mathbf{x}-\mathbf{x}') = \delta(\mathbf{r}-\mathbf{r}')\delta_{\sigma,\sigma'} \quad \text{and} \quad \int \mathrm{d}\mathbf{x} \equiv \int \mathrm{d}\mathbf{r} \sum_{\sigma}$$

The employment of the field operator $\hat{\mathbf{\Psi}}(\mathbf{x})$ allows us to introduce a set of time-dependent orthonormal orbitals

$$\int \mathrm{d}\mathbf{x}\, \phi_k^*(\mathbf{x},t)\phi_q(\mathbf{x},t) = \delta_{kq}$$

in which the particles reside. This is simply done by expanding the (time-independent) field operator as follows:

$$\hat{\mathbf{\Psi}}(\mathbf{x}) = \sum_k \hat{a}_k(t)\phi_k(\mathbf{x},t), \qquad \hat{a}_k(t) = \int \phi_k^*(\mathbf{x},t)\hat{\mathbf{\Psi}}(\mathbf{x})\,\mathrm{d}\mathbf{x} \tag{17.2}$$

where the $\hat{a}_k(t)$ are time-dependent annihilation operators. The index k refers to both spatial and spin quantum numbers. The orbitals $\phi_k(\mathbf{x},t)$ are, in general, spin orbitals. In what follows we will also use the term 'orbitals' to describe spin orbitals whenever unambiguous. The time-dependent annihilation and corresponding creation operators obey the usual commutation (anticommutation) relations for bosons (fermions) at any time,

$$\hat{a}_k(t)\hat{a}_q^\dagger(t) \mp \hat{a}_q^\dagger(t)\hat{a}_k(t) = \delta_{kq}$$

The many-body Hamiltonian describing the N identical particles comprises one-body $\hat{h}(\mathbf{x})$ and two-body $\hat{W}(\mathbf{x},\mathbf{x}')$ terms, which may, in the most general case, be time-dependent quantities. The formulation provided here holds in this generic case. With the annihilation and creation operators, the many-body Hamiltonian is conventionally written as follows:

$$\begin{aligned}
\hat{H} &= \sum_{k,q} h_{kq}\hat{a}_k^\dagger\hat{a}_q + \tfrac{1}{2}\sum_{k,s,l,q} W_{ksql}\hat{a}_k^\dagger\hat{a}_s^\dagger\hat{a}_l\hat{a}_q \\
h_{kq} &= \int \phi_k^*(\mathbf{x},t)\hat{h}(\mathbf{x})\phi_q(\mathbf{x},t)\,\mathrm{d}\mathbf{x} \\
W_{ksql} &= \iint \phi_k^*(\mathbf{x},t)\phi_s^*(\mathbf{x}',t)\hat{W}(\mathbf{x},\mathbf{x}')\phi_q(\mathbf{x},t)\phi_l(\mathbf{x}',t)\,\mathrm{d}\mathbf{x}\,\mathrm{d}\mathbf{x}'
\end{aligned} \tag{17.3}$$

Note that the matrix elements h_{kq} and W_{ksql} are generally time-dependent quantities (even if $\hat{h}(\mathbf{x})$ and $\hat{W}(\mathbf{x},\mathbf{x}')$ are time-independent operators) since they are evaluated with time-dependent orbitals. Here and throughout this chapter the explicit time dependence of quantities is not shown whenever unambiguous.

In the MCTDH approach for identical particles, the *Ansatz* for the time-dependent many-body wavefunction is written as a linear combination of time-dependent permutational symmetry-adapted configurations (permanents for bosons or determinants for fermions),

$$\begin{aligned}
|\Psi(t)\rangle &= \sum_{\mathbf{n}} C_{\mathbf{n}}(t)|\mathbf{n};t\rangle \\
|\mathbf{n};t\rangle &\equiv |n_1,n_2,\ldots,n_M;t\rangle \\
&= \frac{1}{\sqrt{n_1!n_2!\cdots n_M!}}[\hat{a}_1^\dagger(t)]^{n_1}[\hat{a}_2^\dagger(t)]^{n_2}\cdots[\hat{a}_M^\dagger(t)]^{n_M}|\mathrm{vac}\rangle
\end{aligned} \tag{17.4}$$

The occupation numbers n_k are collected in the vector $\mathbf{n}$ and satisfy the conservation-of-particles condition $n_1 + n_2 + \cdots + n_M = N$. The summation over $\mathbf{n}$ runs over all possible configurations (permanents or determinants) generated by distributing N particles over M orbitals. The configurations $|\mathbf{n};t\rangle$ satisfy, of course, the orthonormalization condition $\langle \mathbf{n};t|\mathbf{n}';t\rangle = \delta_{\mathbf{n},\mathbf{n}'}$. Their normalization to unity is readily verified from the basic action of creation and annihilation operators in Fock space:

$$\begin{aligned}
&\hat{a}_k(t)|n_1,n_2,\ldots,n_k,\ldots,n_M;t\rangle \\
&\qquad = (\pm 1)^{\sum_{l=1}^{k-1} n_l}\sqrt{n_k}\,|n_1,n_2,\ldots,n_k-1,\ldots,n_M;t\rangle \\
&\hat{a}_q^{\dagger}(t)|n_1,n_2,\ldots,n_q,\ldots,n_M;t\rangle \\
&\qquad = (\pm 1)^{\sum_{l=1}^{q-1} n_l}\sqrt{n_q+1}\,|n_1,n_2,\ldots,n_q+1,\ldots,n_M;t\rangle
\end{aligned} \tag{17.5}$$

For bosons each occupation number n_k can take any integer value between 0 and N in accordance with Bose–Einstein statistics, whereas for fermions each n_k can be either 0 or 1 in accordance with Fermi–Dirac statistics. These restrictions on the occupation numbers are implicitly assumed in Equation (17.5).

So far, we have used the indistinguishability of particles (either bosons or fermions) to express the many-body Hamiltonian $\hat{H}$ and multiconfigurational wavefunction $\Psi(t)$ in a unified form based on second quantization. In turn, the indistinguishability of particles has much deeper implications for the multiconfigurational treatment of the many-body problem. Specifically, it allows one to introduce reduced density matrices into the theory. Since identical particles commonly interact via two-body interactions (see the standard many-body Hamiltonian in Equation (17.3)), the equations of motion in this case involve reduced one- and two-body density matrices only. The reduced one-body density matrix of the indistinguishable-particle multiconfigurational wavefunction $\Psi(t)$ is given by

$$\begin{aligned}
\rho(\mathbf{x}_1|\mathbf{x}_1';t) &= N\int \Psi^*(\mathbf{x}_1',\mathbf{x}_2,\ldots,\mathbf{x}_N;t)\Psi(\mathbf{x}_1,\mathbf{x}_2,\ldots,\mathbf{x}_N;t)\,\mathrm{d}\mathbf{x}_2\,\mathrm{d}\mathbf{x}_3\ldots\mathrm{d}\mathbf{x}_N \\
&= \langle\Psi(t)|\hat{\mathbf{\Psi}}^{\dagger}(\mathbf{x}_1')\hat{\mathbf{\Psi}}(\mathbf{x}_1)|\Psi(t)\rangle \\
&= \sum_{k,q=1}^{M}\rho_{kq}(t)\phi_k^*(\mathbf{x}_1',t)\phi_q(\mathbf{x}_1,t)
\end{aligned} \tag{17.6}$$

where

$$\rho_{kq}(t) = \langle\Psi(t)|\hat{a}_k^{\dagger}(t)\hat{a}_q(t)|\Psi(t)\rangle$$

are its matrix elements in the orbital basis. We collect these matrix elements as $\boldsymbol{\rho}(t) = \{\rho_{kq}(t)\}$. Similarly, the reduced two-body density matrix of the indistinguishable-particle multiconfigurational wavefunction $\Psi(t)$ reads

$$\begin{aligned}
&\rho(\mathbf{x}_1, \mathbf{x}_2|\mathbf{x}_1', \mathbf{x}_2'; t) \\
&\quad = N(N-1)\int \Psi^*(\mathbf{x}_1', \mathbf{x}_2', \mathbf{x}_3, \ldots, \mathbf{x}_N; t)\Psi(\mathbf{x}_1, \mathbf{x}_2, \mathbf{x}_3, \ldots, \mathbf{x}_N; t)\, \mathrm{d}\mathbf{x}_3 \cdots \mathrm{d}\mathbf{x}_N \\
&\quad = \langle \Psi(t)|\hat{\mathbf{\Psi}}^\dagger(\mathbf{x}_1')\hat{\mathbf{\Psi}}^\dagger(\mathbf{x}_2')\hat{\mathbf{\Psi}}(\mathbf{x}_2)\hat{\mathbf{\Psi}}(\mathbf{x}_1)|\Psi(t)\rangle \\
&\quad = \sum_{k,s,l,q=1}^{M} \rho_{kslq}(t)\phi_k^*(\mathbf{x}_1', t)\phi_s^*(\mathbf{x}_2', t)\phi_l(\mathbf{x}_2, t)\phi_q(\mathbf{x}_1, t)
\end{aligned} \tag{17.7}$$

where

$$\rho_{kslq}(t) = \langle \Psi(t)|\hat{a}_k^\dagger(t)\hat{a}_s^\dagger(t)\hat{a}_l(t)\hat{a}_q(t)|\Psi(t)\rangle$$

are its matrix elements in the orbital basis.

17.2.2
Equations of Motion with Reduced Density Matrices

Combining the above ingredients, we are ready to specify the equations of motion of MCTDH to systems of identical particles. In what follows, we prescribe the essentials of the derivation. For more details, the reader is referred to Refs. [266,268]. We employ the Lagrangian formulation of the time-dependent variational principle [269, 270] and hence write the functional action of the time-dependent Schrödinger equation as follows:

$$\begin{aligned}
&S[\{C_\mathbf{n}(t)\}, \{\phi_k(\mathbf{x}, t)\}] \\
&\quad = \int \mathrm{d}t \left\{ \left\langle \Psi(t) \left| \hat{H} - \mathrm{i}\frac{\partial}{\partial t} \right| \Psi(t) \right\rangle - \sum_{k,j}^{M} \mu_{kj}(t)[\langle \phi_k|\phi_j\rangle - \delta_{kj}] \right\}
\end{aligned} \tag{17.8}$$

This allows us first to evaluate the expectation value $\langle \Psi(t)|\hat{H} - \mathrm{i}\partial/\partial t|\Psi(t)\rangle$ and only afterwards to perform the variation with respect to the orbitals $\{\phi_k(\mathbf{x}, t)\}$ and coefficients $\{C_\mathbf{n}(t)\}$. The time-dependent Lagrangian multipliers $\mu_{kj}(t)$ take on a dual role: (i) they ensure that the time-dependent orbitals $\{\phi_k(\mathbf{x}, t)\}$ remain normalized and orthogonal to one another at all times throughout the propagation; and (ii) they exactly 'compensate' for those terms appearing within the Dirac–Frenkel formulation of the time-dependent variational principle [32, 33], that is, when the variation of $\Psi(t)$ is performed before matrix elements $\langle \delta\Psi(t)|\hat{H} - \mathrm{i}\partial/\partial t|\Psi(t)\rangle$ are evaluated (also see previous chapters and [266,270]).

Thus, to obtain the equations of motion for $\{\phi_k(\mathbf{x}, t)\}$ and $\{C_\mathbf{n}(t)\}$, we need to evaluate the expectation value $\langle \Psi(t)|\hat{H} - \mathrm{i}\partial/\partial t|\Psi(t)\rangle$ in Equation (17.8) and then to perform the variation with respect to the orbitals and coefficients. We must remember that the orbitals and coefficients are independent arguments of $S[\{C_{\mathbf{n}}(t)\},\{\phi_k(\mathbf{x},t)\}]$. To perform the variation with respect to $\{\phi_k(\mathbf{x},t)\}$, we express $\langle\Psi(t)|\hat{H}-\mathrm{i}\partial/\partial t|\Psi(t)\rangle$ in a form that explicitly depends on the orbitals:

$$\left\langle\Psi\left|\hat{H}-\mathrm{i}\frac{\partial}{\partial t}\right|\Psi\right\rangle = \sum_{k,q=1}^{M}\rho_{kq}\left[h_{kq}-\left(\mathrm{i}\frac{\partial}{\partial t}\right)_{kq}\right] + \frac{1}{2}\sum_{k,s,l,q=1}^{M}\rho_{kslq}W_{ksql}-\mathrm{i}\sum_{\mathbf{n}}C_{\mathbf{n}}^{*}\frac{\partial C_{\mathbf{n}}}{\partial t} \tag{17.9}$$

$$\left(\mathrm{i}\frac{\partial}{\partial t}\right)_{kq} = \mathrm{i}\int\phi_k^{*}(\mathbf{x},t)\frac{\partial\phi_q(\mathbf{x},t)}{\partial t}\,\mathrm{d}\mathbf{x}$$

The matrix elements of $(\mathrm{i}\partial/\partial t)$ with respect to the $\{\phi_k(\mathbf{x},t)\}$ reflect the fact that, when acting on the orbitals, the time derivative can be viewed as a one-body operator. Examining representation (17.9) shows that the only explicit dependence on the orbitals is grouped into the matrix elements h_{kq}, $(\mathrm{i}\partial/\partial t)_{kq}$ and W_{ksql}, whereas the elements ρ_{kq} and ρ_{kslq} of the reduced one- and two-body density matrices do not depend explicitly on the orbitals.

It is now straightforward to perform the variation of $S[\{C_{\mathbf{n}}(t)\},\{\phi_k(\mathbf{x},t)\}]$ with respect to $\{\phi_k(\mathbf{x},t)\}$. Making use of the orbitals' orthonormality condition to solve for the Lagrange multipliers and then eliminating them by introducing a projection operator,

$$\mu_{kj}(t) = \sum_{q=1}^{M}\left\{\rho_{kq}\left[h_{jq}-\left(\mathrm{i}\frac{\partial}{\partial t}\right)_{jq}\right]+\sum_{s,l=1}^{M}\rho_{kslq}W_{jsql}\right\}$$

$$\hat{\mathbf{P}} = \sum_{j'=1}^{M}|\phi_{j'}\rangle\langle\phi_{j'}| \tag{17.10}$$

we arrive at the following set of equations of motion for the orbitals, with $j=1,\ldots,M$:

$$(1-\hat{\mathbf{P}})\mathrm{i}|\dot{\phi}_j\rangle = (1-\hat{\mathbf{P}})\left(\hat{h}|\phi_j\rangle+\sum_{k,s,l,q=1}^{M}\{\boldsymbol{\rho}(t)\}_{jk}^{-1}\rho_{kslq}\hat{W}_{sl}|\phi_q\rangle\right) \tag{17.11}$$

where

$$\hat{W}_{sl}(\mathbf{x},t) = \int\phi_s^{*}(\mathbf{x}',t)\hat{W}(\mathbf{x},\mathbf{x}')\phi_l(\mathbf{x}',t)\,\mathrm{d}\mathbf{x}'$$

are time-dependent, local (for spin-independent interactions) one-body potentials and $\dot{\phi}_j\equiv\partial\phi_j/\partial t$. Note that at this point the projection operator $(1-\hat{\mathbf{P}})$ appears on both sides of the equations of motion (17.11). We remark that the projection operator $(1-\hat{\mathbf{P}})$ is denoted in our Refs. [265, 266, 268] by $\hat{\mathbf{P}}$.

The equation of motion for the MCTDH expansion coefficients was discussed in previous chapters. For completeness, in the context of the MCTDH approach for identical particles, we discuss it here in brief. The variation of $S[\{C_{\mathbf{n}}(t)\}, \{\phi_k(\mathbf{x}, t)\}]$ with respect to the $\{C_{\mathbf{n}}(t)\}$ is easily done after expressing the expectation value $\langle\Psi(t)|\hat{H} - \mathrm{i}\partial/\partial t|\Psi(t)\rangle$ in a form that explicitly depends on the coefficients:

$$\left\langle \Psi \middle| \hat{H} - \mathrm{i}\frac{\partial}{\partial t} \middle| \Psi \right\rangle = \sum_{\mathbf{n}} C_{\mathbf{n}}^* \left[\sum_{\mathbf{n}'} \left\langle \mathbf{n}; t \middle| \hat{H} - \mathrm{i}\frac{\partial}{\partial t} \middle| \mathbf{n}'; t \right\rangle C_{\mathbf{n}'} - \mathrm{i}\frac{\partial C_{\mathbf{n}}}{\partial t} \right] \tag{17.12}$$

The following equation of motion then emerges straightforwardly:

$$\boldsymbol{\mathcal{H}}(t)\mathbf{C}(t) = \mathrm{i}\frac{\partial \mathbf{C}(t)}{\partial t}, \qquad \mathcal{H}_{\mathbf{n}\mathbf{n}'}(t) = \left\langle \mathbf{n}; t \middle| \hat{H} - \mathrm{i}\frac{\partial}{\partial t} \middle| \mathbf{n}'; t \right\rangle \tag{17.13}$$

where the vector $\mathbf{C}(t)$ collects together the coefficients $\{C_{\mathbf{n}}(t)\}$.

So far, only the orthonormality of the orbitals has been utilized to obtain the equations of motion for the orbitals and coefficients. It is well known and discussed in the previous chapters that the multiconfigurational wavefunction $\Psi(t)$ is invariant to unitary transformations of the orbitals, compensated by the 'reverse' transformations of the expansion coefficients. This invariance allows one to simplify the equations of motion and, in particular, their numerical integration by introducing the differential condition [30]:

$$\langle \phi_k | \dot{\phi}_q \rangle = 0 \qquad \text{for } k, q = 1, \ldots, M \tag{17.14}$$

(For the more general form of the differential condition, see [34, 37] and the previous chapters. For an explicit proof that the differential condition (17.14) can always be satisfied given an initial set of orbitals satisfying $\int \mathrm{d}\mathbf{x}\, \phi_k^*(\mathbf{x}, 0)\phi_q(\mathbf{x}, 0) = \delta_{kq}$, see [266].) With the differential condition (17.14), the equations of motion for the orbitals and expansion coefficients can be simplified to read, with $j = 1, \ldots, M$:

$$\mathrm{i}|\dot{\phi}_j\rangle = (1 - \hat{\mathbf{P}}) \left[\hat{h}|\phi_j\rangle + \sum_{k,s,l,q=1}^{M} \{\boldsymbol{\rho}(t)\}_{jk}^{-1} \rho_{kslq} \hat{W}_{sl} |\phi_q\rangle \right]$$

$$\mathbf{H}(t)\mathbf{C}(t) = \mathrm{i}\frac{\partial \mathbf{C}(t)}{\partial t}, \qquad H_{\mathbf{n}\mathbf{n}'}(t) = \langle \mathbf{n}; t | \hat{H} | \mathbf{n}'; t \rangle \tag{17.15}$$

The coupled equation sets (17.11) for the orbitals $\{\phi_j(\mathbf{x}, t)\}$ and (17.13) for the expansion coefficients $\{C_{\mathbf{n}}(t)\}$, or, respectively, Equation (17.15) constitute a unified and compact representation of the MCTDH approach for identical particles (also see [268]), either bosons (MCTDHB; [265,266]) or fermions (MCTDHF). We note that the form of MCTDHF in the literature [242,247,267] did not exploit *reduced two-body density matrices*.

Before we proceed to apply MCTDH to other systems, it is instructive to consider the stationary, self-consistent version of MCTDH for identical particles. In the spirit of Refs. [38, 261], the latter is readily arrived at by setting $t \to -it$ in the equations of motion (17.11) and (17.13) or in (17.15). When this is done, one arrives at the multiconfigurational self-consistent (time-independent) working equations, with $k = 1, \ldots, M$:

$$\sum_{j=1}^{M} \left[\rho_{kj}\hat{h} + \sum_{s,l=1}^{M} \rho_{kslj}\hat{W}_{sl} \right] |\phi_j\rangle = \sum_{j=1}^{M} \mu_{kj}|\phi_j\rangle$$

$$\mathbf{HC} = \varepsilon\mathbf{C} \tag{17.16}$$

where $\varepsilon = \langle\Psi|\hat{H}|\Psi\rangle$ is the eigenenergy of the system. The values of the Lagrange multipliers can be read from (17.10). Making use of the fact that the matrix of Lagrange multipliers $\{\mu_{kj}\}$ is Hermitian (for stationary states) and of the invariance property of multiconfigurational wavefunctions discussed in the previous chapters and above, one can transform (17.16) to a representation where $\{\mu_{kj}\}$ is a diagonal matrix. Equation (17.16) constitutes a unified representation of the literature multiconfigurational self-consistent field theories for bosons [261] and for fermions [271, 272].

17.3 Bose–Bose, Fermi–Fermi and Bose–Fermi Mixtures: MCTDH-BB, MCTDH-FF and MCTDH-BF

In this section we consider a mixture of two kinds of identical particles, and discuss the corresponding specification of the MCTDH approach. There are three possible mixtures (Bose–Bose, Fermi–Fermi and Bose–Fermi), and the resulting MCTDH-BB, MCTDH-FF and MCTDH-BF approaches are presented below in a unified manner based, as in the previous section, on the second-quantization formalism and reduced density matrices [273]. Aiming at an economical account, we present here only the additional ingredients to the single-species case, followed directly by the resulting equations of motion. The reader is referred to Ref. [273] for more details.

17.3.1 Ingredients for Mixtures

We consider a mixture with $N = N_\mathrm{A} + N_\mathrm{B}$ particles. The mixture consists of N_A identical particles (bosons or fermions) of type A and N_B identical particles (bosons or fermions) of type B. In what follows, when needed, we denote quantities in the mixture by A, B or AB superscripts.

For the A and B species we need two field operators, each expanded by a complete set of time-dependent orbitals:

$$\hat{\mathbf{\Psi}}^{(A)}(\mathbf{x}) = \sum_k \hat{a}_k(t)\phi_k(\mathbf{x},t), \qquad \hat{\mathbf{\Psi}}^{(B)}(\mathbf{y}) = \sum_{\bar{k}} \hat{b}_{\bar{k}}(t)\psi_{\bar{k}}(\mathbf{y},t) \tag{17.17}$$

Note that orbitals belonging to different species need not be orthogonal or *a priori* have any relation with respect to one another. Also, field operators, annihilation and creation operators corresponding to different species commute. The many-body Hamiltonian of the mixture is conveniently written as a sum of three terms:

$$\begin{aligned}
\hat{H}^{(AB)} &= \hat{H}^{(A)} + \hat{H}^{(B)} + \hat{W}^{(AB)} \\
\hat{W}^{(AB)} &= \sum_{k,\bar{k},q,\bar{q}} W^{(AB)}_{k\bar{k}q\bar{q}} \hat{a}_k^{\dagger}\hat{a}_q\hat{b}_{\bar{k}}^{\dagger}\hat{b}_{\bar{q}} \\
W^{(AB)}_{k\bar{k}q\bar{q}} &= \iint \phi_k^*(\mathbf{x},t)\psi_{\bar{k}}^*(\mathbf{y},t)\hat{W}^{(AB)}(\mathbf{x},\mathbf{y})\phi_q(\mathbf{x},t)\psi_{\bar{q}}(\mathbf{y},t)\,\mathbf{dx}\,\mathbf{dy}
\end{aligned} \tag{17.18}$$

The first two terms of $\hat{H}^{(AB)}$ are the A and B single-species Hamiltonians and can be read directly from Equation (17.3). The third term of $\hat{H}^{(AB)}$ is the interaction between the two species.

In the MCTDH approach for mixtures of identical particles, the *Ansatz* for the many-body wavefunction $\Psi^{(AB)}(t)$ is taken as a linear combination of products of time-dependent permutational symmetry-adapted configurations:

$$|\Psi^{(AB)}(t)\rangle = \sum_{\mathbf{n},\mathbf{m}} C_{\mathbf{nm}}(t)|\mathbf{n};t\rangle \times |\mathbf{m};t\rangle \equiv \sum_{\mathbf{n},\mathbf{m}} C_{\mathbf{nm}}(t)|\mathbf{n},\mathbf{m};t\rangle \tag{17.19}$$

The configurations $\{|\mathbf{n};t\rangle\}$ and $\{|\mathbf{m};t\rangle\}$ are either permanents or determinants depending on whether the mixture is a Bose–Bose, Fermi–Fermi or Bose–Fermi mixture, and can be read directly from Equation (17.4). The summation over **n**,**m** runs, respectively, over all possible configurations (permanents or determinants) generated by distributing N_A, N_B particles over M, $\bar{M}$ orbitals.

To derive the equations of motion, we will make use of the reduced density matrices of the mixture. In a mixture, see Equation (17.18), two types of reduced density matrices appear:

(i) the intra-species (A and B) reduced one- and two-body density matrices,

$$\rho^{(A)}(\mathbf{x}_1|\mathbf{x}_1';t), \qquad \rho^{(B)}(\mathbf{y}_1|\mathbf{y}_1';t)$$

and

$$\rho^{(A)}(\mathbf{x}_1,\mathbf{x}_2|\mathbf{x}_1',\mathbf{x}_2';t), \qquad \rho^{(B)}(\mathbf{y}_1,\mathbf{y}_2|\mathbf{y}_1',\mathbf{y}_2';t)$$

which can be read directly from Equations (17.6) and (17.7), respectively; and

(ii) the inter-species (AB) reduced two-body density matrix of $\Psi^{(AB)}(t)$, which is given by

$$\begin{aligned}
&\rho^{(AB)}(\mathbf{x}_1, \mathbf{y}_1|\mathbf{x}_1', \mathbf{y}_1'; t) \\
&= N_A N_B \int \Psi^{(AB)^*}(\mathbf{x}_1', \mathbf{x}_2, \ldots, \mathbf{x}_{N_A}, \mathbf{y}_1', \mathbf{y}_2, \ldots, \mathbf{y}_{N_B}; t) \\
&\quad \times \Psi^{(AB)}(\mathbf{x}_1, \mathbf{x}_2, \ldots, \mathbf{x}_{N_A}, \mathbf{y}_1, \mathbf{y}_2, \ldots, \mathbf{y}_{N_B}; t) \\
&\quad \times d\mathbf{x}_2\, d\mathbf{x}_3 \ldots d\mathbf{x}_{N_A}\, d\mathbf{y}_2\, d\mathbf{y}_3 \ldots d\mathbf{y}_{N_B} \\
&= \langle \Psi^{(AB)}(t)|\{\hat{\mathbf{\Psi}}^{(A)}(\mathbf{x}_1')\}^\dagger \hat{\mathbf{\Psi}}^{(A)}(\mathbf{x}_1)\{\hat{\mathbf{\Psi}}^{(B)}(\mathbf{y}_1')\}^\dagger \hat{\mathbf{\Psi}}^{(B)}(\mathbf{y}_1)|\Psi^{(AB)}(t)\rangle \\
&= \sum_{k,q=1}^{M} \sum_{\bar{k},\bar{q}=1}^{\bar{M}} \rho^{(AB)}_{k\bar{k}q\bar{q}}(t)\phi_k^*(\mathbf{x}_1', t)\phi_q(\mathbf{x}_1, t)\psi_{\bar{k}}^*(\mathbf{y}_1', t)\psi_{\bar{q}}(\mathbf{y}_1, t)
\end{aligned} \tag{17.20}$$

where

$$\rho^{(AB)}_{k\bar{k}q\bar{q}}(t) = \langle \Psi^{(AB)}(t)|\hat{a}_k^\dagger(t)\hat{a}_q(t)\hat{b}_{\bar{k}}^\dagger(t)\hat{b}_{\bar{q}}(t)|\Psi^{(AB)}(t)\rangle$$

are its matrix elements in the orbital basis.

17.3.2
Equations of Motion With Intra- and Inter-Species Reduced Density Matrices

The equations of motion of MCTDH-XY – the MCTDH approach for mixtures of two kinds of identical particles – utilize the above ingredients and are straightforwardly derived from the corresponding action functional

$$S[\{C_{\mathbf{nm}}(t)\}, \{\phi_k(\mathbf{x}, t)\}, \{\psi_{\bar{k}}(\mathbf{y}, t)\}]$$

and expectation value

$$\begin{aligned}
&\left\langle \Psi^{(AB)} \middle| \hat{H}^{(AB)} - i\frac{\partial}{\partial t} \middle| \Psi^{(AB)} \right\rangle \\
&= \sum_{k,q=1}^{M} \rho^{(A)}_{kq}\left[h^{(A)}_{kq} - \left(i\frac{\partial}{\partial t}\right)_{kq}\right] + \frac{1}{2}\sum_{k,s,l,q=1}^{M} \rho^{(A)}_{kslq} W^{(A)}_{ksql} \\
&\quad + \sum_{\bar{k},\bar{q}=1}^{\bar{M}} \rho^{(B)}_{\bar{k}\bar{q}}\left[h^{(B)}_{\bar{k}\bar{q}} - \left(i\frac{\partial}{\partial t}\right)_{\bar{k}\bar{q}}\right] + \frac{1}{2}\sum_{\bar{k},\bar{s},\bar{l},\bar{q}=1}^{\bar{M}} \rho^{(B)}_{\bar{k}\bar{s}\bar{l}\bar{q}} W^{(B)}_{\bar{k}\bar{s}\bar{q}\bar{l}} \\
&\quad + \sum_{k,q=1}^{M}\sum_{\bar{k},\bar{q}=1}^{\bar{M}} \rho^{(AB)}_{k\bar{k}q\bar{q}} W^{(AB)}_{k\bar{k}q\bar{q}} - i\sum_{\mathbf{nm}} C^*_{\mathbf{nm}} \frac{\partial C_{\mathbf{nm}}}{\partial t} \\
&= \sum_{\mathbf{n},\mathbf{m}} C^*_{\mathbf{nm}}\left[\sum_{\mathbf{n}',\mathbf{m}'}\left\langle \mathbf{n}, \mathbf{m}; t \middle| \hat{H}^{(AB)} - i\frac{\partial}{\partial t} \middle| \mathbf{n}', \mathbf{m}'; t\right\rangle C_{\mathbf{n}'\mathbf{m}'} - i\frac{\partial C_{\mathbf{nm}}}{\partial t}\right]
\end{aligned} \tag{17.21}$$

Eliminating the corresponding Lagrange multipliers by introducing projection operators,

$$\mu_{kj}^{(A)}(t) = \sum_{q=1}^{M} \left\{ \rho_{kq}^{(A)} \left[h_{jq}^{(A)} - \left(\mathrm{i}\frac{\partial}{\partial t} \right)_{jq} \right] + \sum_{s,l=1}^{M} \rho_{kslq}^{(A)} W_{jsql}^{(A)} + \sum_{\bar{k},\bar{q}=1}^{\bar{M}} \rho_{k\bar{k}q\bar{q}}^{(AB)} W_{j\bar{k}q\bar{q}}^{(AB)} \right\}$$

$$\mu_{\bar{k}\bar{j}}^{(B)}(t) = \sum_{\bar{q}=1}^{\bar{M}} \left\{ \rho_{\bar{k}\bar{q}}^{(B)} \left[h_{\bar{j}\bar{q}}^{(B)} - \left(\mathrm{i}\frac{\partial}{\partial t} \right)_{\bar{j}\bar{q}} \right] + \sum_{\bar{s},\bar{l}=1}^{\bar{M}} \rho_{\bar{k}\bar{s}\bar{l}\bar{q}}^{(B)} W_{\bar{j}\bar{s}\bar{q}\bar{l}}^{(B)} + \sum_{k,q=1}^{M} \rho_{k\bar{k}q\bar{q}}^{(AB)} W_{k\bar{j}q\bar{q}}^{(AB)} \right\}$$

$$\hat{\mathbf{P}}^{(A)} = \sum_{j'=1}^{M} |\phi_{j'}\rangle\langle\phi_{j'}|, \qquad \hat{\mathbf{P}}^{(B)} = \sum_{\bar{j}'=1}^{\bar{M}} |\psi_{\bar{j}'}\rangle\langle\psi_{\bar{j}'}| \tag{17.22}$$

and making explicit use of the differential conditions,

$$\langle\phi_k|\dot{\phi}_q\rangle = 0 \quad (k,q = 1,\ldots,M), \qquad \langle\psi_{\bar{k}}|\dot{\psi}_{\bar{q}}\rangle = 0 \quad (\bar{k},\bar{q} = 1,\ldots,\bar{M}) \tag{17.23}$$

the final result for the equations of motion of the orbitals $\{\phi_k(\mathbf{x},t)\}$ and $\{\psi_{\bar{k}}(\mathbf{y},t)\}$ and expansion coefficients $\mathbf{C}^{(AB)}(t) = \{C_{\mathbf{nm}}(t)\}$ reads, with $j = 1,\ldots,M$ and $\bar{j} = 1,\ldots,\bar{M}$:

$$\begin{aligned} \mathrm{i}|\dot{\phi}_j\rangle = (1-\hat{\mathbf{P}}^{(A)}) \Bigg[\hat{h}^{(A)}|\phi_j\rangle + \sum_{k,q=1}^{M} \{\boldsymbol{\rho}^{(A)}(t)\}_{jk}^{-1} \\ \times \Bigg\{ \sum_{s,l=1}^{M} \rho_{kslq}^{(A)} \hat{W}_{sl}^{(A)} + \sum_{\bar{k},\bar{q}=1}^{\bar{M}} \rho_{k\bar{k}q\bar{q}}^{(AB)} \hat{W}_{\bar{k}\bar{q}}^{(AB)} \Bigg\} |\phi_q\rangle \Bigg] \end{aligned} \tag{17.24}$$

$$\begin{aligned} \mathrm{i}|\dot{\psi}_{\bar{j}}\rangle = (1-\hat{\mathbf{P}}^{(B)}) \Bigg[\hat{h}^{(B)}|\psi_{\bar{j}}\rangle + \sum_{\bar{k},\bar{q}=1}^{\bar{M}} \{\boldsymbol{\rho}^{(B)}(t)\}_{\bar{j}\bar{k}}^{-1} \\ \times \Bigg\{ \sum_{\bar{s},\bar{l}=1}^{\bar{M}} \rho_{\bar{k}\bar{s}\bar{l}\bar{q}}^{(B)} \hat{W}_{\bar{s}\bar{l}}^{(B)} + \sum_{k,q=1}^{M} \rho_{k\bar{k}q\bar{q}}^{(AB)} \hat{W}_{kq}^{(BA)} \Bigg\} |\psi_{\bar{q}}\rangle \Bigg] \end{aligned} \tag{17.25}$$

$$\mathbf{H}^{(AB)}(t)\mathbf{C}^{(AB)}(t) = \mathrm{i}\frac{\partial \mathbf{C}^{(AB)}(t)}{\partial t} \tag{17.26}$$

$$H_{\mathbf{nm},\mathbf{n}'\mathbf{m}'}^{(AB)}(t) = \langle \mathbf{n},\mathbf{m};t|\hat{H}^{(AB)}|\mathbf{n}'\mathbf{m}';t\rangle \tag{17.27}$$

where

$$\hat{W}_{\bar{k}\bar{q}}^{(AB)}(\mathbf{x},t) = \int \psi_{\bar{k}}^*(\mathbf{y},t)\hat{W}^{(AB)}(\mathbf{x},\mathbf{y})\psi_{\bar{q}}(\mathbf{y},t)\,\mathrm{d}\mathbf{y}$$

are one-body potentials that the B species exerts on the A species, and

$$\hat{W}_{kq}^{(BA)}(\mathbf{y},t) = \int \phi_k^*(\mathbf{x},t)\hat{W}^{(AB)}(\mathbf{x},\mathbf{y})\phi_q(\mathbf{x},t)\,\mathrm{d}\mathbf{x}$$

are one-body potentials that the A species exerts on the B species. These inter-species potentials are local (for spin-independent interactions), time-dependent potentials. We remark that the projection operators $(1 - \hat{\mathbf{P}}^{(A)})$ and $(1 - \hat{\mathbf{P}}^{(B)})$ are denoted in our Ref. [273] by $\hat{\mathbf{P}}^{(A)}$ and $\hat{\mathbf{P}}^{(B)}$, respectively.

The coupled equations of motion (17.24)–(17.27) for the A species orbitals $\{\phi_k(\mathbf{x},t)\}$, the B species orbitals $\{\psi_{\bar{k}}(\mathbf{y},t)\}$, and the mixture's expansion coefficients $\{C_{\mathbf{nm}}(t)\}$ (also see [273]) constitute the MCTDH method for mixtures of identical particles (MCTDH-XY), which may be Bose–Bose (MCTDH-BB), Fermi–Fermi (MCTDH-FF) or Bose–Fermi (MCTDH-BF) mixtures.

Finally, in the spirit of Refs. [38, 261], the MCTDH-XY equations (17.24)–(17.27) can be transformed by setting $t \to -it$ therein to a multiconfigurational self-consistent theory for time-independent (stationary) mixtures. The resulting working equations readily read, with $k = 1, \ldots, M$ and $\bar{k} = 1, \ldots, \bar{M}$, respectively:

$$\sum_{j=1}^{M} \left[\rho_{kj}^{(A)} \hat{h}^{(A)} + \sum_{s,l=1}^{M} \rho_{kslj}^{(A)} \hat{W}_{sl}^{(A)} + \sum_{\bar{k},\bar{q}=1}^{\bar{M}} \rho_{k\bar{k}j\bar{q}}^{(AB)} \hat{W}_{\bar{k}\bar{q}}^{(AB)} \right] |\phi_j\rangle = \sum_{j=1}^{M} \mu_{kj}^{(A)} |\phi_j\rangle$$

$$\sum_{\bar{j}=1}^{\bar{M}} \left[\rho_{\bar{k}\bar{j}}^{(B)} \hat{h}^{(B)} + \sum_{\bar{s},\bar{l}=1}^{\bar{M}} \rho_{\bar{k}\bar{s}\bar{l}\bar{j}}^{(B)} \hat{W}_{\bar{s}\bar{l}}^{(B)} + \sum_{k,q=1}^{M} \rho_{k\bar{k}q\bar{j}}^{(AB)} \hat{W}_{kq}^{(BA)} \right] |\psi_{\bar{j}}\rangle = \sum_{\bar{j}=1}^{\bar{M}} \mu_{\bar{k}\bar{j}}^{(B)} |\psi_{\bar{j}}\rangle$$

$$\mathbf{H}^{(AB)} \mathbf{C}^{(AB)} = \varepsilon^{(AB)} \mathbf{C}^{(AB)} \tag{17.28}$$

where $\varepsilon^{(AB)} = \langle \Psi^{(AB)} | \hat{H}^{(AB)} | \Psi^{(AB)} \rangle$ is the eigenenergy of the mixture. The self-consistent equations (17.28) (also see [273]) generalize to mixtures the single-species equations (17.16), that is, the self-consistent field theories for bosons [261] and for fermions [271, 272].

17.4
Higher-Order Forces and Reduced Density Matrices: Three-Body Interactions

In the previous two sections we considered identical particles and mixtures of two kinds of identical particles interacting via two-body interactions. We note that the available formulations of MCTDHB [265,266] and MCTDHF [242,247, 267], as well as that of Ref. [268], are for identical particles interacting via two-body interactions. A two-body interaction is the most basic interaction in an interacting (quantum) system. When the particles comprising the quantum system have internal structure, higher-order interactions (forces) may come into play. For instance, in nuclear physics it has been accepted that three-body interactions are necessary to fully understand the structure of nuclei (see, for example, [274]). Much more recently, and in the context of another field, the proposition to utilize cold polar molecules to engineer (condensed-matter) systems with three-body interactions has been put forward [275].

Envisioning the interest in multiconfigurational time-dependent dynamics of systems with three-body forces, we extend the unified formulation of MCT-DHB and MCTDHF of Ref. [268] and Section 17.2 to include three-body interactions. Aiming at a short, yet concise account, we present here only the addition to the single-species, two-body case.

17.4.1 Ingredients for Three-Body Interactions

The many-body Hamiltonian $\hat{H}$ describing the N indistinguishable particles now comprises one-body $\hat{h}(\mathbf{x})$, two-body $\hat{W}(\mathbf{x},\mathbf{x}')$ and three-body $\hat{U}(\mathbf{x},\mathbf{x}',\mathbf{x}'')$ terms. In second quantization it is written as follows [276]:

$$\hat{H}^{(3B)} = \sum_{k,q} h_{kq}\hat{a}_k^\dagger\hat{a}_q + \frac{1}{2}\sum_{k,s,l,q} W_{ksql}\hat{a}_k^\dagger\hat{a}_s^\dagger\hat{a}_l\hat{a}_q + \frac{1}{6}\sum_{k,s,p,r,l,q} U_{kspqlr}\hat{a}_k^\dagger\hat{a}_s^\dagger\hat{a}_p^\dagger\hat{a}_r\hat{a}_l\hat{a}_q$$

$$U_{kspqlr} = \iiint \phi_k^*(\mathbf{x},t)\phi_s^*(\mathbf{x}',t)\phi_p^*(\mathbf{x}'',t)\hat{U}(\mathbf{x},\mathbf{x}',\mathbf{x}'') \times \phi_q(\mathbf{x},t)\phi_l(\mathbf{x}',t)\phi_r(\mathbf{x}'',t)\,d\mathbf{x}\,d\mathbf{x}'\,d\mathbf{x}'' \tag{17.29}$$

The matrix elements U_{kspqlr} are generally time-dependent quantities even if the three-body interaction $\hat{U}(\mathbf{x},\mathbf{x}',\mathbf{x}'')$ is time independent. Note also that the interaction terms $\hat{W}(\mathbf{x},\mathbf{x}')$ and $\hat{U}(\mathbf{x},\mathbf{x}',\mathbf{x}'')$ are required to be *symmetric* to permutations of any two particles, *independently* of whether the particles are bosons or fermions.

To derive the equations of motion, we will again make use of the reduced density matrices of the interacting system. In addition to the reduced one- and two-body density matrices, the reduced three-body density matrix is required. The reduced three-body density matrix of the indistinguishable-particle multiconfigurational wavefunction $\Psi(t)$, see Equation (17.4), reads

$$\begin{aligned}
&\rho(\mathbf{x}_1,\mathbf{x}_2,\mathbf{x}_3|\mathbf{x}_1',\mathbf{x}_2',\mathbf{x}_3';t) \\
&= N(N-1)(N-2)\int \Psi^*(\mathbf{x}_1',\mathbf{x}_2',\mathbf{x}_3',\mathbf{x}_4,\ldots,\mathbf{x}_N;t) \\
&\quad \times \Psi(\mathbf{x}_1,\mathbf{x}_2,\mathbf{x}_3,\mathbf{x}_4,\ldots,\mathbf{x}_N;t)\,d\mathbf{x}_4\cdots d\mathbf{x}_N \\
&= \langle\Psi(t)|\hat{\mathbf{\Psi}}^\dagger(\mathbf{x}_1')\hat{\mathbf{\Psi}}^\dagger(\mathbf{x}_2')\hat{\mathbf{\Psi}}^\dagger(\mathbf{x}_3')\hat{\mathbf{\Psi}}(\mathbf{x}_3)\hat{\mathbf{\Psi}}(\mathbf{x}_2)\hat{\mathbf{\Psi}}(\mathbf{x}_1)|\Psi(t)\rangle \\
&= \sum_{k,s,p,r,l,q=1}^{M} \rho_{ksprlq}(t)\phi_k^*(\mathbf{x}_1',t)\phi_s^*(\mathbf{x}_2',t)\phi_p^*(\mathbf{x}_3',t)\phi_r(\mathbf{x}_3,t)\phi_l(\mathbf{x}_2,t)\phi_q(\mathbf{x}_1,t)
\end{aligned} \tag{17.30}$$

where

$$\rho_{ksprlq}(t) = \langle\Psi(t)|\hat{a}_k^\dagger(t)\hat{a}_s^\dagger(t)a_p^\dagger(t)\hat{a}_r(t)\hat{a}_l(t)\hat{a}_q(t)|\Psi(t)\rangle$$

are its matrix elements in the orbital basis.

17.4.2 Equations of Motion With Three-Body Reduced Density Matrix

The equations of motion of MCTDHB and MCTDHF in the presence of three-body interactions can now be derived, and in a unified form. We start from the action functional $S[\{C_{\mathbf{n}}(t)\},\{\phi_k(\mathbf{x},t)\}]$, see Equation (17.8), and expectation value

$$\left\langle\Psi\left|\hat{H}^{(3B)} - \mathrm{i}\frac{\partial}{\partial t}\right|\Psi\right\rangle$$

$$= \sum_{k,q=1}^{M} \rho_{kq}\left[h_{kq} - \left(\mathrm{i}\frac{\partial}{\partial t}\right)_{kq}\right] + \frac{1}{2}\sum_{k,s,l,q=1}^{M} \rho_{kslq}W_{ksql}$$

$$+ \frac{1}{6}\sum_{k,s,p,r,l,q=1}^{M} \rho_{ksprlq}U_{kspqlr} - \mathrm{i}\sum_{\mathbf{n}} C_{\mathbf{n}}^* \frac{\partial C_{\mathbf{n}}}{\partial t} \tag{17.31}$$

Eliminating the Lagrange multipliers,

$$\mu_{kj}^{(3B)}(t)$$

$$= \sum_{q=1}^{M}\left\{\rho_{kq}\left[h_{jq} - \left(\mathrm{i}\frac{\partial}{\partial t}\right)_{jq}\right] + \sum_{s,l=1}^{M}\rho_{kslq}W_{jsql} + \frac{1}{2}\sum_{s,p,r,l=1}^{M}\rho_{ksprlq}U_{jspqlr}\right\} \tag{17.32}$$

and making use of the differential condition (17.14), the final result for the equations of motion of the the orbitals $\{\phi_k(\mathbf{x},t)\}$ and expansion coefficients $\{C_{\mathbf{nm}}(t)\}$ read, with $j = 1,\ldots,M$:

$$\mathrm{i}|\dot{\phi}_j\rangle = (1 - \hat{\mathbf{P}})$$

$$\times\left[\hat{h}|\phi_j\rangle + \sum_{k,s,l,q=1}^{M}\{\boldsymbol{\rho}(t)\}_{jk}^{-1}\left(\rho_{kslq}\hat{W}_{sl} + \frac{1}{2}\sum_{p,r=1}^{M}\rho_{ksprlq}\hat{U}_{splr}\right)|\phi_q\rangle\right] \tag{17.33}$$

$$\mathbf{H}^{(3B)}(t)\mathbf{C}(t) = \mathrm{i}\frac{\partial\mathbf{C}(t)}{\partial t} \tag{17.34}$$

$$H_{\mathbf{nn}'}^{(3B)}(t) = \langle\mathbf{n};t|\hat{H}^{(3B)}|\mathbf{n}';t\rangle \tag{17.35}$$

where

$$\hat{U}_{splr}(\mathbf{x}) = \iint \phi_s^*(\mathbf{x}',t)\phi_p^*(\mathbf{x}'',t)\hat{U}(\mathbf{x},\mathbf{x}',\mathbf{x}'')\phi_l(\mathbf{x}',t)\phi_r(\mathbf{x}'',t)\,\mathrm{d}\mathbf{x}'\,\mathrm{d}\mathbf{x}''$$

are local (for spin-independent interactions), time-dependent one-body potentials.

The coupled equation sets (17.33)–(17.35) for the orbitals $\{\phi_j(\mathbf{x},t)\}$ and expansion coefficients $\{C_{\mathbf{n}}(t)\}$ constitute a unified and compact representation of MCTDHB and MCTDHF for the dynamics of identical particles interacting via three-body forces.

Finally, by setting $t \to -it$ in (17.33)–(17.35), we arrive at a multiconfigurational self-consistent theory for stationary properties of identical-particle systems with three-body forces. The resulting working equations take on the form, with $k = 1, \ldots, M$:

$$\sum_{j=1}^{M}\left[\rho_{kj}\hat{h} + \sum_{s,l=1}^{M}\left(\rho_{kslj}\hat{W}_{sl} + \frac{1}{2}\sum_{p,r=1}^{M}\rho_{ksprlj}\hat{U}_{splr}\right)\right]|\phi_j\rangle = \sum_{j=1}^{M}\mu_{kj}^{(3B)}|\phi_j\rangle$$

$$\mathbf{H}^{(3B)}\mathbf{C} = \varepsilon^{(3B)}\mathbf{C} \tag{17.36}$$

where $\varepsilon^{(3B)} = \langle\Psi|\hat{H}^{(3B)}|\Psi\rangle$ is the eigenenergy of the system. The self-consistent equations (17.4.2) generalize the standard self-consistent field theories for bosons [261] and for fermions [271, 272], see Equations (17.16), to include three-body interactions.

17.5 Illustrative Numerical Examples for Bosons: MCTDHB

MCTDHB [265, 266] has so far been employed to explore several interesting problems dealing with the dynamics of interacting many-boson systems. These include: splitting of a Bose–Einstein condensate by a time-dependent barrier [265]; dynamics of fragmented condensates in double-well potentials [266]; formation of many-body fragmented states in attractive one-dimensional condensates [277]; and demonstration of build-up of coherence between initially independent subsystems (condensates) [278]. We would like to stress that MCTDHB has been implemented numerically as a separate code, independently of the Heidelberg MCTDH package [13]. For more details, see Ref. [266]. We mention that the MCTDH approach itself has been employed successfully and fruitfully to study dynamics [279, 280] as well as statics [281–283] of few-boson systems; also see Chapter 26.

In this section we would like to present two numerical applications of MCTDHB to (spinless) bosons in one dimension. It is convenient to choose a length scale L and to rescale the problem to dimensionless units. This is formally done by dividing the many-body Hamiltonian (17.3) by the energy unit $\hbar^2/mL^2$, where m is the mass of a boson. The one-body Hamiltonian then reads:

$$\hat{h}(x) = -\frac{1}{2}\frac{\partial^2}{\partial x^2} + \hat{V}(x)$$

As the two-body interaction, we take the popular contact interaction [260,284]

$$\hat{W}(x, x') = \lambda_0 \delta(x - x')$$

The first example deals with a comparative study of the dynamics of coherent and fragmented condensates in a double well; also see Ref. [266]. N bosons are prepared in the ground state of the double-well potential (with $\sigma = 2.6$)

$$\hat{V}_0(x) = \frac{x^2}{2\sigma^2} + 8\exp\left(-\frac{x^2}{2\sigma^2}\right)$$

Two systems are considered: the first with $N = 100$ bosons and interaction strength $\lambda_0 = 9.99/99 \approx 0.1$; and the second with $N = 1000$ bosons and interaction strength $\lambda_0 = 9.99/999 = 0.01$. The two systems are characterized by the same 'mean-field' factor $\lambda_0(N-1) = 9.99$, meaning that on the Gross–Pitaevskii mean-field level the dynamics of both systems are identical [260, 284].

The ground-state wavefunction $\Psi(0)$ is computed by imaginary-time propagation of the Gross–Pitaevskii equation and of the MCTDHB(2) equations of motion (17.15) (MCTDHB(2) means MCTDHB with $M = 2$ orbitals). At the Gross–Pitaevskii mean-field level, the condensate is always coherent, and all N bosons occupy one and the same orbital. At the many-body level, we find for both systems that the condensate is fully two-fold fragmented. In a fully two-fold fragmented system, the bosons occupy two orbitals (fragments). Here, due to the reflection symmetry of the initial potential, there are $\frac{1}{2}N$ bosons in each fragment. At time $t = 0$ we lower the barrier between the two wells and translate the whole potential to the left, $x \to x + 2$. The resulting double-well potential is given by (with $\sigma = 2.6$):

$$\hat{V}(x) = \frac{(x+2)^2}{2\sigma^2} + 4\exp\left(-\frac{(x+2)^2}{2\sigma^2}\right)$$

The many-boson wavefunction $\Psi(0)$ is allowed to evolve in time. The dynamics is computed by real-time propagation of the Gross–Pitaevskii and MCTDHB(2) equations of motion.

In Figure 17.1 we plot the time-dependent density $\rho(x, t) = \rho(x \mid x = x'; t)$ of each system computed from the time-dependent wavefunction $\Psi(t)$, see

Equation (17.6). Let us analyse the results. At $t = 0$ the corresponding stationary densities coincide because the barrier between the two potentials wells is initially quite high. At $t = 3.0$ differences start to mount. The mean-field density acquires density wiggles as a result of collisions with the double-well walls. The many-body densities for $N = 100$ and $N = 1000$ bosons, on the other hand, develop almost no density wiggles. Apart from being broader, the many-body densities at this moment in time resemble in shape the initial conditions, see Figure 17.1. Finally, at $t = 50.0$ the three densities are different from one another and from the initial conditions. In particular, the two systems with $N = 100$ and $N = 1000$ bosons, which at the mean-field level have the same dynamics, exhibit at the many-body level different time evolution. For more on the time evolution of fragmented condensates in this problem, see Ref. [266].

The second example that we present deals with splitting of a condensate by a time-dependent barrier; also see Ref. [265]. We consider $N = 200$ bosons with $\lambda_0 = 0.1$ initially prepared in the harmonic potential (with $\sigma = 2.6$):

$$\hat{V}_0(x) = \frac{x^2}{2\sigma^2}$$

At time $t = 0$ we ramp up a Gaussian barrier linearly in time up to a certain barrier height $V_{\text{max}} = 30.0$. The time-dependent trap potential is given by

$$\hat{V}(x,t) = \frac{x^2}{2\sigma^2} + V_{\text{max}} \exp\left(-\frac{x^2}{2\sigma^2}\right) \times \begin{cases} t/T_{\text{ramp}} & t \leqslant T_{\text{ramp}} \\ 1 & t > T_{\text{ramp}} \end{cases}$$

The resulting time evolution of the system is integrated by the MCTDHB(2) equations of motion. The MCTDHB with $M = 2$ orbitals was found to describe accurately the dynamics for the parameters considered here [265]. By fully optimizing the time-dependent expansion coefficients and orbitals in the multiconfigurational time-dependent expansion of $\Psi(t)$, we are able to go much farther in treating the problem of splitting a condensate than in Ref. [285].

To analyse the results, we resort to the reduced one-body density matrix of $\Psi(t)$. Diagonalizing the reduced one-body density matrix, we obtain its natural orbitals and natural occupation numbers:

$$\rho(x \mid x'; t) = \sum_{j=1}^{2} \rho_j(t) \phi_j^{*\text{NO}}(x', t) \phi_j^{\text{NO}}(x, t)$$

The corresponding time-dependent natural occupation numbers $\rho_1(t)$ and $\rho_2(t)$ are plotted in Figure 17.2 as a function of time. Two reference points of the system under investigation are: (i) the initial state is a slightly depleted condensate, $\rho_1(0) = 99.62\%$ and $\rho_2(0) = 0.38\%$; and (ii) at full barrier height,

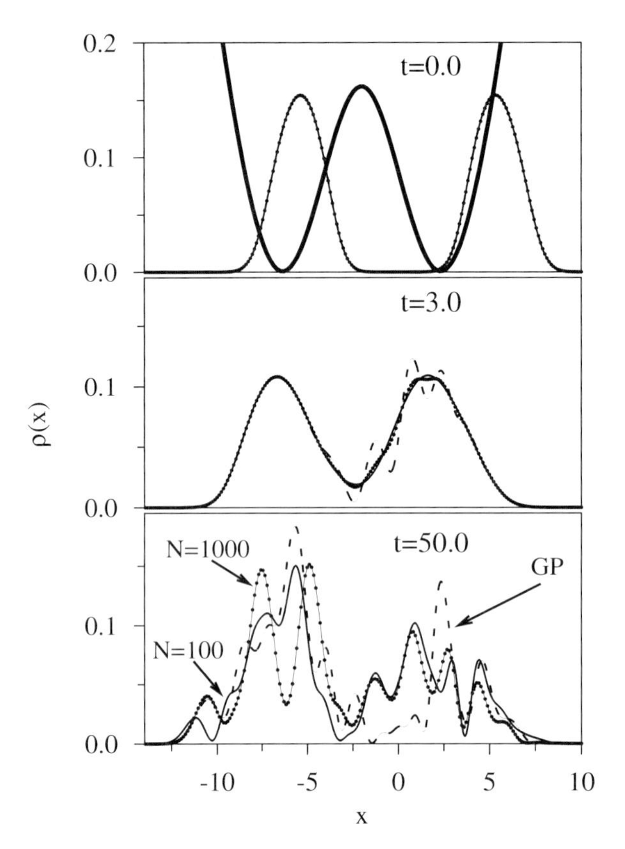

Figure 17.1 Comparison of the mean-field and many-body time evolution of two condensates made of $N = 100$ and $N = 1000$ bosons in a double-well potential with corresponding interaction strengths $\lambda_0 = 9.99/(N-1)$ computed by the Gross–Pitaevskii and MCTDHB(2) theories. Shown are three time snapshots of the density $\rho(x,t) = \rho(x \mid x = x'; t)$, normalized to 1: $N = 100$ *and* $N = 1000$ mean-field (dashed line); $N = 100$ many-body (solid line); and $N = 1000$ many-body (dotted–solid line). At $t = 0$ all densities coincide on the scale shown. To guide the eye, the double-well trap potential in which the condensates evolve is illustrated in the top panel (thick solid line). The quantities shown are dimensionless.

the ground state of the *stationary* system is a fully two-fold fragmented condensate, that is, $\rho_1 = 50.0\%$ and $\rho_2 = 50.0\%$.

The three panels shown in Figure 17.2 represent three numerical experiments for ramping-up times of $T_{\text{ramp}} = 25$, 3000 and 10 000. It is helpful to compare these ramping-up times with the period of the initial harmonic trap, which is given by $2\pi\sigma = 16.34$. Examining Figure 17.2 we see strong oscillations of the natural occupation numbers in time. The oscillations of $\rho_1(t)$ and $\rho_2(t)$ around the fully fragmented value of 50.0% indicate that, even when the barrier has reached its final height, the system *is not* in the ground state

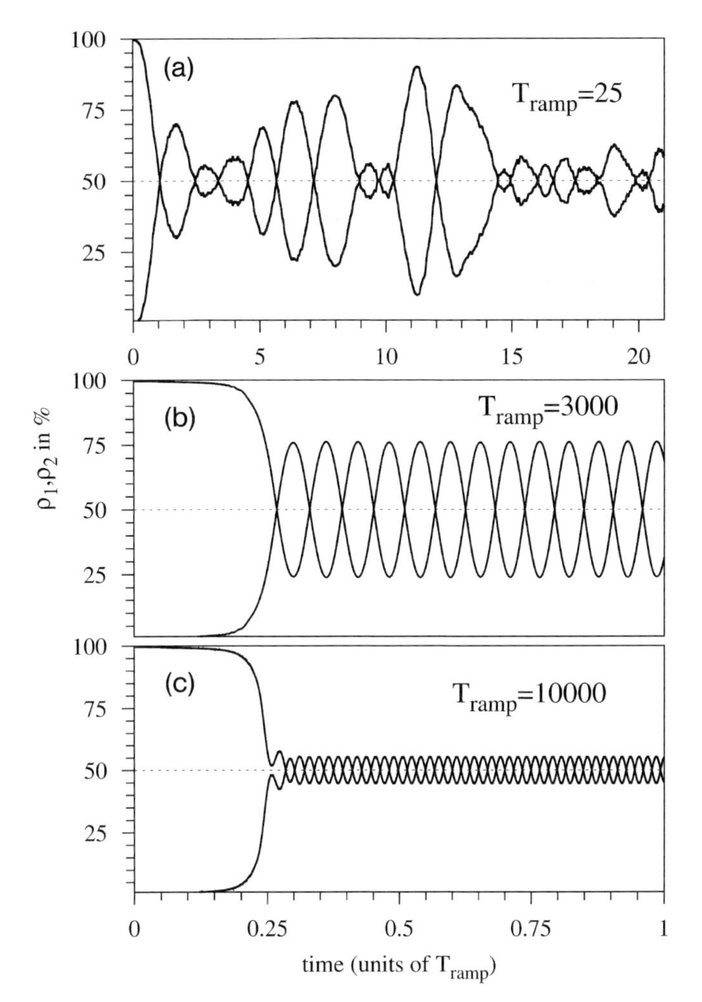

Figure 17.2 How difficult is it to fragment a condensate fully by ramping up a time-dependent barrier? Shown are the natural occupation numbers $\rho_1(t)$ and $\rho_2(t)$ of $N = 200$ bosons with $\lambda_0 = 0.1$ and for three ramp-up times T_{ramp}. The period of the initial harmonic trap is $2\pi\sigma = 16.34$. On increasing T_{ramp}, the time-dependent many-body state approaches the ground state of the double-well potential, which is a fully fragmented condensate, $\rho_1 = \rho_2 = 50.0\%$. When the ramp-up time is longer, the process of splitting a condensate can be made more adiabatic. Still, even for $T_{\text{ramp}} = 10\,000$, which is more than 600 times the period of the initial harmonic potential (and is of the order of the lifetime of a condensate), the amplitude of oscillations is about 10%. The quantities shown are dimensionless.

of the double well. Generally, the oscillations of $\rho_1(t)$ and $\rho_2(t)$ indicate that the many-body state $\Psi(t)$ involves the ground state *and* excited state(s) of the trapped many-body system.

Figure 17.2 tells us how difficult it is to split a condensate adiabatically. Clearly, when the ramping-up time is longer, the splitting process is made 'more' adiabatic. This is evident from the three panels of Figure 17.2, which demonstrate that: (i) the amplitude of oscillations becomes smaller for longer ramp-up times; and (ii) for longer ramp-up times, a single main frequency contributes to the oscillations, indicating that essentially only the lowest many-body excited state is coupled to the ground state. Yet, even for $T_{\mathrm{ramp}} = 10\,000$, which is more than 600 times the period of the initial harmonic potential (and is, in fact, of the order of the lifetime of a condensate), the final state deviates noticeably from the ground state, and the dynamical process of splitting a condensate is still not 'ideally adiabatic'. For more on the time evolution of a condensate halved by a time-dependent barrier and the role of excited states in this process, see Ref. [265].

17.6
Discussion and Perspectives

When the permutational symmetry of identical particles is taken into account, the equations of motion of the MCTDH approach simplify considerably. In particular, making explicit use of the fact that particle–particle interactions are commonly two-body interactions introduces into the time-dependent multi-configurational treatment of the dynamics *reduced two-body density matrices.* These developments have allowed us to solve numerically the many-body dynamics of 1000 identical bosons – a much larger number of particles than MCTDH is capable of.

We have presented in this chapter, in a unified form, the MCTDH approach for systems that comprise identical particles – bosons (MCTDHB) or fermions (MCTDHF). Similarly, the MCTDH approach for mixtures of two types of identical particles, Bose–Bose (MCTDH-BB), Fermi–Fermi (MCTDH-FF) and Bose–Fermi (MCTDH-BF) mixtures, has been presented in a unified form. We have explicitly considered the cases of two-body and three-body particle–particle interactions, which translate – with the help of second-quantization formalism – to reduced two-body and three-body density matrices, respectively. Continuing and combining these lines, if one thinks of a general system of interacting particles as a mixture of different kinds of particles interacting via several- or even many-body forces, one can reformulate MCTDH itself using reduced density matrices. Such an extension of the present formulation is straightforward. In the most general case of N distinguishable degrees of freedom coupled by N-body forces, the resulting equations of motion contain reduced N-body inter-species density matrices, such as Equation (17.20).

Returning to identical particles and mixtures thereof, the difference in the particles' statistics (Bose–Einstein or Fermi–Dirac) appears explicitly in our derivation in the first of the above equations, Equation (17.1), namely in the commutation or anticommutation relations between the field operators, annihilation and creation operators. The final equations of motion, for example, Equation (17.15) for identical bosons or fermions, do not depend explicitly on the particles' statistics. The difference in the particles' statistics translate only implicitly into the matrix elements appearing in the theory, which are: (i) the matrix elements of the reduced density matrices in the orbital basis; and (ii) the matrix elements of the Hamiltonian in the configuration basis. It turns out that even these matrix elements, the matrix elements of the Hamiltonian and separately the matrix elements of the reduced density matrices, can be formulated in a unified manner for bosons and fermions; see Refs. [268] and [273] for the respective treatment of single-species systems and mixtures. In particular, the famous Slater–Condon rules for evaluating matrix elements with determinants [272] and their bosonic analogue for permanents [261] can be written in a unified form [268].

The above formulations for single-species systems and mixtures of identical particles were made for particles with spin, that is, in terms of spin orbitals. This allows us to examine the relation between spin degrees of freedom and distinguishability, and point out its relevance to multiconfigurational dynamics (also see Ref. [273]). Specifically, it is well known that identical particles with different spin projections can be viewed and treated as distinguishable particles [276]. Consequently, and in our context, given a system of identical spin-S particles interacting via *spin-independent* forces (for simplicity, the initial-condition wavefunction should be an eigenfunction of the total spin operator and total-spin projection operator), we can treat the dynamics of the system in two different ways:

(i) with MCTDHB or MCTDHF, as a single-species spin-S particle system; or

(ii) as a mixture of, in general, $(2S+1)$ spinless species, with the corresponding MCTDH approach for mixtures.

The second possibility employs purely spatial orbitals, which saves unnecessary overheads in terms of summation over spin-S degrees of freedom.

The unified formulation of the MCTDH approach for systems of identical particles goes beyond the appearance and structure of the equations of motion. Let us concentrate for brevity on the single-species case. As is well known, the multiconfigurational *Ansatz* becomes exact when the number of orbitals M goes to ∞ (the orbitals then become time-independent; see previous chapters). In practical calculations, one has, of course, to restrict the number of orbitals included. Suppose that we take M' virtual orbitals on top of the minimal number of orbitals needed to accommodate N identical particles. For bosons $M = M' + 1$, whereas for fermions $M = M' + N$. With these observations, it can be shown that the sizes of the multiconfigurational Hilbert spaces of MCTDHB(M) *and* MCTDHF(M) are the same and given by [268,286]:

$$\text{number of configurations} = \binom{N + M'}{N} \tag{17.37}$$

Thus, in view of all the above unified properties, it is deductive and even practical to think of a single multiconfigurational code for systems of identical particles.

The motivation to devise strategies for Hilbert-space truncation methods beyond the multiconfigurational *Ansatz* of the many-body wavefunction, where *all* configurations generated by distributing N identical particles over M orbitals are taken into account, should be obvious. From the methodological point of view, configuration-interaction expansions scale very badly with the number of particles N and number of orbitals M, even when the orbitals in use are determined variationally or self-consistently. Therefore, devising strategies for how to go beyond the common multiconfigurational reasoning and further truncate the unnecessary or less necessary parts of the many-body Hilbert space is of practical importance as well as of basic interest.

Two natural strategies to devise approximation schemes stem from the above formulation of the equations of motion (see also [268]). The first strategy is simply to limit the number of configurations taken into account in the multiconfigurational expansion (17.4). (This strategy applies in particular to Equations (17.11) and (17.13), before the differential condition (17.14) is applied.) The second strategy stems from the appearance of reduced density matrices in the equations of motion, and amounts to replacing the equation of motion for the coefficients with equations of motion for the reduced density matrices themselves (in this context, see Ref. [257]), and solving the latter equations of motion *directly* with those of the orbitals. When the entire hierarchy of equations of motion for the reduced density matrices is included, no additional approximations have been introduced. When truncations of the equations of motion for the reduced density matrices are employed, one arrives at a new type of self-consistent-like propagation scheme. The merit of these propagation schemes can be evaluated by comparison to the 'exact' solution obtained within the 'N particles in M orbitals' multiconfigurational subspace.

Finally, in our opinion, the motivation to apply multiconfigurational treatments and Hilbert-space truncation schemes to the identical-particle quantum many-body problem should be high. From the perspective of the underlying physics, in particular for bosonic systems, even severe truncations of the many-body Hilbert space have led to very many valuable and substantial physical results. Gross–Pitaevskii theory, the simplest of such approximations, which employs only a single configuration where all the bosons of the same kind reside in the same orbital, is the most familiar and best-documented example [260, 284]. More recently, going beyond the Gross–Pitaevskii mean field by allowing bosons to occupy a single, general configuration of the many-boson Hilbert space, the multi-orbital (best mean-field) approach [287–289] has generated several fundamental and interesting physical results [290–295]. All the above provide a strong and solid ground to continue attacking the quantum many-boson and many-body problems with multiconfigurational approaches in combination with reduced density matrices.

Part 3 Applications

Multidimensional Quantum Dynamics: MCTDH Theory and Applications.
Edited by Hans-Dieter Meyer, Fabien Gatti, and Graham A. Worth

ISBN: 978-3-527-32018-9

18
Multidimensional Non-Adiabatic Dynamics

Graham A. Worth, Horst Köppel, Etienne Gindensperger and Lorenz S. Cederbaum

18.1
Introduction

A class of problems that have been treated with particular success by the MCTDH method are those in which a conical intersection between potential energy surfaces dominates the dynamics. These non-adiabatic systems – so-called as they cannot be described by a single adiabatic potential energy surface – are able to undergo radiationless electronic state crossing on an ultrafast (femtosecond) time-scale [4,73]. The signature of conical intersections is found in many spectra, particularly photoelectron spectra [173, 205, 296], and they provide important pathways in photochemistry [171,297] in systems ranging from H_3 [298, 299] to biologically active chromophores such as retinal [300, 301] and DNA bases [302, 303].

Wavepacket dynamics simulations have a natural connection to detailed experiments using femtochemistry laser spectroscopy, and simulations are routinely required to aid the interpretation of these studies. The main hurdles for simulations are the size of the systems studied. Non-adiabatic phenomena are inherently multidimensional in nature, often with a number of vibrational modes coupled strongly to the electronic degree of freedom. A typical example is provided by the absorption spectrum of pyrazine ($C_4N_2H_4$). The conical intersection connecting the S_1 and S_2 states strongly couples five vibrational modes, and weakly couples the remaining 19 [62]. Not only is such a system too large to simulate using standard wavepacket dynamics methods, but there is also the problem of obtaining suitable potential energy surfaces.

A powerful, yet simple, description of coupled potential surfaces is provided by the vibronic coupling model Hamiltonian, which uses the correspondence between the adiabatic and diabatic pictures to full effect [173]. The adiabatic picture is that given by the clamped-nucleus Hamiltonian, with sets of energy-ordered potential energy surfaces provided by the electronic states. These surfaces are obtained directly from electronic structure calcula-

Multidimensional Quantum Dynamics: MCTDH Theory and Applications.
Edited by Hans-Dieter Meyer, Fabien Gatti, and Graham A. Worth

ISBN: 978-3-527-32018-9

tions. Coupling between these states is provided by nuclear momentum-like operators. The diabatic picture is one in which couplings are provided by potential-like operators. The potential energy surfaces can then be related to an electronic configuration, and so to chemical entities. The surfaces in the diabatic picture, as they are smooth, can be described by a low-order Taylor expansion.

The vibronic coupling model is described in Section 18.2. In addition to its simplicity, it is in the product form vital for the efficient application of the MCTDH algorithm (Section 3.3). In Section 18.3 details are given of why the MCTDH method is so suited for solving problems using the model Hamiltonian. After that, in Section 18.4 examples are given of problems treated. Finally, in Section 18.5 an extension of the basic model is provided that shows how to include a bath correctly as a set of effective modes.

18.2
The Vibronic Coupling Hamiltonian

The vibronic coupling model adopted uses the well-known concept of diabatic electronic states [304–306]. Contrary to the usual adiabatic electronic states, they are not – except for isolated points in nuclear coordinate space – eigenfunctions of the electronic Hamiltonian. Adiabatic electronic wavefunctions may have singular first derivatives of the nuclear coordinates, for example, at conical intersections of potential energy surfaces [4, 73, 173]. These important topological features have emerged as paradigms for non-adiabatic excited-state dynamics [4, 73, 171, 297]. They are thus difficult, if not impossible, to deal with in a quantum dynamics treatment in the adiabatic basis, because of diverging non-adiabatic – or derivative – coupling terms.

These singularities are removed by switching to a diabatic electronic basis, by a suitable orthogonal transformation. This is thus the method of choice for quantum dynamics calculations. To be sure, the derivative couplings cannot be entirely removed in this way [307], but the remaining terms are non-singular and usually considered negligible for practical purposes. Also, for our purposes they are neglected, which may be considered as part of the model assumptions adopted. The potential coupling terms appearing instead in the diabatic basis are expanded in a low-order Taylor series in some suitable displacement coordinates. This constitutes the multimode vibronic coupling approach [173], which is used here. For the general case of n interacting electronic states, we decompose the Hamiltonian into kinetic and potential energy parts, T_N and V_0, of some reference electronic state, and an $n \times n$ potential energy matrix $\mathbf{W}$, describing the changes in potential energy with respect to V_0

in the interacting manifold (**1** is the $n \times n$ unit matrix):

$$\hat{\mathbf{H}} = (T_N + V_0)\mathbf{1} + \mathbf{W} \tag{18.1}$$

The matrix elements of **W** are written as follows:

$$W_{nn}(\mathbf{Q}) = E_n + \sum_i k_i^{(n)} Q_i + \sum_{i,j} \gamma_{ij}^{(n)} Q_i Q_j + \cdots \tag{18.2}$$

$$W_{nn'}(\mathbf{Q}) = \sum_i \lambda_i^{(nn')} Q_i + \sum_{i,j} \mu_{ij}^{(nn')} Q_i Q_j + \cdots \quad (n \neq n') \tag{18.3}$$

The truncation of the Taylor series after the first-order or second-order terms (the latter being shown here) is coined the linear or quadratic vibronic coupling approach (LVC or QVC, respectively) [4,73,173]

In typical applications we consider a photoexcitation or photoionization process where T_N and V_0 relate to the initial electronic state (usually the ground state), described in the harmonic approximation. The Q_i in Equations (18.2, 18.3) are then the relevant dimensionless normal coordinates (harmonic frequencies ω_i) and we have

$$T_N = -\sum_i \frac{\omega_i}{2} \frac{\partial^2}{\partial Q_i^2}, \qquad V_0 = \sum_i \frac{\omega_i}{2} Q_i^2 \tag{18.4}$$

The quantities E_n appearing in Equation (18.2) have the meaning of vertical excitation or ionization energies, referring to the centre of the Franck–Condon zone, $\mathbf{Q} = \mathbf{0}$ (boldface denotes the vector of all coordinates). Because we take the diabatic and adiabatic basis states to coincide at this geometry, the E_n have no counterpart in the off-diagonal elements of Equation (18.3). The other parameters appearing in these expressions are called linear or quadratic coupling constants, in an obvious notation, either intra-state (for $n = n'$) or inter-state (for $n \neq n'$).

In molecules with symmetry elements, the latter can impose important restrictions on the modes appearing in the various summations of Equations (18.2, 18.3). These are relevant, in particular, for the linear coupling terms, for which they read:

$$\Gamma_n \otimes \Gamma_Q \otimes \Gamma_{n'} \supset \Gamma_A \tag{18.5}$$

Explicitly, a given vibrational mode with symmetry Γ_Q can couple electronic states with symmetries Γ_n and $\Gamma_{n'}$ in first order only if the direct product on the left-hand side of Equation (18.5) comprises the totally symmetric irreducible representation Γ_A of the point group in question. The generalization to the second-order terms should be apparent, though it is less restrictive. From Equation (18.5) one immediately deduces (given an Abelian point group) that for $n = n'$ only totally symmetric modes enter the Hamiltonian in first order. Thus – for electronic states of different symmetries – the intra-state and inter-state linear couplings are caused by different sets of modes [173]. This will indeed be the case for the examples below, as far as Abelian point groups are concerned. For non-Abelian point groups there may be electronic states degenerate by symmetry, and the above discussion has to be suitably generalized. That is, the direct product $\Gamma_n \otimes \Gamma_n$ has to be replaced by its symmetric counterpart, and the indices appearing in Equations (18.2) and (18.3) should be extended to cover also the various components of degenerate irreducible representations. Consequently, also non-totally symmetric modes may appear in the diagonal elements of Equation (18.2) in first order. This amounts to the Jahn–Teller effect, which is dominated by symmetry restrictions even more than for the case of Abelian point groups discussed above. For details, we refer to the large amount of literature in the field [308, 309].

Despite the importance of the diabatic basis for dynamical calculations, the adiabatic representation is useful at least in two different respects. First, the key features of the adiabatic potential energy surfaces, such as minima of crossing seams, double minima occurring at a reduced symmetry, and so on, are vital to interpreting essential features of the nuclear dynamics such as spectra and electronic populations [4, 173]. Second, as already mentioned in the introduction, the adiabatic surfaces are also needed to determine the various coupling constants entering Equations (18.2) and (18.3) from *ab initio* electronic structure calculations. The latter necessarily give adiabatic quantities, at least in a direct sense. The comparison of the adiabatic surfaces underlying Equations (18.2) and (18.3) with *ab initio* results thus allows the parameters such as coupling constants to be determined by requiring that the corresponding model surfaces reproduce the *ab initio* data as well as possible. For the linear intra-state couplings, particularly simple expressions can be given [173], since these are just the gradients of the potential energy surface with respect to the normal coordinates of the modes in question:

$$k_i^{(n)} = (\partial V_n / \partial Q_i)\big|_{\mathbf{Q}=0} \tag{18.6}$$

Similarly, for a two-state problem with a non-totally symmetric active mode (coordinate Q_u, frequency ω_u), the parabolic plus hyperbolic shape of the resulting adiabatic potential curves V_1 and V_2 [173]

$$V_{1,2} = \tfrac{1}{2}(E_1 + E_2) + \tfrac{1}{2}\omega_u Q_u^2 \pm \sqrt{\tfrac{1}{4}(E_1 - E_2)^2 + (\lambda Q_u)^2} \tag{18.7}$$

readily gives the following expression for the inter-state coupling constant:

$$\lambda = \sqrt{\frac{1}{8}\frac{\partial^2 (V_1 - V_2)^2}{\partial Q_u^2}\bigg|_{\mathbf{Q}=0}} \tag{18.8}$$

In more general situations, such as three states interacting through the same vibrational mode, the coupling constants may be determined by a least-squares fit of the model eigenvalues to electronic structure data [310]. A general fitting procedure for any size of system is also described below in Section 18.4.

We conclude this section by pointing out that the model nature of the Hamiltonian, Equations (18.2) and (18.3), and its potential energy surfaces, apparently introduces restrictions on the type of problem to be treated, for example, photochemical transformations [171,297]. More recently, an extension has been proposed and successfully applied, where the model has been used only for the adiabatic-to-diabatic mixing angle [4]. This so-called concept of regularized diabatic states [201, 202, 311] allows the treatment of general potential energy surfaces, but at the expense of losing the structural simplicity of the Hamiltonian. As pointed out above, and will become further apparent below, it is this structural simplicity, where all operators entering the Hamiltonian are simple products of the coordinates, which brings the MCTDH algorithm to full power. This would apparently no longer be the case with general potential energy surfaces appearing within the concept of regularized diabatic states. Therefore, in the applications presented below, we use the vibronic model in the original, direct form as expressed by the Hamiltonian (18.1)–(18.3). Despite the restricted form, it will become clear below that the model covers a rich variety of phenomena and can be applied to truly multidimensional problems.

18.3 Combining the Vibronic Coupling Model with MCTDH

Perhaps the easiest way to show why MCTDH and the vibronic coupling model Hamiltonian fit so well together is to look at an example. The calculation that really proved the potential of the MCTDH method was the calculation of the absorption spectrum of pyrazine explicitly including all 24 vibrational modes [62]. The first two bands of the absorption spectrum of this molecule provide a classic example of a conical intersection. The lower band has a well-defined vibrational structure, as expected for a bound state. The upper band is intense and fairly featureless [312]. This lack of structure was shown to be due to a conical intersection between the S_1 and S_2 states, which results in the short lifetime in the upper electronic state [313].

The pyrazine molecule has 24 vibrational modes. Its equilibrium geometry has a point group D_{2h} and the coupled S_1 and S_2 states have B_{3u} and B_{2u} symmetry, respectively. Thus the quadratic vibronic coupling model Hamiltonian can be written as

$$\hat{\mathbf{H}} = \sum_i \frac{\omega_i}{2}\left(-\frac{\partial^2}{\partial Q_i^2} + Q_i^2\right)\mathbf{1} + \begin{pmatrix} -\Delta & 0 \\ 0 & \Delta \end{pmatrix} + \sum_{i\in G_1} \begin{pmatrix} \kappa_i^{(1)} & 0 \\ 0 & \kappa_i^{(2)} \end{pmatrix} Q_i$$
$$+ \sum_{(i,j)\in G_2} \begin{pmatrix} \gamma_{i,j}^{(1)} & 0 \\ 0 & \gamma_{i,j}^{(2)} \end{pmatrix} Q_i Q_j + \sum_{i\in G_3} \begin{pmatrix} 0 & \lambda_i \\ \lambda_i & 0 \end{pmatrix} Q_i$$
$$+ \sum_{(i,j)\in G_4} \begin{pmatrix} 0 & \mu_{i,j} \\ \mu_{i,j} & 0 \end{pmatrix} Q_i Q_j \tag{18.9}$$

where G_1 are the five symmetric modes that appear linearly on the diagonal and G_3 is the b_{1g} mode that provides linear coupling between the two states. G_2 are the pairs of modes whose product is totally symmetric and so appear with quadratic and bilinear terms on the diagonal, and finally G_4 are the pairs of modes whose product has symmetry b_{1g} and thus provide bilinear coupling terms.

A four-mode model, including the coupling mode ν_{10a} and three of the a_{1g} modes, ν_{6a}, ν_1 and ν_{9a}, was shown by Domcke and co-workers to be able to reproduce the features of the S_2 spectrum using standard wavepacket dynamics calculations [313]. The envelope, however, was only reproduced by adding a phenomenological broadening to the spectrum, damping the autocorrelation function with a fast relaxation time of 30 fs. This must be due to the coupling between the four-mode 'system' and the 'bath' provided by the remaining 20 modes. Thus, a full calculation of the spectrum requires the model to be extended to second order, including quadratic and bilinear terms, and thus all 24 modes are involved.

The 174 parameters required for the second-order model were calculated by Raab *et al.* [62] using the simplest possible method for calculating electronic states, configuration interaction with single excitations (CIS). The spectrum was then obtained from the Fourier transform of the autocorrelation function (see Section 6.2.1), calculated using the MCTDH method. After minor adjustment of key parameters, the agreement with the experimental spectrum is seen to be very good (Figure 18.1(a)). Note that a small, 150 fs, damping has been added to the autocorrelation function to produce this spectrum to allow for the finite propagation time of the simulation.

The power of the MCTDH method can be seen in the fact that this calculation was possible at all using the hardware available at the time. The technical details of the basis sets used are summarized in Table 18.1. The rows correspond to different models studied: the Domcke four-mode model; a 12-mode model that augments these four modes with the remaining eight modes with g symmetry; and the full 24-mode system. Two 24-mode calculations are listed, with different numbers of SPFs. The second column details how the degrees

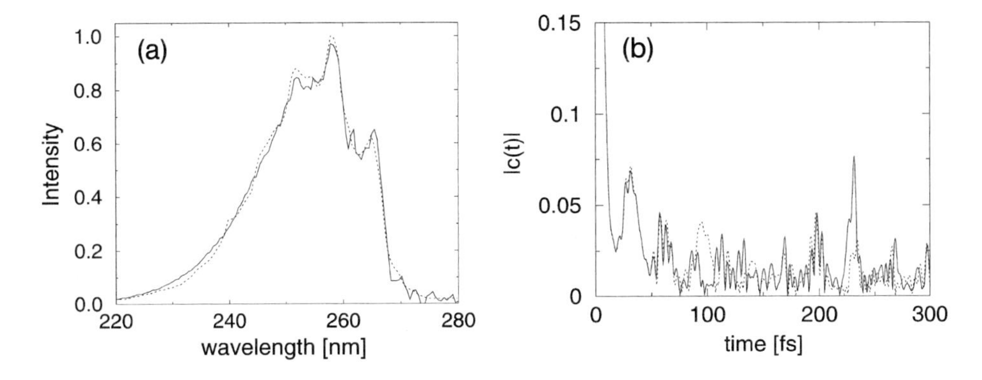

Figure 18.1 The absorption spectrum of pyrazine. (a) The experimental spectrum (dotted line) compared to that calculated using a 24-mode vibronic coupling model Hamiltonian and the MCTDH method with a large basis set (full line). (b) The calculated absolute value of the pyrazine autocorrelation function from two 24-mode MCTDH calculations using different basis sizes: large basis set (full line), small basis set (dotted line). See Table 18.1 for details.

of freedom were combined together to form multidimensional 'particles'. As discussed in Section 3.3, this keeps the length of the wavefunction expansion short. The four-mode calculation used four one-dimensional particles, that is, four sets of one-dimensional functions were used as the SPFs. The 12-mode calculation used five particles with, for example, the ν_{10a} and ν_{6a} combined together to give a two-dimensional particle. The 24-mode calculation used eight particles.

The wavefunction expansion length is the total single-particle function (SPF) basis size, given by the product of the number of SPFs per particle, summed over the two states. The numbers are given in column 3. For the four-mode and 12-mode calculations, the expansion length is 10 720 and 45 240, respectively. For the two 24-mode calculations, calculation I has a length of 502 200 and calculation II a length of 2771 440. The four-mode and 12-mode calculations are both converged with respect to the autocorrelation function, and hence the spectrum. A full test of convergence could not be made for the 24-mode calculations, but the number of SPFs for 24-mode II were chosen so that the population of the highest natural orbital was less than 0.01, suitable for averaged quantities. A comparison with the smaller 24-mode I calculation supports this. The autocorrelation function is shown in Figure 18.1(b), and the all-important first two peaks are nearly identical.

The SPFs need to be described by a primitive basis set. For this, a harmonic oscillator discrete variable representation (DVR) basis was used [34], which has been found to be very efficient for such bound-state problems. The numbers of functions required for each degree of freedom are given in column 4.

Table 18.1 Technical details of the MCTDH calculations of the absorption spectrum of pyrazine. From Ref. [62]. The round brackets denote the combination of vibrational modes; the square brackets denote the number of single-particle functions (SPFs) used for the representation of the wavefunction in the S_1 and S_2 states. The number of modes in one combination defines the dimensionality of the corresponding SPFs. These SPFs are represented on a grid whose size is given by the product of the number of grid points used for each mode of the corresponding combination.

Model	Combination of modes	Number of SPFs $[S_1, S_2]$	Number of grid points
4 mode	$\nu_{10a}, \nu_{6a}, \nu_1, \nu_{9a}$	[10, 8], [16, 10], [7, 6], [7, 6]	40, 32, 16, 12
12 mode	$(\nu_{10a}, \nu_{6a}), (\nu_1, \nu_{9a}),$ $(\nu_2, \nu_{6b}, \nu_{8b}), (\nu_4, \nu_5),$ $(\nu_{7b}, \nu_{8a}, \nu_3)$	[14, 11], [10, 8], [6, 6], [7, 6], [5, 5]	(40, 32), (20, 12), (4, 12, 24), (24, 8), (4, 8, 12)
24 mode I	$(\nu_{10a}, \nu_{6a}), (\nu_1, \nu_{9a}, \nu_{8a}),$ $(\nu_2, \nu_{6b}, \nu_{8b}), (\nu_4, \nu_5, \nu_3),$ $(\nu_{16a}, \nu_{12}, \nu_{13}), (\nu_{19b}, \nu_{18b}),$ $(\nu_{18a}, \nu_{14}, \nu_{19a}, \nu_{17a}),$ $(\nu_{20b}, \nu_{16b}, \nu_{11}, \nu_{7b})$	[12, 9], [6, 5], [4, 3], [5, 3] [4, 3], [6, 6] [4, 4], [3, 3]	(40, 32), (20, 12, 8), (4, 8, 24), (24, 8, 8), (24, 20, 4), (72, 80), (6, 20, 6, 6) (6, 32, 6, 4)
24 mode II	same as I	[14, 11], [8, 7], [6, 5], [6, 4], [4, 5], [7, 7], [5, 5], [3, 4]	same as I

Thus 40 DVR functions were used for the ν_{10a} mode, 32 for the ν_{6a} mode, and so on. The primitive basis size for each particle is given by the bracketed numbers. So, for example, the primitive basis size of the two-dimensional particle containing the ν_{10a} and ν_{6a} modes in the 12-mode calculation is 1280. For efficiency, it is important to keep the primitive basis sizes for the various particles similar in length.

The total primitive basis, that which would be required in a standard wave-packet calculation, is given by the product of the number of grid points for all modes. For the four-mode, 12-mode and 24-mode problems, these were, respectively, 245 760, 2.6×10^{13} and 6.4×10^{26}. The contraction efficiency of the MCTDH method is then the ratio of the MCTDH wavefunction expansion length to the primitive basis size. For these large calculations, the expansion length is clearly a much smaller number than the primitive basis.

Finally, we should mention the resources required for these calculations. For propagation lengths of 150 fs, the four-mode calculation required only 20 min on an IBM RS/6000 workstation and 16 MB memory. These are very cheap for a full four-dimensional quantum dynamics calculation. For the 12-

mode calculation on the same machine, this rose to 10 h and 45 MB memory. For the large 24-mode II calculation, a CRAY T90 vector machine was used and 485 h of CPU time were required with 650 MB memory. This is a substantial, but manageable, amount of time. The power of the method again can be seen in that the smaller, 24-mode I, calculation required only 100 h and 205 MB to produce a spectrum that is of good quality. Note that these calculations were performed in 1999. Repeating the 24-mode calculation on a modern PC takes a couple of days.

18.4 Examples

A number of systems have by now been treated using the vibronic coupling model combined with the MCTDH method. In the following, a few calculations are used to demonstrate the work.

18.4.1 Allene Cation

The calculation of the pyrazine absorption spectrum detailed above showed the importance of including second-order terms for a complete treatment. A further example where second-order terms must be included into the model to describe a spectrum correctly is found in the photoelectron spectrum of allene. The equilibrium structure of allene has the point group D_{2d}. Doubly degenerate states of the ion, labelled ^2E, are thus subject to $E \otimes \beta$ Jahn–Teller coupling, where the symmetry of the state is lowered by coupling to pairs of modes, one with B_1 and one with B_2 symmetry. The $\tilde{A}^2E$ state is further pseudo-Jahn–Teller coupled to the $\tilde{B}^2B_2$ state via the doubly degenerate E modes.

The photoelectron spectrum [314] for this coupled band shows a well-structured lower energy portion that could be explained by the Jahn–Teller coupled $\tilde{A}$ band with progressions from one symmetric stretch and one Jahn–Teller active mode [315]. Later work then assigned the diffuse higher part of the spectrum to the pseudo-Jahn–Teller coupled system [316]. However, the assignment of the lower part of the spectrum was found to be incompatible with the coupling when looking at all the possible modes, as there are three strongly coupled modes with relevant frequencies. The answer was that the second-order coupling between these modes leads to significant changes in the frequencies, by what is termed Duschinsky rotation. A simulation with all 15 modes and three states, while still not in perfect agreement with experiment, supports this [63].

An interesting feature of the allene cation system is that the doubly degenerate ground-state wavefunction can be written so that each component has a hole at different ends of the molecule. This molecule thus provides a simple model for charge transfer along a conjugated chain – starting in one component of the ground state is equivalent to removing an electron from one end of the molecule, and population transfer between the components then monitors the transfer. Owing to the vibronic coupling, this is found to be an ultrafast process [317].

The problem when including second-order terms is not only the increase in system size, but also the number of parameters that need to be determined. The linear model for pyrazine has 13 parameters and the second-order model has 174. Similarly, the linear model for allene has 25 parameters, and a further 16 second-order parameters, thought to be the most important, were added from the many possible. In these examples, the parameters were calculated by hand from information obtained at a few points on the potential energy surfaces using the formulae given in Section 18.2. This quickly becomes a very laborious task for more modes, the more so if many states are involved.

To deal with this fitting problem, an automated scheme has been set up and implemented as the VCHAM program [319], which is distributed with the MCTDH package [13]. This was first used to calculated the 79 parameters in a quadratic model of the butatriene cation [206]. The program sets up appropriate geometries for calculating the energies along cuts through the potential surfaces, collates the information, and then fits the parameters so that the model matches the calculated adiabatic surfaces. In a recent example, the VCHAM procedure has been used to obtain parameters for a vibronic coupling model of the lowest six excited states of benzene at the complete active space self-consistent field (CASSCF) level, revealing the different types of coupling present in these states [320].

The allene radical cation demonstrates the utility of this approach [318]. The surfaces for the $\tilde{A}^2E/\tilde{B}^2B_2$ coupled states, together with the $\tilde{C}^2A_1$, have been studied [318]. All four states are required for a good fit. Furthermore, it was found during the fitting process that satellite states also had to be included to get the form of the potentials along the low-frequency doubly degenerate modes that are important in the pseudo-Jahn–Teller coupling. Fourth-order terms were also required along some modes. The procedure also allowed the use of electronic structure methods for which analytic gradients are not available as only single-point energies are required.

The quality of the fits along the most important modes is shown in Figure 18.2. It is clear that, despite the simplicity of the model, it is able to describe the anharmonicity of the adiabatic surfaces extremely well. The model for the related, but larger, pentatetraene system has also been calculated and used to interpret the experimental spectrum [321].

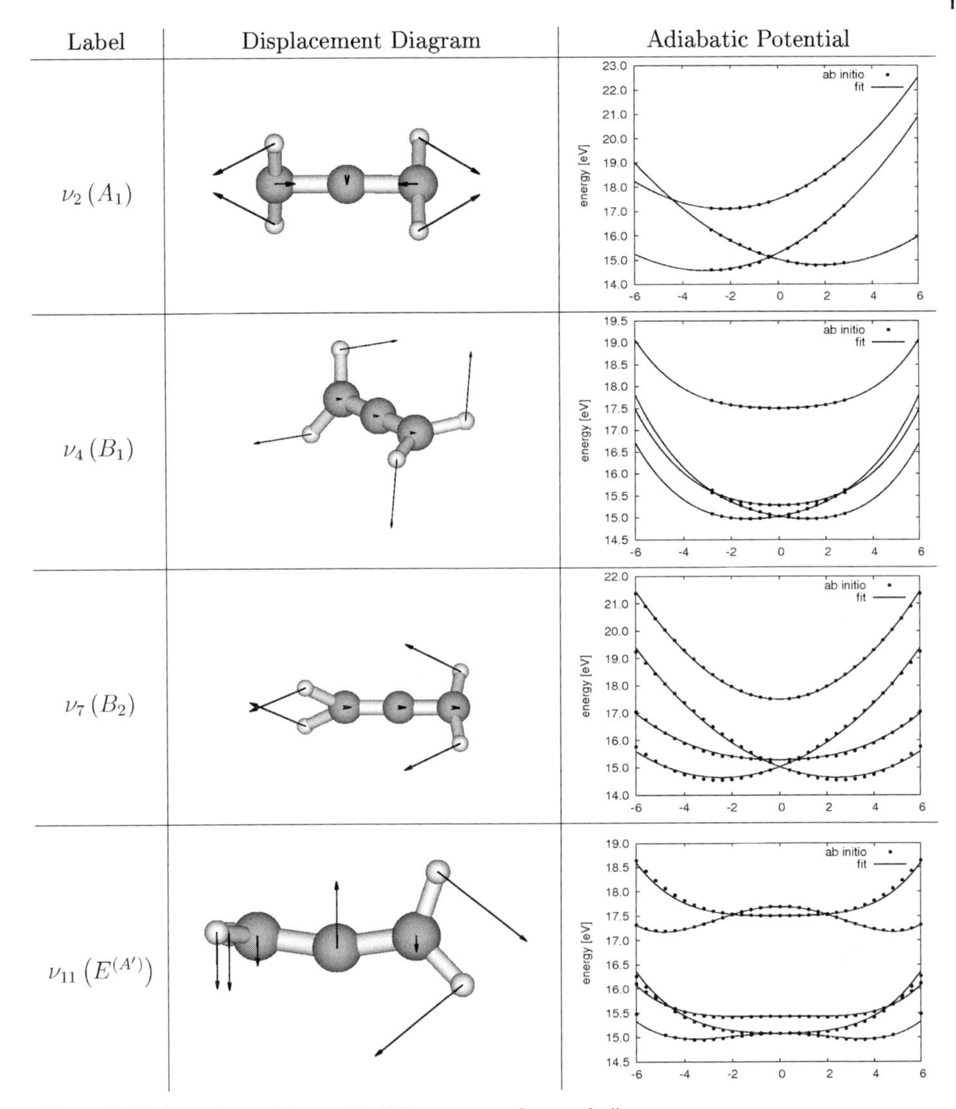

Figure 18.2 Cuts through the potential energy surfaces of allene along normal modes important for the non-adiabatic dynamics. The potentials show the data from *ab initio* calculations as points. The lines are the adiabatic surfaces from the vibronic coupling model Hamiltonian. Taken from [318].

18.4.2
$Cr(CO)_5$

By not focusing on just the intersection region, the fitting procedure also allows a better analysis of the global surfaces, and can lead to new findings. For example, the ground-state adiabatic surface of $Cr(CO)_5$ shows the moat and three minima typical of a second-order Jahn–Teller interaction in the ground state that is doubly degenerate at D_{3h} geometries [185]. On fitting the surfaces globally, however, the topology was actually found to be predominantly due to an $(E \oplus A) \otimes e$ pseudo-Jahn–Teller interaction between the ground state and the lowest singly degenerate excited state [322]. The dynamics in a pseudo-Jahn–Teller system are distinct from those in a Jahn–Teller system: in the latter a wavefunction propagated on the lower adiabatic surface is subject to the geometric phase effect, while in the former it is not [323]. This has consequences for the shape of the evolving wavepacket.

Calculations were performed using the three states and the five most important vibrations, namely the two doubly degenerate pairs that account for both the Jahn–Teller and the pseudo-Jahn–Teller coupling in addition to the symmetric breathing mode. The dynamics after forming the $Cr(CO)_5$ molecule by photodissociation of $Cr(CO)_6$ is shown in Figure 18.3. The results of two calculations are shown: including just the two strongest coupling modes, and including the five most important ones. The calculation of the adiabatic populations of the five-mode model is a huge job – requiring the multimode transformation operator on the full primitive grid (see Section 6.4), which has 1.02×10^{11} points. A Monte Carlo integration scheme was used for this.

The diabatic populations represent the population of the states with electronic wavefunctions dominated by chromium d-electron configurations $\Phi_{\tilde{X}} = d_{xz}^2 d_{yz}^2 d_{xy}^2$, $\Phi_{\tilde{A}} = d_{xz}^2 d_{yz}^2 d_{xy}^1 d_{x^2-y^2}^1$ and $\Phi_{\tilde{B}} = d_{xz}^2 d_{yz}^2 d_{x^2-y^2}^2$, respectively. The system starts in the diabatic $\tilde{A}$ state. In the two-mode calculation, after 100 fs there is a large transfer of population to both the other states. After another 200 fs there is a further transfer, after which little is left in the initial state. The transfer is similar, but less smooth, in the five-mode calculation. Earlier transfer is also seen, and the second transfer is weaker due to the spreading of the wavepacket in the larger available space reducing the effect of the recurrence.

The adiabatic states in this system are effectively $\tilde{\Phi}_{S_1} = \Phi_{\tilde{X}} - \Phi_{\tilde{B}}$, $\tilde{\Phi}_{S_2} = \Phi_{\tilde{A}}$ and $\tilde{\Phi}_{S_3} = \Phi_{\tilde{X}} + \Phi_{\tilde{B}}$. In these states, the population transfer in the two-mode model is more dramatic: it is effectively finished after 100 fs, having transferred 90% of the population to the ground state. In the five-mode model, the transfer out of S_2 is less, and the S_3 state becomes more populated.

Figure 18.4 plots snapshots of the adiabatic wavepacket motion over the ground and first excited states for this system. The plot is in the space of the doubly degenerate vibrational mode that has the strongest coupling. The potential energy surfaces for the lowest two adiabatic states are shown in the figure, with three minima on the lower state at C_{4v} symmetry, and three narrow minima on the upper state all due to the pseudo-Jahn–Teller coupling between the three diabatic states. The intersection between the states is at the centre of the plot at the D_{3h} geometry. This plane corresponds to pseudo-

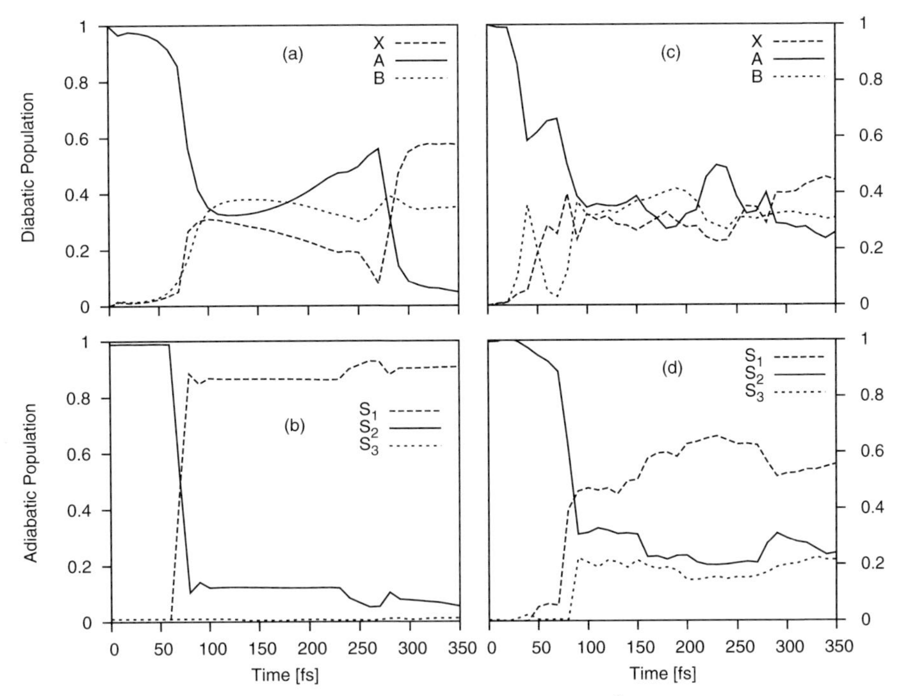

Figure 18.3 State populations of $Cr(CO)_5$ after formation in the $\tilde{A}$ state by photodissociation of $Cr(CO)_6$. The model included the lowest three electronic states and the most strongly coupled modes. (a) The diabatic state populations and (b) the adiabatic state populations including only two modes. (c) The diabatic state populations and (d) the adiabatic state populations including five modes.

rotation of the molecule: moving from minimum to minimum corresponds to a rearrangement of the three equatorial carbonyl groups [185].

The dynamics start on the first adiabatic excited state with a C_{4v} structure, distorted along the Q_2 mode. This initial condition is that formed by the sudden removal of a single carbonyl group, and the wavepacket at this time is taken to have the form of the undisturbed ground-state vibrational eigenfunction for the vibrations. After 80 fs the wavepacket has reached the D_{3h} geometry and population transfer to the ground state takes place. This bifurcates and returns to the D_{3h} centre after 240 fs. There is a small recurrence to the upper state seen in the adiabatic populations at this time. Finally, the wavepacket on the ground state reaches the right-hand side of the well after 340 fs. This time-scale fits the time-scale of coherent motion measured by Trushin *et al.* for this system [324]. Note that the wavepacket on the ground state in Figure 18.4 is symmetrical, in contrast to the plot in the original paper (Figure 8 in Ref. [322]). This was due to a plotting error in the analysis.

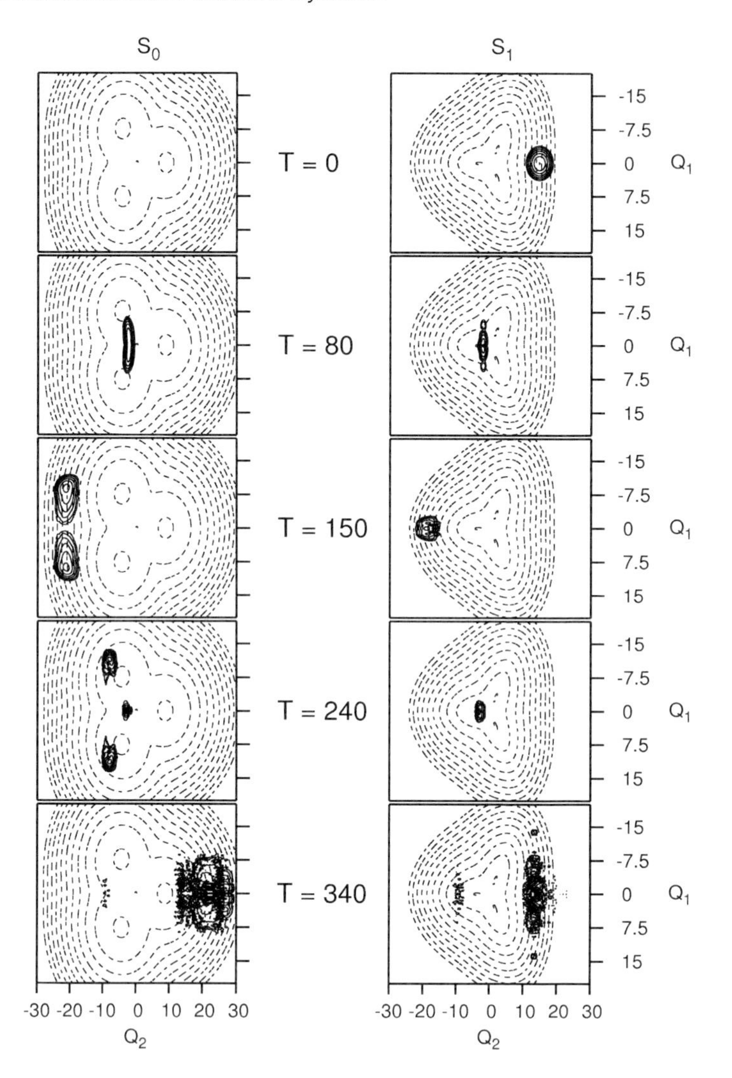

Figure 18.4 Snapshots of the adiabatic wavepacket of $Cr(CO)_5$ after formation from the photodissociation of $Cr(CO)_6$. The right-hand panel shows the upper state, S_1, and the left-hand panel the lower, S_0. The dotted lines are contours representing the adiabatic potential energy surfaces. The full contours represent the wavepacket density. The coordinates are the strongest coupling doubly degenerate modes.

18.4.3 Benzene Cation

The ability to follow the dynamics in a manifold of coupled states is exemplified by calculations on the benzene cation [220, 310]. The photoelectron spectrum of benzene has a number of bands in the region 9–20 eV [325]. The surfaces for the lowest five bands (eight states) have been fitted using the linear vibronic coupling model [310]. These states are all vibronically coupled and Figure 18.5 shows the coupling along an effective mode.

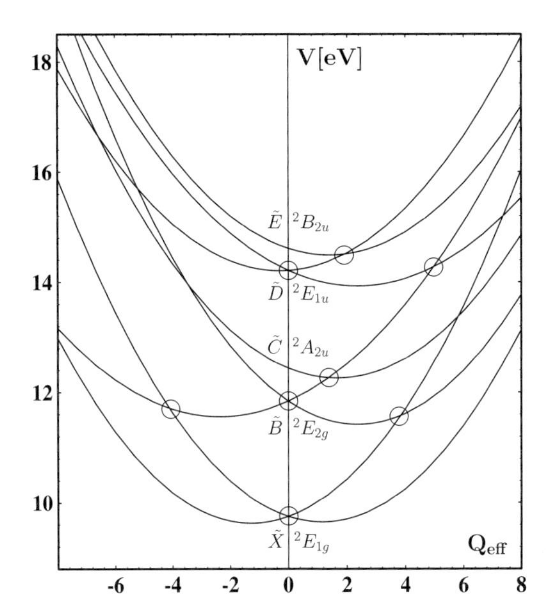

Figure 18.5 A schematic diagram of the lowest eight electronic states in the benzene radical cation shown as a cut along an effective mode. Conical intersections between the states are circled.

Large MCTDH calculations have shown that the model is able to reproduce the experimental spectrum [220]. These then allow a detailed analysis of the modes important for the system dynamics. Figure 18.6 shows the state populations after starting in the non-degenerate $\tilde{C}$ state. The modes required are the symmetric breathing mode and the doubly degenerate modes with e_{2g} symmetry that provide the Jahn–Teller coupling within the $\tilde{X}$ and $\tilde{B}$ states. The pseudo-Jahn–Teller coupling between the $\tilde{B}$ and $\tilde{C}$ states is provided by modes with e_{2u} symmetry, and that between the $\tilde{X}$ and $\tilde{B}$ states by modes with b_{2g} symmetry. An effective mode was used to model the coupling provided by the pair of modes with this symmetry.

The population is seen to decay rapidly from the $\tilde{C}$ state and after 200 fs the population is shared equally by the $\tilde{B}$ and $\tilde{X}$ states. Initial transfer occurs to the $\tilde{B}$ state, followed by transfer to the ground state. Similar results were obtained ignoring the degeneracy of the modes (Figure 18.6(b)). These findings are of relevance for the fluorescence dynamics of the benzene cation. They provide a pathway for ultrafast $\tilde{C} \rightarrow \tilde{X}$ non-radiative relaxation, and

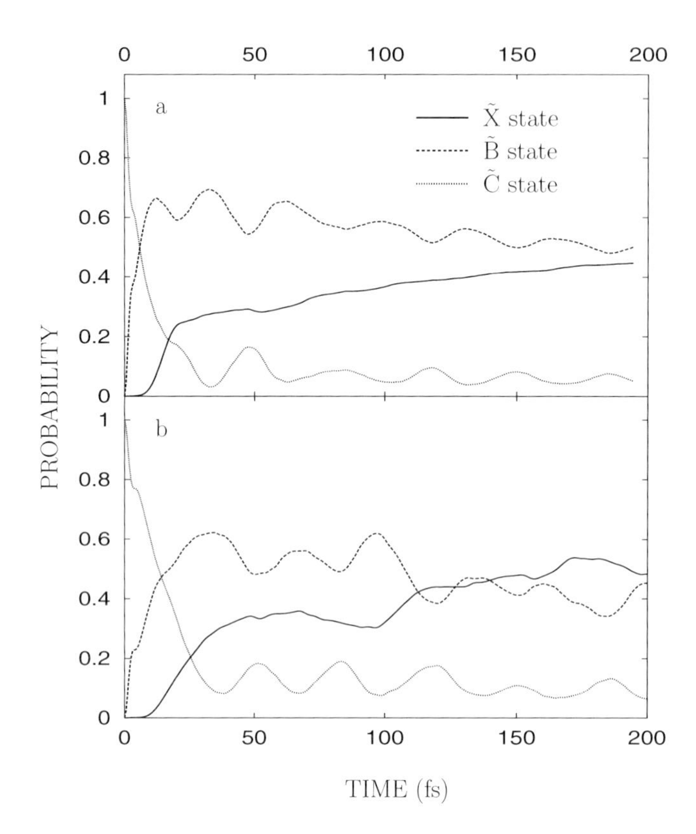

Figure 18.6 The population dynamics of the $\tilde{X}$–$\tilde{B}$–$\tilde{C}$ system of the benzene radical cation. (a) The symmetric mode ν_2 along with the degenerate modes ν_{16}, ν_{18} and ν_{19} (Herzberg numbering) as well as an effective mode with b_{2g} symmetry were included along with all five electronic states. The populations of $\tilde{B}$ and $\tilde{X}$ are the sum of the two components. (b) The same calculation, but treating all modes and states as non-degenerate (five modes and three states). Taken from [220]

thus explain the absence of emission in this system. Similar calculations have also been performed for the higher excited states of Figure 18.5 and have been related to the fragmentation dynamics of the cation [326]. Finally, the studies have been extended to the monofluoro derivative [327, 328] and also the three difluorobenzene isomers [329], and the characteristic changes observed experimentally for fluorination been reproduced and interpreted in this way.

18.5 Effective Modes

The vibronic coupling model is well suited for use with MCTDH. One can successfully treat problems involving roughly up to 30 modes in vibronic coupling situations with two electronic states, and even more when using the ML-MCTDH variant (see Chapter 14). However, the numerical effort becomes unaffordable when truly many modes participate in the dynamics, as is the case in extended molecular species or system–environment complexes subject to vibronic interactions. A way to reduce this effort is to construct suitable effective modes out of the original modes. By using only a few of these effective modes, the problem can be attacked.

The idea of constructing effective modes goes back to the late 1970s to the study of multimode Jahn–Teller situations [330, 331]. It has been introduced only recently for the general case of conical intersections [332] and used in quantum dynamical calculations with MCTDH. While some particular modes of the molecular complex, hereafter referred to as system modes, may require a treatment beyond the linear approximation (large-amplitude motion, strong anharmonicity), the linear vibronic coupling (LVC) model is generally sufficient for the vast majority of the other modes, so-called bath modes. These many bath modes render the explicit quantum dynamical treatment impossible and will be transformed to effective modes. The procedure for constructing the effective modes is detailed in Refs. [332–335].

The original Hamiltonian of the molecular complex reads

$$\hat{\mathbf{H}} = \hat{\mathbf{H}}_S + \hat{\mathbf{H}}_{LVC} \tag{18.10}$$

where $\hat{\mathbf{H}}_S$ stands for the Hamiltonian of the system part not restricted to the LVC model. The many (N) bath modes are collected in $\hat{\mathbf{H}}_{LVC}$, which has the form of Equation (18.1), neglecting the second-order coupling terms (the matrix of constant energy splitting is further transferred to the system part.) In this Hamiltonian, the bath modes tune or couple the electronic states, or may even do both in the absence of symmetry (assumed in the following). While the coupling strength varies from one bath mode to the other, they all play formally the same role and are coupled via the electronic subsystem only.

The idea underlying the effective-mode approach is to define a few modes that collect cumulative effects due to all the original modes, thus allowing pertinent approximations. These modes are basically defined by a rotation of the original modes, and the LVC Hamiltonian is transformed accordingly, leaving the physics unchanged. This has been done for the two-state LVC model by showing that only three effective modes participate directly in the coupling between the electronic states [332, 333]. These three modes can be shown to couple to three additional effective modes out of the $(N-3)$ remaining modes, which in turn couple to three other effective modes, and so on [334]. One can therefore build a *hierarchy of LVC effective Hamiltonians*, the members of which are made up of three effective modes only and chained up by bilinear couplings [334, 335]:

$$\hat{\mathbf{H}}_{\mathrm{LVC}} = \hat{\mathbf{H}}_1 + \sum_{k=2}^{n} H_k \hat{\mathbf{1}} \tag{18.11}$$

with the first effective Hamiltonian

$$\hat{\mathbf{H}}_1 = \sum_{i=1}^{3} \frac{\tilde{\omega}_i}{2}(\tilde{P}_i^2 + \tilde{X}_i^2)\hat{\mathbf{1}} + \sum_{i=1}^{3} \begin{pmatrix} K_i^{(1)}\tilde{Q}_i & \Lambda_i\tilde{Q}_i \\ \Lambda_i\tilde{Q}_i & K_i^{(2)}\tilde{Q}_i \end{pmatrix} \tag{18.12}$$

and the higher-order members, $k = 2, \ldots, n$,

$$H_k = \sum_{j=3(k-1)+1}^{3(k-1)+3} \frac{\tilde{\omega}_j}{2}(\tilde{P}_j^2 + \tilde{Q}_j^2) + \sum_{i=3(k-2)+1}^{3(k-2)+3} \sum_{j=3(k-1)+1}^{3(k-1)+3} d_{ij}(\tilde{P}_i\tilde{P}_j + \tilde{Q}_i\tilde{Q}_j) \tag{18.13}$$

The tildes over capital letters reflect the changes in the quantities upon rotation of the original modes – see Refs. [333–335] for details. Here n is the total number of effective Hamiltonians. It is readily seen that each effective Hamiltonian couples only to the nearest neighbours, like in a chain, in contrast with the original LVC Hamiltonian. Only $\hat{\mathbf{H}}_1$ (and $\hat{\mathbf{H}}_{\mathrm{S}}$) couples the electronic states; all the higher members of the hierarchy are scalar operators and do not couple directly to the system. See also Refs. [336–339] for a closely related perspective.

Having constructed the complete hierarchy, all modes are still present and the dynamics still cannot be computed. We shall then neglect some effective Hamiltonians, and keep only the number we want or can include in the calculation. While the transformation of the Hamiltonian leaves the physics unchanged, the truncation of the hierarchy apparently constitutes an approximation. In fact, it is not on a given time-scale! Indeed, it has been shown by a moment analysis of the autocorrelation function that truncating the hierarchy at the order k, inclusive, suffices to reproduce *numerically exactly* all the moments up to and including the order $2k + 1$ [335]. Thus, using only $\hat{\mathbf{H}}_{\mathrm{S}} + \hat{\mathbf{H}}_1$ suffices to reproduce the short-time dynamics of the entire complex, that is, using three modes instead of N for the bath. In the frequency domain, one gains access to the width and main asymmetry of the spectrum. It is worth noting that even these low-order moments are not exact if one neglects some modes in the original LVC Hamiltonian.

If time-scales beyond short times are of interest, one may include a few more effective Hamiltonians, extending the time-scale of accuracy and the affordable resolution of spectra. Since the dynamics in the presence of vibronic coupling is typically fast, the use of a few members of the hierarchy will suffice to describe the dynamics correctly. Regarding spectra, they are often so dense that they are difficult to resolve, and one mainly observes broad bands. Again, with only a few members of the hierarchy, these band shapes can be accurately calculated. The properties of the effective-mode approach have been shown mathematically [333,335], and also exemplified numerically for several cases [334, 336–338, 340, 341].

The new form of the LVC Hamiltonian is, like the original form, well suited for calculations with MCTDH. The product form is preserved, although now bilinear kinetic terms appear between the members of the hierarchy. Combined with the effective-mode approach of the LVC problem, MCTDH allows the efficient computation of the full quantum dynamics of complexes with an arbitrarily large initial number of modes over a given time-scale. Finally, this approach is not restricted to two electronic states, but remains valid when more states are involved, allowing the treatment of multistate conical intersections and cascading situations [335]. For n_{el} coupled electronic states, a maximum of $n_{el}(n_{el}+1)/2$ effective modes enters in each effective Hamiltonian.

18.6 Summary

The vibronic coupling model Hamiltonian is an excellent starting point for the study of photoinduced dynamics in which non-adiabatic effects play an important role. In its simplest form, the linear vibronic coupling model, it has been shown to describe correctly the dynamics of a system as it passes close to and through a conical intersection connecting different electronic states. Extensions to higher orders then add more details, important for describing the dynamics at longer time-scales and at geometries away from the intersection.

Fitting the model to the adiabatic surfaces from electronic structure calculations provides a suitable way of obtaining the diabatic surfaces and couplings required without the necessity of defining the diabatic functions themselves. In fact, the model is a good way of providing these functions. The use of fitting routines can produce the many required parameters in a semi-automatic way from a small number of single-point *ab initio* calculations at suitable points.

Non-adiabatic dynamics is by its nature multidimensional: often the motion of a number of modes is coupled. Here the MCTDH method has proved very successful in studying these systems. The form of the vibronic coupling model is automatically in the form required, and calculations including 10 or more modes and a number of electronic states are presently feasible. The systems studied to date have provided an invaluable insight into the dynamics of these systems. For other systems treated recently along similar lines, see, for example, singlet excited furan [342], the cyclobutadiene radical cation [343] and the cyclopropane radical cation [344].

On-going work will enable more modes to be treated with greater accuracy. The development of the effective-mode formalism will certainly help by providing a framework for the reduction of a huge system. By dividing the modes into a 'system' and a 'bath', the vibronic coupling model can be reformulated in a hierarchy of Hamiltonians, enabling an analysis of the important dynamics with a limited effort. A complementary development, not dealt with in this chapter, is the use of a Gaussian wavepacket basis in place of the flexible SPFs of the full MCTDH method (see Section 3.5). These are particularly effective for the bath modes. A recent study on the pyrazine absorption spectrum using this method shows its potential [194].

19
MCTDH Calculation of Flux Correlation Functions: Rates and Reaction Probabilities for Polyatomic Chemical Reactions

Fermín Huarte-Larrañaga and Uwe Manthe

19.1
Introduction

Probably the magnitude that interests most chemists when investigating a reaction is its thermal rate constant, $k(T)$. It is the basic observable linking the macroscopic variation of partial pressures or concentrations and the microscopic molecular collision domain. For an elementary reaction

$$\mathrm{A} + \mathrm{BC} \rightarrow \mathrm{AB} + \mathrm{C}$$

it is found from

$$\frac{\mathrm{d}[\mathrm{AB}]}{\mathrm{d}t} = k(T) \cdot [\mathrm{A}][\mathrm{BC}]. \tag{19.1}$$

The rate constant of such an elementary reaction can be obtained from first principles as the thermal average of the *cumulative reaction probability*, $N(E)$:

$$k(T) = \frac{1}{2\pi Q_\mathrm{r}(T)} \int_{-\infty}^{\infty} \mathrm{d}E \, e^{-E/k_\mathrm{B}T} N(E) \tag{19.2}$$

where $Q_\mathrm{r}(T)$ is the reactant partition function per unit volume (atomic units, $\hbar = 1$). The cumulative reaction probability (CRP) is the sum of all probabilities for any energy-accessible reactant state to react and end up in any energy-accessible product state. This magnitude can generally be obtained from a complete reactive scattering simulation of the particular reactive process. Such a strategy implies performing scattering calculations to obtain the complete S-matrix for a continuum of energies and then averaging all the state-to-state probabilities in the CRP. Accurate scattering calculations of the S-matrix can be a demanding and extremely delicate task, since it requires solving the Schrödinger equation starting from a situation where reactants are asymptotically apart and ending in the equivalent situation on the product side.

Multidimensional Quantum Dynamics: MCTDH Theory and Applications.
Edited by Hans-Dieter Meyer, Fabien Gatti, and Graham A. Worth

ISBN: 978-3-527-32018-9

This state-to-state based approach is inefficient if the purpose of the simulation is to obtain $N(E)$ or $k(T)$. In this case, it is the possibility of overcoming the reaction barrier that matters, and any information about the asymptotic states is irrelevant. If the reaction occurs through a reaction barrier and no long-lived complexes are present, the dynamics in the vicinity of the reaction barrier will determine the value of $k(T)$. Then, a dynamics simulation in this region can be employed to calculate the rate constant. This can be achieved by employing flux correlation functions, as will be explained in the following section. Such an approach requires a much smaller computational effort with respect to the full scattering calculations. Avoiding the propagation from the reaction barrier to the asymptotic channels has significant computational advantages: the region where the dynamics is simulated is significantly reduced and the simulation time is much shorter.

An example of this fast time convergence is the flux correlation function for the $CH_4 + H \rightarrow CH_3 + H_2$ reaction, shown in Figure 19.1. Regardless of the computational details, which are not given here, one clearly sees that the function reaches a plateau at about 20 fs. Thus, dynamical simulations on a short time-scale in the 10 fs range are sufficient to converge the flux correlation function rigorously and later to obtain the reaction rate.

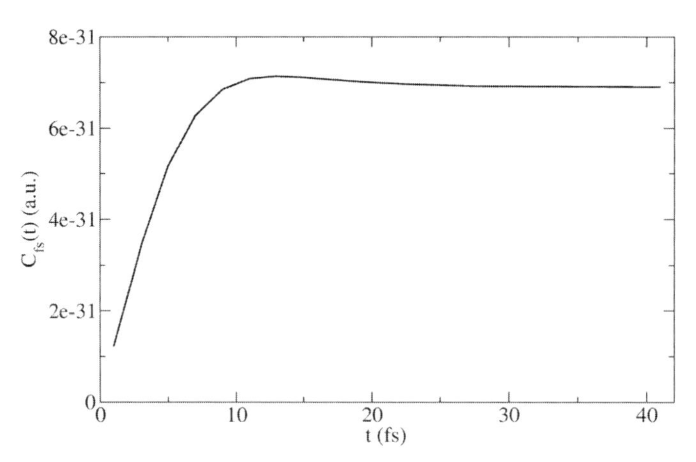

Figure 19.1 Time evolution of the flux position correlation function for the $H + CH_4 \rightarrow H_2 + CH_3$ reaction.

This short-time dynamics simulation scheme is particularly well suited for the MCTDH scheme. If a suitable coordinate system is chosen, the short simulation time often limits the correlation between the different degrees of freedom. In return, the capability of the MCTDH approach to treat multidimensional systems represents a key element in the efficiency of the method, which is reviewed here.

These numerical advantages have enabled *direct* quantum dynamics calculations of reaction rates to be the first step in the extension of the accurate first-principles theory of reactions to more complex, even polyatomic, systems. In this respect, the first quantum dynamics simulation of a four-atom reaction in its full dimensionality consisted in the *direct* calculation of the $H_2 + OH \rightarrow H_2O + H$ cumulative reaction probability [345]. Equivalent full-dimensional initial-state-selected [346, 347] and state-to-state [348] quantum simulations appeared shortly after. At that time, direct rate constant calculations were still based on the use of time-independent grids and did not utilize MCTDH propagation.

The combination with the MCTDH approach allowed a significant step further in terms of the complexity of the systems that could be accurately treated. For instance, this combined scheme enabled the accurate full-dimensional all-J thermal rate calculation of the $H_2 + OH$ benchmark reaction [349]. Probably the most significant achievement of the methodology reviewed here is the calculation of thermal rates for polyatomic reactions, such as hydrogen abstraction from methane: $X + CH_4 \rightarrow CH_3 + HX$. The dimensionality of these systems (12 internal degrees of freedom) is still beyond the capability of standard reactive scattering approaches. While the first accurate full-dimensional ($J = 0$) $k(T)$ calculations for $H + CH_4$ were published in 2000 [217], the first state-specific full-dimensional simulation of a five-atom reaction, $H_2 + C_2H$, did not appear until 2006 [350]. Thus, while the reach of accurate state-selective quantum simulations is still more limited, alternative approaches focusing on the accurate calculation of thermal reaction rates have advanced towards the treatment of more complex systems.

The purpose of this chapter is to give a detailed overview of the theory of direct and accurate reaction rate calculations, to discuss the use of the MCTDH approach within this context, to present recent applications studying polyatomic reactions, and to try to outline results of general relevance.

19.2 Flux Correlation Functions and Quantum Transition-State Concept

19.2.1 Thermal Rates From Flux Correlation Functions

Flux correlation functions represent an attractive, efficient and theoretically robust means of obtaining *directly* reaction rate constants or cumulative reaction probabilities, avoiding scattering (or half scattering) calculations. Originally mentioned by Yamamoto [351], the foundations of the present theory of flux correlation functions were grounded in 1974 in an article by Miller [352] in which the flux–position correlation function was derived from scattering theory and the quantum transition-state concept was discussed. Miller, Schwarz and Tromp [15] later extended this initial work and introduced a set of different flux correlation functions, from which the $k(T)$ of an elementary reaction can be calculated. Among them, the flux–flux autocorrelation function is particularly useful:

$$C_{\mathrm{f}}(t) = \mathrm{tr}(\hat{F}\, \mathrm{e}^{\mathrm{i}(\hat{H}t+\mathrm{i}\hat{H}\beta/2)}\, \hat{F}\, \mathrm{e}^{-\mathrm{i}(\hat{H}t-\mathrm{i}\hat{H}\beta/2)}) \tag{19.3}$$

(with $\beta = 1/k_{\mathrm{B}}T$). In the equation, $\hat{H}$ stands for the Hamiltonian, h is a function that equals unity on the product side of the dividing surface and vanishes on the reactant side, and $\hat{F} = \mathrm{i}[\hat{H}, h]$ is a flux operator that measures the flux through the dividing surface.

The thermal rate constant can be calculated from this flux correlation function as [15]:

$$k(T) = \frac{1}{Q_{\mathrm{r}}(T)} \int_0^{\infty} \mathrm{d}t\, C_{\mathrm{f}}(t) \tag{19.4}$$

Here $Q_{\mathrm{r}}(T)$ denotes the partition function of the reactants. The connection between this approach and the transition-state concept can be easily understood by considering the equivalent equation in classical mechanics (see, for example, [353]).

Combining the expression of $k(T)$ in terms of the energy-dependent $N(E)$ and Equation (19.4) in terms of the time-dependent $C_{\mathrm{f}}(t)$, one can derive an expression that allows the *direct* calculation of the CRP, that is, without any reference to the asymptotic states [15]. A convenient form of this expression reads

$$N(E) = \tfrac{1}{2}\, \mathrm{e}^{E/kT_0} \int_{-\infty}^{\infty} \mathrm{d}t \int_{-\infty}^{\infty} \mathrm{d}t'\, \mathrm{e}^{-\mathrm{i}E(t'-t)} C_{\mathrm{f}}(t, t'; T_0) \tag{19.5}$$

where $C_{\mathrm{f}}(t, t'; T_0)$ is a *generalized* flux correlation function [354]. This function can be written in its asymmetric form

$$C_{\mathrm{f}}(t, t'; T_0) = \mathrm{tr}(\hat{F}_{T_0}\, \mathrm{e}^{\mathrm{i}\hat{H}t'}\, \hat{F}'\, \mathrm{e}^{-\mathrm{i}\hat{H}t}) \tag{19.6}$$

or symmetric form

$$C_{\mathrm{f}}(t, t'; T_0) = \mathrm{tr}(\hat{F}_{2T_0}\, \mathrm{e}^{\mathrm{i}\hat{H}t'}\, \hat{F}'_{2T_0}\, \mathrm{e}^{-\mathrm{i}\hat{H}t}) \tag{19.7}$$

In the above expressions $\hat{F}_{T_0}$ is the thermal flux operator [110],

$$\hat{F}_{T_0} = \mathrm{e}^{-\hat{H}/2kT_0}\, \hat{F}\, \mathrm{e}^{-\hat{H}/2kT_0}$$

At this point, it is worth mentioning that $\hat{F}$ and $\hat{F}'$ are not necessarily defined on the same dividing surface. While all the above expressions are rigorously valid for any dividing surface chosen, the time required until the integral converges to its $t \to \infty$ limit depends on the choice of the dividing surface. Consequently, the efficiency of a numerical application will vary depending on this choice. If one is aiming at the calculation of $k(T)$ or $N(E)$, then it is usually most efficient to employ flux operators defined on one single dividing surface that includes the transition-state geometry. However, whenever flux correlation functions are employed to obtain state-selected reaction probabilities and rates, then one of the dividing surfaces is located in the asymptotic area [355–357].

19.2.2
Thermal Flux Operator: Properties and Physical Interpretation

As we have just shown, the calculation of rate constants via flux correlation functions requires the evaluation of the trace of some operator or combination of operators. Consequently, an efficient scheme to evaluate this trace is a prerequisite for the actual computation of flux correlation functions, and therefore $N(E)$ and $k(T)$. A straightforward computation of the trace employing a complete basis set is unfeasible even for medium-sized triatomic systems. A successful strategy most commonly employed in these cases is to employ the eigenstate representation of an appropriate operator. Here we will focus on a particular approach provided by the properties of the thermal flux operator.

The work by Park and Light [110] describes some important characteristics of the flux and thermal flux operators. Given that the flux operator is a singular operator, we will henceforth focus on its regularized version, the thermal flux operator. The thermal flux operator in a one-dimensional (1D) system has only two non-zero eigenvalues. For a 1D system with $h = \theta(x - x_{\mathrm{DS}})$ one finds

$$\begin{aligned}
\hat{F}_T &= \mathrm{e}^{-\beta\hat{H}/2}\mathrm{i}[\hat{H}, \theta(\hat{x} - x_{\mathrm{DS}})]\,\mathrm{e}^{-\beta\hat{H}/2} \\
&= \mathrm{e}^{-\beta\hat{H}/2}\frac{1}{2m}\{\hat{p}\delta(x - x_{\mathrm{DS}}) + \delta(x - x_{\mathrm{DS}})\hat{p}\}\,\mathrm{e}^{-\beta\hat{H}/2} \\
&= \mathrm{e}^{-\beta\hat{H}/2}\frac{1}{2m}\{\hat{p}|x_{\mathrm{DS}}\rangle\langle x_{\mathrm{DS}}| + |x_{\mathrm{DS}}\rangle\langle x_{\mathrm{DS}}|\hat{p}\}\,\mathrm{e}^{-\beta\hat{H}/2}
\end{aligned} \tag{19.8}$$

Thus, $\hat{F}_T$ is of rank two and its eigenvectors span the two-dimensional space $\{\mathrm{e}^{-\beta\hat{H}/2}|x_{\mathrm{DS}}\rangle, \mathrm{e}^{-\beta\hat{H}/2}\hat{p}|x_{\mathrm{DS}}\rangle\}$. The eigenvalues have equal size and opposite sign, $f_{T-} = -f_{T+}$, and their corresponding eigenvectors are the complex conjugate of one another.

Let us next consider a multidimensional system with a separable Hamiltonian, $\hat{H} = \hat{H}_R + \hat{H}_\perp$. Here $\hat{H}_R$ is a 1D operator that describes the motion along the reaction coordinate R, and $\hat{H}_\perp$ contains the motion in the remaining (bound) degrees of freedom. The dividing surface in this system will therefore be only a function of R, $h = \theta(R - R_{DS})$, and the thermal flux operator can be factorized into two components:

$$\hat{F}_T = e^{-\hat{H}_\perp \beta} \hat{F}_{T,R} \tag{19.9}$$

with

$$\hat{F}_{T,R} = e^{-\hat{H}_R \beta/2} i[\hat{H}_R, \theta(R - R_{DS})] \, e^{-\hat{H}_R \beta/2} \tag{19.10}$$

The operator can thus be seen as the product of a 1D thermal flux operator, $\hat{F}_{T,R}$, and the thermal weighting in the degrees of freedom orthogonal to R. In this separable system, each eigenstate of $\hat{H}_\perp$, ϕ_i, can be interpreted as a vibrational eigenstate. Moreover, if the dividing surface is set on top of the reaction barrier, the ϕ_i eigenstates can actually be seen as vibrational states of the activated complex. Each state, combined with the two $\hat{F}_{T,R}$ eigenstates, gives rise to a pair of non-zero eigenvalues of the overall $\hat{F}_T$, weighted according to the thermal occupation of the vibrational state: $f_{i\pm} = \pm f_{R,T} \, e^{-\beta E_{\perp,i}}$. Therefore, if the vibrational spacing is larger than the thermal energy, the exponential term will rapidly decrease and only a small number of eigenvalues of $\hat{F}_T$ should be relevant.

Although chemical reactions are actually non-separable systems, the previous statement is qualitatively generally valid. That is, the spectrum of non-zero eigenvalues of the thermal flux operator is rather small and its eigenstates have a physical interpretation in terms of vibrational states of the activated complex. Numerical results for the $H + H_2$ benchmark reaction for vanishing total angular momentum, $J = 0$, can corroborate this statement. The eigenvalues of the thermal flux operator for this reaction obtained at a reference temperature of 1000 K are listed in Table 19.1 (first column).

Table 19.1 Thermal flux eigenvalues (1000 K) for the $H + H_2 \rightarrow H_2 + H$ reaction.

f_n (a.u.)	$\ln(f_0/f_n)$	$\beta(E_n - E_0)$	(n_s, n_b)
$\pm 6.78 \times 10^{-5}$	0	0	(0, 0)
$\pm 0.80 \times 10^{-5}$	2.14	$2\beta\omega_b = 2.59$	(0, 2)
$\pm 0.40 \times 10^{-5}$	2.83	$\beta\omega_s = 2.96$	(1, 0)
$\pm 0.10 \times 10^{-5}$	4.22	$4\beta\omega_b = 5.18$	(0, 4)
$\pm 0.05 \times 10^{-5}$	4.91	$\beta(\omega_s + 2\omega_b) = 5.55$	(1, 2)
$\pm 0.03 \times 10^{-5}$	5.42	$2\beta\omega_s = 5.92$	(2, 0)
$\pm 0.01 \times 10^{-5}$	6.51	$6\beta\omega_b = 7.78$	(0, 6)

If a normal-mode analysis is performed at the transition-state geometry, the contributing eigenstates can be labelled according to the vibrational (n_s, n_b) states of the activated complex. Here n_s and n_b denote the quantum numbers of the symmetric stretch mode ($\omega_s = 0.256$ eV) and the bending mode

($\omega_b = 0.112\,\text{eV}$). The asymmetric stretch corresponds naturally to the reaction coordinate. If the system were strictly separable, one should be able to obtain all these eigenvalues as a vibrational progression starting from the first (ground-state) eigenvalue f_0, that is, $f_n = \mathrm{e}^{-\beta(E_n - E_0)} f_0$, where E_n is the energy of the nth vibrational level. This is the same as saying that

$$\ln(f_0/f_n) = \beta(E_n - E_0) \tag{19.11}$$

These two terms are listed in Table 19.1 for comparison purposes: the $\ln(f_0/f_n)$ obtained from the calculated thermal flux eigenvalues are compared to the $\beta(E_n - E_0)$ obtained from the corresponding vibrational state. The two columns in the table would be identical for a strictly separable system and harmonic in the non-reactive coordinates. Although this is not the case, the values are indeed quite similar, and the idea of a vibrational progression within the flux eigenvalues is therefore confirmed.

The physical interpretation of the flux eigenstates as vibrational states of the activated complex can be illustrated by representing the corresponding wavefunctions. In Figure 19.2 the contour plots of the F_T eigenfunctions for the $H + H_2$ collinear reaction are shown together with the potential energy surface, using Jacobi coordinates. The first thing that one may notice from the figure is that the wavefunctions cannot be described as a product of 1D functions and therefore the $H + H_2$ system is not strictly separable. The assumption of separability is thus not accurate enough to be used in numerical calculations. However, a picture based on vibrational states of the activated complex is nevertheless very helpful for interpretation. The upper panel corresponds to the F_T eigenstate pair with the largest eigenvalue. By looking at the contour plot, this state can easily be interpreted as a ground wavefunction since it shows no nodes in either direction. Instead, the function in the lower panel corresponds to the second eigenpair and shows one node. This function can therefore be interpreted as a first excited state.

19.2.3 Calculation of *N*(*E*) and *k*(*T*)

The fact that the thermal flux has a small number of non-zero eigenvalues is extremely useful in order to evaluate the trace in the flux correlation function efficiently. Employing the spectral decomposition of the thermal flux operator, one may straightforwardly transform the expression for the asymmetric flux correlation function in Equation (19.6) as

$$C_\mathrm{f}(t, t'; T_0) = \sum_{f_{T_0}} f_{T_0} \langle f_{T_0} | \mathrm{e}^{\mathrm{i}\hat{H}t'} \hat{F}' \, \mathrm{e}^{-\mathrm{i}\hat{H}t} | f_{T_0} \rangle . \tag{19.12}$$

The symmetric version of the correlation function (Equation (19.7)) offers the additional advantage that both flux operators can be expressed in terms of the

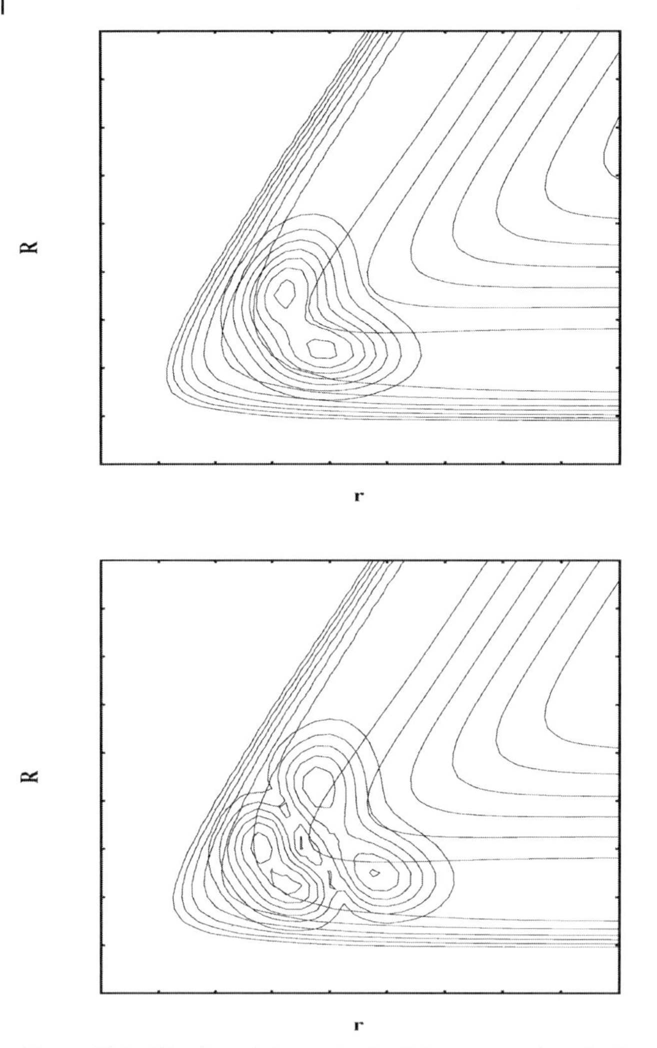

Figure 19.2 FF_T eigenstates and potential energy surface for the collinear $H + H_2$ reaction. The contour plots correspond to the two eigenpairs with largest eigenvalues (top to bottom, decreasing absolute value).

eigenstates:

$$C_f(t, t'; T_0) = \sum_{f_{2T_0}} \sum_{f'_{2T_0}} f_{2T_0} f'_{2T_0} \langle f_{2T_0} | e^{i\hat{H}t'} | f'_{2T_0} \rangle \langle f'_{2T_0} | e^{-i\hat{H}t} | f_{2T_0} \rangle \ , \tag{19.13}$$

then Equation (19.5) can be used to finally compute the corresponding cumulative reaction probability. If the symmetric version is used, the resulting equation takes a particularly simple form [354]:

$$N(E) = \tfrac{1}{2} e^{\beta_0 E} \sum_{f_{2T_0}} \sum_{f'_{2T_0}} f_{2T_0} f'_{2T_0} \left| \int_{-\infty}^{\infty} dt \, e^{iEt} \langle f_{2T_0} | e^{-i\hat{H}t} | f'_{2T_0} \rangle \right|^2 \tag{19.14}$$

Thus, to compute cumulative reaction probabilities, one first has to calculate the eigenstates of the thermal flux operator. Next, these states have to be propagated, and the resulting matrix elements in the correlation function are Fourier transformed. In some cases, the numerical stability of the method can be improved by diagonalizing the flux operator at a higher reference temperature T_1 and propagating the resulting thermal flux for the remaining imaginary time until the desired reference temperature value, T_0.

19.3 Rate Constant Calculations

19.3.1 Propagating All F_T Eigenstates

The previously shown features for the thermal flux operator, namely the small spectrum of non-vanishing eigenvalues and the correspondence of the eigenstates to vibrational states of the activated complex, are not restricted to small prototypical systems. Similar features have been observed for larger systems [222, 358–360] and vanishing total angular momentum, $J = 0$. In these cases, iterative diagonalization techniques will be particularly efficient for the computation of the thermal flux eigenstates. Calculations employing the MCTDH approach use the techniques presented in Chapter 9 for this purpose. Some technical aspects of MCTDH calculations of reaction rates may be mentioned. Typically, the potential energy surfaces employed show a comparatively complicated structure and therefore the correlation discrete variable representation (CDVR) scheme (Chapter 10) is employed in the potential evaluation. The integration schemes employed are based on the constant mean-field concept [50] and are described in detail in Ref. [51] (see also Chapter 4). To improve numerical stability, $\hat{F}_{T_0}\psi_i$ functions are propagated instead of the ψ_i functions that span the basis set of the $\hat{F}_{T_0}$ representation [354].

19.3.2 Statistical Sampling

In contrast to the previous $H+H_2$ case, whenever a reaction occurs through a transition state with very low vibrational frequencies or if an accurate treatment of the overall rotation is required, one can no longer assume that the number of non-vanishing thermal flux eigenstates will be small. Rotational and low-frequency vibrational spacings are typically much smaller than thermal energies and an enormous number of rovibrational states need to be considered. Having to include a huge number of thermal flux eigenstates significantly complicates the direct calculation of $k(T)$ or $N(E)$. First, the number of iterations required becomes very large and can converge slowly (more iterations are needed than actually contributing states). Second, a larger number of eigenstates need to be propagated. Both issues cause a significant increase in the computational effort involved.

However, these difficulties cannot be ignored if one aims at an accurate description of reactive processes. In order to obtain experimentally measurable observables, overall rotation might need to be included explicitly in the simulation, and a large number of total angular momenta generally have to be considered in the quantum mechanical calculation of reaction cross-sections, rate constants, or cumulative reaction probabilities. In addition to this, whenever the transition-state geometry presents low-frequency vibrational states, an accurate treatment will necessarily encounter a large number of contributing states. This is generally the case for polyatomic reactions involving rather heavy atoms or transition states with a floppy geometry.

A statistical sampling procedure [361] was devised specifically for situations in which the number of non-vanishing flux eigenstates is large and it is clearly inefficient to obtain and propagate them all explicitly. In this scheme, random sample functions are introduced in the coordinates that correspond to the degrees of freedom with low energy spacing:

$$|\psi_j\rangle = \sum_i (-1)^{\alpha_i(j)} |\phi_i\rangle \tag{19.15}$$

with $\alpha_i(j)$ being an integral random number and $\{|\phi_i\rangle\}$ the basis set for the respective degree of freedom. These random sample functions satisfy the following completeness relation:

$$\lim_{M\to\infty} \frac{1}{M} \sum_{j=1}^{M} |\psi_j\rangle\langle\psi_j| = 1 \tag{19.16}$$

Given this completeness relation, the thermal flux operator $\hat{F}_T$ can be rewritten as

$$\begin{aligned} \hat{F}_T &= \sum_{j=1}^{M} \mathrm{e}^{-\hat{H}\beta/2} \left(\lim_{M\to\infty} \frac{1}{M} |\psi_j\rangle\langle\psi_j|\hat{F} \right) \mathrm{e}^{-\hat{H}\beta/2} \\ &= \lim_{M\to\infty} \frac{1}{M} \sum_{j=1}^{M} \hat{F}_{Tj} \end{aligned} \tag{19.17}$$

where

$$\hat{F}_{Tj} = \mathrm{e}^{-\hat{H}\beta/2} |\psi_j\rangle\langle\psi_j|\hat{F}\, \mathrm{e}^{-\hat{H}\beta/2}$$

can be called a statistical thermal flux operator. Since the operator $\hat{F}_{Tj}$ contains a projection onto the random sample state, no progression corresponding to the low-frequency coordinates appears in the spectrum of $\hat{F}_{Tj}$. The number of non-vanishing eigenvalues of the statistical thermal flux operator is therefore determined exclusively by the coordinates that are not treated statistically (high-frequency vibrations) and only a small number of states will contribute to the eigenstate representation of $\hat{F}_{Tj}$:

$$\hat{F}_{Tj} = \sum_{f_{Tj}} |f_{Tj}\rangle f_{Tj} \langle f_{Tj}| \tag{19.18}$$

This decomposition of the statistical thermal flux operator may be used in Equation (19.12) to calculate a statistical sample of the cumulative reaction probability, $N_j(E)$. The scheme followed is then analogous to the one just described in Section 19.3.1, only a projection onto the random sample function follows the application of the flux operator [349, 362]. The final CRP can be obtained as the statistical average of the random samples:

$$N(E) = \lim_{M\to\infty} \frac{1}{M} \sum_{j=1}^{M} N_j(E) \tag{19.19}$$

The scheme described above is formally identical in the case of accurate all-J and low-frequency vibrational degrees of freedom. The only difference lies in the primitive basis used in Equation (19.15): for accurate all-J flux correlation function calculations, the random sample functions are built from Wigner rotation functions (that is, three-dimensional rigid rotor eigenfunctions), while Hermite associated functions are employed in the case of low-frequency vibrational modes [222]. Finally, it should be emphasized that this theoretical treatment is rigorously correct and does not invoke any approximation.

It is interesting to note that the MCTDH approach is ideally suited to the statistical sampling scheme. A 1D random function as given in Equation (19.15) can be represented by one single-particle function (SPF). Moreover, the projection operator $|\psi_j\rangle\langle\psi_j|$ is a one-particle operator. Therefore, the large primitive grids required in the representation of the rotational motion by expansion in Wigner functions can be used without major computational effort in the MCTDH approach.

19.4 Application to Polyatomic Reactions

The $CH_4 + H \rightarrow CH_3 + H_2$ reaction was the first reaction with more than four atoms for which accurate cumulative reaction probabilities were calculated for zero total angular momentum [217, 363]. Thermal rate constants were obtained by employing the method reviewed here and applying the J-shifting scheme [364] to account for the overall rotation. The Jordan and Gilbert potential energy surface (JG-PES), given in [365], employed in these first works has a reaction barrier that is significantly smaller than suggested by recent high-level *ab initio* results. The $k(T)$ values obtained from these first converged full-dimensional quantum dynamics calculations [217, 360, 363] differed significantly from experiment.

Recently, accurate theoretical rate constants could be calculated on a new accurate PES [225,366] developed using the Shepard interpolation [213,367] of high-quality *ab initio* data. In that work, converged values of the thermal rate constant were obtained for temperatures between 250 and 500 K. Details of the calculation have been published in Ref. [366], and only the most relevant computational features are commented upon here in order to highlight the efficiency of the flux correlation method and the MCTDH approach. But first it should be mentioned that 12 iterations of the Lanczos-based diagonalization scheme (Chapter 9) are needed to collect all the non-vanishing contributions from the flux eigenstates. The resulting eigenstates need to be propagated over 40 fs to converge the flux correlation function and $k(T)$.

It is important to emphasize the fact that the efficiency of the MCTDH scheme has been a prerequisite for the 12-dimensional (12D) quantum simulations. A qualitative, though non-rigorous, indication of the saving introduced by the MCTDH scheme is given in Table 19.2, where the numerical parameters of the MCTDH wavefunction are listed. The simulation was performed employing normal-mode coordinates defined at the transition-state geometry. These coordinates minimize the correlation effects between the nuclear degrees of freedom in the strong interaction region. Table 19.2 contains the 12 normal coordinates, grouped when they correspond to degenerate modes. For each coordinate, the number of SPFs and the number of primitive basis functions employed in the calculation are given, as well as the type of primitive function used. The sizes of the simple direct products of the time-dependent basis (272 160 functions) and the time-independent basis ($\sim 2.9 \times 10^{16}$) give a very clear idea of the reduction in computational effort resulting from the MCTDH contraction.

The calculated $k(T)$ values are compared in Figure 19.3 with experimental results from different groups [368, 369] in an Arrhenius plot. The agreement between theory and experiment is significantly improved with respect to the older JG-PES calculation. However, a rigorous comparison of theoretical and experimental results is rather difficult due to the spread of the experimental values. Probably, the availability of room- and lower-temperature experimental results would be very helpful in assessing the accuracy of the Shepard interpolated PES. However, the fact that the accuracy of theoretical prediction can now rival the accuracy of measured rate constants, even for reactions in-

Table 19.2 Parameters for the MCTDH wavefunction in the $H + CH_4 \rightarrow H_2 + CH_3$ dynamics simulation of Refs. [366]. (FFT stands for fast Fourier transform basis and H-DVR for the DVR functions based on Hermite polynomials.)

Coordinate	SPF (n_i)	Grid points (N_i)	Scheme
Q_1	7	128	FFT
Q_2, Q_3	3	25	H-DVR
Q_4	3	30	FFT
Q_5, Q_6	3	25	H-DVR
Q_7, Q_8	2	10	H-DVR
Q_9	5	192	FFT
Q_{10}	2	10	H-DVR
Q_{11}, Q_{12}	10	8	H-DVR

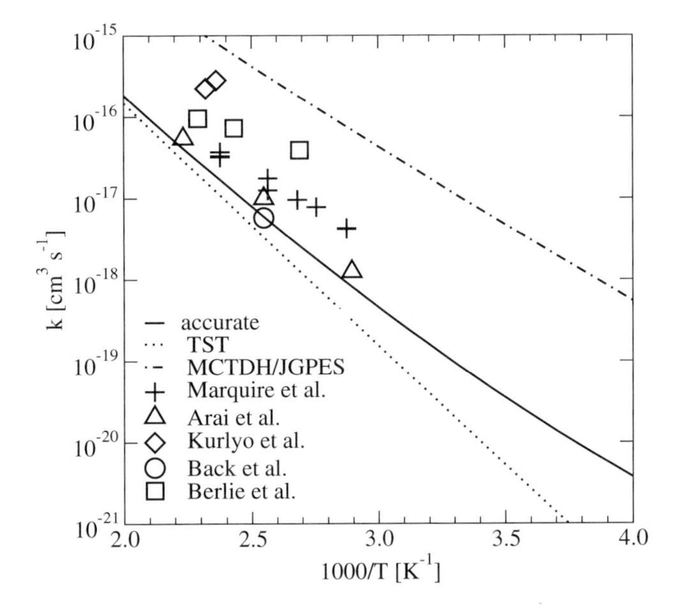

Figure 19.3 Thermal rate constant of $CH_4 + H \rightarrow CH_3 + H_2$ in an Arrhenius plot. The accurate theoretical data are given as a solid line, and classical transition-state theory results on the same PES are shown as a dotted line.

volving six atoms, can be seen as a significant step forwards in the area of theoretical reaction dynamics.

The importance of the tunnelling effect on the $CH_4 + H \rightarrow CH_3 + H_2$ reaction rate was examined in the same work. Classical transition-state theory (TST) results (dotted line in Figure 19.3) are significantly smaller than the accurate quantum dynamics data (solid line) in the lower-temperature range, confirming the importance of tunnelling at lower temperatures. However, already at temperatures of about 500 K, the TST results agree with the quantum results and tunnelling no longer seems to be particularly important.

Besides the actual agreement of the accurate dynamics with the experiment, an important purpose of accurate benchmark data is testing approximate models on such complex systems. The reliability of the data produced by any dynamics approach is intimately bound to the quality of the potential energy surface employed. Therefore, agreement between approximate theoretical data and experimental results may sometimes be fortuitous. The rigorous assessment of the reliability of an approximate method requires comparison not to experiment but to accurate theoretical data obtained on the same potential energy surface, if these are available.

Prior to the publication of the Shepard interpolated PES, most of the theoretical studies of the $CH_4 + H \rightarrow CH_3 + H_2$ reaction had been performed on the JG-PES. In particular, several approximate reduced-dimensionality quantum dynamics methods have been applied, such as the rotating bond umbrella approach of Yu and Nyman [370] and the semi-rigid vibrating rotor target of Zhang and co-workers [371]. Although both schemes consider explicitly four (out of 12) degrees of freedom of the reactive system, dramatic discrepancies arise between them. This finding points out the fact that the choice of the coordinates explicitly treated in a reduced-dimensionality calculation is of crucial importance and not an easy subject. While the above-mentioned reduced-dimensionality results appeared before the 12D MCTDH results were published, more reduced-dimensionality quantum calculations have been presented since then. Wang and Bowman [372] have studied the system on the same surface employing a six-dimensional model. Zhang and co-workers [373] have presented very recently seven- (7D) and eight-dimensional (8D) transition-state wavepacket calculations showing significantly lower rate constant values than the 12D result. A reasonable explanation for these differences is still a matter of debate. Variational transition-state approach calculations incorporating tunnelling corrections [374] have also been performed on the JG-PES, obtaining a good agreement with the accurate results, even at low temperatures.

Similar studies have been performed for the methane combustion reaction, $CH_4 + O \rightarrow CH_3 + OH$. Concerning the simulation parameters, the same techniques were employed as in the $H + CH_4$ reaction. In this case, the contributing eigenstates at a reference temperature of 300 K were converged after 10 Lanczos iterations and later propagated for 120 fs. The presence of the oxygen atom makes the system significantly heavier than the $H + CH_4$ one, and consequently the energy spacing of the activated complex vibrational states is comparable to thermal energies. A slightly different approach was employed in this case, estimating the flux contributions not explicitly included in the di-

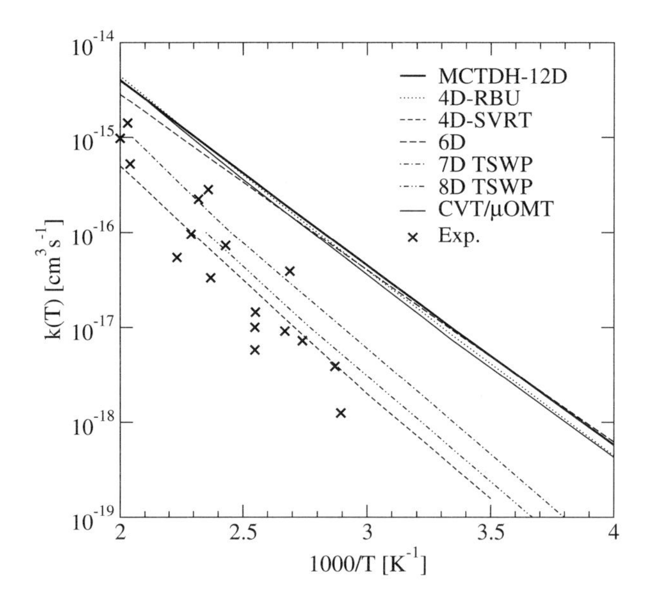

Figure 19.4 Arrhenius plot of computed thermal rate constants for $H + CH_4 \rightarrow H_2 + CH_3$, from the accurate calculation, reduced-dimensionality calculations (RBU-4D [370], SVRT-4D [371], 6D [372], 7D and 8D TSWP [373]), and improved variational transition state with tunnelling correction (CVT/μOMT) [374]. Crosses indicate experimental results taken from Ref. [368].

agonalization by a harmonic progression. For a more detailed discussion, the reader is referred to Ref. [222].

19.5 The Effect of Rotation–Vibration Coupling on Rate Constants

MCTDH rate constant calculations have also been used to study the effect of the coupling between overall rotation and the internal degrees of freedom in the $O + HCl \rightarrow OH + Cl$ [362], $H_2 + Cl \rightarrow HCl + H$ [358], $H_2 + OH \rightarrow H_2O + H$ [349], and $OH + HCl \rightarrow H_2O + Cl$ [375] reactions. For these systems, the importance of rotation–vibration coupling was investigated by comparing approximate results to accurate all-J ones, calculated with the statistical sampling approach. Here we do not intend to discuss in detail the individual systems, but rather present the general picture that has emerged from these studies.

The rotation–vibration decoupling approximation, most commonly known as *J*-shifting, is based on the standard, separable, rigid-rotor analysis of the transition state. This approximation assumes that the internal motion and the overall rotation are uncoupled and that the contribution of the overall rotation to $k(T)$ is determined by the rotational partition function of the activated complex. Such an approximation is only reasonable if the reaction is a direct process and the transition state (TS) has a well-defined geometry. It also implicitly assumes that the moments of inertia of the complete systems do not change significantly on the time-scale relevant to the determination of the rate constant.

An illustrative case can be the comparison of the $OH + H_2$ and $OH + HCl$ reactions. While the *J*-shifting approximation provides satisfactory results in the $OH + HCl$ [375] reaction, comparison with accurate results in the $OH + H_2$ case [349] proved a significant overestimation of the thermal rate constant. The difference can be explained by the fact that the reaction barrier occurs later in the $OH + HCl$ case, the H–O–H structure at the TS is therefore more robust, and the vibration–rotation coupling is consequently smaller. In the case of $OH + H_2$, the TS normal mode correlating with the product water molecule bending mode couples rather strongly to the smallest moment of inertia of the TS, I_a. Moreover, this normal mode presents a small vibrational frequency (572 cm^{-1}) and can therefore be significantly excited at thermal energies. Neglecting the vibration–rotation coupling will in this case give rise to inaccuracies in the 30% range. Instead, in the $OH + HCl$ reaction, the same mode couples less efficiently to the overall rotation and has an oscillatory frequency of 1669 cm^{-1}. The excitation of this mode therefore contributes much less to the thermal ensemble and the *J*-shifting approximation is suitable.

Thus it may be inferred that, whenever the reaction occurs through a rigid TS and all moments of inertia are large at the TS geometry, the *J*-shifting approximation is expected to give reliable results. This is the case for the $X + CH_4$ family of reactions, where the TS usually presents large moments of inertia, for example 21 423 and 57 370($\times$2) atomic units in the case of $H + CH_4$.

19.6
Concluding Remarks and Outlook

The combination of the flux correlation function formalism with the MCTDH approach has enabled the calculation of accurate quantum mechanical thermal rates for polyatomic systems. The achievements of the last eight years open the perspective for future accurate simulations of reactions relevant to atmospheric chemistry as well as simple organic reactions. Having accurate theoretical data at hand is extremely valuable in a three-fold way: first, comparison with the experiment gives a rigorous indication of the accuracy of the PES employed and the quantum chemical calculations on which the PES is based; second, approximate methodologies can be confronted prior to their application on larger systems; and, finally, accurate theoretical results might support, or even replace, experimental data whenever the kinetic measurement faces experimental difficulties.

Acknowledgments

Financial support by the the Spanish 'Ministerio de Educación y Ciencia' (CTQ2005-3721, UNBA05-33-001), 'Direcció General de Universitats Recerca i Societat de la Informació' (2005 PEIR 0051/69) (F.H.L.), the 'Deutsche Forschungsgemeinschaft' and the 'Fond der Chemischen Industrie' (U.M.) are gratefully acknowledged. The Spanish 'Ministerio de Educación y Ciencia' is also acknowledged for a Ramón y Cajal fellowship (F.H.L.).

20
Reactive and Non-Reactive Scattering of Molecules From Surfaces

Geert-Jan Kroes, Rob van Harrevelt and Cédric Crespos

20.1
Introduction

Chemical reactions at surfaces are of tremendous practical interest. To give an example of the practical importance of chemical processes at surfaces, about 90% of the chemical manufacturing processes employed throughout the world use catalysts in one form or another. Examples of such processes are the production of ammonia (a raw material for artificial fertilizer) and the current commercial process for the production of hydrogen (steam reforming of methane). In an illustration of the importance of the field, the 2007 Nobel Price for Chemistry was recently awarded to Professor G. Ertl for unravelling the reactions producing NH_3, involving nitrogen and hydrogen molecules and atoms, and taking place at an iron surface [376]. In ammonia production and steam reforming of methane alike, the chemical reactions are made up of several elementary steps, one of which is rate limiting (the dissociative chemisorption of N_2 for ammonia production, and the dissociative adsorption of CH_4 on Ni into surface-adsorbed $H + CH_3$ for steam reforming [377]).

Molecule–surface scattering also displays a number of unique phenomena. The earliest experiments on diffractive scattering of He and H_2 from LiF(001) helped to demonstrate the wave nature of atomic and molecular motion [378]. Compared to gas-phase dissociation, a unique feature of the photodissociation of molecules adsorbed onto surfaces is that it may display surface-induced rainbow scattering [379]. The interaction of a molecule with a metal surface may lead to electron–hole pair excitation, to the point that scattering of molecules from surfaces with low work-functions may even lead to electron emission [380]. These phenomena serve to make the field of scattering of molecules from surfaces an interesting playground, but they also pose challenges. For instance, it is clear that processes in which electron–hole pair excitation plays an important role should take dissipation (of electronic energy) into account, and this complicates the dynamic treatment of such systems.

Multidimensional Quantum Dynamics: MCTDH Theory and Applications.
Edited by Hans-Dieter Meyer, Fabien Gatti, and Graham A. Worth

ISBN: 978-3-527-32018-9

A variety of elementary processes can occur in reactive and non-reactive interactions of atoms and molecules with surfaces. Their detailed understanding through computational and experimental studies can provide a thorough understanding of several aspects of the interaction of atoms and molecules with surfaces. Perhaps the simplest scattering process from a surface is atomic diffraction. This process yields information on the corrugation of the atom–surface interaction (its dependence on the coordinates of motion of the atom in the two directions parallel to the surface). In the scattering of a molecule from the surface, rotationally inelastic scattering of the molecule can also occur. Rotationally inelastic scattering was first verified experimentally in conjunction with diffractively inelastic scattering for H_2 + LiF(001) [381] and for HD and H_2 + Mg(001) [382] in the mid-1970s. Unambiguous experimental evidence for direct vibrational excitation (whereby energy transfer occurs in essentially a single gas–surface encounter) in molecule–surface scattering was first reported for NO scattering from Ag(111) [383].

The dissociation of H_2 on metal surfaces is relevant to several heterogeneously catalysed reactions, and is of fundamental importance as a model system. Experiments [384, 385] and theoretical work [386] have shown that phonons do not affect the reaction in a major way, and that the effect of surface temperature on reaction can be easily modelled. Recent calculations on H_2 + Pt(111) have shown that the effect of electron–hole pair excitation on the reaction of initially non-vibrating hydrogen on metal surfaces can likewise be ignored [387]. Furthermore, in dynamical studies of dissociation of H_2 on surfaces, it has been possible for some time now to treat the motion in the six molecular degrees of freedom fully quantum mechanically, without any approximations [388–390]. This makes the system ideal for testing the quality of electronic structure theory for predicting molecule–surface interaction energies. A wide range of experimental results is available for systems such as H_2 + Cu(111) [391, 392] and Cu(100) [393–395] and H_2 + Pt(111) [396]. Other dissociative chemisorption systems of great interest include N_2 + Ru(0001), about which there has been speculation that the reaction should be affected to a large extent by electron–hole pairs [397], and CH_4 + Ni(111), for which experiments have shown that the pre-excitation of the molecular vibrations can promote the reaction in a mode-specific way [398, 399].

Most of the molecule–surface scattering processes described above occur on short time-scales, making them amenable to calculations employing the multiconfiguration time-dependent Hartree (MCTDH) method.

MCTDH calculations have addressed:

(i) the rotationally and diffractionally inelastic scattering of H_2 from LiF(001) [400] and from a model rectangular surface [401], and of N_2 from LiF(001) [34, 80, 401, 402]; and

(ii) the dissociative chemisorption of H_2 on a model transition-metal surface [403], on Cu(100) [404,405] and on Pt(111) [406], of N_2 on Ru(0001) [407], and of CH_4 on Ni [408–411].

MCTDH calculations have also addressed other processes, and the following have all been studied recently:

(iii) the photodissociation of CH_3I on MgO [412–414] and of HCl on ice [415, 416];

(iv) the vibrational relaxation of H adsorbed onto Si(100) and the laser-induced desorption of H from Si(100) [417, 418];

(v) the vibrational relaxation and the sticking of CO on Cu(100) [97];

(vi) the coherent control of hydroxyl formed due to HCl adsorption onto MgO(001) [419]; and

(vii) the femtosecond time-scale photoinduced electron transfer from a dye molecule (coumarin 343) to a semiconductor surface (of TiO_2) in the presence of solvent molecules [229, 232].

Finally, methods from the more general class of time-dependent multiconfiguration self-consistent field methods have addressed:

(viii) surface energy transfer in the single-phonon limit in scattering of He from a Cu surface [29]; and

(ix) ultraviolet laser-induced desorption of CO from LiF(001) [420].

A discussion of the last two papers falls outside the scope of this review; instead, below, this review will focus on papers addressing molecule–surface scattering and employing the MCTDH method proper.

20.2 Theory

In this section we will discuss some of the salient features of the theory for molecule–surface scattering, as used in MCTDH applications. We will focus on the molecule–surface problems to which MCTDH has been most applied, that is, the reactive and non-reactive scattering of a diatomic molecule from a periodic surface. We will formulate the theory for the dissociative chemisorption problem, treating all six degrees of freedom of the molecule.

Calculations on the scattering of diatomic molecules from surfaces are usually carried out within the Born–Oppenheimer approximation [387]. Also neglecting the phonons, and using atomic units as done throughout this chapter, the Hamiltonian may be written [421]:

$$\hat{H} = -\frac{1}{2M}\frac{\partial^2}{\partial Z^2} - \frac{1}{2\mu}\frac{\partial^2}{\partial r^2} - \frac{1}{2M\sin^2\gamma}\left[\frac{\partial^2}{\partial x^2} - 2\cos\gamma\frac{\partial}{\partial x}\frac{\partial}{\partial y} + \frac{\partial^2}{\partial y^2}\right] + \frac{\hat{\jmath}^2}{2\mu r^2} + V_{6D}(Z,r,x,y,z,\theta,\phi) \tag{20.1}$$

In Equation (20.1), M is the mass of the molecule, μ is its reduced mass, $\hat{\jmath}$ is the angular momentum operator, and V_{6D} is the potential energy of the two atoms of the molecule interacting with one another and with the surface. The coordinates are defined as follows: Z is the molecule–surface distance; x and y are the coordinates for the molecule's motion parallel to the surface, measuring the distances associated with the projection of the centre of mass of the molecule on the surface along the lattice vectors $\mathbf{x}$ and $\mathbf{y}$ spanning the surface unit cell; r is the intramolecular distance of the molecule; and the polar and azimuthal angles θ and ϕ define the orientation of the molecule with respect to the surface normal and $\mathbf{x}$. Furthermore, γ is the angle between $\mathbf{x}$ and $\mathbf{y}$, which is 90° for square and rectangular surface unit cells ((100) and (110) surfaces of fcc materials) and 60° for surfaces with hexagonal symmetry ((111) surfaces of fcc materials). In calculations on non-reactive scattering, the motion in r can often be neglected: then, the term involving the second derivative with respect to r can be omitted from Equation (20.1), and in the rotational energy operator r is best set to the value corresponding to the molecule's rotational constant for its ground vibrational state.

A unique feature of surface scattering is that, owing to the periodicity of the surface, the translational momentum of the molecule parallel to the surface can only take on discrete values. To express the implications of this, we first define the primitive lattice vectors expressing the periodicity of the surface. Let $\mathbf{s}_1$ (in the direction of $\mathbf{x}$) and $\mathbf{s}_2$ (in the direction of $\mathbf{y}$) be the primitive vectors in real space spanning the surface unit cell. The reciprocal lattice vectors are then given by

$$\mathbf{k}_i \cdot \mathbf{s}_j = 2\pi\delta_{ij}, \qquad i = 1,2 \tag{20.2}$$

The restrictions on the changes that the parallel translational momentum of the molecule can undergo can be expressed as follows. If the molecule comes in with an initial parallel translational momentum defined by the two-dimensional vector $\mathbf{K}^0$, then after scattering the final translational parallel momentum obeys

$$\mathbf{K}^{\mathrm{f}}_{nm} = \mathbf{K}^0 + n\Delta\mathbf{k}_1 + m\Delta\mathbf{k}_2 \tag{20.3}$$

In Equation (20.3), the length of $\Delta\mathbf{k}_i$ is the minimum amount of momentum that can be gained or lost in the direction of $\mathbf{k}_i$, and is given by

$$\Delta\mathbf{k}_i = \frac{2\pi}{|\mathbf{s}_i| \sin\gamma} \tag{20.4}$$

In view of the above, in time-dependent quantum dynamics calculations, the initial wavefunction is best expressed as

$$\begin{aligned}\Psi_0(Z,r,x,y,\theta,\phi) &= \phi_{vj}(r) Y_{jm_j}(\theta,\phi) \frac{1}{\sqrt{A}} \exp[\mathrm{i} \sin\gamma (K_1^0 x + K_2^0 y)] \\ &\times \int \mathrm{d}k_z\, b(k_z) \frac{1}{\sqrt{2\pi}} \exp(\mathrm{i}k_z Z)\end{aligned} \tag{20.5}$$

and the primitive basis for motion in x and y is best taken to consist of basis functions

$$\eta_{nm}(x,y) = \frac{1}{\sqrt{A}} \exp[\mathrm{i}\sin\gamma(K_1^0 x + K_2^0 y)] \exp[\mathrm{i}\sin\gamma(n\Delta k_1 x + m\Delta k_2 y)] \tag{20.6}$$

In Equation (20.5), A is the surface of the unit cell, and the prefactor $1/\sqrt{A}$ normalizes the part of the wavefunction describing parallel translational motion. Furthermore, $\phi_{vj}(r)$ is the wavefunction describing the initial vibrational state of the molecule with vibrational quantum number v and rotational quantum number j, $Y_{jm_j}(\theta,\phi)$ is the spherical harmonic describing the initial rotational state of the molecule, m_j being the projection of $\mathbf{j}$ on the surface normal, and the integral describes a Gaussian wavepacket travelling to the surface with an appropriate momentum distribution $b(k_z)$, k_z being the translational momentum in the direction along Z. Finally, n and m are the quantum numbers for diffraction.

As suggested by Equations (20.5) and (20.6), it is best to use a Fourier basis as primitive functions to represent motion along x and y. Because the motions in the angles θ and ϕ are strongly coupled by the rotational energy operator, it is best to use the spherical harmonics $Y_{jm_j}(\theta,\phi)$ as a primitive basis for two-dimensional single-particle functions (SPFs) describing the simultaneous motion in these two degrees of freedom [80]. A convenient choice for the motion in Z and r is to use a Fourier basis as primitive basis; for long-time propagations this is best combined with the use of a complex absorbing potential (CAP) to avoid artefacts associated with the periodicity of Fourier basis functions.

For dissociative chemisorption, initial-state-selected reaction probabilities can be computed with the help of a CAP located in the exit channel, starting at an appropriately chosen value of $r = r_{cut}$, in combination with a flux analysis approach [422] (see Section 6.5). State-to-state scattering probabilities for rotationally and diffractionally inelastic scattering, or partial state-to-state scattering probabilities (considering only rotational transitions, or only diffractive transitions), can be computed by defining a CAP in the entrance channel and employing projection operators to project onto the final states of interest, as is also discussed in Section 6.5. Including also the rotational and/or diffractive energies of the states projected on, it is also possible to compute values of the average rotational energy transfer (ARET) or the average energy transferred to motion parallel to the surface (APET) [80].

20.3 Applications of the MCTDH Method to Molecule–Surface Scattering

20.3.1 Rotationally and Diffractionally Inelastic Scattering

The first application of the MCTDH method to rotationally and diffractionally inelastic scattering concerned the scattering of H_2 from LiF(001) [400]. The goal was to investigate the convergence behaviour of the MCTDH method with respect to the number of SPFs required to solve this type of problem. Calculations were performed for a collision energy of 0.2 eV, using a model potential depending on four degrees of freedom (Z, x, y and θ) of factorizable form [423]. State-to-state probabilities for rotationally and diffractively inelastic scattering were compared with time-independent (close-coupling, CC) and time-dependent (close-coupling wavepacket, CCWP) quantum scattering calculations [424]. The comparison showed that, for normal incidence, convergence to within 1% could be achieved using (5, 5, 15, 4) SPFs for (x, y, Z, θ), employing Fourier primitive basis sets for x, y and Z, and a Gauss–Legendre polynomial primitive basis set for θ.

In the same year, results of MCTDH calculations on rotationally and diffractionally inelastic scattering of H_2 from a model rectangular surface and of N_2 from LiF(001) were published [401]. The calculations used a so-called surface dumbbell model [425], in a parametrization for H_2 interacting with a rectangular lattice [425] and for N_2 interacting with LiF(001) [426]. The dumbbell potential is likewise of factorizable form, and depends on all five molecular degrees of freedom of the diatom considered as a rigid rotor. The goal was to examine critically the performance of the MCTDH algorithm with respect to different primitive basis representations and mode combination schemes. Rotational excitation probabilities for N_2 + LiF(001) were in good agreement with CCWP results. The paper established that it is useful to mode-combine θ and ϕ in one mode, and to mode-combine x and y in one mode.

The $N_2 + LiF(001)$ problem was revisited in subsequent five-dimensional (5D) MCTDH calculations employing the same potential as used in the previous calculations [401]. The new work [80] and the work discussed above [401] have already been reviewed earlier [34], so that we will not discuss it further here.

The $N_2 + LiF(001)$ dumbbell model potential was also used in 5D MCTDH calculations that served as a benchmark to test the performance of the mixed quantum–classical Bohmian (MQCB) trajectory method [402]. In the MQCB calculations, the motion in x and y was treated quantum mechanically, while the motion in Z and the angles was treated classically. Rotationally averaged diffraction probabilities computed with the MQCB method were in excellent agreement with MCTDH results for the collision energy range investigated (0.1–0.3 eV). The fact that MCTDH results were used as benchmark in this study may be viewed as signifying that the MCTDH method has matured, to the point that the scientific community now accepts that it can provide very accurate and well-converged results.

20.3.2 Dissociative Chemisorption

20.3.2.1 Dissociative Chemisorption of H_2 on Metal Surfaces

The earliest application of the MCTDH method to a molecule–surface scattering problem [403] modelled dissociative chemisorption of H_2 on a transition-metal surface, using model potentials developed earlier for standard wave-packet studies [427]. The calculations modelled motion in two degrees of freedom, that is, r and Z. It was shown that numerically exact solutions could be obtained with the MCTDH method.

MCTDH calculations modelling motion in all six molecular degrees of freedom were recently done for dissociative chemisorption of H_2 on Cu(100) [404, 405]. The potential energy surface (PES) of Ref. [428] was used, which is based on density functional theory (DFT) calculations. This potential has the form:

$$V(r, Z, x, y, \theta, \phi) = \sum_{i=1}^{20} V_i^{(rZ)}(r, Z) V_i^{(xy)}(x, y) V_i^{(\theta\phi)}(\theta, \phi) \tag{20.7}$$

Equation (20.7) has the form of Equation (3.34) if the composite coordinates $Q_{rZ} = (r, Z)$, $Q_{xy} = (x, y)$ and $Q_{\theta\phi} = (\theta, \phi)$ are used in the MCTDH calculations. The potential then has a form that is suitable for MCTDH calculations. The $rZ/xy/\theta\phi$ mode combination scheme is also advantageous for other reasons. For instance, an advantage of combining θ and ϕ using spherical harmonics as a primitive basis set is that the use of this basis set avoids the singularities associated with the $\hat{\jmath}^2$ operator that exist for $\theta = 0°$ and $180°$.

A converged calculation required 26 SPFs for the rZ mode, 34 SPFs for the xy mode, and 23 SPFs for the $\theta\phi$ mode. With these numbers of SPFs, for each mode the smallest natural population (see Section 3.1) is smaller than the threshold of 10^{-4}, which was found to lead to accurate results, and which threshold was therefore used in all calculations on $H_2 + Cu(100)$.

For normal incidence, Figure 20.1 shows that the reaction probabilities computed with the MCTDH method are in good agreement with symmetry-adapted wavepacket (SAWP) results [429]. The slight discrepancies that are observable are presumably caused by resonances. An accurate description of the details of the resonance structure would require a very large number of SPFs, which tends to cause technical problems. If smaller numbers of SPFs are used, then the structures are averaged out.

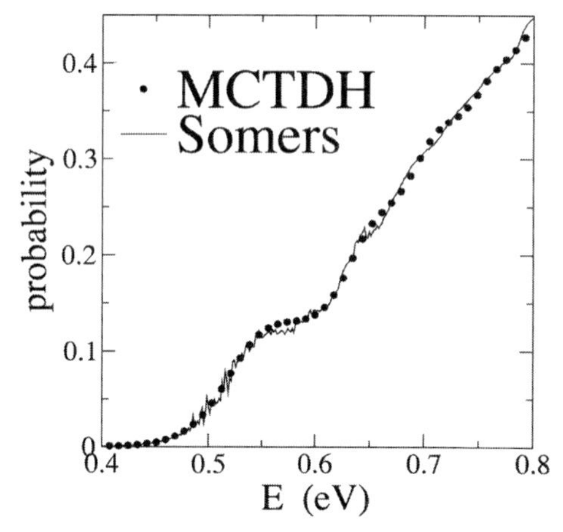

Figure 20.1 Sticking probability of $H_2(v=0, j=0)$ on Cu(100) as a function of the initial translational energy E, for normal incidence: comparison of the MCTDH results [404] (circles) with the results of the standard wavepacket calculations of Somers *et al.* [429] (full line). From Ref. [404].

The MCTDH approach was found to be an efficient alternative for the SAWP or other wavepacket approaches for this type of six-dimensional (6D) problem. The calculation for normal incidence took only 19 h on a 500 MHz EV6 Alpha processor. However, mode combination is essential for the efficiency of the MCTDH approach. This is the consequence of the large numbers of SPFs required to obtain convergence for this type of problem. The high efficiency of the calculations is partly due to the special form of the potential (see Equation (20.7)). In principle, any potential can be fitted to an expression similar to Equation (20.7), for example using the potfit method, which fits the PES to an expansion in natural potentials (see Section 11.1). However, the expansion would generally require many more than 20 terms to represent the

PES sufficiently accurately. For general PESs, the required CPU time would probably be a factor of 5–10 larger than for the PES used.

To demonstrate the importance of mode combination, the required CPU time was investigated for a calculation where r and Z were treated as separate modes. As the potential of Equation (20.7) then does not have a product form compatible with the mode combination scheme, the potfit method was employed to expand the functions $v_i^{(r,Z)}$ in a set of one-dimensional functions:

$$v_i^{(r,Z)} = \sum_j a_{ij} v_{ij}^{(r)}(r) v_{ij}^{(Z)}(Z) \tag{20.8}$$

Details of the potfit procedure used were given in [404]. The calculation not using mode combination for r and Z required seven SPFs for r and 22 SPFs for Z, almost as many as needed for r and Z in the calculation in which these degrees of freedom were mode-combined (26 SPFs). Consequently, the calculation using mode combination of r and Z used much less CPU time (19 h) and used a much more compact representation of the wavepacket than the calculation not using mode combination for these degrees of freedom (110 h on a 500 MHz EV6 Alpha processor).

The calculations on $H_2 + Cu(100)$ discussed so far describe collisions with normal incidence. It is also interesting to study the sticking probability for off-normal incidence. If the PESs were independent of x and y, then the sticking probability would be independent of the translational energy parallel to the surface. However, for most PESs the barrier height and position depend strongly on x and y. Darling and Holloway [430] have shown that in that case the parallel translational energy can either increase or decrease the sticking probability, depending on the PES.

Off-normal-incidence calculations require more computational effort than normal-incidence calculations because of the increase in the total energy. Furthermore, for off-normal incidence the symmetry of the metal surface cannot be used to the same advantage in the SAWP approach as for normal incidence. (At normal incidence, the use of symmetry can save a factor of 8 in computational expense for $j = 0$; at off-normal incidence, for incidence along a symmetry direction of the crystal, this is decreased to a factor of 2). The MCTDH approach is well suited to off-normal-incidence calculations, because of its high efficiency. For details of these calculations, see Ref. [405].

Figure 20.2 presents the sticking probabilities for different values of the initial translational energy for motion along x. For $E_Z < 0.5$ eV, the parallel translational energy increases the sticking probability. (If the reaction probability did not depend on the parallel translational energy, but only on E_Z, the curves in Figure 20.2(a) would fall on top of each other; this case is called normal energy scaling of the reaction.) However, the parallel translational energy is less effective than the normal translational energy E_Z. If it were as efficient, then the sticking probability plotted as a function of the total translational energy E would be independent of E_x (total energy scaling). Figure 20.2(b) shows that this is not the case. This shows that, according to the theory, and with the PES used, the reaction does not obey either normal energy scaling or total energy scaling. Experimentally, the situation is not really clear: for H_2 on Cu(100), the sticking probabilities measured in molecular beam experiments could be well fitted to contributions from the initial vibrational ground state and the first excited state, which obey normal energy scaling. However, good fits to experiment could also be obtained with $v = 0$ and $v = 1$ reaction probabilities, which showed deviations from normal energy scaling [431].

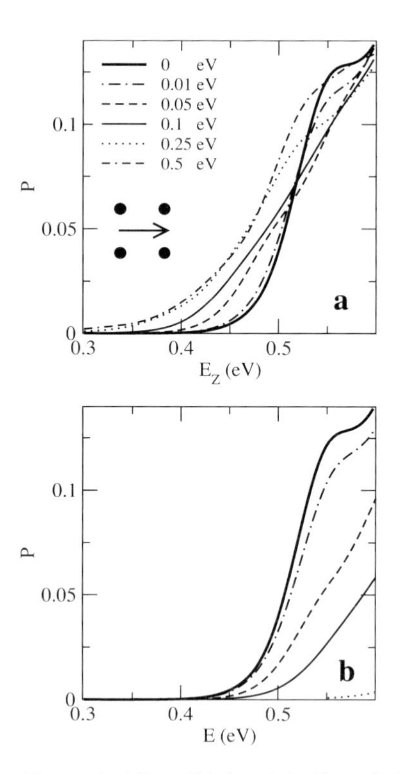

Figure 20.2 The sticking probability of $H_2(v=0, j=0)$ on Cu(100) as a function of (a) the normal translational energy E_Z and (b) the total translational energy E, for different values of the initial parallel translational energy along x, E_x ($E_y = 0$). From Ref. [405].

The performance of the MCTDH method compared to that of the time-dependent wavepacket (TDWP) method was also considered for a different H_2–surface reaction, that is, $H_2 + Pt(111)$ [406]. A new aspect for molecule–surface reactions considered for this system was how potentials should be handled that are of a general form, that is, that are not expressed as sums of products of low-dimensional functions. This limitation can be overcome by using the potfit method. Four-dimensional (4D) quantum dynamics calculations were performed on the dissociative chemisorption of H_2 on Pt(111) and the initial-state-resolved dissociation probability of H_2 was calculated within two 4D models.

The first set of calculations on H_2 dissociating on the Pt(111) surface was performed within a 4D model treating explicitly the molecular rotations. The 4D model involved the molecular centre-of-mass motion normal to the surface, the H–H distance r, and the polar and azimuthal angles θ and ϕ. Two different 4D PESs have been used for the calculations. These PESs are 4D cuts through the 6D H_2 + Pt(111) PES [432] obtained by interpolation of DFT points. The first 4D PES was obtained by fixing the centre-of-mass position on top of a surface atom. The second 4D PES was obtained by fixing the impact site above a bridge site, that is, between two nearest-neighbour surface atoms.

The H_2 dissociation probability was computed for the $(v_0 = 0, j_0 = 0)$ state. The MCTDH and TDWP (CCWP) reaction probabilities obtained for the top site and the bridge site are plotted in Figure 20.3(a) and (b), respectively. In both cases, the agreement is very good, especially for the bridge site, for which the curves are almost identical. A comparison between the computer time required by the MCTDH and TDWP standard methods for both calculations is made in Table 20.1. The MCTDH calculation for the top site has been split in two: the first propagation run corresponds to the low-energy range and the second one to a higher-energy regime. The total propagation time compares well with the time used in the TDWP standard method (assuming that the processors used in the MCTDH calculations are five times faster). The same remark can be made for the bridge site calculation. The overall conclusion for the 4D fixed site model is that MCTDH can give an accurate description of the dissociative chemisorption of H_2 on a metal surface for a PES of general form, at a computational cost comparable to that of the TDWP method. For numerical details, see Ref. [406].

Table 20.1 Comparison between the computer time required by the MCTDH and TDWP standard methods for 4D calculations within the fixed site model [406].

	MCTDH 'top' (0.02–0.13 eV) / (0.13–0.41 eV)	MCTDH 'bridge'	TDWP 'top'	TDWP 'bridge'
Processor	2.67 GHz	2.67 GHz	500 MHz	500 MHz
CPU time	0 h 38 min / 0 h 36 min	2 h 57 min	4 h 59 min	22 h 13 min

In the second 4D model [433], the molecule's centre-of-mass motion was fully represented, modelling motion in x, y, Z and r. The 4D PES built in Ref. [433] was used in MCTDH and in analogous 4D TDWP studies of dis-

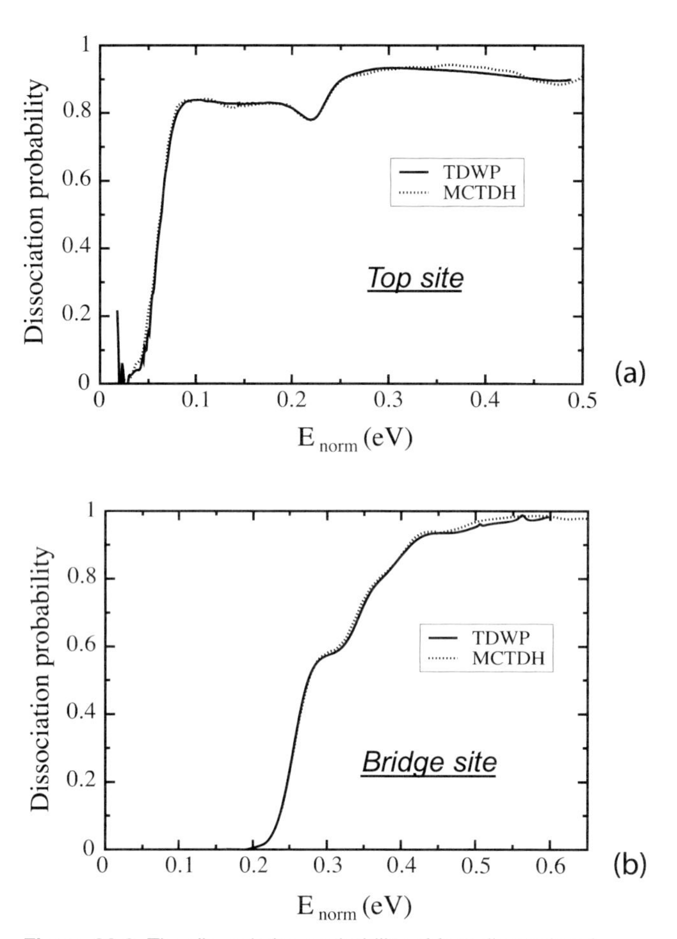

Figure 20.3 The dissociation probability of four-dimensional (a) 'on top' and (b) 'on bridge' quantum dynamics calculations for $H_2(v{=}0, j{=}0)$ on Pt(111) plotted as a function of the translational energy E_{norm}. From Ref. [406].

sociative and diffractive scattering of H_2 from Pt(111). One calculation was performed for the $H_2(v{=}0)$ initial state (Figure 20.4(a)) and a second one for $H_2(v{=}1)$ (Figure 20.4(b)). The agreement was very good for the $v = 0$ case, where the two curves are are almost superimposed. The $v = 1$ case shows a small discrepancy at high energy, which is probably a convergence issue. The CPU time compared well between the two methods for the $v = 0$ case but less so for the $v = 1$ case, where the MCTDH method was a bit slower (Table 20.2). The CPU time comparison should be more meaningful for a full 6D calculation. For numerical details, see Ref. [406].

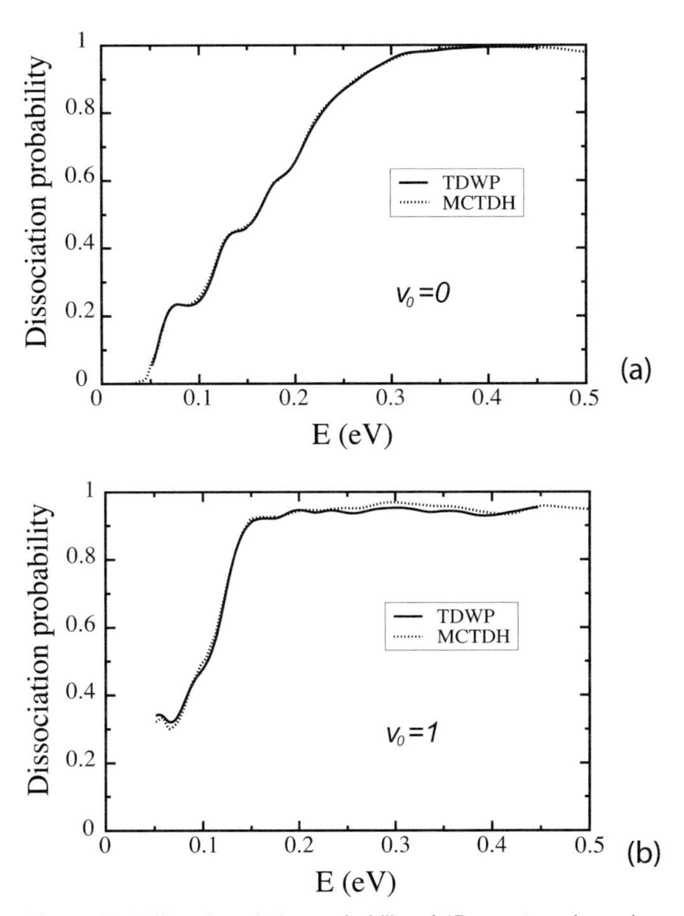

Figure 20.4 The dissociation probability of 4D quantum dynamics calculations for (a) $H_2(v=0)$ and (b) $H_2(v=1)$ on Pt(111) plotted as a function of the translational energy, E. From Ref. [406].

Table 20.2 Comparison between the computer time required by the MCTDH and TDWP standard methods for 4D calculations within the (Z, r, x, y) model [406].

	MCTDH $(v=0)$ (0.02–0.18 eV) / (0.18–0.45 eV)	MCTDH $(v=1)$ (0.02–0.18 eV) / (0.18–0.50 eV)	TDWP $(v=0)$ (0.05–0.15 eV) / (0.15–0.45 eV)	TDWP $(v=1)$ (0.05–0.15 eV) / (0.15–0.45 eV)
Processor	500 MHz	500 MHz	500 MHz	500 MHz
CPU time	0 h 17 min / 0 h 22 min	0 h 37 min / 1 h 26 min	3 h 29 min / 2 h 59 min	9 h 59 min / 5 h 15 min

In summary, the study of the H_2 + Pt(111) reaction system has shown that the MCTDH method is clearly an efficient method for modelling the dissociative chemisorption of hydrogen on a metal surface within a 4D model. The quantum dynamics of gas–surface processes involving dissociation is well reproduced in spite of the strong coupling between all the degrees of freedom in the dissociation region. The calculations on H_2 + Pt(111) have shown that it is possible to use a generic PES form as the basis of accurate MCTDH calculations.

20.3.2.2 Dissociative Chemisorption of N_2 on Metal Surfaces

The dissociation of N_2 on Ru is the rate-limiting step in the industrial synthesis of ammonia on Ru catalysts [434]. As a result, there has been much interest in this system. The problem has been studied in a number of low-dimensional dynamics studies, employing model PESs. The results of some of these model studies suggested that at room temperature the reaction should be completely dominated by a tunnelling mechanism [435], which of course should be somewhat surprising for such a 'heavy-atom system'.

Rate constants were computed for N_2 dissociating on stepped Ru(0001) using the MCTDH method [407], on the basis of a PES computed with DFT, using the revised Perdew–Burke–Ernzerhof (RPBE) generalized gradient approximation (GGA). A flux correlation approach [436] was used, in which the cumulative reaction probability was computed by calculating the eigenstates of the thermal flux operator using diagonalization, and subsequently propagating these eigenstates in time. The MCTDH study, the full details of which are given in Ref. [407], employed six degrees of freedom, that is, those of the N_2 molecule.

The major outcome of the calculations was the tunnelling correction κ, given by $\kappa = k(T)/k^{\mathrm{TST}}(T)$, where $k^{\mathrm{TST}}(T)$ is the rate constant computed using harmonic, classical transition-state theory. As Figure 20.5 shows, the tunnelling correction is quite moderate, being no larger than 1.4 at 200 K. As noted by the authors, the obstinacy of tunnelling theories in the literature for N_2 + Ru(0001) (see, for instance, Ref. [435]) is probably due to the experimental observation of significant reaction probabilities at energies well below the reaction barriers computed for low-index surfaces of Ru. However, it has subsequently become known that at low energies the reaction is dominated by steps that have much lower associated barriers [437]. As Figure 20.5 also shows, the PES could be calculated using only two reference points, where a reference point contained the energy of that point, the gradient at that point and the Hessian (modified Shepard interpolation was used to obtain the PES directly from DFT data). The tunnelling correction was found to be well described with a Wigner tunnelling correction factor.

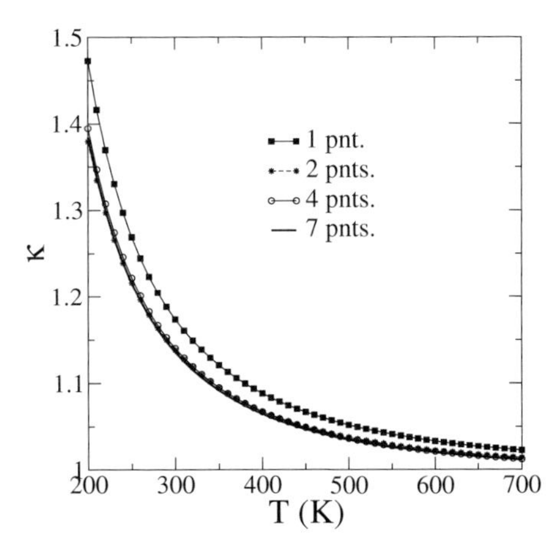

Figure 20.5 The tunnelling correction computed with MCTDH for the dissociative chemisorption of N_2 on Ru(0001), for modified Shepard PESs based on different numbers of reference points. From Ref. [407].

20.3.2.3 **Dissociative Chemisorption of CH_4 on Ni(111)**

Jansen and Burghgraef carried out three-dimensional MCTDH calculations on the dissociation of CH_4 on Ni(111) [408]. The methane molecule was treated as a pseudo-diatomic molecule. The three degrees of freedom modelled were: the distance of the molecule to the surface, Z; one CH stretching mode, s; and the angle of the three-fold axis of the CH_3 moiety treated as being non-reactive with the surface normal, τ. A PES was constructed on the basis of DFT calculations on CH_4 interaction with Ni clusters representing Ni(111) surfaces. The barrier geometry was taken from these DFT calculations, but the barrier height was adjusted to obtain agreement with experimental measurements of the reaction probability for a collision energy of 64 kJ mol^{-1}.

The reaction probability computed for the rovibrational ground state of CH_4 with the three-dimensional (3D) model as well as a two-dimensional (2D) model that ignored the rotation of the umbrella rises too steeply with the collision energy when compared to experiment. According to the authors, this implied that the barrier of the PES used was too wide and too low. The effect of pre-exciting the CH_3–H stretch was also investigated. Owing to the lateness of the barrier obtained from the DFT calculations, this has the effect of shifting the reaction probability curve to lower energies by 22–26 kJ mol^{-1} [408].

Jansen and co-workers also studied the translational to vibrational energy (T–V) and vibrational to translational energy (V–T) transfers in collisions of CH_4 with Ni(111) employing model PESs [409–411]. These calculations were motivated by early experiments showing that pre-exciting the ν_3 mode with one quantum leads to a large increase in the reaction probability [398], but suggested that the effect of exciting this mode should be too small to account for the vibrational effects observed in seeded beam experiments on CH_4 + Ni [438], suggesting that pre-exciting other vibrational modes should enhance the reaction even more effectively.

In the papers, it was argued that the efficiency of energy flow between a particular vibration and translation should be a measure of how effective pre-exciting that mode should be in enhancing the reactivity. Although the 10-dimensional calculations were undoubtedly quite advanced at the time they were carried out, the results of the calculations were somewhat contradictory. For instance, in the calculations on the scattering of methane in its initial vibrational ground state, most energy was found to be transferred to the ν_3 mode for geometries where methane had one or two H atoms pointed down to the surface [410]. In contrast, in calculations on vibrational de-excitation, more energy was generally found to be transferred from the ν_1 mode to translation than from the ν_3 mode to translation [411]. By the reasoning applied in both papers, then either pre-exciting the ν_3 mode should be most effective for promoting reaction (from the calculations on vibrational excitation) or pre-exciting the ν_1 mode should be most effective for promoting reaction (from the calculations on vibrational de-excitation). Experiments later found [439] that pre-exciting the ν_1 mode is 10 times more effective for promoting reaction than pre-exciting the ν_3 mode.

20.3.3
Photodissociation of Molecules on Insulator Surfaces

The photodissociation of CH_3I on MgO was studied in 2D [412] and 4D [413] MCTDH calculations. In the 2D calculations, the C–I axis was always taken perpendicular to the surface, and the degrees of freedom modelled were the distance of the centre of mass to the surface, R, and the C–I distance, r. The photodissociation could proceed on two spin–orbit coupled excited states, that is, the $^3Q_{0+}$ state corresponding to CH_3 and the spin–orbit excited state of I, I*, and the $^1Q_1$ state. A model PES was used to model the interaction of CH_3I with MgO; see Ref. [412].

The 2D calculations focused on differences in the outcomes between the initial situations where CH_3 initially points up to the gas phase or down to the surface. These calculations nicely illustrated the importance of going beyond the time-dependent Hartree (TDH) single-configuration approximation. The case where CH_3 initially points up to the gas phase was well described with the TDH method (the overlap of the TDH and exact wavefunctions being 95% at 2000 a.u. of time): the dissociation is direct, because CH_3 freely moves away from I and the surface. The case for CH_3 initially pointing down to the surface is different. Here, the CH_3 bounces in between the surface and the heavy I atom, as the I atom attempts to move away. The difficulty of describing this with a single configuration is illustrated in Figure 20.6, where the probability of finding the system in the electronic state corresponding to I* is plotted as a function of time. The first transition around 1000 a.u. of time, where the system first moves through the potential crossing point, is still rather well described with the TDH method ($n = 1$), but the second transition is not. The calculations found that, for an accurate description of the chattering dynamics and the non-adiabatic photodissociation, four configurations were required.

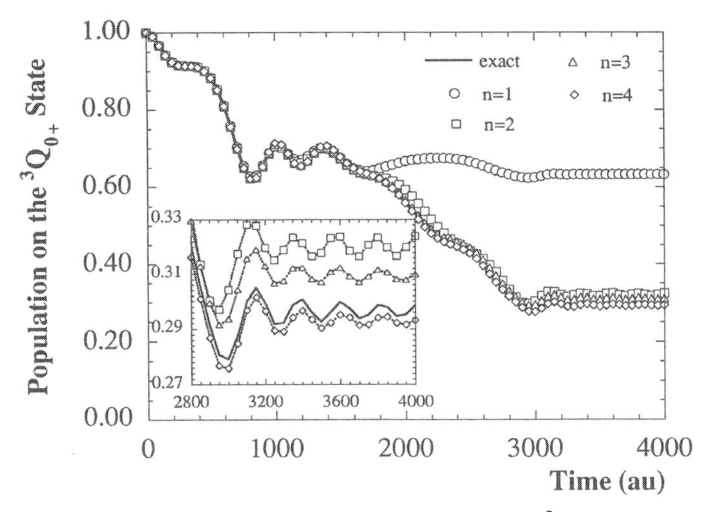

Figure 20.6 Time-dependent population of the excited $^3Q_{0+}$ state of CH_3I for the 2D MCTDH calculations. The single-configuration TDH ($n = 1$) fails to describe the second transition. A more detailed comparison for the time period 2800–4000 a.u. is given in the insert. Reused with permission from Ref. [412]. Copyright 1994, American Institute of Physics.

In the 4D MCTDH calculations [413], the degrees of freedom modelled were the Cartesian coordinates x and Z of I and of the CH_3 moiety considered as a pseudo-atom. The calculations only considered the less complicated (but also less interesting) case where the lighter CH_3 moiety of the physisorbed CH_3I initially points up to the gas phase.

Woittequand *et al.* have used the MCTDH method to model the photodissociation of HCl (X→A) adsorbed onto the (0001) face of ice [415, 416]. In their first study, motion in two degrees of freedom was studied, that is, the Z- and x-coordinates of the H atom of HCl, while the Cl atom of HCl and the water molecules were all kept fixed, the water molecules at positions and orientations corresponding to a perfectly ordered ice surface [415]. The potential used was a sum of the intramolecular potential of HCl and a somewhat modified HCl–ice interaction potential based on pair potentials [440]. The results showed the dynamics to be heavily affected by the fact that in the initial geometry the hydrogen-bonded H atom of HCl points down to the surface. The time evolution of the wavepacket shows a part that collides with the surface and then reflects from it, while another part of the wavepacket representing the motion of H was found to move into an ice cavity (about 1%). The HCl–ice absorption spectrum was found to be shifted to the blue by about 0.36 eV relative to the gas-phase HCl spectrum. The computed angular distribution describing the motion of the H atom contains a surface rainbow peak, and is affected by a shadow effect due to the presence of a Cl atom on top of the H atom in the initial HCl–ice adsorption geometry.

In their second study [416], Woittequand also took into account motion in a third molecular degree of freedom (the second coordinate for motion of H along the surface, that is, y), while the motion of the water molecules was modelled in an approximate way, that is, by performing the MCTDH calculations for 25 initial geometries representing different initial HCl–ice configurations. These configurations were obtained by performing molecular dynamics simulations of HCl interacting with an ice surface at $T = 210\,\mathrm{K}$, treating both the HCl and the H_2O molecules as rigid rotors. The MCTDH results obtained for the 25 initial configurations were then averaged to obtain final results for photodissociation of HCl on ice at $T = 210\,\mathrm{K}$. The same PES was used as in the earlier calculations. Compared to the earlier 2D calculations, even at 0 K the 3D spectrum was found to exhibit less structure than the 2D spectrum. The reason for this is that the extra motion in y allowed in the 3D model allows the H atom to escape from under the Cl atom more easily, leading to less trapping between Cl and ice. The averaging over 25 initial configurations used to take surface temperature effects into account served to diminish the structure in the spectrum even more.

20.3.4
Molecule–Surface Dynamics With Dissipation

In molecule–surface dynamics problems, there will usually be an effect of either the phonons or electron–hole pair excitations (for metal surfaces) on the dynamics one is interested in. However, if one is only interested in the dynamics of the molecule interacting with the surface, then a logical and convenient way to proceed is to divide the system up into a small system of interest (the primary system, often taken as the molecule only) and a so-called bath (the environment, often taken as the surface). Such a formalism allows one to 'trace out' the bath. The equation of motion to be solved is

$$\dot{\rho} = L(\rho) \tag{20.9}$$

where ρ is the reduced density, that is, the density associated with the primary system.

An MCTDH formalism to deal with so-called open systems (consisting of a primary system coupled to an external heat bath) was developed by Raab and Meyer [95] and applied by Cattarius and Meyer [97]. In the formalism, the assumption is made that the correlation times of the external heat bath system are small compared to those of the primary system. In this case, the Markov approximation, which neglects memory effects and is therefore local in time, is valid. The linear super-operator L was taken to be of the so-called Lindblad form, such that there can be an energy flow from the system to the bath, but not the other way around. The MCTDH method for wavefunctions was reformulated as a method for propagating the density of the system, using single-particle density operators rather than SPFs. Full details of the method used are presented in Ref. [97].

The MCTDH density-operator formalism was subsequently applied to the vibrational relaxation of CO on Cu(100) and the sticking of CO on Cu(100), using six and three degrees of freedom for these problems, respectively. The calculations presented served mainly to show that the newly developed formalism was functioning correctly.

Saalfrank and co-workers have studied the vibrational relaxation of H adsorbed on Si(100) using both the MCTDH [417] and the time-dependent self-consistent field (TDSCF) [418] methods. Their MCTDH calculations divided up the system of interest into two vibrations involving H (the primary system) and up to 50 phonons (the bath). An MCTDH wavefunction-based approach was taken similar to that described in Ref. [441]. The two H–Si vibrations considered were the H–Si stretching vibration and Si–Si–H bending vibration. The phonon coordinates and frequencies were obtained from a normal-mode calculation for a Si cluster covered by hydrogen atoms, consisting of 144 Si atoms and 36 H atoms. The normal-mode analysis employed a semi-empirical potential to describe the Si–Si bonds. Pairs of phonons were included in the calculation if the sum of their frequencies nearly coincided with differences in primary system energy levels, and if the phonon modes were strongly coupled to the primary system vibrations. In the system–bath coupling term of the Hamiltonian, it was necessary to include both single-phonon and two-phonon coupling terms.

The lifetime of the first excited Si–H stretching vibrational state, which is of the order of nanoseconds, was outside the realm that could be simulated with the MCTDH method. The decay of the Si–Si–H bending excited states was found to be non-exponential, but the half-life $T_{1/2}$ could be computed. The computed $T_{1/2}$ values (3.54, 1.13, 0.87 and 0.73 ps, for vibrational quantum numbers of the bending mode of 1, 2, 3 and 4, respectively) were too large compared with those obtained from perturbation theory (0.94, 0.48, 0.33 and 0.26 ps, respectively) [442] by factors in the range 2–4, which was attributed to the finite bath size (only 50 phonons were considered). However, a semi-quantitative agreement with the results of perturbation theory was observed, in the sense that simple scaling laws observed with perturbation theory concerning the dependence of the half-life on the bending quantum numbers were also found to be obeyed in the MCTDH calculations. Later TD-SCF calculations by the same group that considered a much larger number of phonons (534 instead of 50) obtained very good quantitative agreement with the results of perturbation theory, suggesting that for this kind of problem it is more important to include a large amount of bath modes than to include correlation using the MCTDH approach [418].

Perhaps the most impressive MCTDH calculations on a molecule–surface dynamics problem incorporating dissipation have been performed on photoinduced, ultrafast electron transfer between a dye molecule, coumarin 343, and a semiconductor surface, TiO_2 [229, 232]. To model the electron transfer following photoexcitation, a Hamiltonian was used that was represented in charge-localized diabatic electronic states, that is, one electronic state $|\phi_d\rangle$ representing the photoexcited dye molecule interacting with the electronic ground-state TiO_2 surface (the donor state), and a quasi-continuum of acceptor states $|\phi_k\rangle$ representing the situation where an electron has been transferred from the dye molecule (which has become a cation) to a level in the conduction band of the semiconductor. In the first calculation [229], a semi-empirical method based on the Newns model of chemisorption [443] and parametrized using a tight-binding model [444] was used to calculate donor and acceptor electronic energies and coupling matrix elements.

In the second calculation [232], some details of which are given below, an approach was taken based on electronic structure calculations and motivated by a projection operator approach previously used to describe resonant electron–molecule scattering [445]. A partitioning scheme was used based on DFT calculations on a complex of the dye molecule with a TiO_2 cluster. The scheme for describing the $|\phi_d\rangle$ and $|\phi_k\rangle$ states was based on the following three steps: (i) using localized basis sets, the Hilbert space was partitioned in a donor and acceptor group; (ii) the Hamiltonian was partitioned according to the donor–acceptor separation; and (iii) the donor and acceptor blocks of the partitioned Hamiltonian were diagonalized separately. More details are provided in Ref. [232].

To simulate the quantum dynamics associated with the electron transfer, the multilayer (ML) formulation [119, 228] of the MCTDH method was used (see Chapter 14). More specifically, the calculation [232] was done by modelling the vibrational motion in 39 vibrational modes of the coumarin 343 molecule, 1184 acceptor electronic states, and 20 oscillators to simulate the effect of the solvent molecules. The time-scale of the electron transfer predicted by the calculations (about 13 fs, see Figure 20.7) was faster than the experimental results for the system investigated, which were in the range 20–200 fs. The calculations predicted a slower injection component (by about a factor of 2) than the previous calculations based on the semi-empirical Newns model (see Figure 20.7). Furthermore, the first-principles model for the donor and acceptor energies and coupling strengths was found to lead to electronic coherences not observed in the calculations based on the semi-empirical model (see Figure 20.7), because the donor–acceptor coupling strengths depend significantly on the energy of the acceptor states. Finally, the comparison showed that the effect of the coupling to the nuclear degrees of freedom on the electron injection time is more pronounced in the semi-empirical model than in the more sophisticated electronic structure model (see Figure 20.7).

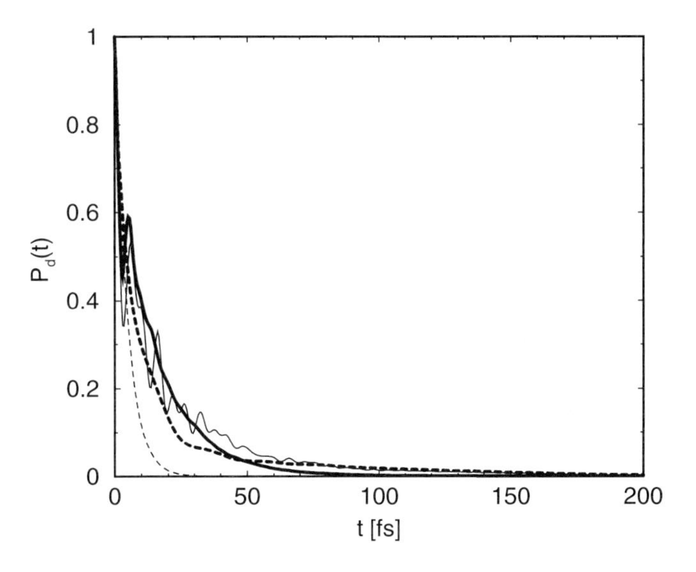

Figure 20.7 The population of the donor state after photoexcitation as a function of time, for the coumarin 343–TiO_2 system. In the calculations represented by solid lines, donor and acceptor energies and coupling strengths were calculated explicitly using an electronic structure approach detailed in [232]. In the calculations represented by dashed lines, donor and acceptor energies and coupling strengths were calculated by a semi-empirical theory explained in [229]. The thick lines present results of calculations including vibrational modes of coumarin 343, while the thin lines are results of purely electronic calculations. Reprinted with permission from Ref. [232]. Copyright (2007) American Chemical Society.

20.4 Summary and Outlook

Many molecule–surface scattering processes occur on short time-scales, allowing successful studies of these processes using the MCTDH method. MCTDH calculations have successfully addressed the rotationally and diffractionally inelastic scattering of H_2 from LiF(001) and from a model rectangular surface, and of N_2 from LiF(001). For the last system the MCTDH method was even used as a benchmark method to test an approximate mixed quantum–classical method, illustrating the maturity of the MCTDH method. The MCTDH method has also been using to study the dissociative chemisorption of H_2 on a model transition-metal surface, on Cu(100) and on Pt(111), of N_2 on Ru(0001), and of CH_4 on Ni. In the most challenging applications, six-dimensional calculations were carried out of the reaction probabilities of H_2 dissociating on Cu(100), and of rate constants for N_2 dissociating on a stepped Ru surface. MCTDH calculations have also addressed the photodissociation of molecules on insulator surfaces, for example, CH_3I on MgO, and HCl on ice(0001).

The MCTDH method has also been applied successfully to molecule–surface scattering problems affected by dissipation. Examples include the vibrational relaxation of H adsorbed onto Si(100), the vibrational relaxation and the sticking of CO on Cu(100), and the femtosecond time-scale photoinduced electron transfer from a dye molecule to a semiconductor surface in the presence of solvent molecules.

While applications of the MCTDH method to molecule–surface scattering problems have been quite successful, there is no reason for complacency. For instance, the MCTDH method has been used successfully in 6D quantum dynamical calculations of probabilities for the dissociation of H_2 on Cu(100). However, the PES employed was ideally suited to the MCTDH method, consisting of an expansion of products of two-dimensional functions containing only 20 terms. The form used cannot provide a very accurate fit describing dissociative chemisorption of a diatomic molecule on a metal surface [446]. Accurate reaction probabilities can be obtained using PESs fitted with the corrugation-reducing procedure [447] (available for $H_2 + Pt(111)$ as well as Cu(100) [432]) and using PESs interpolated with the GROW method [204, 448] (available for $N_2 + Ru(0001)$ [397]). However, accurate 6D MCTDH calculations on these systems employing these PESs or refitted forms of these PESs using the potfit method have yet to be demonstrated. The present situation is brighter for the computation of rate constants. It was recently demonstrated that accurate quantum mechanical values of rate constants can be obtained using an approach based on Shepard interpolation of the PES in the vicinity of the reaction barrier for the dissociation of N_2 on stepped Ru, employing just a few data points (see Section 20.3.2.2).

A new challenge to theorists working on the dynamics of molecule–surface reactions is to treat the dissociative chemisorption of polyatomic molecules on metal surfaces using quantum dynamics and describing the motion in all molecular degrees of freedom. For instance, many interesting experiments have been done on the dissociative chemisorption of methane on metal surfaces. Statistical theories have been used to argue that this reaction should be equally affected by energy in the molecule's translational motion towards the surface, in its vibrations (with no distinction between the different normal modes), and in the surface phonons [449]. This view has recently been contradicted by experiments on CH_2D_2 [399] and CH_4 [439, 450, 451] dissociating on Ni surfaces, as explained below.

In experiments on the dissociation of CH_4 on Ni(111), the reaction probabilities were measured for methane initially in its $3\nu_4$ state (containing 47 kJ mol^{-1} of vibrational energy) and for methane initially in its ν_3 state (containing 36 kJ mol^{-1} of energy). Based on statistical theories, one would expect the former state to be most reactive, but experiment shows that the latter state is more reactive [451]. This is shown in Figure 20.8. In an even more dramatic example, the reactivity of the ν_1 and ν_3 states (which have comparable energies, the energy of the former state however being somewhat lower) of methane on Ni(100) was measured. The reactivity of the ν_1 state was found to be higher than that of the ν_3 state by an order of magnitude, even though statistical theory predicts them to be about the same! See Figure 20.9 [439].

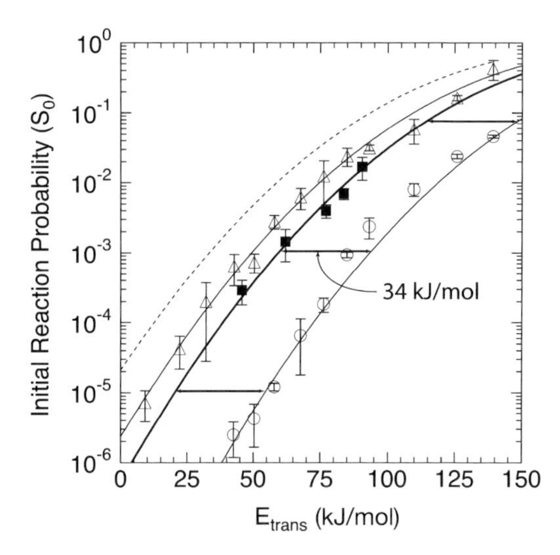

Figure 20.8 State-resolved experimental reaction probabilities for dissociation of CH_4 incident on Ni(111) in $v = 0$ (circles), ν_3 (triangles) and $3\nu_4$ (squares). Arrows indicate the shift in the reaction probability measured for the $3\nu_4$ state relative to that measured for $v = 0$. The dashed line indicates the reaction probability that would be expected for the $3\nu_4$ state if vibrational energy in the ν_4 mode were as effective in promoting reaction as vibrational energy in the ν_3 mode. Reprinted with permission from Ref. [451]. Copyright (2005) by the American Physical Society. See also http://link.aps.org/abstract/PRL/e208303.

It would be nice if MCTDH calculations could reproduce the trends in Figures 20.8 and 20.9, explain them, and make predictions about the reactivity of methane in excited states not yet investigated experimentally, such

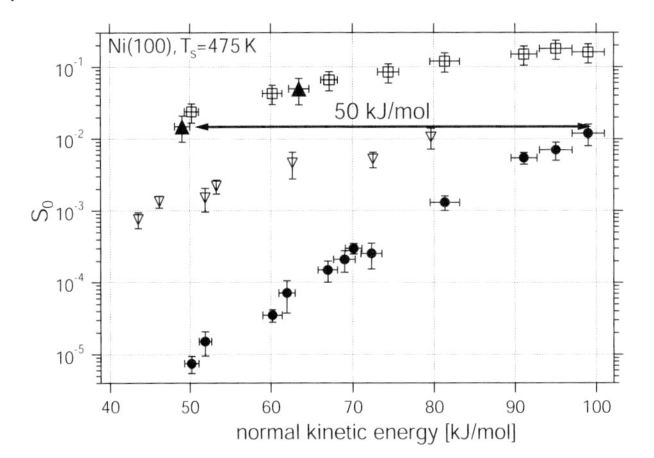

Figure 20.9 State-resolved dissociation probabilities for CH_4 in the ν_1 (filled triangles), ν_3 (empty inverted triangles), $2\nu_3$ (empty squares) and ground (filled circles) vibrational states incident on Ni(100). The error bars represent 95% confidence intervals. Reprinted with permission from Ref. [439]. Copyright (2005) by the American Physical Society. See also http://link.aps.org/abstract/PRL/v94/e246104.

as states involving excitation of the ν_2 mode only. So far, calculations using the MCTDH method and explicitly modelling dissociation of CH_4 have only treated up to three degrees of freedom [408], while calculations using the standard wavepacket method have modelled up to four degrees of freedom [452, 453]. Nevertheless, MCTDH calculations on the dissociative chemisorption treating all molecular degrees of freedom are likely to become possible in the near future, given recent developments of parallel computer hardware and recent improvements of the MCTDH method [119, 228]. Such calculations should be able to explain the mode selectivity of the reaction of methane on Ni surfaces (the trends in Figures 20.8 and 20.9). A major challenge that will have to be met concerns the representation of the PES, as a potfit representation in 15 degrees of freedom is unlikely to be effective or even feasible. Recent 6D calculations on H_2 + Pt(111) suggest that the use of n-mode potentials may represent a route to success [454], but further tests on late barrier systems (like H_2 + Cu) and ultimately polyatomic molecules reacting with surfaces are necessary to confirm this.

Computing accurate reaction probabilities for CH_4 + Ni is also a challenge for MCTDH calculations also modelling dissipation: recent 4D calculations on CH_4 + Ni(111) predict that during reaction the Ni lattice reconstructs, effectively lowering the barrier [453]. Another observation requiring explanation is the loss of energy observed in rotationally inelastic scattering of H_2 from Cu(100) [395], which could be due either to phonons or to electron–hole pair excitation. Other observations requiring accurate modelling with dissipation to electron–hole pairs include the observation of multi-quantum vibrational relaxation of NO on metal surfaces [455], and of electron emission from a surface with low work-function due to scattering of highly vibrationally excited NO from it [380]. In conclusion, the future for the MCTDH method looks bright in applications to molecule–surface scattering problems, with ample scope for exciting and challenging applications in this area.

21 Intramolecular Vibrational Energy Redistribution and Infrared Spectroscopy

Fabien Gatti and Christophe Iung

21.1 Introduction

Knowledge about the competition and the time-scales of intramolecular vibrational energy redistribution (IVR) of selectively vibrationally excited molecules on a molecular level is important for the understanding of rates, pathways and efficiencies of chemical transformations [456–460]. IVR studies aim to answer questions such as: Starting from a well-defined initial excitation of a molecule, where does the energy go? How long does it take? In the statistical theories of Rice and Marcus [461], IVR is assumed to be rapid and complete. It is indeed sometimes observed that the IVR time-scale is linked to the state density according to the Fermi golden rule and that the energy will be rapidly redistributed in a statistical way through all vibrational modes.

However, since then, our understanding of IVR has been substantially refined by the success of numerous experimental investigations. New tools (among which are lasers, molecular beams and double-resonance techniques) have indeed allowed the difficulties that plagued early experiments to be overcome, most notably the impossibility of preparing and studying a well-defined initial state [462, 463]. Short-pulse lasers can directly follow this flow of energy by exciting a specific vibrational motion and monitoring the transient response of the molecule. In the light of the many experimental findings for various molecules in the gas phase, it appears that neither the fast nor the statistical character of the IVR process should be taken for granted [464–467]. It can even happen that IVR proceeds via very specific pathways. If these specific channels lead to important processes, it then opens the way to the induction of chemical reactions by means of lasers. Indeed, one dream of chemists [468] is to employ laser sources to drive selective chemical reactions in such a way that only the desired products occur. This approach is apparently becoming a reality [469–472]. This field also offers new possibilities for the understanding of biological phenomena such as the elementary steps of

Multidimensional Quantum Dynamics: MCTDH Theory and Applications.
Edited by Hans-Dieter Meyer, Fabien Gatti, and Graham A. Worth

ISBN: 978-3-527-32018-9

vision and photosynthesis [468]. Indeed, life seems to exploit extremely efficient and selective pathways to induce the conversion of light into chemical energy.

Much theoretical effort must thus be directed towards the development of more refined models to study systems with less drastic approximations than before. Such general methods of molecular quantum dynamics coupled with quantum chemistry calculations could predict the vibrational states leading to a desired reaction path. Pioneering works were carried out in the 1990s and are technically based on the determination of an active space in which the dynamics is studied. The active space is extracted from an underlying primitive space by the wave operator method based on the Bloch formalism [456,473,474]. This approach has been applied by Wyatt, Iung and co-workers [475–483] to perform full nine- and 30-dimensional studies of the spectroscopy and the IVR of HCF_3 and benzene, respectively.

More recently, it was demonstrated in several examples that the Heidelberg package [13] of the multiconfiguration time-dependent Hartree (MCTDH) algorithm [30, 34, 37, 38] is an efficient tool to investigate IVR. Here, we confine ourselves to IVR studies in molecules in their ground electronic state. In order to begin a systematic study of the IVR process, there are reasons for focusing on the electronic ground state [484]. First, more spectroscopic data for an individual molecule are, in general, available than for electronic excited states, and these data are very important to interpret the IVR process correctly. Second, in electronic excited states, there are several other competing processes in the overall energy redistribution (involving the internal conversion to other states), which drastically complicates the interpretation of the IVR simulations. Moreover, in the ground state, certain vibrations can be related to specific chemical bonds (in particular, the stretching modes of vibration), in contrast to the vibrational modes in excited states, which involve the motion of many atoms and so already start out spatially delocalized. For all these reasons, it is more reasonable to start a systematic prospect of laser-enhanced, mode-selective chemistry with molecules in their electronic ground state.

As mentioned above, high-resolution vibrational spectroscopy is essential for understanding flow in molecules. Consequently, IVR studies are directly linked to the simulations of infrared (IR) spectra. The natural way to generate eigenvalues from the knowledge of a propagated wavepacket is to Fourier-transform the autocorrelation function. However, this approach is useful when a high resolution of individual lines is not desired. When individual lines are to be resolved, the energy levels may be converged with MCTDH by analysing the autocorrelation function with the aid of the filter diagonalization method [53–55, 64, 485]. Recently, a very important advance has been made with the development of the *improved relaxation* method (Chapter 8). The latter allows convergence with a high accuracy of the eigenvalues and eigenstates for rather large systems. For instance, a comprehensive calculation of an IR spectrum by improved relaxation has recently been achieved for six-dimensional (6D) problems, namely H_2CS [101] and HONO [65, 107]. Several 15-dimensional eigenstates have also been converged for the Zundel cation $H_5O_2{}^+$ [59, 156, 486].

Consequently, the Heidelberg MCTDH package [13] could offer a precious framework for a synergy between experimenters and theoreticians in the field of IVR and IR spectroscopy. The goal of the present chapter is to review some applications in this domain.

21.2
Local-Mode Excitation of the CH Stretch in Fluoroform and Toluene

Molecules containing a light-atom (XH) stretching vibration are ubiquitous. For such XH vibrations, the oscillator strength is almost exclusively carried by the light-atom local stretching motion. Numerous experimental studies have shown that, if some energy is deposited in this XH bond, the IVR of the system is not statistical: see, for instance, CF_3H [487–491], HNCO [492], HOCl [493], CH_3OH [494], H_2O_2 [495] and benzene [496–498]. Since the XH local mode of vibration is close to a normal mode of vibration, it is natural to discuss the physics of the system in terms of the number of quanta in this bond.

A very popular system has been the CH 'chromophore', since it can be probed in a large variety of molecular environments and has pronounced absorption signals. In our studies with MCTDH, we have considered [499] the CH overtones of fluoroform, which exhibit distinct spectral features indicating a pronounced separation of the IVR time-scales and therefore multiple IVR pathways. The dominant feature in the CH overtone spectra of fluoroform is the strong Fermi resonance ($\nu_{\text{CH stretch}} = 2\nu_{\text{FCH bend}}$) between the CH stretch and the two FCH bends. This characteristic was recognized in 1948 by Bernstein and Herzberg [500]. The CH stretch and the two FCH bends constitute the so-called three-dimensional (3D) CH chromophore and, in high-resolution overtone spectroscopy [487, 491, 501–505], several polyad band structures appear.

In Ref. [499], we have simulated the dynamics during several picoseconds after excitation of a local mode of the CH stretch. The system was described by means of nine rectilinear normal coordinates, and we have used the potential surface of Maynard *et al.* [506]. For $v_{\text{CH}} = 1$, we observe, first, a rapid and reversible exchange between the CH stretch and the two FCH bends, and, second, a reversible exchange of the energy between the chromophore and the other modes of vibration on a longer time-scale. The positions of the main bands of the fundamental and the first CH overtone are shown in Table 21.1.

In order to avoid very long propagations, we have obtained these levels by filter diagonalization [53, 64, 507]. For $v_{\text{CH}} = 1$, we see in Table 21.1 that two

Table 21.1 Energies (in cm^{-1}) of those eigenstates whose projections on $|1\nu_{CH}\rangle^0$ (or $|2\nu_{CH}\rangle^0$) are larger than 0.05.

Line	WOSA [482]	MCTDH	Expt [487–491,501]
$1\nu_{CH}$	3050	3050	3035
$2\nu_{bend}$	2750	2750	2710
Satellite lines	3046	3046	3031
—	3069	3068	3040
$2\nu_{CH}$	5986	5989	5959
$1\nu_{CH} + 2\nu_{bend}$	5760	5759	5710
$4\nu_{bend}$	5438	5436	5337

main levels crop up corresponding to the Fermi resonance between the CH stretch and the FCH bends. Moreover, two satellite lines appear corresponding to background resonances between the CH stretch and the other modes of vibration. The comparison between our values and those obtained with the wave operator sorting algorithm (WOSA) method [482] demonstrates the reliability of both calculations. For $v_{CH} = 2$, we observe the same exchange of energy between the CH stretch and the two FCH bends, but the energy flow to the other modes of vibration is irreversible and very slow. The other vibrational modes now play the role of a 'bath'. IVR is not statistical and it is likely that a full equilibrium of the energy between the chromophore and bath modes takes at least 100 ps. For $v_{CH} = 3, 4$, we observe similar trends: there is a very fast exchange of energy between the CH stretch and the two FCH bends. However, the coupling with the bath modes accelerates this energy redistribution. More importantly, a large part of the energy remains in the chromophore. Consequently, and as expected experimentally, the CH chromophore acts as an energy reservoir, especially the FCH bends, although the state density becomes very large. Very recently, we have implemented the nine-dimensional (9D) Hamiltonian operator in polyspherical coordinates using a kinetic energy operator coming from Equation (12.41) in Section 12.4.2 and a new and more accurate potential surface from Ramesh and Sibert [508, 509]. We hope that we will be in a position to investigate the dynamics higher in energy. HCF_3 was the first application of MCTDH in the field of IVR and was chosen as benchmark results were available (with the WOSA method of Ref. [482]). These results were important, as they proved that MCTDH can be very accurate in this new domain of applications.

Moreover, we have also investigated [152] IVR in the rotating methyl group in toluene $C_6H_5CH_3$, which can be viewed as a model study on methyl groups, which are ubiquitous in organic molecules. Experiments [510–517] have shown that the excited CH stretching spectra of toluene exhibit a complex structure. However, it is not clear whether this structure can be ascribed to a coupling between the CH stretching and the almost free internal rotation of the methyl group, or to a perturbation of these spectral profiles by the CH stretch–bend Fermi resonances [518]. The work in Ref. [152] was important because it was the first calculation with the improved relaxation method (Chapter 8) and also the first application of *potfit* to a 9D potential energy surface (PES). However, the empirical potential surface used [518] is not likely to be accurate enough to reproduce correctly all the details of the investigated energy transfer. A new PES should be calculated and could be very important for fully understanding the relative importance of the rotation of the CH_3 subsystem and of the Fermi resonances between the CH stretching and the HCH bending modes.

21.3 Study of Highly Excited States in HFCO and DFCO

We focus now on HFCO and DFCO, which display a relatively low dissociation threshold (about 14 000 cm^{-1} above the ground state). Sophisticated experiments developed by Moore and co-workers [519–523] seem to indicate that the IVR dynamics in DFCO is very different from the strong mode-specific IVR in HFCO when the out-of-plane mode is highly excited in the range 13 000–23 000 cm^{-1}. HFCO is an excellent prototype to study mode specificity in unimolecular dissociation, since the out-of-plane bending mode is weakly coupled to the reaction coordinate, which lies entirely in the molecular plane (the transition state is planar, with an energy of approximately 17 900 cm^{-1} above the minimum). According to the experimental study [519–521], the characterized states for HFCO in the range 13 000–23 000 cm^{-1} are extremely stable with respect to state mixing, in contrast to the general belief that the vibrational quantum states above the dissociation threshold are chaotic. The experimental studies have stimulated Kato and co-workers [524] to determine a sophisticated *ab initio* global 6D PES that includes the dissociation pathway (HFCO $\rightarrow$ HF + CO). An energy cut-off at 3.5 eV (about 28 230 cm^{-1}) has been introduced in this PES. Hence the isomerization HFCO $\rightarrow$ ^{t}FCOH is not described by this surface. The high quality of this PES has recently been demonstrated by comparing experimental and numerical energies up to 9000 cm^{-1} in DFCO [525].

IVR in HFCO [154] and DFCO [155], whose out-of-plane mode is highly excited, has been the subject of numerical simulations using the MCTDH package to answer the following questions:

(i) How efficient is the MCTDH package in providing highly excited energies of a system described by a global PES whose expression is not factorized and not ideally suited to an MCTDH calculation?

(ii) What are the main features (selectivity, time-scale of the energy flow) of IVR in XFCO (X = H or D) whose out-of-plane mode is initially excited by n quanta ($n = 2, \ldots, 20$)?

(iii) How different is the energy flow in DFCO and HFCO?

First, the efficiency of MCTDH coupled to a filtered diagonalization and a fit of the PES has been established by calculating for HFCO [154] the energies of the out-of-plane overtones $|n\nu_6\rangle$ ($n = 2, \ldots, 18$) and by comparing MCTDH data with some energy levels obtained by a time-independent approach based on a Davidson algorithm [526–529] and with the available experimental data [519–521]. Comparison with experimental data and time-independent data for HFCO is provided in Table 21.2.

Table 21.2 Energies (in cm^{-1}) of HFCO out-of-plane overtones.

$\|n\nu_6\rangle$	Time-independent[a]	MCTDH [154]	Experiment [519]	MCTDH$^{corr\ b}$
$n = 1$	1019	1019	1011	1011
$n = 2$	2031	2031		
$n = 4$	4036	4034		
$n = 6$	6018	6012		
$n = 8$	7985	7968		
$n = 10$	9949	9906		
$n = 12$		11 826		
$n = 14$		13 723	13 622	13 611
$n = 16$		15 609	15 486	15 481
$n = 18$		17 477	17 319	17 333
$n = 20$		19 240	19 125	19 080

[a] These values have been obtained by a time-independent approach based on a Davidson scheme, which provides selectively eigenstates and eigenvalues [526–528, 530].
[b] These corrected values have been obtained using a potential that was modified to correct for the error observed in the fundamental frequency ($1\nu_{\text{out-of-plane}}$).

The last column of Table 21.2 gives corrected values of the energies by taking into account the error generated by the PES for the fundamental transition. An error of about 8 cm^{-1} is observed for $1\nu_6$. Consequently, the corrected values for the out-of-plane mode overtones ($n\nu_6$) given in column 5 are obtained by subtracting $n \times 8\,cm^{-1}$ from the MCTDH energy. A recent study [525] shows that such a correction provides energies with a very good accuracy. Comparison between columns 4 and 5 demonstrates the great quality of the PES used [524] even for describing highly excited states. For the lower levels, a comparison has been performed between the MCTDH energies and the values provided by a time-independent approach [528, 529]. The difference between columns 2 and 3 is generated by the fact that the existence of an energy cut-off introduced in the PES (for $E = 3.5\,\text{eV}$) has not been treated in the same way in these two studies [154, 528]. In the Davidson calculation [528], the potential is considered as infinite for energies larger than 3.5 eV, whereas the MCTDH study [154] assumes a constant potential energy equal to 3.5 eV. It is important to note that such a difference for energies larger than 3.5 eV generates some small but not negligible differences for the first overtones and for the low excited part of the spectrum. This remark shows how the calculation of the energy levels is very sensitive to the features of the PES even when the difference occurs for very high excited energies. Consequently, MCTDH is an excellent tool for extracting some highly excited states with a very good accuracy. Time-independent methods are not, at the moment, able to provide such highly excited overtones using a global PES. This study demonstrates also the quality of the PES used [524].

Let us focus now on the very beginning of the IVR in an XFCO (X = H or D) molecule whose out-of-plane mode is initially excited by n quanta. Using the MCTDH package, the evolution of the wavepacket, $\Psi(t)$, has been calculated and analysed. The fraction of energy in the normal mode Q_i of the molecule whose out-of-plane mode has been initially excited by n quanta is denoted $F_i^n(t)$ and is estimated by:

(i) for in-plane modes

$$F_i^n(t) = \frac{E_i(t) - E_i(t{=}0)}{E_\varphi^n - E_\varphi^0}$$

(ii) for the initially excited out-of-plane mode

$$F_\varphi^n(t) = \frac{E_\varphi(t) - E_\varphi^0}{E_\varphi^n - E_\varphi^0}$$

with $E_i(t) = \langle \Psi(t) | h_i^0 | \Psi(t) \rangle$, where h_i^0 is a 1D anharmonic zero-order Hamiltonian describing mode Q_i and E_φ^n is the energy of the nth excited state of h_φ^0.

Figures 21.1 and 21.2 provide the time evolution of the fraction of energy for three different modes and for different excitation energies from 2000 to 20 000 cm^{-1} in HFCO and DFCO, respectively. The three modes have been selected for the following reasons: the out-of-plane mode is the initially excited mode, the CO stretch exhibits a Fermi resonance in DFCO ($\nu_{\text{CO}} \simeq 2\nu_\varphi$), while the FCO bend has a dynamical behaviour similar to the other in-plane modes. Figure 21.1 shows that IVR in HFCO is not strongly affected by the increase in the state density. The structure of the IVR during the first step of the dynamics is very stable with respect to the number of quanta in the out-of-plane mode from $2\nu_6$ up to $20\nu_6$. This means that no new efficient couplings crop up when n increases and that the effect of the huge increase in the state density is very limited.

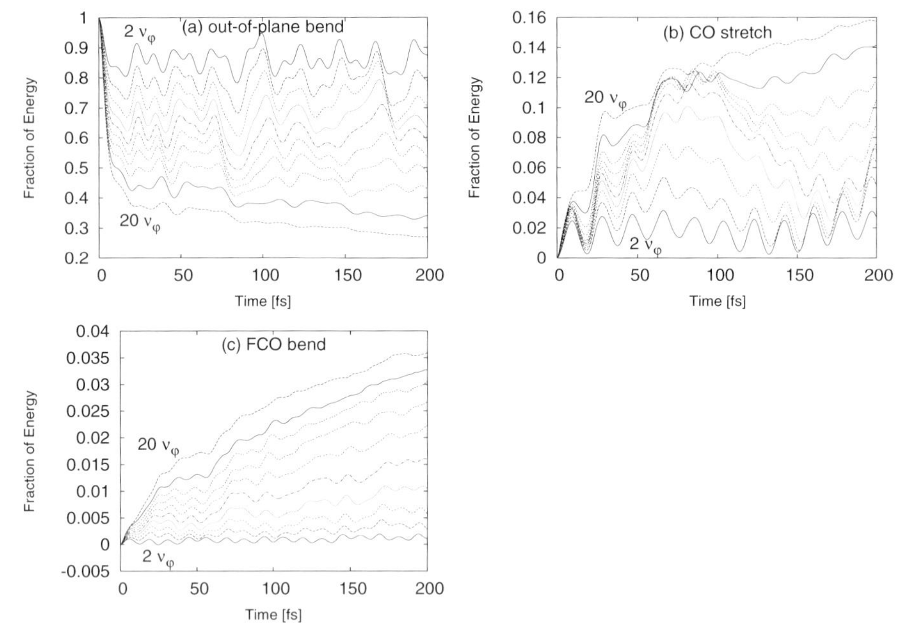

Figure 21.1 Evolution of the fraction of the energy in different modes in an HFCO molecule whose out-of-plane mode has been excited by n quanta ($n = 2$–20). The figures correspond to (a) the out-of-plane mode, (b) the CO stretch and (c) the FCO bend. The trends of the figures that provide the fraction of energies in the other in-plane modes (CF and CH stretches and HCO bend) are similar to (c).

What happens in DFCO, in which a Fermi resonance exists between the out-of-plane mode and the CO stretch (Table 21.3)? Figure 21.2(a) and (b) convey a very different message from the similar figures for HFCO (Figure 21.1(a) and (b)). They show a reversible exchange between the out-of-plane mode and the CO stretch, disclosing a strong Fermi resonance. Moreover, the remarkable quasi-stability of the short-time dynamics initiated by the out-of-plane mode excitation observed for HFCO is totally destroyed. This is in perfect agreement with the experimental predictions. Moreover, the global qualitative behaviour of the exchange of the energy between these two modes is strongly affected by the increase of the excitation energy. The structure of the energy flow quickly disappears and becomes hardly visible for $14\nu_6$. Finally, the Fermi resonance seems to become less effective for high-energy excitation. Figure 21.2(c) shows that the dynamical behaviour of the in-plane modes (which are not the CO stretch) is similar to their behaviour in HFCO (Figure 21.2(c) reproduces only the FCO bend behaviour but a similar trend is obtained for the CD and CF stretches and the FCO bend). This means that the existence of the Fermi resonance does not significantly modify the dynamical behaviour of the molecule.

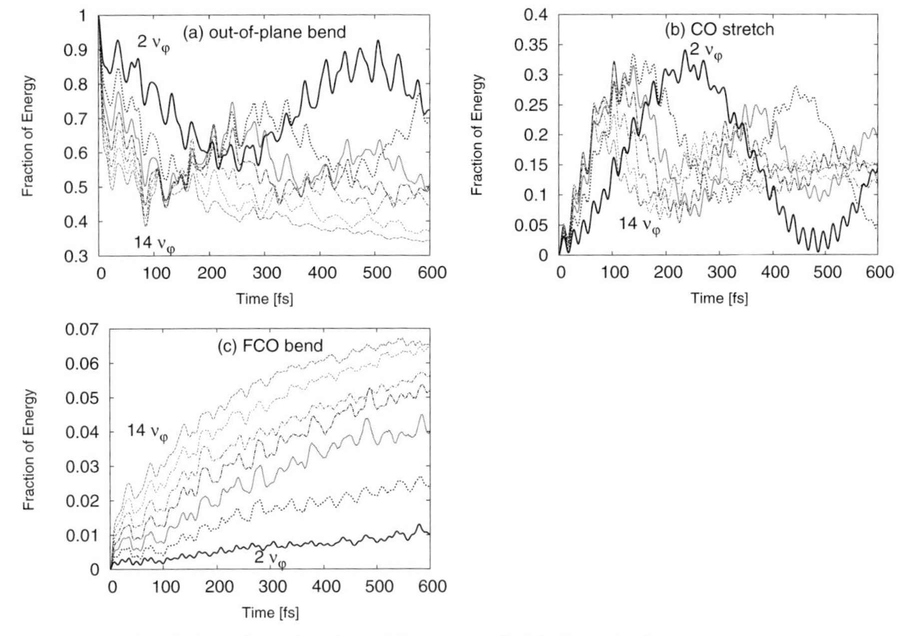

Figure 21.2 Evolution of the fraction of the energy in (a) the out-of-plane mode, (b) the CO stretch and (c) the FCO bend in a DFCO molecule whose out-of-plane mode has been excited by n quanta ($n = 2$–14). The trends of the figures that provide the fraction of energies in the other in-plane modes (CF and CD stretches and DCO bend) are similar to (c).

Table 21.3 Energies (in cm^{-1}) of HFCO out-of-plane overtones $|n\nu\rangle$.

	HFCO $X = H$	*DFCO* $X = D$
ν_1 (CX stretch)	2981	2262
ν_2 (CO stretch)	1837	1797
ν_3 (XCO bend)	1343	968
ν_4 (CF stretch)	1065	1073
ν_5 (FCO bend)	662	657
ν_6 (τ, out-of-plane)	1011	857

Figure 21.3 provides the fraction of energy in the different normal modes after an initial excitation energy of about 8000 cm^{-1} (Figure 21.3(a) and (c)) and 18 000 cm^{-1} (Figure 21.3(b) and (d)) for HFCO (Figure 21.3(a) and (b)) and DFCO (Figure 21.3(c) and (d)). The structure of the energy flow to the other modes in DFCO is not very different from HFCO. For instance, the energy flow to the modes that are linked to the dissociation, such as the CF stretch and the DCO bend, is still very small. The XCO bend and CX stretch (X = H

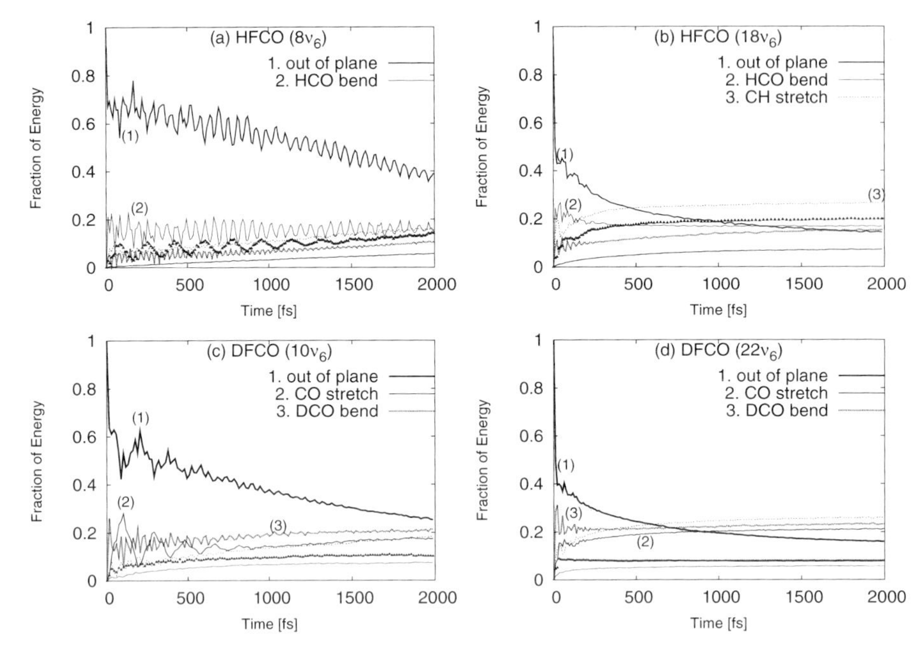

Figure 21.3 Evolution of the fractions of energy in the different modes when the initial excited state is (a) $8\nu_6$ in HFCO, (b) $10\nu_6$ in DFCO (about 8000 cm^{-1} of excitation), (c) $18\nu_6$ in HFCO and (d) $22\nu_6$ in DFCO (about 18 000 cm^{-1} of excitation).

and D) are the modes that receive most of the energy from the out-of-plane mode at the beginning of the dynamics. It is remarkable that the CX (X = H or D) stretch plays a similar role in these two molecules, while ω_{CD} in DFCO is much smaller than ω_{CH} in HFCO. Interestingly enough, no strong resonance between the CO stretch and the DCO bend occurs in DFCO, whereas the experimental work supposed that such resonance might play an important role. Our results seem to indicate a rather different behaviour: the periods of the energy flows into the CO stretch and the DCO bend are totally different. We do observe an increase of the energy flux from the out-of-plane bend to the DCO bend with the number of quanta. However, this increase is due to a direct coupling with the out-of-plane mode exactly like in HFCO and is not due to an indirect coupling through the CO stretch. Such a conclusion does not agree with the experimental predictions, which consider that the Fermi resonance should drastically modify the DFCO dynamical behaviour. Our simulation shows that the CO stretch is not efficiently coupled with the other in-plane modes. Consequently, the dynamical behaviours of HFCO and DFCO are rather similar in this respect.

For the sake of completeness, it could be important in the future to study the effect of rotational motion on the mode specificity of HFCO. It is indeed possible that the Coriolis forces can couple the out-of-plane motions to the reaction coordinate, leading to dissociation. We are now adding the dipole moments to carry out simulations in the presence of an external field. Our preliminary [531] results show that the time evolution of an XFCO molecule excited in the out-of-plane mode by an external field is very similar to the results provided by Figures 21.1 and 21.2.

21.4 Selective Population of Vibrational Levels in H_2CS in the Presence of an External Field

In previous work, the initial states were chosen to be excitations of local modes and the energy redistribution was investigated up to a few picoseconds. These preliminary studies are essential to highlight the IVR selectivity and the qualitative features of the processes. However, these excitations of local modes are somewhat artificial, since they can be different from those which are obtained by irradiation with laser light, the standard way to excite molecules experimentally. The next step is thus to explicitly implement the dipole moment surfaces and the external fields in the quantum mechanical simulations. We have used H_2CS as a first application because an accurate potential function is available for the electronic ground state [532] as well as the dipole moment function [533]. Moreover, previous calculations of eigenvalues, transition moments and selective population of the vibrational levels have been reported [533] using a different approach, that is, a development of the time-dependent wavefunction in stationary vibrational eigenstates.

In order to mimic a laser pulse, we used pulse parameters similar to those investigated by Leonard *et al.* [533] (non-chirped pulse):

$$E(t) = E_0 \cos(\omega t) \sin^2\left(\frac{\pi t}{t_p}\right) \theta(t)\theta(t_p - t) \tag{21.1}$$

where E_0 is the field strength, ω the field frequency, t_p the pulse duration and θ the step function. We use the Loudon [534] presentation, which describes the molecule quantum mechanically, the electric field classically and their interaction by first-order perturbation theory, that is,

$$\hat{H}_{\text{tot}}(t) = \hat{H}_0 + \hat{h}(t) \tag{21.2}$$

where $\hat{H}_0$ is the Hamiltonian of the H_2CS molecule without an external field, and $\hat{h}(t)$ is the interaction between the molecule and the electromagnetic field, which reads

$$\hat{h}(t) = -\boldsymbol{\mu} \cdot \mathbf{e}\, E(t)$$
$$= -\mu \cos(\theta) E_0 \cos(\omega t) \sin^2\left(\frac{\pi t}{t_\mathrm{p}}\right) \theta(t)\theta(t_\mathrm{p} - t) \qquad (21.3)$$

In this equation, $\boldsymbol{\mu}$ is the total dipole moment vector of the molecule, $\mathbf{e}$ is a unitary vector parallel to the electric field polarization, and θ is the angle between the electric field polarization and the dipole moment vector. We consider the molecule as oriented and assume that the direction of the dipole moment vector does not change during the interaction with the laser field.

A comprehensive calculation of an IR spectrum by improved relaxation was achieved for the first time for H_2CS [101]. Moreover, we have selectively populated given vibrational eigenstates with the parameters optimized by Leonard *et al.* [533] using the downhill simplex method. Table 21.4 gives the final populations and compares them with the results of Ref. [533]. We recall that, in Ref. [533], the approach to propagate the wavepacket is totally different since the latter calculation used an expansion of the wavefunction in a basis set of bound eigenstates.

Table 21.4 Comparison of the populations reached at the end of the pulse for selected target states. E_0, ω and t_p are the three parameters of the optimized pulses as in Equation (21.1). When the maximum of the population with MCTDH is reached before t_p, this time is given in parentheses. ΔE is the difference between the eigenenergy of this work and that of Ref. [533] with MULTIMODE.

Target	Energy[a]	ω (cm^{-1})	E_0/E_h[b]	t_p (fs)	Pol.[c]	Pop.[d]	Population[a]	ΔE
$1\nu_2$	1457.4	1457.58	0.00455	2419	z	1.00	1.00	−0.2
$2\nu_2$	2876.0	1441.84	0.0086	3437.2	z	0.97	0.97	−0.8
$\nu_1 + \nu_2$	4441.9	1471.3	0.01477	1355.6	z	0.37	0.09 (740 fs)	−6.7
$2\nu_3$	2101.8	1056.73	0.00455	2007.0	z	0.82	0.83	−0.6
$2\nu_3$	3128.3	1052.1	0.00856	1123.86	z	0.72	0.58	−1.0
$2\nu_5$	6013.8	6011.8	0.010	6000	z	1.00	0.14 (3075 fs)	+1.8
$2\nu_6$	1979.1	1974.76	0.01396	2209.45	z	0.84	0.31	−0.3
$2\nu_1$	5789.0	2904.49	0.03179	220.63	z	0.66	0.16 (135 fs)	−20.8
$2\nu_1$	5789.0	5804.26	0.03952	1829.1	z	1.00	0.52 (1115 fs)	−20.8
$1\nu_4$	990.5	996.61	0.00485	2816.87	x	0.91	0.69 (2010 fs)	−1.2
$2\nu_4$	1939.5	965.05	0.00260	2775.43	x	0.76	0.23 (2060 fs)	−4.5
$1\nu_6$	1979.1	988.22	0.00207	5435.15	y	0.75	0.53 (6540 fs)	−0.3

[a] From MCTDH. [b] Energy in hartrees. [c] Polarization. [d] Population from Ref. [533].

For some of the calculations, the agreement with the MCTDH calculation and the value obtained in [533] is excellent. However, for other calculations, the deviation is rather large. This difference is not due to a lack of convergence in either of the two dynamical approaches but must be traced back to the great sensitivity of the dynamics with respect to the eigenstates: the pulses (above all, the frequencies) are optimized for the eigenstates calculated in [533]. If these eigenstates are not perfectly converged, the MCTDH calculations with these pulses partially miss the targets. It should be emphasized that this sensitivity is aggravated for multiphoton excitations (see, for instance, Ref. [535, 536] for a discussion regarding this point). This tendency directs us towards using shorter pulses in our IVR simulations in the future.

21.5 *Cis–Trans* Isomerization of HONO

HONO is one of the smallest molecules that exhibits a *cis–trans* conformational equilibrium, and the corresponding isomerization possesses a strong mode selectivity. Consequently, it constitutes an ideal prototype for theoreticians to investigate IVR leading to a chemical process.

A new six-dimensional *cis–trans* double-minimum PES of the electronic ground state was determined using a coupled cluster approach [537]. More recently, we have calculated a six-dimensional dipole moment function [107]. The calculated and experimental transition energies and moments for the *cis* geometry are given in Table 21.5. They are in good agreement with the experimental values especially for the 000100 level corresponding to the ON central stretching mode, which is the most important one for our studies, as explained below.

Table 21.5 Comparison of calculated and experimental transition energies in cm^{-1} for the *cis* geometry of HONO (τ is the torsional angle). Given in parentheses are the transition moments in $km\,mol^{-1}$.

	ν_{exp} [a]	Calc. [b]	Calc. [c]	Calc. [d]	Var. [e]	CCSD(T)/DFT [f]
000010 (ONO)	609	606(42)	592(37)	584(40)	616.3	616.7(33.7)
000100 (ON)	852(291)	854(296)	830(346)	841(214)	849.7	850.1(281)
001000 (HON)	—	1336(9)	1269(15)	1308(88)	1312.6	1311.6(7.5)
010000 (N=O)	1641	1641 (157)	1576(195)	1694(77)	1639.5	1632.0(177)
100000 (OH)	3426	3653(38)	3404(69)	3405(6)	3438.5	3438.1(18.0)
000002 (τ)	—	—	1130(9)	1219(18)	1213.4	1212.7 (0.2)

[a] From Ref. [538]. [b] CCSD(T) calculations from Ref. [539].
[c] (MP2/CC-VSCF)/HF 2D anharmonic calculations from Ref. [540].
[d] DFT, 6D anharmonic calculations from [541].
[e] From [537]. [f] Our MCTDH results.

The interconversion barrier was found to be relatively low at 4105 cm^{-1} above the *trans* ground state. Even though the *cis* ground state lies only 94.0 cm^{-1} above the *trans* one, the two conformers present a very different behaviour due to Fermi resonances, which are more efficient in the *cis* conformer than in the *trans* one. In particular, in agreement with previous experimental [538, 540, 542–547] and theoretical work [548–553], our first MCTDH calculations (without a laser field) [65] show that the *cis* $\rightarrow$ *trans* process proceeds much faster than in the opposite direction. Moreover, we have shown that there are very large differences between the energy redistributions after different excitations of the local modes [65], stressing the strong mode selectivity of HONO.

In Ref. [107], we triggered isomerization via a laser field. Since the *cis* $\rightarrow$ *trans* isomerization is much faster than in the opposite direction, we started from the *cis* ground state. A preliminary study showed that an efficient local mode for triggering *cis*$\rightarrow$ *trans* isomerization is the ON middle stretching. Following the analysis in [537], the isomerization after ON stretching mode excitation is essentially driven by an efficient potential coupling with the torsional angle. It was indeed shown that the barrier height is considerably lowered when the ON stretching mode is allowed to adjust. We have used an external field parallel to the ON bond. A more precise analysis of the process has shown that an efficient *selective* IVR pathway leading to the *cis* $\rightarrow$ *trans* isomerization is expected when a wavepacket motion is induced by introducing a simultaneous increase of the central ON stretching and the ONO bending. This exactly corresponds to an excitation of the eigenstates corresponding to the $000n000$ series and to a laser with a frequency of about $850\,\text{cm}^{-1}$. Consequently, we have used a z-polarized laser field with a frequency of about $850\,\text{cm}^{-1}$, the z-axis being parallel to the ON bond. The irradiation duration was 500 fs.

Figure 21.4 summarizes the whole process. We have plotted a two-dimensional section of the potential hypersurface of HONO, including the two minima and the saddle point. The abscissa corresponds to τ (the torsional angle) and the ordinate to R_3 (the central ON distance). The lowest contour line corresponds to 0.017 eV and the highest one to 2.0 eV. Also depicted are snapshots of the evolution of reduced-dimensionality probability densities of HONO between 290 and 400 fs after excitation. Figure 21.4 shows that the wavepacket performs a localized motion in the initial phase of the excitation with the laser pulse ($t = 290$ and 305 fs): excitation along the z-axis leads to regular oscillations along the optimal direction. The averaged value of R_3 is smaller than the value at the equilibrium geometry for 290 fs and larger for 305 fs. When the averaged value of R_3 becomes larger, the wavepacket starts to delocalize in the τ direction towards the *trans* geometry ($t = 315$ fs). The absorbed energy is thus sufficient to allow motion of the wavepacket over the inversion barrier ($t = 325$ and 350 fs). After 350 fs, the system has absorbed a larger amount of energy and the population in the *trans* grows considerably ($t = 370$ and 400 fs). The part of the wavepacket in the *trans* potential well is spread out and several maxima of probability appear.

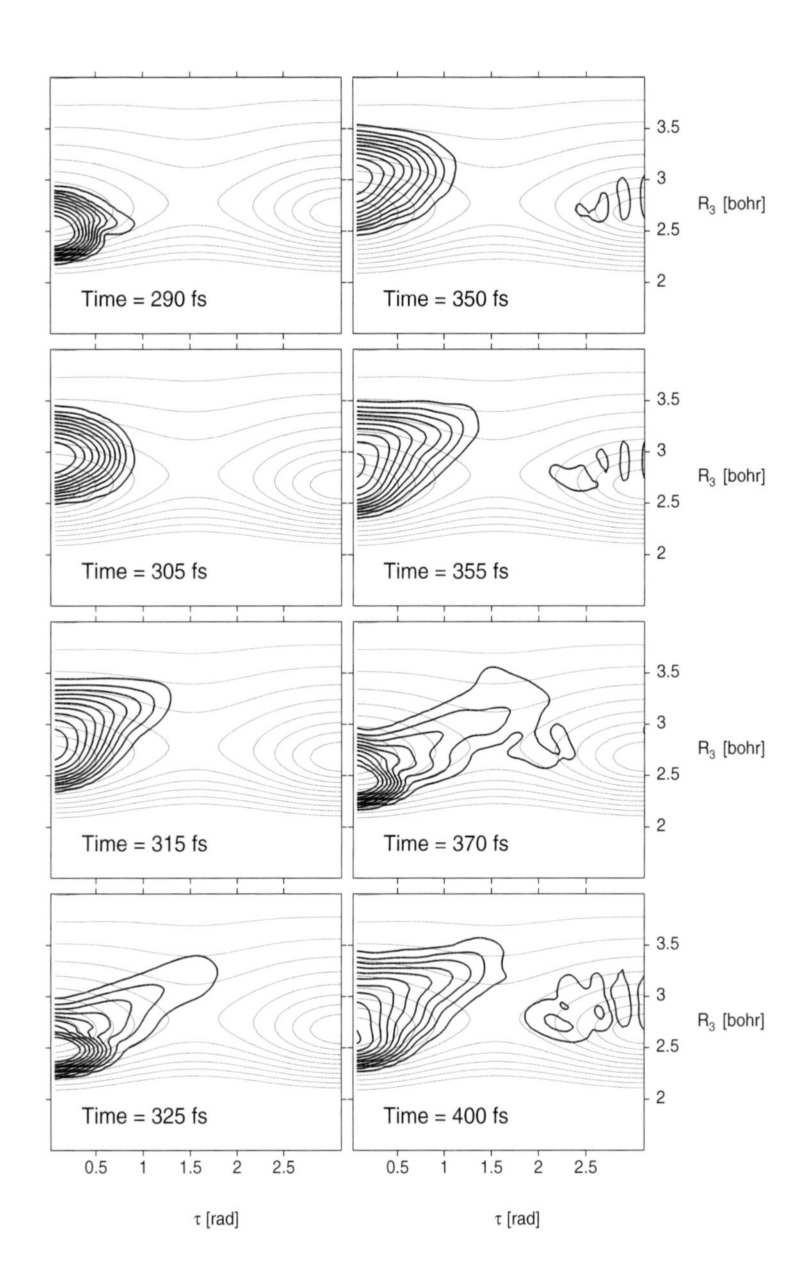

Figure 21.4 Reduced density logarithmic contour plots of the wavepacket in the R_3, τ plane during laser excitation using a carrier frequency corresponding to 850 cm^{-1}, a maximal field strength of 0.0035 a.u. and a $\sin^2$ pulse envelope with a duration of 500 fs. Underlaid is the PES with the other coordinates fixed at *cis* equilibrium.

After 500 fs the laser is off but the induced process continues during the IVR. We have estimated the *trans* probability after 2.5 ps in Table 21.6 for different laser intensities. The final *trans* probability P^{F}_{trans} increases roughly linearly with increasing absorbed energy within the energy range considered. Higher yields are likely if higher intensities or longer irradiation times are used. We have also observed that the process is relatively robust against variation of the excitation frequency. To give an estimate of the expected final *trans* probability for an experiment with free orientation, we have also propagated for orientations other than z. We found no isomerization for x and y polarizations for 0.0035 a.u. field strength. Orientation of 45° between the x- and z-axes yields about the half of the final P_{trans} of the z-polarized case. Thus, assuming isotropic orientation during the laser irradiation, one can roughly estimate that one-third of the calculated P_{trans} will be observed.

Table 21.6 Comparison of the different values of the final *trans* probability P^{F}_{trans} for different total energies and laser intensities. The laser frequency is equal to 850 cm^{-1}.

Laser intensity (a.u.)	Total energy (eV)	P^{F}_{trans} ($t = 2500$ fs)
0.0035	1.045	0.17
0.004	1.128	0.22
0.005	1.260	0.28
0.006	1.363	0.34

21.6 Conclusion

In this chapter, we have shown how it is possible with MCTDH to develop a systematic study of the IVR and laser control for several molecular systems. This could offer an efficient framework for a synergy between experimenters and theoreticians in this field. HONO is a good example of what could be feasible to spur new experimental investigations. Such an approach is similar to 'intelligent control of reactivity' by laser pulses in organic chemistry as developed by Robb and co-workers [199,554], in which excited electronic states coupled by conical intersections are involved. Such joint experimental–theoretical studies are essential to establish unambiguously the connection between the experimental signals that are measured and the molecular dynamics that is to be observed.

For the sake of completeness, it could be important to study in future the effect of Coriolis coupling on the mode specificity of the IVR. It is indeed sometimes observed experimentally that the Coriolis forces can influence the dynamics in molecular systems. Finally, it should also be emphasized that the present IVR studies must be seen as complementary to the MCTDH studies focusing on the vibronic coupling problems involving conical intersections. The latter provide the microscopic mechanism for ultrafast relaxation processes in polyatomic molecules and are particularly relevant for our understanding of mechanisms of photochemistry. Consequently, both of them (IVR and vibronic coupling studies) can provide a better understanding of very complex processes involving competition between pure IVR and transfer of the energy between different electronic states.

Acknowledgments

Financial support by the NT05-3_42315 Projet Blanc of the ANR of the French 'Centre National de la Recherche Scientifique' (CNRS) is gratefully acknowledged.

22
Open System Quantum Dynamics With Discretized Environments

Mathias Nest

22.1
Introduction

In most chemical reactions, it is possible to discriminate between a small number of degrees of freedom that are important for the process, and a large number of environmental degrees of freedom that have only a weak effect. This is the case, for example, for reactions in condensed-phase, catalytic and enzymatic reactions, and many more. In particular, the example of catalytic reactions shows that the influence of the environment cannot entirely be neglected. This discrimination of primary modes of interest and secondary modes that can be described in an approximate fashion leads to the notion of system–bath type quantum dynamics. In this section, recent developments in the quantum dynamical description of this kind of situation will be presented.

Formally, we assume that the Hamiltonian of the full system can meaningfully be divided into three parts, which describe the system, the environment (bath) and the interaction between the two:

$$H = H_{\mathrm{S}} + H_{\mathrm{I}} + H_{\mathrm{B}} \tag{22.1}$$

We will use the terms 'environment' and 'bath' in the following mostly interchangeably. Details of the three Hamiltonians will be discussed later, after the general subject of this section has been outlined.

During the 1990s, attempts were made to treat situations of this kind with methods of open system quantum dynamics. The system degree of freedom, usually the reaction path coordinate plus one or two other important modes, were described accurately with a reduced density matrix, ρ. The environmental degrees of freedom were integrated out. This approach leads to an equation of motion for the reduced density matrix, a so-called generalized master equation,

$$\dot{\rho} = -\mathrm{i}[H_{\mathrm{S}}^{\mathrm{ren}}, \rho] + \mathcal{L}_D[\rho] \tag{22.2}$$

Multidimensional Quantum Dynamics: MCTDH Theory and Applications.
Edited by Hans-Dieter Meyer, Fabien Gatti, and Graham A. Worth

ISBN: 978-3-527-32018-9

Here, H_S^{ren} is the (possibly renormalized) system Hamiltonian, which generates the system dynamics in the absence of the environment. The second term on the right-hand side is the so-called dissipative Liouvillian, which is supposed to describe the net effect of the environment on the system. A very large number of dissipative functionals have been developed over the years – Refs. [555–563] give a small and certainly biased overview.

However, this approach has not been as successful as hoped. The problem is that the derivation of the dissipative Liouvillian often requires drastic approximations. Briefly, the main problems have been as follows:

(i) *Strong coupling.* Often, reduced equations of motion are derived using perturbation theory. This prohibits the treatment of medium and strong interactions.

(ii) *Non-Markovian effects.* Real environments have non-vanishing response and memory times, which makes a Markovian description questionable.

(iii) *Nonlinearity.* Nonlinearity and anharmonicity in the bath or interaction Hamiltonian are difficult to include.

(iv) *Non-equilibrium states.* Many finite environments do not behave like an ideal bath, which is always in the same equilibrium state.

(v) *Initial correlation.* Most derivations assume factorizing initial conditions for the system and bath degrees of freedom. But if both parts are already in contact when the propagation starts, this is unphysical.

Even if some of these problems can be solved, it is impossible to solve all of them in a single generalized master equation. Therefore, a different approach to system–bath type quantum dynamics has appeared in recent years. The idea is simply to include the environmental degrees of freedom in the wavepacket dynamics. The resulting wavefunction is very large, but can be treated with the MCTDH method. The earliest calculations of this kind have been on pyrazine [221,564]. One way to explain the efficiency of MCTDH is to describe it as a scheme to restrict the amount of correlation that can be represented with the *Ansatz* for the wavefunction. A single Hartree product is, by definition, an uncorrelated wavefunction, while more and more correlation is included when the expansion length is increased. Because the bath degrees of freedom are usually mostly uncorrelated, at least at time $t = 0$, this approach works especially well for system–bath type wavefunctions.

In the following section, some more details of the system–bath *Ansatz* are given, and some general properties of the systems to be studied later are discussed. Section 22.3 introduces the basic effects due to an environment, and provides comparisons between reduced equations of motion (REOMs) and exact dynamics. Section 22.4 extends the treatment to finite temperatures. Finally, Section 22.6 summarizes and provides a brief outlook to ongoing and future work.

22.2 The System–Bath *Ansatz*

We can now give the general forms of the three Hamiltonians, Equation (22.1), without selecting a specific physical problem yet. The system will be described by a Hamiltonian of the form

$$H_\mathrm{S} = \frac{p^2}{2M} + V(z) \tag{22.3}$$

with a general potential $V(z)$, which is typically, but not necessarily, a Born–Oppenheimer potential energy surface. If the environment constitutes a true bath, then it is fully defined by its spectral density, and can thus be modelled by a collection of harmonic oscillators:

$$H_\mathrm{B} = \sum_{\kappa=1}^{N} \frac{p_\kappa^2}{2m} + \frac{1}{2} m\omega_\kappa^2 x_\kappa^2 \tag{22.4}$$

This replacement of the actual, but often unknown, Hamiltonian of the environment is called the *surrogate Hamiltonian* approach [565,566]. The choice of mass m will be discussed later. The interaction between the system and the bath is described by H_I. If this is taken from *ab initio* data, it can take on arbitrarily complicated functional forms, with very different physical interpretations. In the following, we will restrict our attention to interaction Hamiltonians of the form

$$H_\mathrm{I} = \sum_{\kappa=1}^{N} c_\kappa f(z) x_\kappa \tag{22.5}$$

Linearity in the bath coordinate can be assumed if the effect of the system on the bath is only small, and only small deviations from the bath equilibrium state occur. The function $f(z)$ already provides significant flexibility to describe a large number of different situations.

The bath that has just been defined can be characterized by a spectral function [567,568]

$$J(\omega) = \frac{1}{2}\pi \sum_{\kappa=1}^{N} \frac{c_\kappa^2}{m\omega_\kappa} \delta(\omega - \omega_\kappa) \tag{22.6}$$

which is related to the damping kernel $\gamma(t)$ by

$$\gamma(t) = \frac{2}{M\pi} \int_0^\infty \frac{J(\omega)}{\omega} \cos(\omega t)\, \mathrm{d}\omega \tag{22.7}$$

The latter appears, for example, in the generalized Langevin equation of classical mechanics [88, 568, 569],

$$M\ddot{z}(t) + M\int_{-\infty}^{t} \gamma(t-t')\dot{z}(t)\,\mathrm{d}t + \frac{\mathrm{d}V}{\mathrm{d}z} = \zeta(t) \tag{22.8}$$

where $\zeta(t)$ is a fluctuating force related to the damping kernel by the fluctuation-dissipation theorem [555, 568]. If $\gamma(t)$ is local in time, that is, $\gamma(t) = \gamma\delta(t)$, then the system is called Markovian. Otherwise, the damping kernel contains memory effects of the environment. Its cosine transform gives the frequency-dependent damping rate

$$\gamma(\omega) = \int_0^{\infty} \gamma(t)\cos(\omega t)\,\mathrm{d}t \tag{22.9}$$

In this chapter, we will focus on frequency-independent damping $\gamma(\omega) = \gamma$, which is obtained for

$$J(\omega) = M\gamma\omega \tag{22.10}$$

The above spectral density would lead to an 'ultraviolet catastrophe', so that it has to be truncated at a suitably chosen final frequency ω_N (see discussion below). A comparison of Equations (22.6) and (22.10) leads to an expression for the coupling constants c_κ that appear in the interaction Hamiltonian, H_I:

$$c_\kappa = \kappa\left(\frac{2mM\gamma\Delta\omega^3}{\pi}\right)^{1/2} = \kappa\Delta c \tag{22.11}$$

Here we have assumed that the bath is discretized with harmonic oscillators of equidistant frequencies, $\omega_\kappa = \kappa\Delta\omega$. It is not strictly necessary to do so, but it offers the following advantage. A set of harmonic oscillators with equidistant frequencies has a well-defined recurrence time, or Poincaré time, $T_\mathrm{rec} = 2\pi/\Delta\omega$. A propagation to times longer than that would resolve the bath spectrally, and lead to artefacts due to the discretization. A non-equidistant bath would show partial recurrences at earlier times.

Now that the full Hamiltonian has been specified, it is time to comment on the choice of mass for the bath oscillators: The replacements

$$x_\kappa \to \frac{1}{\sqrt{m\omega_\kappa}}X_\kappa \quad \text{and} \quad p_\kappa \to \sqrt{m\omega_\kappa}P_\kappa \tag{22.12}$$

lead to expressions for H_I and H_B that are independent of the mass m. Thus, the bath is indeed specified by the spectral density alone, and the *Ansatz* for the full Hamiltonian models a 'true' system–bath situation.

22.3 Static and Dynamic Effects of the Bath

The previous section gave the general formulation of a system coupled to an environment. Now, we will proceed to discuss the static and dynamic effects of the environment.

22.3.1 Static Effect: Lamb Shift

Static effects are due to a distortion of the system potential $V(z)$ through the interaction Hamiltonian H_I. If we have a look at the full potential energy,

$$\mathcal{V}(z, x_1, \ldots, x_N) = V(z) + \sum_{\kappa=1}^{N} c_\kappa f(z) x_\kappa + \sum_{\kappa=1}^{N} \frac{1}{2} m\omega_\kappa^2 x_\kappa^2 \tag{22.13}$$

then the minimum of the potential, with respect to the bath degrees of freedom, is given by

$$\frac{\mathrm{d}\mathcal{V}}{\mathrm{d}x_\kappa} = 0 \quad \Leftrightarrow \quad x_\kappa^{(0)} = -\frac{c_\kappa}{m\omega_\kappa^2} f(z) \tag{22.14}$$

Insertion into Equation (22.13) gives

$$\mathcal{V}(z, x_1^{(0)}, \ldots, x_N^{(0)}) = V(z) - \sum_{k=1}^{N} \frac{c_k^2}{2m\omega_k^2} f^2(z) = V(z) - \Delta V(z) \tag{22.15}$$

The coupling results in a distortion of the system potential, which is linear in the coupling strength parameter γ (see Equation (22.11)). This distortion is equivalent to the Lamb shift [555, 570–572] in spectroscopy, where the environment is made of the quantized modes of the electromagnetic field.

From Equation (22.15) it also follows that the distortion does not change the *shape* of the potential, if $f^2(z)$ is proportional to $V(z)$, which is surprisingly often the case. Typical examples are:

(i) a harmonic oscillator coupled bilinearly, $f(z) = z$, to a bath;

(ii) a Morse oscillator, with a coupling function $f(z) = 1 - \mathrm{e}^{-\alpha z}$; and

(iii) a quartic double-well potential, with a coupling function that reduces to the bilinear case in *each* of the wells.

This appealing feature and the simplicity of the above formulae are a consequence of the linearity of H_I in x_κ. If the exchange of two quanta (for example, phonons) were taken into account, that is, if H_I depended on terms such as $x_\kappa x_{\kappa'}$, then these relations would be much more complicated.

22.3.2
Small-Amplitude Motion

Dynamic effects of the coupling can best be illustrated by a suitable example [441]. We choose the problem of the interaction between an atom and a surface, where the surface degrees of freedom (phonons, electron–hole pairs) constitute the bath. The system coordinate z is the surface–atom distance, and the system potential

$$V(z) = D(1 - \mathrm{e}^{-\alpha z})^2 \tag{22.16}$$

is a Morse oscillator, which is a typical choice for covalent bonds [573]. The well depth is chosen to be $D = 49\,\mathrm{meV}$, the mass $M = 50\,000\,m_\mathrm{e}$, and the characteristic length $\alpha = 2\,a_0^{-1}$. From this, a harmonic frequency of the potential well of $\Omega = \alpha\sqrt{2D/M} = 1.2\,\mathrm{m}E_\mathrm{h}/\hbar$ can be computed ($\mathrm{m}E_\mathrm{h}$, millihartrees). If small-amplitude motion is studied, this also sets the relevant frequency range for the bath oscillators. We found that $\omega_N = 2.5\Omega$ is sufficient to obtain a fast-acting bath. In all the calculations that are reported in the following, the number of bath oscillators has been adjusted to ensure a sufficiently large recurrence time. In the interaction Hamiltonian we choose

$$f(z) = \frac{1 - \mathrm{e}^{-\alpha z}}{\alpha} \tag{22.17}$$

because it reduces to the (bi)linear coupling limit in the potential well, and decays for large z. The latter will become important for gas–surface scattering later in this section.

We will begin with a rather simple case: the surface (bath) is at 0 K and the system is initially slightly displaced, causing small-amplitude motion. The latter is modelled by a Gaussian wavepacket, while the bath modes are in their ground states. Figure 22.1 illustrates this paradigmatic case, by energy and position expectation values parametrized with time. The thin line in Figure 22.1 shows energy dissipation and damped oscillations of the system, as expected for true system–bath dynamics. Note that the energy relaxation rate is generally highest around $z = 0$, that is, when the kinetic energy is highest, indicating a damping similar to the classical case, proportional to $\dot{z}$ (see also Equation (22.8)).

How is this reproduced by reduced equations of motion? For small-amplitude motion, it is possible to use a harmonic dissipative Liouvillian, namely the 0 K Lindblad functional [89, 90, 574, 575], with harmonic raising/lowering operators [441, 556, 567]. In the following, we will call this REOM 1. Figure 22.2 compares the time-dependent position expectation value for the full-dimensional wavepacket propagation with open system results, with and without the counter term ΔV (Equation (22.15)). The coupling strength γ has a moderate value of $150\,\mathrm{fs}^{-1}$, comparable to the typical time-

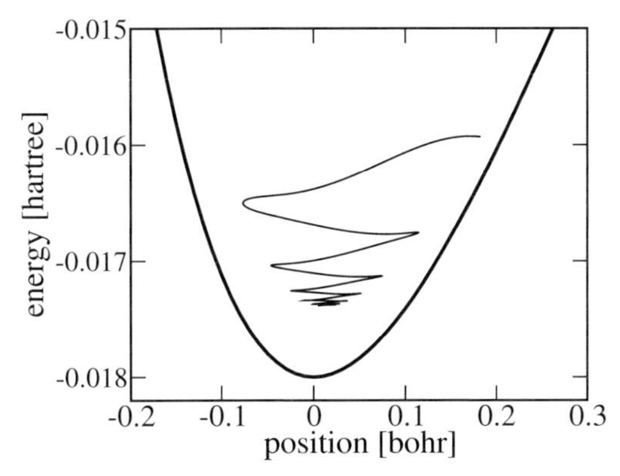

Figure 22.1 Illustration of damped oscillations and energy relaxation in a Morse oscillator coupled to a 0 K bath. The relaxation stops at the ground-state energy.

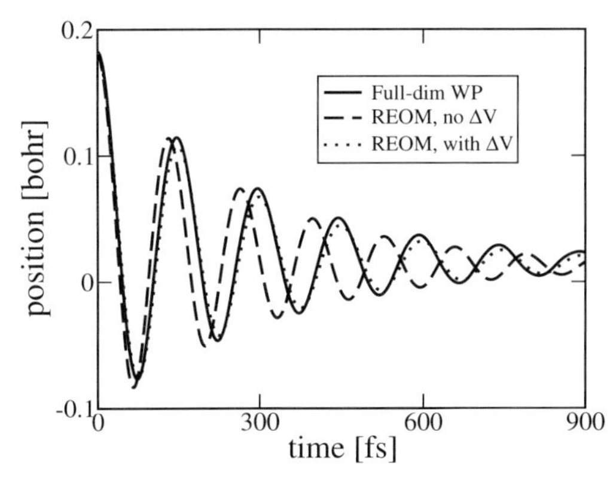

Figure 22.2 Damped oscillations in a Morse oscillator, coupled to a 0 K bath. Only if the counter term ΔV is included in the reduced density matrix propagation is the agreement with exact, full-dimensional calculations very good.

scale of the system. The agreement between the exact solution and REOM 1 *with* the counter term is very good, but a phase shift occurs without it.

A comparison of energy relaxation scenarios is more complicated. While the system energy expectation value is uniquely defined in the reduced density-operator formalism, this is not so for the full-dimensional case, because the full Hamiltonian H (Equation (22.1)) contains the non-separable interaction energy H_{I}. In [441] we could show, by a time-averaged version of the virial theorem, that for harmonic baths, and a coupling linear in the *bath* coordinates, the interaction energy should be split 'democratically',

$$E_{\mathrm{S}} = \langle H_{\mathrm{S}} \rangle + \tfrac{1}{2}\langle H_{\mathrm{I}} \rangle \qquad E_{\mathrm{B}} = \langle H_{\mathrm{B}} \rangle + \tfrac{1}{2}\langle H_{\mathrm{I}} \rangle \tag{22.18}$$

in order to compare well with reduced dynamics. Ref. [441] discusses results for Redfield-type open system dynamics, too. These are not reported here, but they did not differ much from the Lindblad results in the case of small-amplitude motion.

22.3.3 Inelastic Surface Scattering: Adsorption

The inelastic scattering of atoms and molecules at surfaces has attracted considerable attention in the theoretical and physical chemistry community, because it presents an elementary step in heterogeneous catalysis [576–580], and provides a useful testing ground for methods of open system quantum dynamics. For sticking or adsorption to be possible, it is necessary that the surface is non-rigid, and can serve as an energy sink. In this model computation, where we want to compare with a reduced equation of motion, we assume a bath (that is, the spectrum of the surface phonon modes) as in the previous section. This has the advantage that the system frequency is in the middle of the bath spectrum, which leads to somewhat simpler dynamics. In the simulation of a real system, the spectrum would, of course, be taken from *ab initio* information or experimental data [417,442].

The incident atom can be modelled by a wavepacket, which at time $t = 0$ is at some distance from the surface, and has a net momentum towards the surface. In order to have a qualitatively realistic scenario, we have chosen $\gamma = 1\,\mathrm{ps}^{-1}$, which is a typical vibrational lifetime of adsorbates on metal surfaces.

It should be noted that the parameter γ has this meaning strictly only in the harmonic–bilinear case. For the microscopic Hamiltonian and scattering, the nonlinearity becomes important. A golden rule expression can be derived nevertheless, because many properties of the Morse oscillator can be calculated analytically [581–583]. For the anharmonic case, the golden rule gives a lifetime of

$$\gamma_{\mathrm{a}} = \frac{\pi \Delta c^2}{2mM\Delta\omega^3}\left(1 - \frac{2\chi}{\hbar\Omega}\right)\left(1 - \frac{3\chi}{\hbar\Omega}\right) \tag{22.19}$$

where $\chi = \hbar^2\alpha^2/(2M)$ is a measure for the anharmonicity. Obviously, for $\chi \to 0$, the harmonic result is obtained again.

Having established the initial state, we then solve the time-dependent Schrödinger equation, and calculate the sticking probability according to

$$P_{\mathrm{stick}}(t) = \mathrm{Tr}\left\{\left(\sum_n |n\rangle\langle n|\right)|\Psi(t)\rangle\langle\Psi(t)|\right\} \tag{22.20}$$

The sum is over all bound states $|n\rangle$ of the Morse potential.

We compare the exact wavefunction results with reduced density-operator results, where the dissipative Liouvillian consists now of Lindblad operators that are generalized Morse raising or lowering operators (REOM 2) [441, 563, 584, 585]. The resulting energy-resolved sticking probability is shown in Figure 22.3. As expected for this simple model, the sticking probability decreases for higher energies. What is surprising is that the results for both computational methods are so very similar. The reasons for this are the very weak coupling ($\gamma = 1\,\text{ps}^{-1}$), and the absence of non-Markovian effects in the full wavepacket calculation.

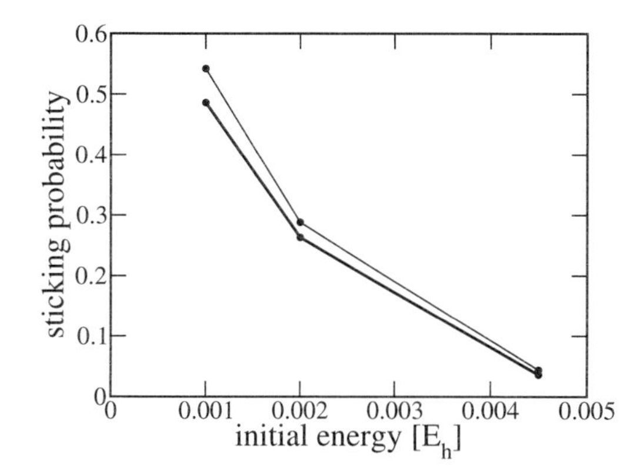

Figure 22.3 Sticking probability as a function of initial kinetic energy of the incident atom: upper curve, reduced density matrix calculation; lower curve, full-dimensional wavepacket calculation.

Differences can be found, however, when one looks at details of the calculations. Figure 22.4 shows the 'flow' of energy, that is, the *change* of the various energy expectation values with time, during such a collision event. While the atom is far from the surface, the system energy is constant (solid lines). Closer to the surface, first the interaction energy rises (dashed line), before energy is transferred to the bath (dotted line). Of course, the latter two quantities are only accessible in the wavepacket picture. For longer times, the interaction energy falls again, but not to zero, because some part of the wavepacket is trapped, and continues to interact with the surface. A qualitative difference between the system energy expectation values $\langle H_S \rangle$ and $\text{Tr}\{H\rho\}$ shows up after 600 fs. There is a small maximum in the wavepacket results, indicating a small energy transfer from the bath back to the system. As the bath has been assumed to be at $T = 0\,\text{K}$, this effect cannot be described within a reduced density matrix picture.

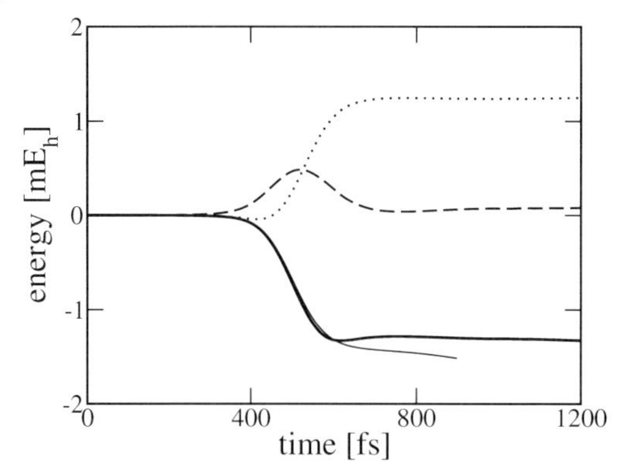

Figure 22.4 Energy flow during a scattering event. The energy difference relative to the initial value is shown as a function of time. Curves: thick solid line, wavepacket subsystem energy; dashed line, wavepacket interaction energy; dotted line, wavepacket bath energy; thin solid line, Morse–Lindblad REOM system energy.

22.4 Finite Temperatures

22.4.1 Random-Phase Wavefunctions

Environments at finite temperatures can be simulated with wavefunction-based methods of open system quantum dynamics, too, if the Boltzmann operator

$$\varrho(\beta) = \frac{1}{Z}\,\mathrm{e}^{-\beta H_0} \tag{22.21}$$

of the system and bath can be sampled. Here, β is as usual the inverse temperature, Z is the partition function, and H_0 is a Hamiltonian whose relation to the full, microscopic Hamiltonian H is discussed below. Generally, the eigenfunctions of H_0 are not known, except for a few model systems. But even if they were known, they would not be helpful because there are too many of them to allow an explicit summation. For example, if in a 20-dimensional system on average five states per dimension are occupied significantly, there are around 10^{14} relevant states, which is numerically intractable.

As an alternative, more practical approach, we suggest using the random-phase thermal wavefunction (RPTWF) method [566,586], which has been used in a similar fashion before in the context of reaction rates in gas-phase reactions [587–589] (see Section 19.3.2). Briefly, the idea is as follows. Given a complete set of basis functions $|\phi_l\rangle$, and a vector of random phases $\boldsymbol{\theta}$, one can construct infinite-temperature wavefunctions according to

$$|\Phi(\boldsymbol{\theta})\rangle = \frac{1}{\sqrt{L}} \sum_{l=1}^{L} \mathrm{e}^{\mathrm{i}\theta_l} |\phi_l\rangle \tag{22.22}$$

and resolve the identity according to

$$\mathbf{1} = \lim_{K\to\infty} \frac{1}{K} \sum_{k=1}^{K} |\Phi(\boldsymbol{\theta}_k)\rangle\langle\Phi(\boldsymbol{\theta}_k)| \tag{22.23}$$

so that the Boltzmann operator can be decomposed into

$$\begin{aligned} \varrho(\beta) &= \frac{1}{Z}\,\mathrm{e}^{-(\beta/2)H_0}\, I\, \mathrm{e}^{-(\beta/2)H_0} \\ &= \lim_{K\to\infty} \frac{1}{Z}\frac{1}{K} \sum_{k=1}^{K} |\Phi(\beta/2,\boldsymbol{\theta}_k)\rangle\langle\Phi(\beta/2,\boldsymbol{\theta}_k)| \end{aligned} \tag{22.24}$$

The random-phase thermal wavefunction $|\Phi(\beta/2,\boldsymbol{\theta}_k)\rangle$ is obtained by propagation in imaginary time up to $\beta/2$. The error of this statistical sampling scheme decreases, like every Monte Carlo method, as $\sigma \propto 1/\sqrt{K}$ with the number of wavefunctions, K.

Thus, in order to obtain a thermal average with the RPTWF method, one has to do the following:

1. Construct a wavefunction with equal amplitude and random phases in the bases of your choice (Equation (22.22)).
2. Propagate in imaginary time up to $\beta/2$.
3. Propagate in real time, so as to simulate the physical process of interest.
4. Repeat steps 1–3 until statistical convergence is achieved.

22.4.2 Inelastic Surface Scattering: Adsorption

We will now take up the problem of inelastic scattering again, and turn to the question of how the temperature of the surface effects the sticking probability [586]. We use basically the same Hamiltonian as in Section 22.3, but with parameters $M = 25\,000\, m_\mathrm{e}$, $D = 50\,\mathrm{meV}$ and $N\Delta\omega = 0.005\, E_\mathrm{h}/\hbar$.

When the RPTWF method is applied to this situation, one has to keep in mind that only the surface is in a thermal state and not the incident atom, which is in a 'microcanonical' state with definite kinetic energy. Therefore, one has to determine the surface Hamiltonian at time $t = 0$, which is equivalent to saying for a scatterer at position $z = +\infty$. Inserting this into the full

Hamiltonian (Equation (22.1)), one finds that

$$H_0 = \sum_{\kappa=1}^{N} \frac{p_\kappa^2}{2m} + \frac{1}{2} m \omega_\kappa^2 x_\kappa^2 - \frac{c_\kappa}{\alpha} x_\kappa \tag{22.25}$$

is the relevant Hamiltonian in the Boltzmann operator, Equation (22.21).

For the application in this subsection, it is necessary to give some more details of the underlying MCTDH wavefunction. Our standard configuration has eight single-particle functions (SPFs) for the system degrees of freedom, with only the first,

$$\varphi_1^{(1)}(z, t{=}0) = \frac{1}{\sqrt{\sigma\sqrt{\pi}}} \exp\left[-\frac{(z-z_0)^2}{2\sigma^2} + \mathrm{i} p_0 z \right] \tag{22.26}$$

initially populated. The centre of the wavepacket is $z_0 = 5\,a_0$, the width $\sigma = 1\,a_0$, and the momentum is chosen such that the kinetic energy is $D/10$. The bath was modelled with 20 harmonic oscillators, which were grouped into five combined modes. The number of SPFs for these combined modes were 5, 5, 5, 4 and 3, with increasing bath mode frequency. In order to build the initial infinite-temperature wavefunction, random numbers of modulus 1 and random phase are assigned to the coefficients $A_{1 j_2 \ldots j_6}$.

After propagation in imaginary time to achieve the desired temperature, and propagation in real time for the scattering event, we can extract the sticking probability for this realization of the thermal ensemble. Figure 22.5 shows the convergence of P_{stick} with increasing number of realizations. After 99 realizations the difference between consecutive points is about a factor of 20 smaller than at the beginning. Achieving another factor of 10 in precision would require 100 times as many realizations, because our scheme converges with the square root of K.

Figure 22.6 shows the results for the temperature-dependent sticking probability. It drops from 20.5% for a surface at 4 K to 10.3% for a surface at 500 K. This temperature range has to be compared to the translational temperature of the incoming particle of $E_{\text{kin}}/k_{\text{B}} = 580\,\text{K}$. Over this range the sticking probability decreases to about half of its value for a cold surface. However, also the accuracy of our values for P_{stick} decreases, because higher temperatures require also a larger number of SPFs. This is indicated by the dashed line, where we used one SPF more for each combined mode of the bath. The sticking probability for the highest temperature is clearly only semi-quantitative. The inset of Figure 22.6 shows the relation between the energy loss of the scattering atom and the sticking probability. For most of the temperature range, it is linear, but at very low temperature, where P_{stick} is almost constant, more energy is transferred to the bath.

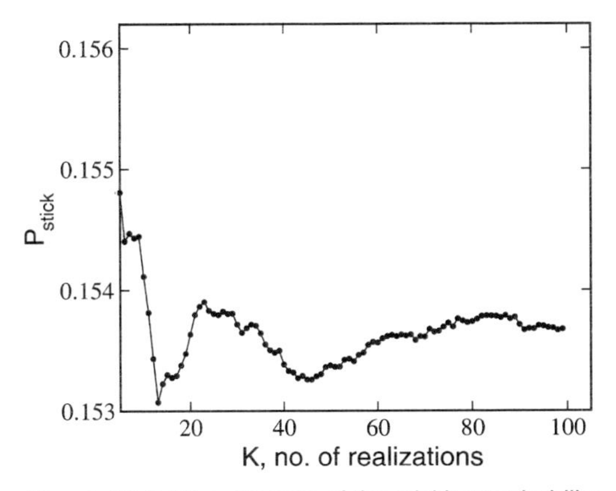

Figure 22.5 Convergence of the sticking probability with the number of realizations of the RPTWF approach, for a surface at $T = 300\,\text{K}$. Other temperatures, except $T = 0\,\text{K}$, show a similar pattern. The error decreases as $1/\sqrt{K}$, as for any method of the Monte Carlo family.

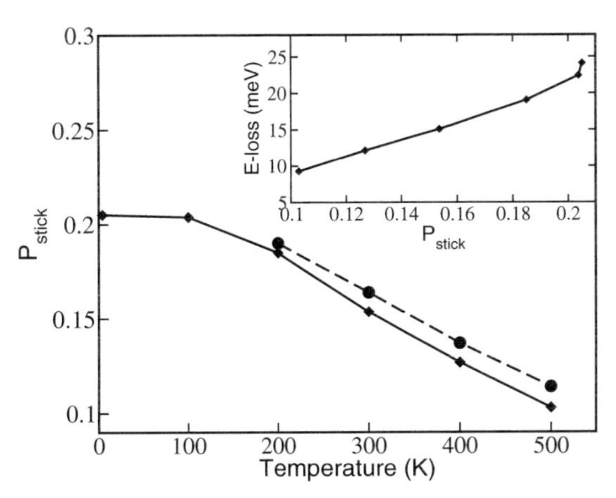

Figure 22.6 The main graph shows the sticking probability as a function of surface temperature. Solid line with diamonds: results of our standard configuration of SPFs. Dashed line with circles: one more SPF per combined mode has been used. The inset shows that the relation between the amount of dissipated energy and sticking probability is largely linear.

Apart from this net energy exchange, one can take a more detailed look at the kinetic energy distributions of the incoming and reflected wavepackets. Figure 22.7 compares the initial (black solid line) and final energies, for scattering at a cold (dashed line) and a hot (dotted line) surface. The area under the final distributions is smaller, because only the unbound part of the wavepackets, which is reflected back into the vacuum, has been considered. Notably, probability density is missing from the low-energy region (30–50 meV). At energies below about 30 meV, where there is almost no density initially, we find

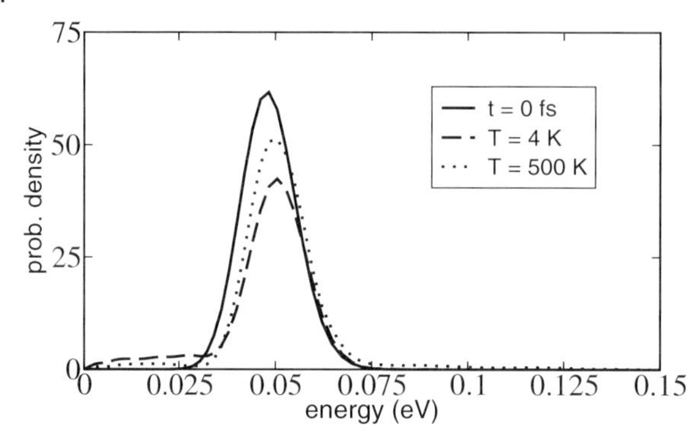

Figure 22.7 Kinetic energy density of the scatterer: at the beginning of the propagation (solid line), and after reflection from a 4 K surface (dashed line) and from a 500 K surface (dotted line). The onset of thermalization can be seen.

a stronger increase for the $T = 4\,\mathrm{K}$ case, while the hot surface redistributes density to higher energies. This is the onset of thermalization, although the contact with the surface is rather short, and the coupling is weak.

22.4.3
Initial Slip and Coupling to Photons

Very often, the initial state of a propagation is assumed to factorize into a system and a bath part,

$$\Psi(0) = \Psi_\mathrm{S}(0) \cdot \Psi_\mathrm{B}(0) \tag{22.27}$$

implying that no correlation between the two is present. If such a state develops with the full Hamiltonian, including H_I, then this state is non-stationary, and on an ultrashort time-scale some violent dynamics can take place before the proper behaviour sets in. This has been a source of trouble for the reduced dynamics community for some time, especially when the Redfield-type master equations [557, 558] have been used. There, this phase of the dynamics is known as *initial slip* [590, 591], and it can lead to negative values for populations or the probability distribution of certain measurements.

Although this initial slip has usually been looked at as an artefact arising from an approximation to the initial state, it also conveys some physics if the interaction between system and bath *is* switched on suddenly [592]. This is the case, for example, when a gas-phase molecule from a jet expansion enters a cavity, so that the photon modes, understood as harmonic oscillators [555], act as a bath.

To model this effect, we chose a two-level system (TLS), represented by the two lowest vibrational levels of a double-minimum potential. The coupling constants c_κ are chosen with a Lorentzian, rather than linear, distribution, but the photon modes are still equidistant. For ease of interpretation, we report temperatures here in units of $\hbar\Delta E/k_B$, where ΔE is the tunnel splitting. Here, we report results for $T_{TLS} = 0.5$, where the second and higher excited states of the double well are not significantly occupied. Figure 22.8 shows the population of the first excited state after the interaction to the photon modes is switched on. Although the temperatures of the two parts are the same (black solid line), the sudden perturbation causes an exchange of energy. In fact, even if the temperature of the radiation (T_B) is larger than the system temperature (dashed line), the oscillations are strong enough to cause a transient energy transfer opposite to what is expected. None of this of course contradicts the laws of thermodynamics, which are recovered at long enough times as steady-state solutions. The effect is damped if the interaction is switched on slowly, but remains present for a large range of parameters. It should be noted that in this study we employed our largest standard MCTDH wavefunction yet, with 101 degrees of freedom.

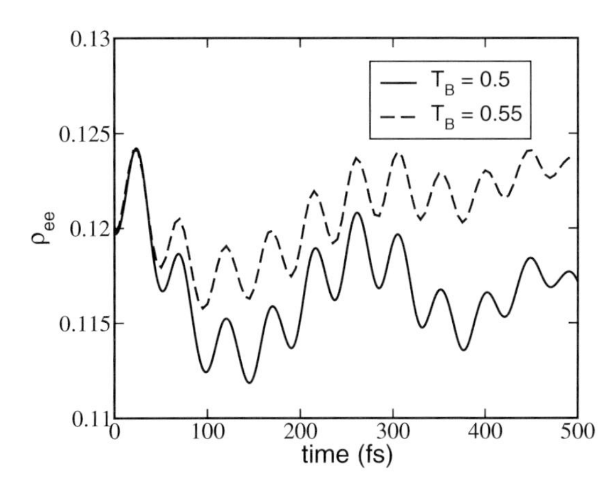

Figure 22.8 Evolution of the population of the excited state, after the interaction between system and bath is switched on. Black line, $T_{TLS} = T_B$: although system and bath are at the same temperature, energy is exchanged. Dashed line, $T_{TLS} < T_B$: at short times, around $t = 100$ fs, there is a transient energy transfer from the colder TLS to the hotter bath.

22.5
Derivatives of MCTDH

Very high-dimensional calculations, which are possible with MCTDH, are required in order to simulate the system and the bath with a wavefunction-based method. Even though highly sophisticated methods and integrators are available, these calculations still take of the order of days or weeks. Therefore, methods have been developed that are adapted to the very large size and/or the system–bath character of the problem. Two possibilities will be mentioned here, briefly. One example is the multilayer MCTDH technique, Chapter 14. In [228] Wang and Thoss used the ML-MCTDH approach to evaluate the thermal Boltzmann operator, in the context of electron transfer reactions in the condensed phase. In contrast to the random-phase thermal wavefunction technique outlined in Section 22.4, they used standard Monte Carlo importance sampling.

Another method that tries to make larger systems accessible is the local coherent-state approximation (LCSA) [593]. It can be seen as a selected configuration MCTDH [42], with the further approximation of single-particle functions being restricted to coherent states. The result is a method that shows *linear scaling* with respect to the dimensionality of the bath. Wavefunctions with several thousand degrees of freedom (comparable to ML-MCTDH) can be propagated. First tests have shown that LCSA can reproduce MCTDH benchmark calculations on energy dissipation very well [593]. As an example, Figure 22.9 shows the damped small-amplitude motion as predicted by LCSA and MCTDH, with very good agreement.

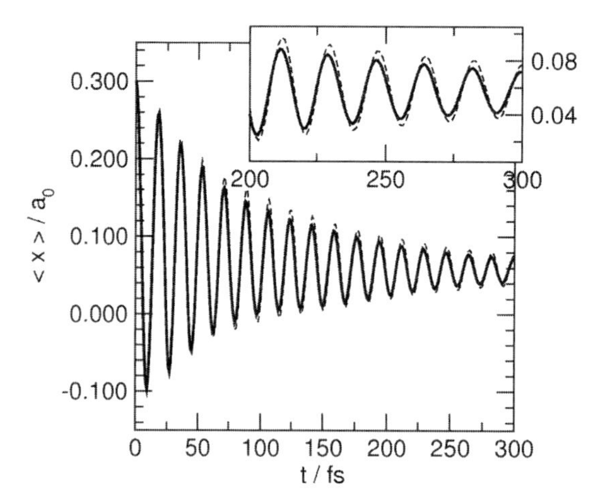

Figure 22.9 Comparison of MCTDH (thick line) and LCSA (dashed line): the position expectation values for $\gamma = 1/50\,\mathrm{fs}^{-1}$ agree very well. The inset shows a small amplitude mismatch at late times.

22.6 Summary and Outlook

The previous studies have shown that it is possible to model the basic effects of an environment on a system. These effects include a distortion of the reaction path/Born–Oppenheimer surface, energy gain or loss, and phase fluctuations. To this end, the spectral density of the environment is discretized, so that it can be replaced by a large number of harmonic oscillators. We could show that this procedure allowed us to reproduce realistically the behaviour of open quantum systems, as long as the total propagation time is too short to resolve the spectral spacing $\Delta\omega$. The limitation that wavefunctions cannot describe finite temperatures can be overcome if the Boltzmann operator is sampled. We have shown that the random-phase technique, applied to the intermediate basis of the Hartree products, is a very efficient sampling scheme. However, the computational effort becomes challenging for high temperatures, because more and more SPFs are required.

All the calculations reported, except the LCSA ones, were performed with the Heidelberg MCTDH package [13].

23
Proton Transfer and Hydrated Proton in Small Water Systems

Oriol Vendrell and Hans-Dieter Meyer

23.1
Introduction

Proton translocation processes are ubiquitous in nature and, together with electronic transfer, constitute the basis of charge transfer mechanisms in chemistry and biology. Let us just mention a few examples: Inter-membrane proteins mediate proton transfer between the two sides of a cellular membrane to generate a potential gradient [594–596]. In fuel cells protons must traverse artificial membranes to generate charge gradients [597,598]. Several proteins rely on proton wires (chains of hydrogen-bonded water molecules) to transfer a charge in the form of an excess proton between acidic and basic centres [599]. In bulk water protons transfer through variants of the Grotthus mechanism [600–602], which lies at the core of acidic chemistry. Proton transfer can also be a photoinitiated process, in which a photoexcited molecule is able to transfer a proton (or a hydrogen atom) to either a neighbouring molecule or a different location in the same molecular unit [603–607].

The fundamental structural unit in which a proton transfer occurs in common chemical and biological systems is the hydrogen bond. A proton is covalently bonded to a *donor* atom and electrostatically bonded to an *acceptor* atom. Eventually, the proton is transferred between the two atoms, which then invert their roles as acceptor and donor. Also hydrogen bonds may not be involved in proton transfer but may play a structural role in shaping materials or biomolecules. It is obvious that proton translocation and the hydrogen bond play a central role in many important areas. However, their study appears as a challenging task, owing to the inherent quantum nature of the lightest stable nucleus. In addition, proton motion is usually strongly coupled to the displacements of surrounding atoms. Although classical [602] or mixed quantum–classical [608] methodologies have been applied to the study of systems containing protons, spectroscopical accuracy in comparing with

Multidimensional Quantum Dynamics: MCTDH Theory and Applications.
Edited by Hans-Dieter Meyer, Fabien Gatti, and Graham A. Worth

ISBN: 978-3-527-32018-9

experiment and a detailed view of the underlying dynamics requires that full-quantum methods be used [226,486,609].

MCTDH is able to treat relatively large systems in terms of coupled nuclear degrees of freedom. It has been quite naturally the method of choice in a number of studies related to proton transfer and hydrogen-bonded systems. Examples can be found of the application of MCTDH to the study of infrared (IR) spectroscopy [59,486,610], intramolecular vibrational energy redistribution (IVR) [610], quantum control [611] or excited-state dynamics [612] of hydrogen-bonded systems. Also, problems of tunnelling splitting in double-well systems have been addressed by MCTDH [226,609,613].

In this chapter we will try to illustrate the power of MCTDH in dealing with proton transfer processes and hydrogen-bonded systems from two very distinct perspectives. In Section 23.2 we discuss how MCTDH was used to simulate a model of the transfer of an excess proton along a chain of hydrogen-bonded water molecules [614]. In this respect, MCTDH constitutes a powerful tool to study model Hamiltonians describing relatively complex processes and to obtain a fundamental insight. Such models are formulated from the beginning such that the required product structure of the potential function (Equation (3.34)) is retained.

The rest of the chapter is devoted to the full-dimensional simulation of the protonated water dimer, a 15-dimensional (15D) problem. In this cationic cluster, a proton is shared between two water molecules forming a strong hydrogen bond. The IR spectroscopy and dynamics of the system are investigated, and results are compared with experiment. This problem, namely the simulation of a high-dimensional system as accurately as possible and in its full dimensionality, requires a quite different approach. Sections 23.3.1 and 23.3.2 discuss briefly how the system Hamiltonian is constructed, especially the construction of a representation of the potential energy surface (PES) for such a high-dimensional problem that can be both accurate and efficient. In Section 23.3.4 the computation of the IR spectrum is described. We also describe here how the assignment of spectral features can be done without relying on the computation of the corresponding vibrational states (which is impracticable for such large systems) but by resorting to time propagation and Fourier analysis.

Finally, Section 23.3.5 discusses the assignment of the spectral features, leading to a concise picture of the dynamics of the cluster and of the hydrated proton in general.

23.2 Proton Transfer Along a Chain of Hydrogen-Bonded Water Molecules

As discussed in Section 23.1, chains of hydrogen-bonded molecules (proton wires) mediate proton transfer over large distances in various contexts, from biological systems to bulk water or materials. Such processes have been extensively simulated by methodologies relying on classical approaches [600, 608, 615, 616]. A purely quantum simulation was, however, missing. In this context, the study of a model of a proton wire was undertaken using MCTDH [614].

23.2.1 Model for the Simulation of a Proton Wire

A minimal model for a proton wire that is able to capture the physics of a multiple-proton translocation between donor and acceptor atoms should contain as dynamical variables the position of each proton with respect to its donor and acceptor atoms and the distance between each donor and acceptor. In our model, these coordinates are r_i, the position of the proton with respect to the midpoint between both heavy atoms, and R_i, the donor–acceptor distance. Such a model is depicted in Figure 23.1. Only the particles depicted with circles are explicitly in the model, while the rest are intended to clarify the meaning of the scheme.

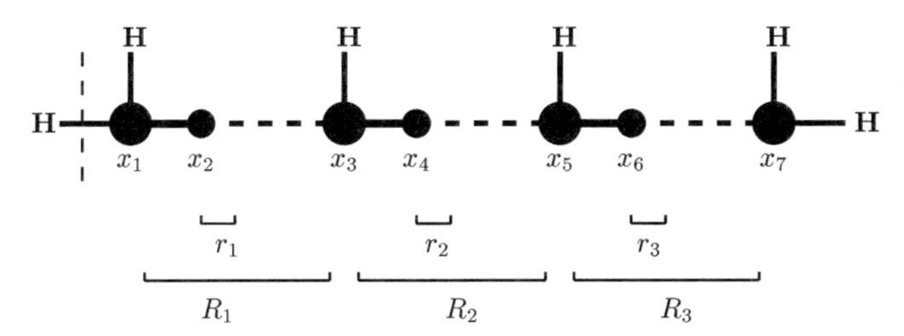

Figure 23.1 Schematic representation of the proton transfer chain model system with three units. The hydrogens denoted by H are only depicted for clarification and do not appear as dynamical variables in the model.

When an excess proton attaches to the leftmost oxygen atom, the excess charge can advance through the chain by successive hops of the protons between each pair of oxygens. At the end, the excess charge can be expelled by proton detachment of the rightmost proton of the chain. The operation of the chain must fulfill certain requirements that will guide the definition of a realistic expression of the PES. For example, the proton denoted x_4 in Figure 23.1 can only transfer to oxygen x_5 as long as x_2 has been transferred to x_3. Otherwise, we would generate a localized positive charge on x_5 (holding an excess proton) and a localized negative charge on x_3 (holding a proton hole). Such considerations lead to the definition of an interaction model in which the local potential felt by each proton, represented by coordinate r_i, is influenced by the position of the two surrounding protons r_{i-1} and r_{i+1}. The definition of the internal coordinates in terms of the Cartesian ones is formulated as:

$$R_i = x_{2i+1} - x_{2i-1} \tag{23.1a}$$

$$r_i = x_{2i} - \frac{1}{2}(x_{2i+1} + x_{2i-1}) \tag{23.1b}$$

$$r_{n+1} = x_{2n+2} - x_{2n+1} \tag{23.1c}$$

In order to define the total potential, we introduce the concept of a unit. A unit is formed by r_i and R_i, the coordinates giving the position of a proton and the distance between its surrounding atoms, respectively, and the position of the two surrounding protons r_{i-1} and r_{i+1}. In this way, each unit is interlaced with each other unit, and the interaction between neighbouring protons can be accounted for. The definition of the potential of a unit is

$$\begin{aligned} V_{\text{unit}}(r_i, R_i, r_{i-1}, r_{i+1}) = {} & V_{\text{exc}}(r_i, R_i)[1 - ls(r_{i-1})][1 - rs(r_{i+1})] \\ & + V^l_{\text{wat}}(r_i, R_i) ls(r_{i-1}) + V^r_{\text{wat}}(r_i, R_i) rs(r_{i+1}) \end{aligned} \tag{23.2}$$

The definition of the PES then follows as the sum of the unit potentials:

$$V_{\text{tot}}(r_1, \ldots, r_n, r_{n+1}, R_1, \ldots, R_n) = \sum_{i=1}^{n} V_{\text{unit}}(r_i, R_i, r_{i-1}, r_{i+1}) \tag{23.3}$$

The function $V_{\text{exc}}(r_i, R_i)$ describes the interaction potential of a unit in the case that both r_{i-1} and r_{i+1} are bonded to donor and acceptor atoms, respectively, while the functions $V^{l,r}_{\text{wat}}(r_i, R_i)$ describe the asymmetric situation of a proton covalently bonded to either donor or acceptor and forming a hydrogen bond with the other atom. The functions $ls(r_{i-1})$ and $rs(r_{i+1})$ are switch functions that control the active potential in each unit as a function of the neighbouring protons' positions. The parametrization of the functions involved in Equation (23.2) as well as their definition is discussed in Ref. [614]. We only mention here that these functions were parametrized with respect to *ab initio* calculations performed on different proton wires. If it is assumed that donor and acceptor atoms and transferring protons lie on a straight line, it is straightforward to obtain the kinetic energy operator (KEO) for this model by using standard differential calculus. More details are found in Ref. [614].

23.2.2
Dynamics of a Proton Wire

The model presented permits the simulation of chains of various sizes and the investigation of the basic mechanisms of operation of a proton wire in a systematic way. Its purpose is oriented more towards obtaining physical insight than to comparison with experimental observations, which is beyond the scope of such an approximation. Various chains have been analysed, and details can be found elsewhere [614]. Here we briefly summarize the results obtained. After protonation of the wire, each proton hop occurs in about 50 fs. This is roughly the time needed for a period of oscillation of the oxygen–oxygen distance R_i. The R_i distances decrease from 2.7 Å to around 2.2 Å during the transfer. Around three transfer events can occur in a row, so that the third proton starts moving 150 fs after the protonation of the first unit of the chain. After that, the fourth proton needs more time to start being transferred since the probability of four transfers in a row is low.

Two important issues were investigated on this model, namely, the correlation between light and heavy particles, and the effect that deuteration has on the operation of the wire. The simulations discussed here correspond to a chain with three units. In order to monitor the rate of transfer under different conditions, a complex absorbing potential (CAP) was set up to absorb the wavefunction as soon as the fourth proton is released from the chain, thus monitoring the amount and rate of overall transfer over time. The results for three cases are depicted in Figure 23.2. The deuterated simulation refers to the substitution of the proton in the second unit (r_2) by a deuterium. When comparing protonated and deuterated chains, it can be seen how, in the first

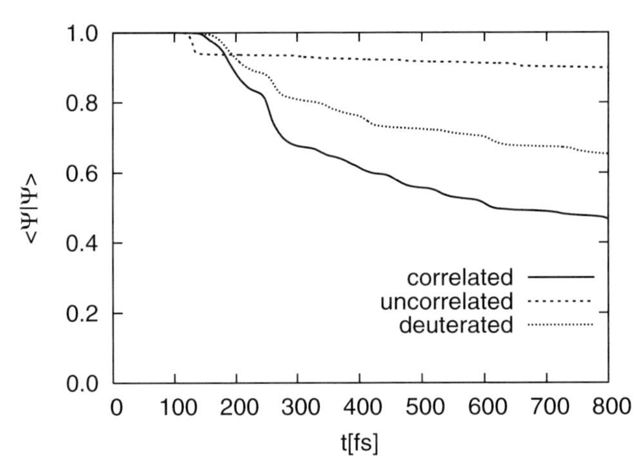

Figure 23.2 Norm-squared of the wavefunction, that is, the survival probability, as a function of time. The wavefunction is absorbed by a CAP as the last proton is released from the chain. The simulations correspond to a chain with three units and a fourth proton that can be expelled from the wire. Protonated, fully correlated chain (full line), deuterated second unit, fully correlated chain (dashed line) and uncorrelated simulation (dotted line) are shown.

stages of the process, up to 250 fs, the transfer rates are quite similar. This corresponds to the impulsive phase, in which the transfer occurs above the potential barriers. Afterwards, the transfer in the protonated case continues over time at a lesser pace, while the transfer in the deuterated chain is almost completely blocked.

This regime corresponds to leaking under the potential barriers, and the transfer occurs mostly by tunnelling. In what are termed 'uncorrelated' simulations, the wavefunction is forced to take the form

$$\Psi = \Psi_r(r_1, \ldots, r_n, t) \prod_{i=1}^{n} \Psi_{R_i}(R_i, t) \tag{23.4}$$

which means that the correlation between light and heavy atoms is only taken into account at the mean-field level. The amount of transfer in the uncorrelated case is dramatically reduced with respect to the correlated case, as depicted in Figure 23.2. This is indicative of the importance of correlated motions in the operation of the wire and of the fact that short-distance and long-distance regimes in the R_i coordinates coexist, rather than the R_i coordinates oscillating between them as a more classical picture would suggest.

This is seen by inspecting Figure 23.3, in which snapshots are depicted of the probability density along coordinates (r_2, R_2). In the case of the fully correlated dynamics (left panel in Figure 23.3), one part of the wavepacket evolves towards short R_i and proton transfer takes place, but an important part stays at long distances. When proton and oxygen motions are correlated only at the mean-field level (wavefunction given by Equation (23.4)), the transfer loses efficiency since the R_i is not able to reach the short-distance regime.

We have shown is this section how a model for a physically complex process such as multiple-proton transfer along a chain of water molecules can be constructed and investigated in detail. The model has been devised from the beginning to be described by a Hamiltonian that has the product structure necessary for efficient MCTDH computations.

23.3 Dynamics and Vibrational Spectroscopy of the Zundel Cation

The Zundel cation ($H_5O_2{}^+$) is the smallest protonated water cluster in which an excess proton is shared between different water molecules. This species has been the target of much research effort in the past owing to its key role as a limiting structure of the hydrated proton in bulk water. The development and improvement of new techniques designed to measure accurate IR spectra of ionic species in the gas phase has led to an accurate characterization of the IR spectrum of $H_5O_2{}^+$ from the experimental side [617–621]. Several the-

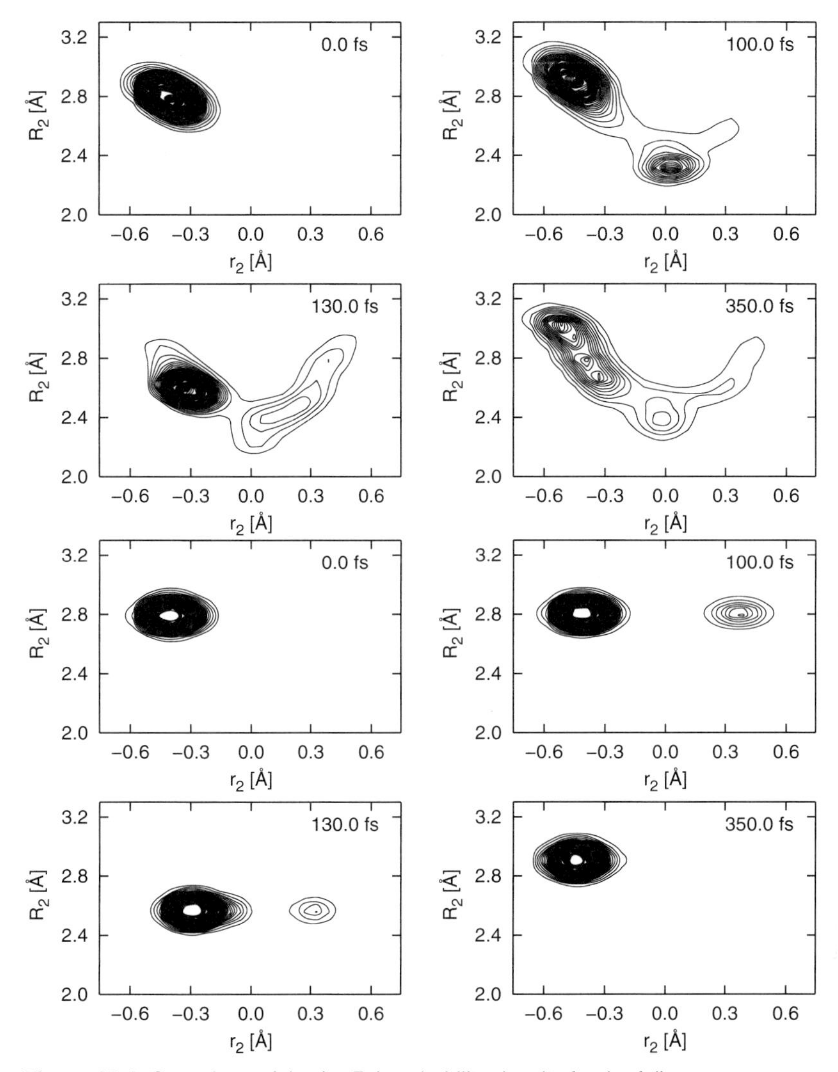

Figure 23.3 Snapshots of the (r_2, R_2) probability density for the fully correlated dynamics (upper four panels) and of the uncorrelated dynamics (lower four panels).

oretical studies that appeared in parallel provide the necessary basis for the rationalization and understanding of the measured spectra.

The IR spectrum of the Zundel cation recorded with the messenger atom technique features a doublet structure in the spectral region around $1000\,\text{cm}^{-1}$ with a splitting of about $120\,\text{cm}^{-1}$. Although it was already clear that the asymmetric proton stretching fundamental (proton transfer motion along the oxygen–oxygen axis) absorbs in that region, the origin of the doublet remained an unsolved issue that could neither be reproduced nor explained by a wide range of computational approaches [210, 620, 622–625]. Moreover, the lowest-frequency region of the spectrum below 800 cm^{-1}, where the fundamental states of the most anharmonic motions are found, has not yet been accessed experimentally, and the assignment of the band at around 1750 cm^{-1}, presumably related to the *ungerade* bending motion of the flanking water molecules, was not completely clear. These difficulties are due to the floppy, anharmonic nature of the inter-atomic potential governing the motion of the cation, in which internal rotation and long-range angular motions take place at relatively low vibrational energies. In this context the study of the IR spectroscopy and dynamics of the Zundel cation was undertaken by the MCTDH method [59, 156, 486].

23.3.1 Set-Up of the Hamiltonian Operator

The simulation of floppy systems like the Zundel cation requires the use of a proper set of coordinates that is able to describe largely anharmonic, angular motions, for example, torsions or internal rotations. To this end a set of polyspherical vectors is used to describe the configuration of the system. The reader is referred to Chapter 12 for further reference on the derivation and usage of KEOs within the polyspherical approach, and more specifically to Section 12.4.2, where the application to semirigid systems is covered. For a detailed derivation of the KEO for $H_5O_2{}^+$, see Ref. [156].

The set of polyspherical vectors that are used to describe the $H_5O_2{}^+$ cation are shown in Figure 23.4. This is a semi-rigid system (at least in the range of vibrational energies involved in the linear IR spectrum), that is, no singularities in the kinetic energy exist for the configurations accessible to the system. This allows the use of standard one-dimensional discrete variable representations (DVRs) (see Appendix in Ref. [156]). The 15 coordinates are defined as follows: the distance between the centres of mass of the two water

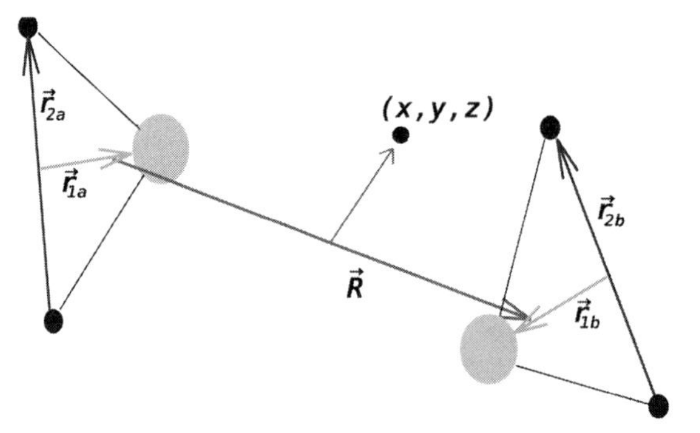

Figure 23.4 Jacobi description of the $H_5O_2{}^+$ system. The vector $\mathbf{R}$ connects the two centres of mass of the water monomers.

molecules (R); the position of the central proton with respect to the centre of mass of the water dimer (x, y, z); the Euler angles defining the relative orientation between the two water molecules (waggings, γ_a, γ_b; rockings, β_a, β_b; internal relative rotation, α); and the Jacobi coordinates that account for the particular configuration of each water molecule $(R_{1(a,b)}, R_{2(a,b)}, \theta_{(a,b)})$, where R_{1x} is the distance between the oxygen atom and the centre of mass of the corresponding H_2 fragment, R_{2x} is the H–H distance and θ_x is the angle between these two vectors.

The representation of the potential energy operator (PEO) for a 15D system is a rather challenging task on its own. In fact, the straightforward use of the PES in the 15D product grid, which would have of the order of 10^{15} points, is today just unthinkable (and probably will be so for quite a long time). However, there are ways to obtain usable representations of such high-dimensional objects. It will not concern us here how the potential energy for each distinct geometry is obtained: this is discussed in the next section. Nonetheless, the availability of a PES fit that is able to provide energies using little computer time will determine to a great extent how and to what degree of precision an accurate PEO can be obtained for the problem at hand.

A way of representing such high-dimensional surfaces is the so-called Cartesian reaction surface (CRS) approach, in which a small number of coordinates (typically two) are selected as defining the reaction surface, while the potential for the rest of the coordinates is given as a second-order Taylor expansion whose coefficients are dependent on the position of the coordinates in the first group. The CRS approach has the required product form, Equation (3.34), meaning that it is an efficient alternative for MCTDH, and in fact it has been used in a number of MCTDH applications [613]. Since the approach is based on orthogonal, rectilinear coordinates, the KEO is simply diagonal. Such an approximation can be used if a large number of coordinates behave almost harmonically and are only weakly coupled among themselves and to the reaction coordinates. Such an approximation breaks down, however, for loosely bonded systems that feature relatively soft, anharmonic potentials. Protonated water clusters belong to this kind of system, and the relatively straightforward CRS is therefore not applicable.

Another possibility to represent our 15D PES $V(\mathbf{Q})$ efficiently is the high-dimensional model representation (HDMR):

$$\tilde{V}(\mathbf{Q}) = V_0 + \sum_i V_i^{(1)}(Q_i) + \sum_{ij} V_{ij}^{(2)}(Q_i, Q_j) + \sum_{ijk} V_{ijk}^{(3)}(Q_i, Q_j, Q_k) + \cdots \quad (23.5)$$

In this approach the PES is given as a sum of terms of increasing order. Each new order n contains corrections to the potential at order $n-1$ with respect to the true potential, that is, correlations between coordinates that had not been accounted for at order $n-1$.

A variant of HDMR that is especially appealing for the straightforward evaluation of the $V^{(n)}$ expansion terms in Equation (23.5) is the so-called cut-HDMR representation [129] – also called n-mode representation [626] – which has already been used to compute accurately the vibrational energy levels of molecular systems [127, 128] and the reaction rates for molecule–surface scattering [454]. In order to define the expansion terms, we first introduce the vector **a**, which represents some reference geometry in the 15D space. In addition, $\mathbf{a}^{\alpha}$ is the same reference geometry but with the αth coordinate removed, while $\mathbf{a}^{\alpha\beta}$ has α and β components removed, and so on. The expansion terms up to order 2 are then defined as:

$$V^{(0)} = V(\mathbf{a}) \tag{23.6a}$$

$$V_i^{(1)}(Q_i) = V(Q_i;\mathbf{a}^i) - V_0 \tag{23.6b}$$

$$V_{ij}^{(2)}(Q_i, Q_j) = V(Q_i, Q_j;\mathbf{a}^{ij}) - V_i^{(1)}(Q_i) - V_j^{(1)}(Q_j) - V_0 \tag{23.6c}$$

where higher orders follow trivially. In a similar way to the definition of the MCTDH *Ansatz*, Equation (3.1), it is advantageous here to work with combined coordinates instead of single (uncombined) coordinates in order to define the PES expansion. Then the Q_κ coordinates refer to composite coordinates of one or more single coordinate(s):

$$Q_\kappa = [q_a, q_b, \ldots] \tag{23.7}$$

In this case we call the $V^{(n)}$ terms combined clusters (CC) and $\mathbf{a}^{\alpha}$ is simply a vector in which all single coordinates of mode α have been removed from the reference geometry. The numerical advantage in the representation of the PES obtained when using combined coordinates instead of single coordinates is analysed in Ref. [156]. The definition of meaningful mode combinations should be possible in most cases. Highly correlated coordinates should be grouped together so that their correlation is taken into account already by the first-order CCs. Such a definition would be based on chemical commonsense, for example, coordinates belonging to the same chemical group or to the same molecule in a cluster are good candidates to be combined.

The application of this scheme does not only lead to numerical advantage in the representation of the PES with respect to the uncombined scheme. Its application in MCTDH will be (much) more efficient if the mode combination used to define the PES is the same as that used in defining the wavefunction. In such a case, the first-order CCs $V^{(1)}(Q_i)$ are full product grids defined in the space of the Q_i combined coordinate. The higher-order terms are given by potfit representations of the total product grids $V^{(n>1)}(Q_i, \ldots)$. A potfit representation can be defined in terms of combined modes (see Equation (11.1) and surrounding text) and in such a case the single-particle potentials (SPPs) are product grids in the space of each Q_i. This strategy can be applied efficiently as long as the full product grid of the nth-order cluster fits into memory. In practice, this means around seven single coordinates, roughly corresponding to CCs of third order.

As a final remark, the convergence of the n-mode representation is highly dependent on the set of coordinates used and the grouping of coordinates into modes. With our set of polyspherical coordinates, which correctly account for the fundamental motions of the cation in terms of rotations, torsions, and so on, and a clever grouping of the coordinates, we obtain a relatively good potential already with a second-order expansion. Details are given in the following section.

23.3.2
Representation of the Potential Energy Surface for $H_5O_2^+$

The PEO for the $H_5O_2{}^+$ system is constructed using the full-dimensional PES developed by Bowman and co-workers. This is based on several tens of thousands of coupled cluster energy calculations, which are combined with a clever fitting algorithm based on polynomials that respect the total permutational symmetry of the system [210]. In order to construct an efficient representation of that PES for the $H_5O_2{}^+$ cation employing the approach described above, one must start by defining the combined modes that are going to be used. In the present case, the following five multidimensional modes are selected: $Q_1 = [z, \alpha, x, y]$, $Q_2 = [R, u_{\beta_a}, u_{\beta_b}]$, $Q_3 = [\gamma_a, \gamma_b]$, $Q_4 = [R_{1a}, R_{2a}, u_{\theta_{1a}}]$ and $Q_5 = [R_{1b}, R_{2b}, u_{\theta_{1b}}]$, where $u_x = \cos x$. It is convenient that coordinates x, y and α are grouped together for symmetry-conserving reasons [156]. Modes Q_2 and Q_3 contain the rocking and wagging coordinates, respectively. Modes Q_4 and Q_5 contain the Jacobi coordinates representing the internal configuration of each water molecule. The definition of the underlying 1D grids is provided in Table I of Ref. [156]. Coordinates z and R would be good candidates to be combined together since they are the two coordinates involved in the description of the intermolecular hydrogen bond. They are not combined here since this would require the definition of a sixth mode.

As a remark, at the time of writing, we have started to do some tests with a six-particle mode combination in which z and R are together in the same mode. This leads to more balanced simulations since all modes are of a similar size. Also the parallel version of the MCTDH performs better with a six-mode set-up since the propagation of the A-vector is efficiently parallelized (Chapter 15).

Following the procedure outlined above, one may select a reference point in coordinate space and proceed straightforwardly to the computation of the clusters defining the cluster expansion. Instead of this, the PES expansion is defined in terms of $M = 10$ reference points whose cluster expansions are averaged with equal weights [156]. The reference points $\mathbf{a}_l$ are located on or very close to stationary points in the lowest energy regions of the PES. The PES expansion is then given by

$$\tilde{V}_{\text{tot}}(\mathbf{Q}) = \frac{1}{M}\sum_{l=1}^{M} \tilde{V}_l(\mathbf{Q}) \tag{23.8}$$

The $\tilde{V}_l(\mathbf{Q})$ terms are given by Equations (23.5) and (23.6). The specific form of $\tilde{V}_l(\mathbf{Q})$ that has been used here is given by

$$\tilde{V}_l(\mathbf{Q}) = V_l^{(0)} + \sum_{i=1}^{5} V_{l,i}^{(1)}(Q_i) + \sum_{i=1}^{4}\sum_{j=i+1}^{5} V_{l,ij}^{(2)}(Q_i, Q_j) + V_{l,z23}^{(3)}(z, Q_2, Q_3) \tag{23.9}$$

where the modes Q_1 to Q_5 have been defined above.

The $V_l^{(0)}$ term is the energy at the reference geometry l. The $V_{l,i}^{(1)}$ terms are the intra-group potentials obtained by keeping the coordinates in other groups at the reference geometry l, while the $V_{l,ij}^{(2)}$ terms account for the group–group correlations. The potential with up to second-order terms gives already a very reasonable description of the system. The $V_{l,z23}^{(3)}$ term accounts for three-mode correlations between the displacement of the central proton, the distance between the two water molecules, and the angular wagging and rocking motions. Note that the primitive grids in each coordinate are the same irrespective of the reference point used to expand the potential. This means that the average, Equation (23.8), can be carried out before the dynamical calculations by summing over all the generated grids of the same coordinates for each reference geometry, involving no extra cost for the dynamics.

The justification for the multi-reference approach lies in the nature of $H_5O_2^+$. The protonated water dimer is a very floppy system featuring several equivalent minima and large-amplitude motions that traverse barriers of low potential energy. Thus, the amount of configurational space available to the system at low vibrational energies is already large and therefore not well covered by a single reference point. One should note that, for a single reference point, the PES expansion is exact at the reference point and hypersurfaces involving the displacement of up to h_m modes (where m is the expansion order). This property is lost after averaging over several reference geometries. However, the overall mean error is reduced by the averaging.

As well as the fact that the configurational space available to the Zundel cation, even at low energies, is quite large and therefore several reference points are necessary to cover it, there are also symmetry-conservation reasons that force us to use several reference geometries. The $H_5O_2^+$ system features eight equivalent absolute minima on the PES that have C_2 symmetry. The system has however (in the range of energies relevant to the IR linear absorption spectrum) $\mathcal{G}_{16}$ symmetry [627]. This symmetry group contains the $\mathcal{D}_{2d}$ group as a subgroup and permits additionally the permutation of the two hydrogens of one of the water monomers (internal rotation). If only one of the minima of the $\mathcal{C}_2$ structure is used as a reference, the total symmetry of the PES is broken. This is due to one $\mathcal{C}_2$ point being exactly described while the others are given only approximately within the corresponding n-mode representation. To avoid this, in fact, for every arbitrary reference geometry, all symmetry-equivalent points generated by the permutations and inversions of the $\mathcal{G}_{16}$ group must also be considered. In the case of taking a $\mathcal{C}_2$ point, this leads to the total eight equivalent absolute minima. Additionally, we also take the two equivalent $\mathcal{D}_{2d}$ points in which both water molecules are planar and located 90° with respect to each other ($\alpha = 90°$ and 270°). These are the 10 reference points used in the construction of the PEO.

The quality of the PES expansion can be assessed by monitoring the expectation values of the different terms of the n-mode representation with respect to the ground vibrational state $|\Psi_0\rangle$. These values are given in cm^{-1} in Table 23.1. The sum of the first-order $\langle\Psi_0|V^{(1)}(Q_i)|\Psi_0\rangle$ terms is close to 6800 cm^{-1}, half the zero-point energy (ZPE), indicating that they carry the major weight in the

Table 23.1 Expectation value of the different terms of the potential expansion (central column) and square root of the expectation value of the potential squared (right column). All energies in cm^{-1}. The combined modes read: $Q_1 = [z, \alpha, x, y]$, $Q_2 = [R, u_{\beta_a}, u_{\beta_b}]$, $Q_3 = [\gamma_a, \gamma_b]$, $Q_4 = [R_{1a}, R_{2a}, u_{\theta_{1a}}]$ and $Q_5 = [R_{1b}, R_{2b}, u_{\theta_{1b}}]$. In addition, $u_x = \cos x$; λ refers to wagging and β refers to rocking motion; and Q_4 and Q_5 contain the Jacobi coordinates of the water molecules.

	$\langle\Psi_0\|V\|\Psi_0\rangle$	$\langle\Psi_0\|V^2\|\Psi_0\rangle^{1/2}$
$V^{(1)}(Q_1)$	1293.6	1807.7
$V^{(1)}(Q_2)$	750.6	966.9
$V^{(1)}(Q_3)$	171.5	266.9
$V^{(1)}(Q_4)$	2293.2	3062.8
$V^{(1)}(Q_5)$	2293.1	3062.8
$V^{(2)}(Q_1, Q_2)$	−526.9	1037.2
$V^{(2)}(Q_1, Q_3)$	−78.8	290.2
$V^{(2)}(Q_1, Q_4)$	−27.5	231.8
$V^{(2)}(Q_1, Q_4)$	−27.4	231.7
$V^{(2)}(Q_2, Q_3)$	−10.5	37.6
$V^{(2)}(Q_2, Q_4)$	−24.7	117.5
$V^{(2)}(Q_2, Q_5)$	−24.7	117.9
$V^{(2)}(Q_3, Q_4)$	−18.8	180.9
$V^{(2)}(Q_3, Q_5)$	−18.8	180.9
$V^{(2)}(Q_4, Q_5)$	1.2	9.9
$V^{(3)}(z, Q_2, Q_3)$	1.0	50.4

description of the PES. The second-order clusters introduce the missing correlation between modes. They have expectation values one order of magnitude smaller than the first-order terms, with one exception, the matrix element arising from the $V^{(2)}(Q_1, Q_2)$ potential. This can be easily understood by noting that modes Q_1 and Q_2 contain coordinates z and R, respectively. These two coordinates are strongly correlated, and indeed they would be good candidates to be put in the same mode in an alternative mode combination scheme. The only third-order term that was introduced presents a rather marginal contribution to the potential energy of the system. These values prove that the PES representation used is of a good quality and rather well converged with respect to the reference PES, at least for the energy domain of interest. The square root of the expectation value of the potential squared is depicted in the third column. It is a measure of the dispersion around the expectation value and should also ideally vanish. The values indicate that the PES representation is good, albeit not yet fully converged.

23.3.3 Ground Vibrational State and Eigenstates in the Low-Frequency Domain

The ground vibrational state and some eigenstates of the vibrational Hamiltonian have been computed using the *improved relaxation* algorithm (Chapter 8). The comparison between the largest, converged MCTDH calculation and other reported results on the same PES is given in Table 23.2. As a reference, we take the given diffusion Monte Carlo (DMC) result [628], which has an associated statistical uncertainty of 5 cm^{-1}. The most comprehensive calculations on the vibrational ground state based on a wavefunction approach to date are those of Bowman and co-workers [628] using the MULTIMODE program [127]. The vibrational configuration-interaction (VCI) results, both using the single reference (SR) and reaction path (RP) variants, are found in Table 23.2. These calculations use a normal-mode-based Hamiltonian. They

Table 23.2 Comparison of the zero-point energy (ZPE) of the $H_5O_2^+$ cation calculated by various approaches on the PES of Huang *et al.* [210]: diffusion Monte Carlo (DMC), normal-mode analysis (harmonic), vibrational CI single reference (VCI-SR) and reaction path (VCI-RP) as published in [628] and MCTDH results. Δ denotes the difference from the DMC result. The converged MCTDH result is obtained with 10 500 000 configurations. (See Table 23.3).

Method	ZPE (cm^{-1})	Δ (cm^{-1})
DMC	12 393	0
Harmonic	12 635	242
VCI-SR	12 590	197
VCI-RP	12 497	104
MCTDH	12 376.3	−16.7

incorporate correlation between the different degrees of freedom due to the cluster expansion of the potential [127] and the use of a CI wavefunction. The best reported VCI result for the ZPE still lies 104 cm^{-1} above the DMC value. It is worth mentioning that, before switching to a Hamiltonian based on polyspherical coordinates, we tried a Hamiltonian expressed in rectilinear coordinates and obtained results similar to those of Bowman and co-workers. The MCTDH converged result for the ZPE is given in Table 23.2. The obtained value for the ZPE is 12 376.3 cm^{-1}, 16.7 cm^{-1} below the DMC value.

Table 23.3 contains ZPE values obtained using an increasing number of configurations. According to these results, the MCTDH reported values are assumed to be fully converged with respect to the number of configurations. The deviation from the DMC result must be attributed to the cluster expansion of the potential in Equation (23.5). Table 23.3 illustrates an interesting property of MCTDH, namely its *early* convergence. The difference in energy between the computation on the top and the most expensive one below is only about 7 cm^{-1}, yet the first calculation can be performed on a laptop while the most expensive one needed several days on a workstation. The variational optimality of the MCTDH equations, in particular the optimality of the SPFs, leads to qualitatively correct results as soon as there are enough SPFs to describe the physics of the problem at hand, while more accurate results can be obtained at the cost of increased computational resources.

Table 23.3 Comparison of the zero-point energy (ZPE) of the $H_5O_2{}^+$ cation for different MCTDH calculations with ascending number of configurations. The Δ values are given with respect to the diffusion Monte Carlo result, 12 393 cm^{-1} [628].

SPFs per mode	No. configs.	ZPE (cm^{-1})	Δ (cm^{-1})
(20,20,12,6,6)	172 800	12 383.7	−9.3
(35,25,15,8,8)	840 000	12 378.5	−14.5
(40,40,20,8,8)	2048 000	12 377.8	−15.2
(60,40,20,8,8)	3072 000	12 376.7	−16.3
(70,50,30,10,10)	10 500 000	12 376.3	−16.7

The probability density of the ground-state wavefunction with respect to some selected coordinates and integration over the remaining coordinates is given in Figure 23.5. Figure 23.5(a) shows the density along the proton transfer coordinate z. The probability density is not negligible in a range spanning about 1 bohr. Figure 23.5(b) depicts the density along the α internal rotation coordinate. Along this coordinate the system interconverts between two equivalent regions of configurational space. The barrier corresponds to planar configurations of the whole system and is about 300 cm^{-1} high depending on the configuration of the rest of the coordinates. The system can interconvert between both halves even when in the ground vibrational state. The dotted curve in Figure 23.5(b) depicts the density at a 10 times enlarged scale and clearly shows a non-vanishing density for $\alpha = 0, \pi$. The splitting state arising from the barrier along α has also been computed, and the splitting energy has been found to be $1\,\text{cm}^{-1}$. Figure 23.5(c) shows the probability density along the wagging coordinates. It consists of four equivalent maxima, each of them centred at around $\pm 30°$ from the planar water configuration, so that both water molecules are in pyramidal configuration. The probability density corresponding to one of the two water molecules in a planar configuration is however quite large and indicates a high probability of exchange between equivalent configurations of the system in which the water molecules switch between pyramidal geometries. Each of these four density maxima corresponds roughly to one of the C_2 equivalent minima on the PES. A total of eight equivalent C_2 minima are present since the barrier along coordinate α divides the configurational space into two equivalent halves. When both monomers are in planar configuration, the system interconverts between both $\mathcal{D}_{2d}$ and $\mathcal{D}_{2h}$ configurations by rotation along α.

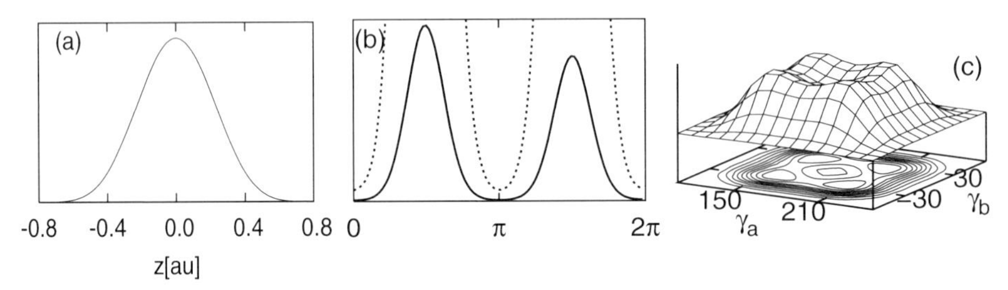

Figure 23.5 For the ground vibrational state, probability density along selected coordinates and integration over the rest: probability density (a) along the z proton transfer coordinate, (b) along the α internal rotation coordinate and (c) on the 2D space spanned by the wagging γ_a and γ_b coordinates. The dotted line in (b) corresponds to a 10 times enlarged scale. It indicates that the probability density at $\alpha = \pi$ is non-vanishing.

In the following we analyse the excited vibrational states related to the wagging motion due to the relevance of some of these modes in shaping the IR spectrum. Further details on the internal rotational motion can be found in Ref. [156]. Figure 23.6 depicts the probability density projected onto the wagging coordinates of the four lowest excited states related to the wagging motion. The shapes of these density plots together with Figure 23.5(c) make it clear that the wagging motion takes place in a very anharmonic region of the PES. State $w_{1a,b}$, which is a doubly degenerate state of an E irreducible representation, is responsible for the absorption at about $100\,\text{cm}^{-1}$ in the IR spectrum. When the system is considered in its full $\mathcal{G}_{16}$ symmetry, the rotational motion around α is also involved in the state corresponding to such absorption [156]. However, in this chapter we consider the symmetry labels as corresponding to the more familiar $\mathcal{D}_{2d}$ point group. The energies of the next three wagging-mode states (w_2, w_3 and w_4) are, respectively, 232, 374 and 422 cm^{-1} and they are shown in Figure 23.6(b), (c) and (d), respectively. These three states correspond to two quanta of excitation in the wagging motions and they can be represented in 'ket' notation by $|11\rangle$, $(|20\rangle - |02\rangle)/\sqrt{2}$ and $(|20\rangle + |02\rangle)/\sqrt{2}$, respectively, where the $|ab\rangle$ notation signifies the quanta of excitation in the wagging motions of monomers A and B. These states have symmetries B_1, B_2 and A_1, respectively. In the harmonic limit, these three states would be degenerate. The w_3 state has four probability density maxima along the 2D space spanned by γ_a and γ_b. They correspond to geometries in which one of the water molecules adopts a pyramidal geometry (H_3O^+ character) and the other adopts a planar geometry (H_2O character). This state transforms according to the B_2 symmetry representation, which is also the symmetry of the proton transfer fundamental. State w_3 will play a major role due to its strong coupling to the proton transfer mode.

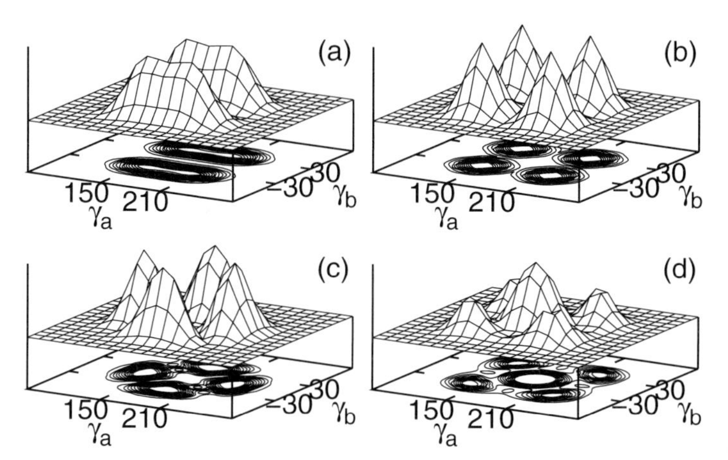

Figure 23.6 Reduced probability density on the wagging coordinates γ_a and γ_b of (a) the first excited ($w_{1a,b}$) wagging-mode state and (b), (c), (d) the excited states w_2, w_3 and w_4, respectively, characterized by two quanta of excitation. In 'ket' notation these four states can be represented as $w_{1a,b} = |10\rangle \pm |01\rangle$, $w_2 = |11\rangle$, $w_3 = |20\rangle - |02\rangle$ and $w_4 = |20\rangle + |02\rangle$, respectively, where the numbers represent the quanta of excitation in the wagging motion of each water molecule in a harmonic limit.

23.3.4
Infrared Absorption Spectrum

The IR absorption cross-section is given by [58]:

$$I(E) = \frac{\pi E}{3c\epsilon_0 \hbar} \sum_n |\langle \Psi_n | \Psi_{\mu,0} \rangle|^2 \, \delta(E + E_0 - E_n) \tag{23.10}$$

where $|\Psi_{\mu,0}\rangle$ is the dipole-operated initial state, that is, $|\Psi_{\mu,0}\rangle \equiv \hat{\mu}|\Psi_0\rangle$, and E_0 is the ground-state energy. The IR spectrum may be equivalently computed in the time-dependent picture. Using a Fourier representation of the delta function it follows that

$$I(E) = \frac{E}{3c\epsilon_0\hbar^2} \operatorname{Re} \int_0^\infty \exp\left(\frac{\mathrm{i}(E+E_0)t}{\hbar}\right) a_\mu(t)\,\mathrm{d}t \tag{23.11}$$

where $a_\mu(t) = \langle\Psi_{\mu,0}|\mathrm{e}^{-\mathrm{i}Ht/\hbar}|\Psi_{\mu,0}\rangle$ denotes the autocorrelation function. Whether a time-dependent or a time-independent approach is more efficient depends on the problem at hand. A time-independent approach relies on the accurate computation of the eigenstates of the Hamiltonian, $|\Psi_n\rangle$, which may be accomplished by iterative-diagonalization methods of the Hamiltonian matrix expressed in some basis, obtaining all the eigenstates up to some desired energy, and then using Equation (23.10) to compute the IR spectrum. A comprehensive calculation of an IR spectrum by improved relaxation has recently been performed for six-dimensional (6D) problems, namely H_2CS [101] and HONO [107]. The lowest 184 states of A' symmetry of the latter molecule were calculated. The improved relaxation approach, however, is not practicable with the current computational capabilities to obtain the whole spectrum for a molecule of the size of $H_5O_2{}^+$, since the density of states at already moderate vibrational energies becomes enormous, while each state must be obtained from a separate calculation. Moreover, the improved relaxation approach contains convergence problems in spectral regions with a high density of states, which makes it unusable for $H_5O_2{}^+$.

Propagation of a wavepacket by MCTDH, on the other hand, is always feasible. A small set of SPFs makes a propagation less accurate, but not impossible. Therefore, the full IR spectrum is efficiently calculated in the time-dependent representation. The resolution at which different peaks of the spectrum are resolved is given by the Fourier transform (FT) of the damping function $g(t)$ [34] by which the autocorrelation $a_\mu(t)$ is to be multiplied when performing the integral (23.11c) to minimize artefacts due to the Gibbs phenomenon. We choose $g(t) = \cos(\pi t/2T)$ and set $g(t) = 0$ for $t > T$, where T denotes the length of the autocorrelation function. Since we make use of the $T/2$ trick, Equation (6.7), T is twice the propagation time. The FT of $g(t)$ is known, Equation (6.11), and its full width at half-maximum (FWHM) is $27\,300\,\mathrm{cm}^{-1}\,\mathrm{fs}/T$.

The IR spectrum calculated with MCTDH is shown in Figure 23.7. The MCTDH spectrum in the full range 0–4000 cm^{-1} is depicted in the left panel and compared with the recent experimental results of Ref. [621] in the right panel. The dipole-moment-operated ground state $\hat{\mu}|\Psi_0\rangle$ was propagated for 500 fs, yielding an autocorrelation of 1000 fs. The spectrum was calculated according to Equation (23.11) and the FWHM resolution of the spectrum is about 30 cm^{-1}.

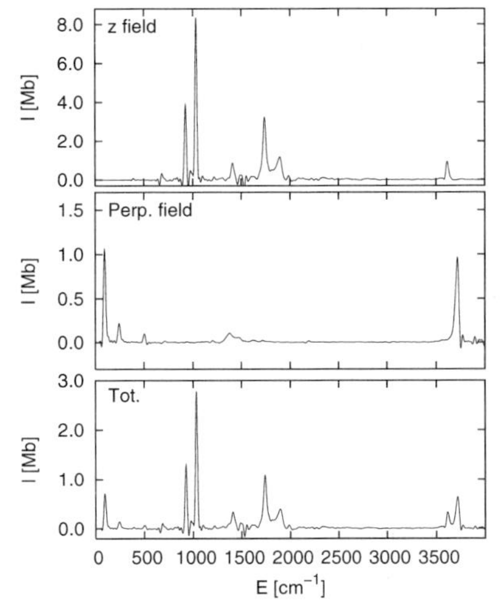

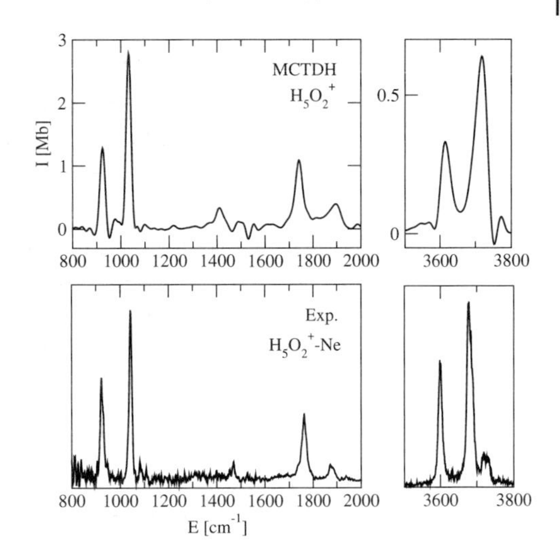

Figure 23.7 Left panel: Simulated MCTDH spectrum in the range 0–4000 cm^{-1}. Excitation in the z-direction (top), perpendicular to the O–H–O axis (middle), and total spectrum, that is, $\frac{1}{3}z + \frac{2}{3}$ perpendicular (bottom). Note the different scale of intensities in the perpendicular-component plot. Autocorrelation time $T = 1000$ fs. Right panel: Comparison between the MCTDH spectrum (top) and the $H_5O_2^+ \cdot$ Ne spectrum of Ref. [620] (bottom). The intensity of the experimental spectrum is adjusted in each spectral region (800–2000 and 3500–3800 cm^{-1}) using the most intense peak of the MCTDH spectrum as a reference. Absorption for the MCTDH spectrum is given in absolute scale in megabarns (1 Mb = 10^{-18} cm^2).

Even if the computation of the IR spectrum in the time-dependent representation is feasible, a means of assigning the different lines to specific motions of the system is still required. Wavefunctions of excited states converged by improved relaxation, Chapter 8, contain all the possible information on that specific state. The intensity of a given excited state $|\Psi_n\rangle$ is readily obtained by computing the dipole moment $|\langle\Psi_n|\hat{\mu}|\Psi_0\rangle|^2$. But even if an excited state of interest $|\Psi_n\rangle$ has been obtained, it is difficult to inspect these mathematical objects directly due to their high dimensionality. Moreover, for the higher excited states, we do not have $|\Psi_n\rangle$ at our disposal but only an autocorrelation function providing spectral lines. In both cases we characterize the eigenstates by their overlaps with carefully chosen test states, that is, by the numbers $|\langle\Phi_{\text{test}}|\Psi_n\rangle|^2$. The following procedures are used.

1. Test states $|\Phi_{\text{test}}\rangle$ are generated as follows:
   (a) Apply some operator $\hat{O}$ to a previously converged eigenfunction,

$$|\Phi_{\text{test}}\rangle = N\hat{O}|\Psi_n\rangle \qquad (23.12)$$

   where N is a normalization constant. For example, $N\hat{z}|\Psi_0\rangle$ generates a test state that in essence differs from the ground state $|\Psi_0\rangle$ by a one-quantum excitation in the proton transfer coordinate z.

(b) Form Hartree products, where the SPFs are obtained through diagonalization of mode Hamiltonians $\hat{h}_\kappa$. The $\hat{h}_\kappa$ are low-dimensional Hamiltonians obtained from the full Hamiltonian by freezing the remaining coordinates. Each $\hat{h}_\kappa$ operates on the coordinate space of an MCTDH particle. Rather than using single Hartree products, one may use linear combinations of products in order to satisfy a symmetry constraint.

2. The overlaps $|\langle\Phi_{\text{test}}|\Psi_n\rangle|^2$ are then computed as follows:
   (a) Directly evaluate the scalar product if $|\Psi_n\rangle$ is available.
   (b) Fourier-transform the autocorrelation function

$$a(t) = \langle\Phi_{\text{test}}|\mathrm{e}^{-\mathrm{i}Ht}|\Phi_{\text{test}}\rangle$$

The overlap is obtained via the formula [101]

$$|\langle\Phi_{\text{test}}|\Psi_n\rangle|^2 = \frac{\pi}{2T}\,\mathrm{Re}\int_0^T \mathrm{e}^{\mathrm{i}E_n t}a(t)\cos\left(\frac{\pi t}{2T}\right)\mathrm{d}t \qquad (23.13)$$

23.3.5
Analysis of the Middle Spectral Region and Dynamics of the Cluster

In the following we apply the techniques described above to analyse the spectral region between 800 and 2000 cm^{-1}. This complicated spectral region is shaped by several couplings that involve collective motions of the whole cation. The complication arises from the floppiness and anharmonicity of the $H_5O_2{}^+$ system. Its complete assignment could only be achieved by the MCTDH method [59, 486], and it constitutes a nice example of the power of this approach to unravel complicated IR signatures. A detailed description of the OH stretching region at about 3600 cm^{-1} as well as a comprehensive list of all fundamentals and various overtones of all vibrational modes may be found in Ref. [59].

The doublet centred at 1000 cm^{-1} is the most characteristic feature of the IR spectrum of $H_5O_2{}^+$. It is depicted in Figure 23.7 (top right). The highest-energy line has been measured to be at 1047 cm^{-1}, while the low-energy component appears at 928 cm^{-1} [620]. These values change slightly to 1042 and 923 cm^{-1}, respectively, in the most recent measurements [621]. There is accumulated evidence in the literature that the absorption of the proton transfer fundamental occurs in the region of 1000 cm^{-1} [210, 620, 623–625]. Specifically, the band at 1047 cm^{-1} in Ref. [620] was assigned to the first excitation of the central proton motion [620] based on MULTIMODE [127] calculations. This band is the most intense band of the spectrum, since the central proton motion along the z-axis induces a large change in the dipole moment of the cation.

The low-energy component has recently been assigned by us [486]. The doublet is seen to arise from coupling between the proton transfer motion, the low-frequency water wagging modes and the water–water stretching motion.

In order to obtain a fundamental understanding of the low-energy ($|\Psi_d^l\rangle$) and high-energy ($|\Psi_d^h\rangle$) components of the doublet, test states were constructed by operating with $\hat{z}$ on the ground state, $|\Phi_{1z}\rangle = \hat{z}|\Psi_0\rangle N$, where N is a normalization constant, and by operating with $(\hat{R} - R_0)$ on the third excited wagging state w_3, $|\Phi_{1R,w_3}\rangle = (\hat{R} - R_0)|\Psi_{w_3}\rangle N$. Note that $|\Phi_{1z}\rangle$ is characterized by one quantum of excitation in the proton transfer coordinate, whereas $|\Phi_{1R,w_3}\rangle$ is characterized by one quantum in the O–O stretch motion and two quanta in the wagging motion. These two test states were propagated and their auto- and cross-correlation functions were used for filter diagonalization (FD) analysis, Section 6.2.2, which yielded an energy of 918 cm^{-1} for $|\Psi_d^l\rangle$ and an energy of 1033 cm^{-1} for $|\Psi_d^h\rangle$. These energies are in good agreement with the peaks in Figure 23.7, which arise from the propagation of $|\Psi_{\mu,0}\rangle$. The spectral intensities were also obtained by FD analysis. The overlaps of the test states with the states making the doublet are: $|\langle\Phi_{1z}|\Psi_d^l\rangle|^2 = 0.09$, $|\langle\Phi_{1R,w_3}|\Psi_d^l\rangle|^2 = 0.83$ and $|\langle\Phi_{1z}|\Psi_d^h\rangle|^2 = 0.46$, $|\langle\Phi_{1R,w_3}|\Psi_d^h\rangle|^2 = 0.10$.

One should take into account that these numbers depend on the exact definition of the test states, which is not unique. However, they provide a clear picture of the nature of the doublet: the low-energy band has the largest contribution from the combination of the symmetric stretch and the third excited wagging (see Figure 23.6(c)), whereas the second largest is the proton transfer motion. For the high-energy band, the importance of these two contributions is reversed. Thus, the doublet may be regarded as a Fermi resonance between two zero-order states that are characterized by $(1R, w_3)$ and $(1z)$ excitations, respectively. The reason why the third wagging excitation plays an important role in the proton transfer doublet is understood by inspecting Figure 23.6(c) and 23.8. The probability density of state w_3 has four maxima, each of which corresponds to a bent conformation of $H_2O–H^+$ (H_3O^+ character) for one of the waters, and a planar conformation (H_2O character) where a lone-pair H_2O orbital forms a hydrogen bond with the central proton. When the proton oscillates between the two waters, the two conformations exchange their characters accordingly. Thus, the asymmetric wagging mode (w_3, 374 cm^{-1}) combines with the water–water stretch motion ($1R$, 550 cm^{-1} [59]) to reach an energy close to the natural absorption frequency of the proton transfer, making these motions coupled. The two states of the doublet transform according to the B_2 irreducible representation of $\mathcal{D}_{2d}$ and hence couple to the z-component of the dipole operator.

The region between the proton transfer doublet and 2000 cm^{-1} features couplings related to the PT and O–O stretching motions. The MCTDH spectrum reported in Figure 23.7 presents three main absorptions in this range, located

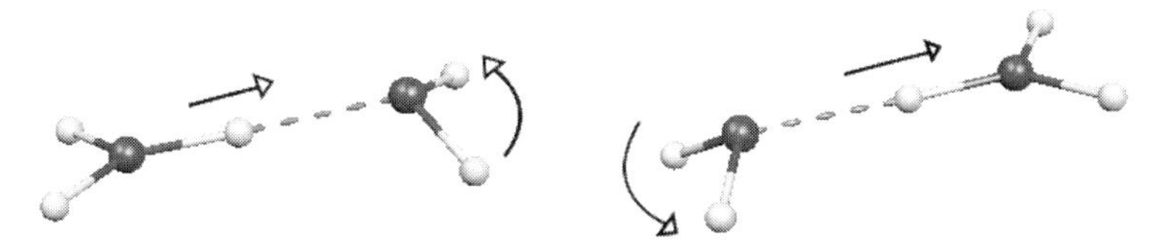

Figure 23.8 Schematic representation of the two most important coupled motions responsible for the doublet peak at 1000 cm^{-1}.

at 1411, 1741 and 1898 cm^{-1}, and we could assign them to $1z1R$, *bu* and $1z2R$ excitations, respectively [59]. Here *bu* stands for bending *ungerade*, that is, one water opens while the other closes. This water bending state deserves particular consideration. It couples very strongly to the proton transfer motion and, in fact, its strong brightness is due to this coupling. In a fashion similar to the coupling to the w_3 wagging motion, as the proton approaches one water molecule, the equilibrium value of the H–O–H angle shifts to a larger value because this water molecule acquires more H_3O^+ character. Conversely, the water molecule at a larger distance from the central proton acquires H_2O character and the angle H–O–H shifts to lower values. That is, the concerted motion is similar to that depicted in Figure 23.8, but with the wagging motion replaced by the bending motion. This coupling shifts the proton transfer line down and the bending line up by at least 120 cm^{-1}. This estimate was obtained by suppressing the correlation between proton transfer and bending by setting the number of SPFs to one, for alternately the proton transfer mode and the water modes. The frequency of the *ungerade* bending when decoupled from the proton transfer motion resembles that of the bare-water bending mode. Thus the shift to a higher frequency is almost entirely due to coupling with the proton transfer mode.

23.4 Conclusion

In this chapter we have discussed the simulation of two systems related to the excess (or hydrated) proton in water and to proton transfer from two very distinct perspectives. First, we described how a model depending on a reduced set of parameters could be used to simulate the operation of a chain of water molecules. Second, the protonated water dimer, or Zundel cation ($H_5O_2{}^+$), which may be looked at as the smallest possible water chain, was investigated in its full dimensionality using an accurate high-quality PES.

We discussed the construction of the system Hamiltonian for $H_5O_2{}^+$, specifically on the PES representation for this 15-dimensional problem. We also described how the IR spectrum could be efficiently calculated and analysed in the time representation.

The reported calculations on $H_5O_2^+$ were in excellent agreement with the experimental measurements of Refs. [620] and [621] on the messenger predissociation spectrum of $H_5O_2^+ \cdot Ne$. Such a remarkable consistency between experiment and theory along the whole spectral range represents, on the one hand, a validation of the underlying potential energy surface of Huang *et al.* [210] and of the mode-based cluster expansion of the potential used in the quantum dynamical simulations, but is also a clear indication that a suitable set of coordinates was selected to tackle the problem [156].

The fact that the reported simulations were successful in obtaining accurate results for a system of the dimensionality of the protonated water dimer is to be attributed in great part to the MCTDH algorithm, in which not only the expansion coefficients, but also the SPFs, are variationally optimal. For a 15-dimensional system, the use of only four basis functions per degree of freedom represents of the order of 10^9 configurations. The largest calculations reported here consist of about 10^7 configurations, while already good results are obtained by using as few as 10^5 configurations (Table 23.2). Such an *early* convergence of the MCTDH method becomes crucial as high-dimensional problems are attempted.

The reported simulations constitute a new example of the ability of the MCTDH method to tackle high-dimensional, complex molecular systems in a rigorous manner. They provide as well important information on the spectroscopy and dynamics of the hydrated proton.

24
Laser-Driven Dynamics and Quantum Control of Molecular Wavepackets

Oliver Kühn

24.1
Introduction

Ultrafast laser spectroscopy provides a real-time look at elementary steps of molecular dynamics such as vibrational energy flow, reactive barrier crossings or charge transfer. This statement holds for few-atom systems in the gas phase as well as for biological complexes. Moreover, recent advances in laser pulse shaping have paved the way from the perspective of merely observing such dynamics to the realm of attaining its control. This gives a new twist to the issue of molecular dynamics, since it opens the opportunity, for instance, to deepen the understanding of elementary reaction steps by guiding the systems into those regions of configuration space that one intends to explore [629].

Given the success of classical trajectory methods in simulating many coupled degrees of freedom (DOFs), one might ask whether a quantum treatment of the dynamics is really necessary, especially if the complexity of the considered systems increases and effects like dephasing may become dominant [630]. Processes involving electronic transitions are definitely of quantum mechanical nature. However, from the perspective of ultrafast spectroscopy, the observation of vibrational coherences with dephasing times of a few picoseconds or less may call for a quantum treatment as well. Furthermore, most strategies for quantum control of wavepacket dynamics are especially designed to make use of the quantum nature of the system's evolution, that is, by exploring phase relations between the states comprising the wavepacket.

Taking another point of view, the validation of classical trajectory methods, such as surface hopping approaches for incorporating electronic transitions, is usually hampered by the bottleneck of performing respective quantum simulations for more than a few coupled DOFs. Fortunately, this situation is changing with the ongoing development of the MCTDH method [13, 34, 38]. Hav-

Multidimensional Quantum Dynamics: MCTDH Theory and Applications.
Edited by Hans-Dieter Meyer, Fabien Gatti, and Graham A. Worth

ISBN: 978-3-527-32018-9

ing at hand an efficient method for propagating wavepackets in many dimensions, it becomes desirable to have an equally efficient way of generating appropriate potential energy surfaces (PESs) using quantum chemistry methods. Moreover, with increasing dimensionality, one should recover system–bath behaviour [630], since not all DOFs will contribute equally to the considered process, which might involve the excitation of a particular chromophore in an extended system. Summarizing these points, the quest goes on for a method with the following properties:

(i) The input for constructing the Hamiltonian should be straightforward to calculate using quantum chemical methods.

(ii) There should be a means for identifying reduced-dimensional Hamiltonians that can be systematically extended to full dimensionality.

(iii) Since many DOFs are involved, the structure of the Hamiltonian should allow factorization in most DOFs without further treatment.

(iv) In the limit of many coupled DOFs, one should recover a Hamiltonian that is of the common system–bath form.

In a number of applications, we have recently shown that the all-Cartesian reaction surface (CRS) Hamiltonian fulfills these criteria and therefore is ideally suited to perform multidimensional wavepacket dynamics using the MCTDH approach. After briefly reviewing important formulae in the next section, applications are given for the laser-driven vibrational dynamics of a model mimicking carboxymyoglobin [631] and of intramolecular hydrogen bonds [610, 611, 632–636]. Subsequently, results on the non-adiabatic dynamics of a photoexcited diatomic molecule embedded in a low-temperature argon matrix are discussed [637–639]. Different extensions of the original CRS approach will also be outlined in the Applications (Section 24.3).

24.2 Theory

The description of laser-driven dynamics is based on a Hamiltonian of the form

$$H = H_{\text{mol}} + H_{\text{field}}(t) \tag{24.1}$$

where the molecular part, H_{mol}, is supplemented by a time-dependent molecular-field coupling part, $H_{\text{field}}(t)$. In the following we will restrict ourselves to the dipole approximation and use

$$H_{\text{field}}(t) = -\mathbf{dE}(t) \tag{24.2}$$

Here, $\mathbf{d}$ is the operator of the dipole moment vector and $\mathbf{E}(t)$ is the electric field vector. This type of operator is straightforwardly incorporated into the MCTDH scheme, provided that the dipole moment operator can be written as a sum of products. This means that, for a set of p MCTDH particles, $\{Q_\kappa\}$, one has (skipping vector notation) in analogy to Equation (3.34),

$$H_{\text{field}}(t) = \sum_{r=1}^{n_s} e_r(t) d_r^{(1)}(Q_1) \cdots d_r^{(p)}(Q_p) \tag{24.3}$$

where the time-dependent coefficients $e_r(t)$ derive from the field $E(t)$. Time-dependent Hamiltonians were first implemented into MCTDH in Ref. [611], where it was also pointed out that the explicit time dependence requires the variable mean-field (VMF) scheme for integration.

In principle, the form of Equation (24.3) as well as of the related molecular Hamiltonian can be obtained by the potfit algorithm. However, since potfit uses the full primitive product grid, its applicability will, in practice, be limited to a few coordinates. Needless to say, the determination of the PES using *ab initio* quantum chemistry faces the same dimensionality bottleneck. Therefore, approximate methods for the *ab initio* determination of the PES are needed, especially for reactive problems involving large-amplitude displacements that cannot be described by local Taylor expansions. A choice that is particularly useful in the context of the MCTDH method is the CRS Hamiltonian approach [640] (for a review including also references to the various blends of reaction surface approaches, see also Ref. [636]). Here, H_{mol} is approximated by

$$H_{\text{mol}} = T_{\mathbf{s}} + V(\mathbf{s}) + T_{\mathbf{q}} - \mathbf{F}(\mathbf{s})\mathbf{q} + \tfrac{1}{2}\mathbf{q}\mathbf{K}(\mathbf{s})\mathbf{q} \tag{24.4}$$

The vector $\mathbf{s}$ comprises large-amplitude coordinates (typically two or three) and $\mathbf{q}$ stands for the majority of coordinates, which are harmonically approximated with respect to a given point $\mathbf{s}$ (mass-weighted normal-mode coordinates). As the choice of $\mathbf{s}$ is not necessarily restricted to minimum-energy configurations, forces, $\mathbf{F}$, appear in Equation (24.4). Further, a certain fixed reference geometry for the definition of the normal modes has to be chosen. Therefore, these coordinates will be mixed by the off-diagonal elements of the Hessian $\mathbf{K}$. While Equation (24.4) provides a full-dimensional description, it is often desirable to restrict the dimensionality of the problem. A convenient measure for the importance of a certain normal mode is its reorganization energy, $\Delta E_i^{\text{reorg}}(\mathbf{s}) = 0.5 K_{ii}(\mathbf{s})[q_i^{(0)}]^2$, with $\mathbf{q}^{(0)} = -\mathbf{K}^{-1}\mathbf{F}$.

As suggested by the name, CRS, all coordinates are taken to be Cartesian, that is, no couplings appear in the kinetic energy operator. Of course, this is an approximation, as the rotational invariance along an arbitrary path on the reaction surface cannot be shown (see, for example, Ref. [613]). However, for large molecules and for a choice of **s** that captures most of the large-amplitude motion, one can expect the error to be small. Different choices of **s** include atomic positions [610, 611, 640, 641] or collective coordinates chosen according to geometries of stationary points on the PES [613, 638] (see Applications in Section 24.3).

The form of the dipole moment surface can be chosen along the same lines, that is, by the expansion

$$d = d(\mathbf{s}, \mathbf{q})\big|_{\mathbf{q}=0} + \nabla_{\mathbf{q}} d(\mathbf{s}, \mathbf{q})\big|_{\mathbf{q}=0}\, \mathbf{q} \tag{24.5}$$

that is, both Equations (24.2) and (24.4) are of factorized form in most DOFs.

Concerning the field $\mathbf{E}(t)$, the applications below will cover cases where it can be assumed that the field is linearly polarized along the dominant dipole direction, that is, the vector character of the field and the dipole can be skipped in the following. Further, analytical pulse forms of the type

$$E(t) = \mathcal{E}(t) \cos[\omega(t)t] \tag{24.6}$$

will be used. Here, $\mathcal{E}(t)$ is the field envelope, which can have, for example, $\sin^2$ shape

$$\mathcal{E}(t) = E_0 \sin^2(t\pi/\tau_\mathrm{p})\theta(t)\theta(\tau_\mathrm{p} - t) \tag{24.7}$$

with τ_p being the pulse length and E_0 its maximum amplitude, or Gaussian shape

$$\mathcal{E}(t) = E_0 \exp[-2 \ln 2 (t/\tau)^2] \tag{24.8}$$

where τ is the full width at half-maximum (FWHM) of the intensity. The frequency in Equation (24.6) has been chosen to be time dependent so as to describe pulse chirping. Specifically, we will use the form [642]

$$\omega(t) = \omega_0 + \frac{\beta(\Delta\omega^2/\ln 2)^2}{8 + 2\beta^2(\Delta\omega^2/\ln 2)^2}\, t \tag{24.9}$$

which yields the FWHM

$$\tau^2 = \tau_0^2 + 4\beta^2\Delta\omega^2 \tag{24.10}$$

Here, β is the chirp rate, $\Delta\omega = 4 \ln 2/\tau_0$, and the frequency and FWHM of the unchirped pulse are ω_0 and τ_0, respectively.

24.3 Applications

24.3.1 Vibrational Ladder Climbing in a Haem–CO Model

Myoglobin is certainly one of the most studied proteins [643] and carboxymyoglobin (MbCO) is particularly well characterized. This characterization includes: a high-resolution X-ray structure [644]; studies of photodissociation dynamics following optical excitation with time-resolved X-ray crystallography of the protein dynamics related to CO migration away from the docking site [645, 646]; investigation of CO [647] as well as haem and protein vibrational motions coupled to the optical transition and bond-breaking dynamics [648–651]; and CO vibrational relaxation in the haem pocket [652]. The vibrational relaxation of the bound CO ligand in MbCO and various model complexes has been investigated thoroughly by Dlott and co-workers [653, 654]. The measured vibrational lifetimes range from 10 to 24 ps. Pure dephasing times were determined from vibrational echo studies to be about 1.5 ps in water at room temperature [655].

Rebinding of CO after photodissociation is a rather slow process requiring microseconds at room temperature and showing signatures of tunnelling below 25 K [656]. This indicates the presence of a barrier, which can be attributed to the fact that, while MbCO is a singlet, deoxy-Mb is a quintet, with the singlet–quintet separation in gas-phase models being about 5 kcal mol^{-1} [657]. For the barrier itself, Harvey obtained a value of about 2.4 kcal mol^{-1}, but it must be stressed that the reaction coordinate for barrier crossing is not simply the Fe–CO centre-of-mass distance, but involves distortion of the haem such that, for example, the central iron moves out of the plane [657]. The question arises whether the *dynamics* in the region of the reaction barrier can be studied using ultrafast spectroscopy. As far as rebinding is concerned, the time-scales and likely stochastic nature of the process are prohibitive. Coming from the reactant's side poses the question of how to break the Fe–CO bond in the electronic ground state using infrared (IR) laser radiation. Compensating for the anharmonicity of the vibrational ladder, chirped pulse excitation is particularly efficient [658]. In fact, coherent vibrational ladder climbing is not restricted to small molecules, as demonstrated by Joffre *et al.* for carboxyhaemoglobin [659]. In this case, however, even though the $\nu = 6$ level of the CO oscillator could be populated ($\sim$12 700 cm^{-1}), which should be well above threshold for Fe–CO bond dissociation (6000–7500 cm^{-1} [660]), no reaction products have been observed.

The excitation process itself has been simulated using a two-dimensional (Fe–CO and C–O coordinates) non-reactive model [662]. However, in order to understand why no dissociation occurs, a reactive model for the Fe–CO

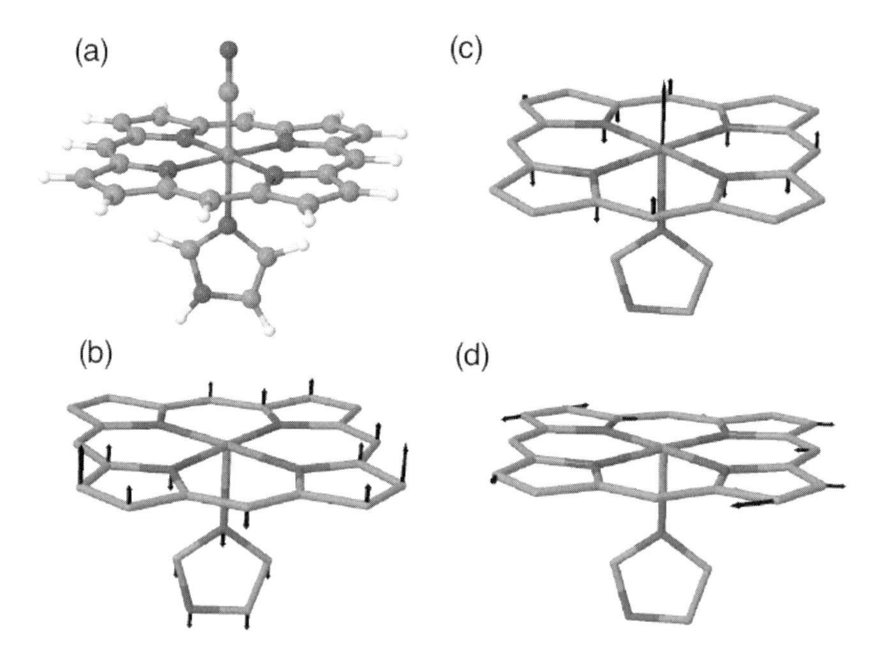

Figure 24.1 (a) Calculated equilibrium geometry of the MbCO model ($FeC_{24}N_6OH_{16}$) using DFT/B3LYP together with a 6-31G basis set for the first-row atoms and the Los Alamos effective core potential for iron as suggested in Ref. [661]. The displacement vectors of the three substrate modes that are used in the present model (H atoms not displayed for visual clarity) are shown in panels (b)–(d). Their frequencies at the geometry shown in panel (a) but with the CO being removed are (b) 129 cm^{-1} (q_1), (c) 400 cm^{-1} (q_2) and (d) 1520 cm^{-1} (q_3).

coordinate is required. Such a model can be designed using the CRS Hamiltonian approach applied to a suitable gas-phase system. In the following, the minimal model shown in Figure 24.1 will be considered, which consists of a CO molecule ligated to an iron porphyrin and an imidazole ligand modelling the effect of the proximal histidine [657,661]. In Ref. [631] a CRS Hamiltonian for this model system has been developed. It comprised two large-amplitude coordinates, that is, the CO bond length, r_{CO}, and the distance between the iron and the CO centre of mass (COM), R_{COM} (with the appropriate reduced masses), that is, $\mathbf{s} = (r_{CO}, R_{COM})$ in Equation (24.4).

As far as important normal modes are concerned, the reorganization energy, ΔE^{reorg}, has been calculated and averaged over the reaction surface. This gave rise to the selection of the two modes shown in Figure 24.1(b) and (c) having frequencies of 129 cm^{-1} (q_1, $\Delta E_1^{av} = 627$ cm^{-1}) and 400 cm^{-1} (q_2, $\Delta E_2^{av} = 779$ cm^{-1}). The collective low-frequency mode q_1 involves some deformation of the haem plane as expected for the process of ligand dissociation [663]. Along mode q_2 the iron atom is moved out of the haem plane, again in accord with the geometry of the deoxy form. A third harmonic mode q_3 with $1520\,\text{cm}^{-1}$, which according to Figure 24.1(d) involves some in-plane deformation of the haem, will be taken into account because in combination with mode q_2 it provides an almost resonant channel for energy flow out of the CO stretching mode. Notice, however, that its average reorganization energy is comparatively small ($16\,\text{cm}^{-1}$). Freezing all other DOFs at their equilibrium geometry, one arrives at a five-dimensional (5D) reactive model for the Fe–CO bond dissociation. As far as the dipole moment is concerned, it is assumed that it does not depend on the harmonic mode coordinates.

Figure 24.2 shows representative cuts of the 5D PES. In panel (a) the large-amplitude part of the coordinate space is shown. Clearly, there is some coupling between the two coordinates, as seen, for example, by the reduction of the equilibrium CO bond length in the dissociation channel. The horizontal solid line marks the position of the predicted singlet–quintet crossing [657] around $R_{\text{COM}} = 5.8\,a_0$, which will be used for defining the dividing surface separating reactants from products. Figure 24.2(b) shows a PES cut for the most strongly coupled mode, q_2. Since this mode involves the out-of-plane motion of the iron centre (see Figure 24.1), it is substantially displaced upon dissociation (not shown). However, its coupling to the CO vibration in the vicinity of the PES minimum is not very pronounced.

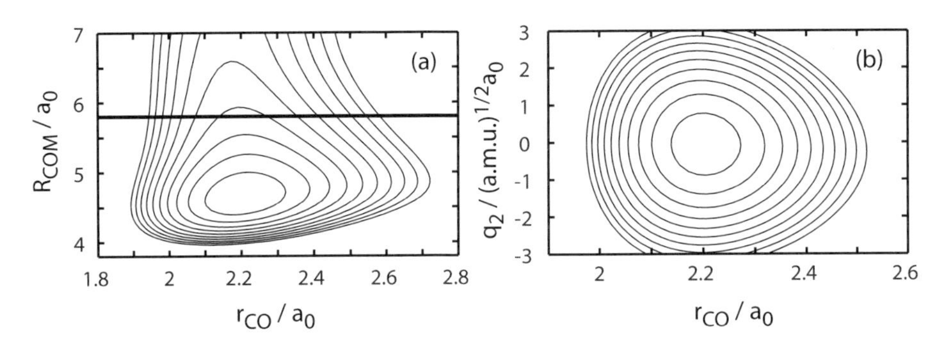

Figure 24.2 Different cuts of the 5D MbCO PES. The other coordinates are fixed at (a) $\{q_i = 0\}$ and (b) $R_{\text{COM}} = 4.67\,a_0$, $q_1 = q_3 = 0$. The contour values are (a) from 0.2 to 2.0 eV in steps of 0.2 eV and (b) from 0.1 to 1.0 eV in steps of 0.1 eV. The horizontal solid line in panel (a) shows the position of the dividing surface used in the reaction probability calculation. The PES as well as the dipole moment surface have been calculated using GAUSSIAN98 [664] on a scattered grid of about 100 points. The data are subsequently fitted to polynomials up to 14th order in R_{COM} and fourth order in r_{CO} for $V(r_{\text{CO}}, R_{\text{COM}})$ and up to sixth order in R_{COM} and r_{CO} for $\mathbf{K}(r_{\text{CO}}, R_{\text{COM}})$ and $\mathbf{f}(r_{\text{CO}}, R_{\text{COM}})$. For the dipole moment, a sixth-order fit has been used. In all calculations, the electronic state has been a singlet.

The following propagation results show the time evolution under the action of a Gaussian pulse, Equation (24.8), with chirped frequency (chirp rate $\beta = 0.032\,\text{ps}^2$ [659]). The two large-amplitude coordinates are combined into a single mode with six single-particle functions (SPFs). For the three normal modes we used $n_1 = n_2 = 5$, and $n_3 = 3$ SPFs. For the representation of these functions, the following primitive grids were employed:

(i) r_{CO} – harmonic oscillator discrete variable representation (HO-DVR) in the range $[1.75\,a_0,\ 2.85\,a_0]$ (50 points);

(ii) R_{COM} – fast Fourier transform (FFT) grid in the range $[3.0\,a_0,\ 7.0\,a_0]$ (64 points); and

(iii) q_i – HO-DVR with $q_1 \in [-4.0,\ 4.0]$ (30 points), $q_2 \in [-3.0,\ 3.0]$ (20 points) and $q_3 \in [-2.0,\ 2.0]$ (15 points) (coordinates in $a_0\sqrt{\text{a.m.u.}}$).

Absorbing boundary conditions were used along the dissociative coordinate R_{COM} realized via a complex potential of the form

$$W(R_{\text{COM}}) = \eta\theta(R_{\text{COM}} - R_{\text{CAP}})(R_{\text{COM}} - R_{\text{CAP}})^2$$

with $\eta = 0.1\,E_{\text{h}}a_0^{-2}$ and $R_{\text{CAP}} = 6.0\,a_0$ [34].

The upper two panels of Figure 24.3 show the field envelope as well as the time-dependent frequency. In the lower panel, the reaction yield is plotted. It is defined by the expectation value of a step function placed at $R_{\text{COM}} = 5.8\,a_0$. Apparently, the reaction yield is increasing during laser driving and keeps increasing in the subsequent field-free propagation interval. On this logarithmic scale the yield arrives at a value of about -5, indicating that only a very small fraction of the wavepacket has reached the exit channel.

In order to analyse the dynamics, we have calculated the population of zero-order states for the CO motion (that is, for the PES cut where all other coordinates are at their equilibrium values). The resulting populations are shown in Figure 24.4. This figure nicely illustrates the ladder climbing in the anharmonic CO oscillator. At the end of the pulse, the highest state with quantum number 7 at energy $14\,380\,\text{cm}^{-1}$ is populated by about 0.03. Energetically, this would be sufficient to overcome the dissociation threshold, which is about $10\,000\,\text{cm}^{-1}$ for the present model. We further note that, after the pulse is over, the population of all considered zero-order states stays approximately constant. In other words, we do not find any sign of vibrational energy redistribution on the considered time-scale.

This is confirmed by inspecting the expectation value of the molecular energy and of the uncoupled CO oscillator in Figure 24.5(a). During most of the pulse duration, both coincide on average; only towards the end of the pulse does a difference become noticeable. Here, energy goes into other DOFs, as seen in Figure 24.5(b), where the energy expectation value for the uncoupled

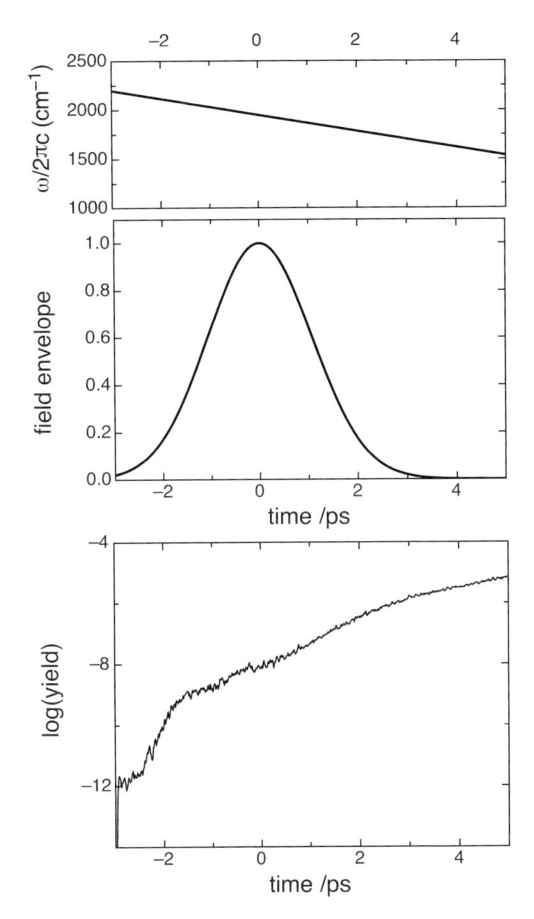

Figure 24.3 Upper panels: Normalized pulse envelope, Equation (24.8), and time-dependent frequency, Equation (24.9), used for the 5D propagation of laser driving the CO coordinate in the MbCO model. The parameters were: $\tau_0 = 0.1$ ps, $\omega_0/2\pi c = 1950\,\text{cm}^{-1}$, $\beta = -0.032\,\text{ps}^2$ and $E_0 = 0.38\,\text{m}E_\text{h}/ea_0$ (mE_h, millihartrees). The lower panel shows the logarithmic reaction yield defined as the expectation value of a step function at $R_\text{COM} = 5.8\,a_0$ (see Figure 24.2).

R_COM coordinate is shown. Apparently, only upon excitation of the $\nu = 6$–7 CO levels does anharmonic coupling start to play a role on this time-scale.

One can conclude that the present model allows for a description of chirped pulse excitation of the CO vibration in a gas-phase MbCO model system. In accord with the experiment [659], we find that, despite controlled CO excitation above the dissociation threshold, there is no noticeable probability for bond breaking. We also observe that on the time-scale of a few picoseconds the effect of the oscillators coupling to the reaction coordinates is negligible. As the wavepacket stays confined along the CO coordinate, the respective

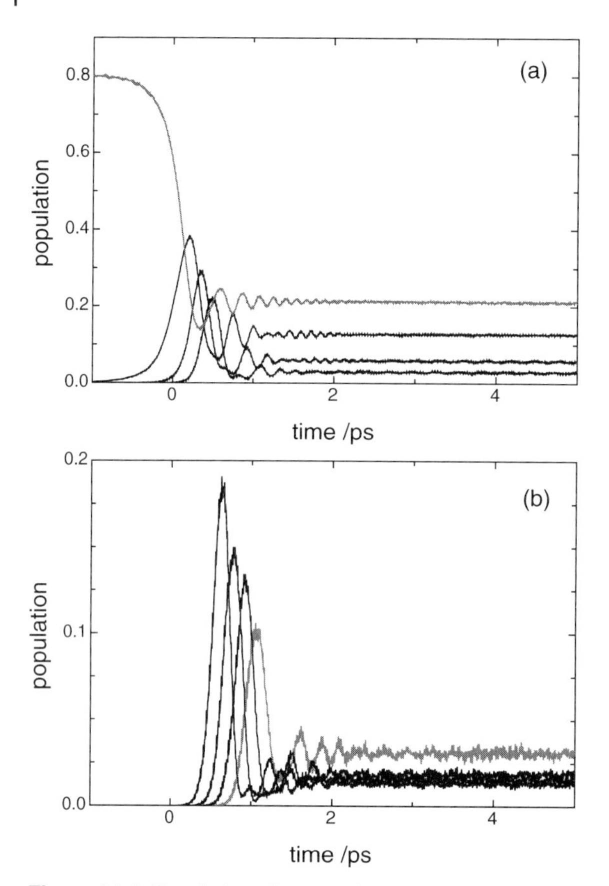

Figure 24.4 Population of zero-order states defined for the CO coordinate with the other coordinates fixed at their equilibrium values. (a) P_0 (in grey) to P_3, (b) P_4 to P_7 (in grey); ordering according to initial increase.

couplings are not very pronounced (see Figure 24.2(b)). Whether the goal of controlled bond breaking in the electronic ground state can be achieved by more complex pulse shapes remains to be investigated.

24.3.2
Hydrogen-Bond Dynamics

Hydrogen bonds A–H· · · B combine two features that make them interesting for MCTDH studies. First, owing to the coupling among the stretching and bending A–H vibrations as well as their interaction with heavy-atom motions, A· · · B, the vibrational dynamics shows pronounced multidimensionality. Second, as one is dealing with the lightest nuclei, quantum effects are likely to be of great importance (for a review, see Ref. [636]). Further-

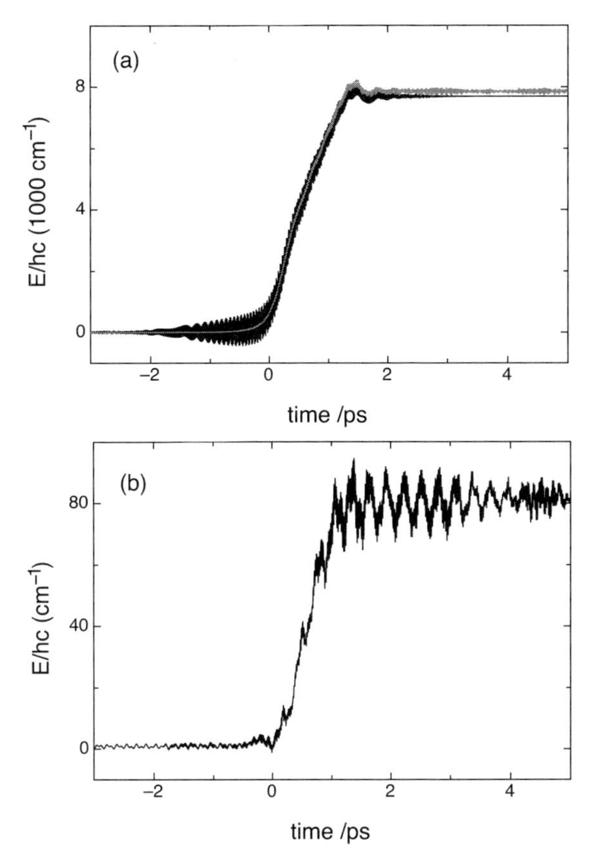

Figure 24.5 Expectation values for the driven 5D MbCO model and the field specified in Figure 24.3. Shown are the results for (a) the total molecular Hamiltonian (in grey) and its uncoupled CO part and (b) the uncoupled R_{COM} part.

more, this is the type of system for which the CRS Hamiltonian was originally developed – light particles are moving in the field of a scaffold of heavier atoms. Motivation also comes from the experimental side, where femtosecond IR laser spectroscopy enables triggering and subsequent observation of coherent wavepacket dynamics in the electronic ground state (see, for example, [459, 665]). The combination of the CRS Hamiltonian and MCTDH wavepacket propagation has been applied in the context of laser driving to two systems, phthalic acid monomethyl ester (PMME) [611, 632, 633] and salicylaldimine (SA) [610, 634, 635], for both the normal and the deuterated species.

PMME is a single-minimum system, that is, the vibrational dynamics is anharmonic, but there is *no* H atom transfer. In fact, it was the first system for which O–H$\cdots$O hydrogen-bond wavepacket dynamics could be initiated upon excitation of the O–H stretching vibration [666]. This finding was first explained in Ref. [632] using a 55-dimensional time-dependent self-consistent field (TDSCF) description. Later, the propagation was refined to include multiconfiguration effects in a nine- (9D) [611] and 19-dimensional (19D) model [633]. Although the basic mechanism of wavepacket excitation due to anharmonic coupling between fast (O–H) and slow (O···O) coordinates does not change, the enhanced flexibility of the MCTDH wavepacket allows for the onset of irreversible energy flow. Representative results of laser-driven wavepacket dynamics for the deuterated form are shown in Figure 24.6. From the decay of the expectation value describing the energy of the excited OD stretching vibration, one can estimate a relaxation time of the order of 20 ps. This is, however, still substantially longer than the observed subpicosecond T_1 time [665]. A proper description of this experimentally observed time-scale requires the interaction with the solvent to be taken into account [667].

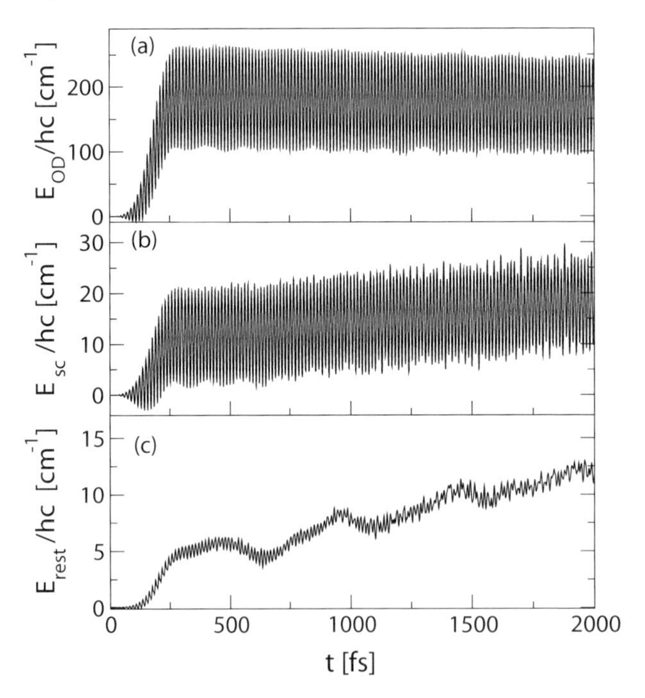

Figure 24.6 Expectation values of single-mode Hamiltonians obtained from an MCTDH propagation on a 9D reaction surface describing the hydrogen-bond dynamics in deuterated phthalic acid monomethyl ester [611]. The driving pulse is resonant with the OD stretching vibration (2430 cm^{-1}) and has a $\sin^2$ shape (Equation (24.7)) with $\tau_p = 300\,\text{fs}$ and $E_0 = 0.5\,\text{m}E_h/ea_B$. The different panels show the expectation values for (a) the OD stretching mode, which corresponds to the reaction coordinate, (b) the most strongly coupled harmonic mode, and (c) the sum of the remaining seven modes. In this simulation, the chirp rate was zero.

Salicylaldimine has an O–H···O hydrogen bond and a double-minimum PES with a barrier of about 2000 cm^{-1} [610]. Treating the atomic coordinates of the in-plane motion of the H/D atom as the large-amplitude reaction coordinate, a seven-dimensional (7D) model of the enol–keto tautomerization reaction was developed in Ref. [610]. A crucial point was the introduction of a (crude) diabatic basis for the reaction coordinate DOFs. Since the diabatic H/D atom states are rather localized, the grid on which the first and second derivatives of the PES has to be calculated is considerably reduced. Fixing the normal-mode coordinates at the more stable enol minimum, the diabatic H/D states have been obtained from the solution of the Schrödinger equation for the wavefunctions $\phi_\alpha(\mathbf{s}{=}(x,y)) \equiv \langle \mathbf{s}|\alpha\rangle$,

$$[T_\mathbf{s} + V(\mathbf{s})]\phi_\alpha(\mathbf{s}) = E_\alpha \phi_\alpha(\mathbf{s}) \tag{24.11}$$

The diabatic representation of the CRS Hamiltonian then becomes

$$H^{\text{diab}} = \sum_{\alpha,\beta}[(\tfrac{1}{2}\mathbf{p}^2 + V_{\alpha\alpha}(\mathbf{q}))\delta_{\alpha\beta} + V_{\alpha\beta}(\mathbf{q})(1-\delta_{\alpha\beta}) - E(t)d_{\alpha\beta}]|\alpha\rangle\langle\beta| \tag{24.12}$$

where $d_{\alpha\beta}$ are the matrix elements of the dipole moment operator, the diabatic PESs are given by

$$V_{\alpha\alpha}(\mathbf{q}) = E_\alpha - \mathbf{F}_{\alpha\alpha}\mathbf{q} + \tfrac{1}{2}\mathbf{q}\mathbf{K}_{\alpha\alpha}\mathbf{q} \tag{24.13}$$

and the coupling between the diabatic states is

$$V_{\alpha\beta}(\mathbf{q}) = -\mathbf{F}_{\alpha\beta}\mathbf{q} + \tfrac{1}{2}\mathbf{q}\mathbf{K}_{\alpha\beta}\mathbf{q} \tag{24.14}$$

This representation offers the additional advantage that one has well-defined zero-order states for the discussion of spectral features and wave-packet dynamics. In Figure 24.7 the population dynamics of the diabatic states after OH (panel (a)) and OD (panel (b)) stretching excitation is shown. Most notable is the difference between OH and OD vibrational stretching dynamics (state $\alpha = 3$), which is mostly due to the different densities of vibrational states. While there is a long-lasting population exchange between the OD stretching state and the ground state (via excited oscillator modes), the excited OH population decays almost to zero within the given interval of 2 ps. This demonstrates that the anharmonicity provided by the PES of the molecule alone suffices to observe subpicosecond population decay.

24.3.3 Predissociation Dynamics of Matrix-Isolated Diatomics

Matrix-isolated diatomics, such as rare gases hosting dihalogens, constitute a seemingly simple model system to study condensed-phase dynamics [668, 669]. In particular, surface hopping classical trajectory methods have been extensively used to unravel details of non-adiabatic dynamics after photoexcitation [670,671]. Here the diatomics-in-molecules (DIM) [672] method plays

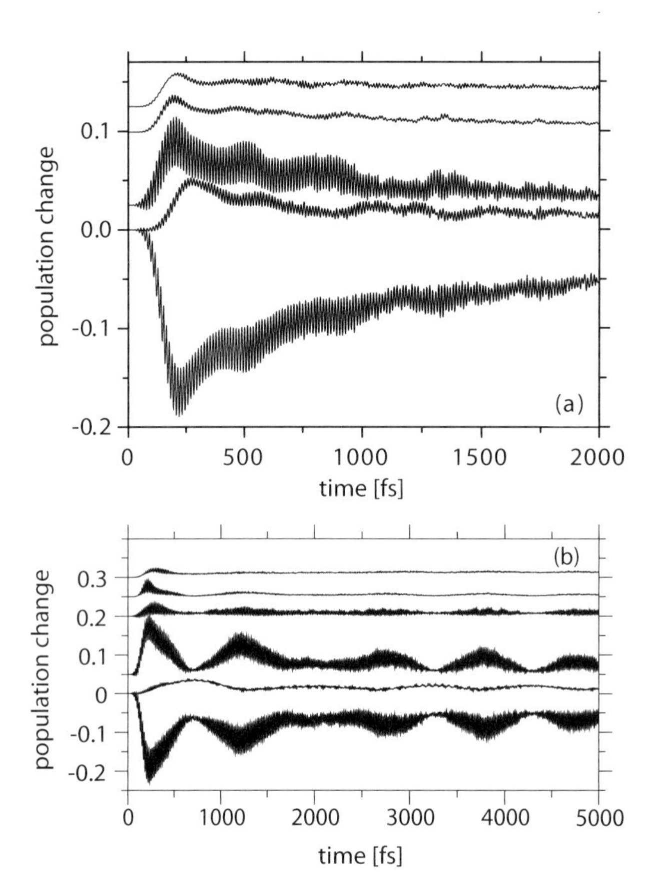

Figure 24.7 Population change of the diabatic H (panel (a)) and D (panel (b)) states calculated for the 7D model of salicylaldimine [610, 634, 635]. (a) Difference populations: ΔP_1, ΔP_2, ΔP_3, ΔP_5 and $\sum_{\alpha=4,6-10} \Delta P_\alpha$ (curves from bottom to top, upper curves are vertically offset). The $\sin^2$-shaped laser pulse was tuned to resonance with the stretching fundamental transition region with pulse parameters: $E_0 = 4\,\mathrm{m}E_\mathrm{h}/ea_0$, $\omega_0/2\pi c = 2680\,\mathrm{cm}^{-1}$, $\tau_\mathrm{p} = 260\,\mathrm{fs}$. (b) Difference populations: ΔP_1, ΔP_2, ΔP_3, ΔP_4, ΔP_7 and $\sum_{\alpha=5,6,8-10} \Delta P_\alpha$ (curves from bottom to top, vertically offset). The pulse frequency has been changed to the OD resonance at $\omega_0/2\pi c = 1920\,\mathrm{cm}^{-1}$. In both cases, the chirp rate was zero.

an essential role for the determination of the forces on the nuclei and the non-adiabatic couplings. Ultrafast spectroscopy and the ability to shape laser light to control molecular dynamics [673, 674], however, requires a quantum mechanical description accounting for the (coherent) evolution of the wavepacket. Reduced-dimensionality models have been shown to be useful for studying the short-time dynamics only [675–677].

Recently, we have made an effort to establish a reaction surface model for quantum dynamics simulations for the system Br_2 in solid argon [637–639].

Our model is based on the DIM description, which provides a diabatic-type representation of the Hamiltonian in terms of the valence-bond basis for Br_2 [637]. In principle, this includes information on the full-dimensional (700 atoms with periodic boundary conditions) PESs and their couplings. In practice, one has to start by focusing on a subset of electronic states. In Ref. [637] it has been shown that, by restricting the model to a specific electronic transition such as $X \to B$, the dimension of the DIM Hamiltonian matrix can be reduced to 17 states (see Figure 24.8). Subsequently, in order to incorporate nuclear DOFs, we have further reduced the model by targeting it to include the $B \to C$ state predissociation channel [638]. This requires the following four states to be taken into account (according to Hund's case c labelling): $X(0_g^+)$, $B(0_u^+)$ and $C(1_u)$, the latter being doubly degenerate.

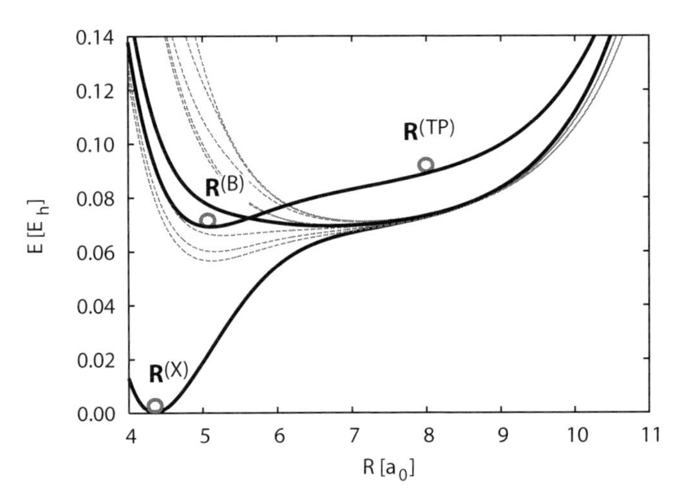

Figure 24.8 Potential energy curves along the Br–Br distance and for the case of Ar atoms fixed at their ground-state equilibrium geometry. The states that are included in the vibronic coupling model (X, B and the doubly degenerate C state) are given by solid lines. Other valence states are shown as dashed lines [638]. The reference geometries considered for the construction of a reaction surface Hamiltonian are shown as grey circles ($\mathbf{R}^{(X)}$, ground-state equilibrium; $\mathbf{R}^{(B)}$, B-state equilibrium; and $\mathbf{R}^{(\mathrm{TP})}$, outer turning point).

Reduced-dimensional models for the nuclear DOFs are typically based on an intuitive choice of one or two coordinates such as the dihalogen's bond distance and a collective matrix coordinate describing cage motion [675]. A more systematic way of designing a reduced model is based on the reaction surface idea, with the choice of large-amplitude coordinates being adopted to the particular process under consideration (similar to the case of H bonds [613]). For the present case of $X \to B$ excitation and subsequent transition to the C states, one can start from the coordinate vectors describing the equilibrium geometry in the ground state and in the B state, that is, $\mathbf{R}^{(X)}$ and $\mathbf{R}^{(B)}$, respectively (see Figure 24.8). Trivially, this change in configuration requires movement along the Br–Br bond distance, which after mass-weighting is the first large-amplitude coordinate, ζ_R. The second coordinate, ζ_{X-B}, can be obtained from the mass-weighted difference vector $\mathbf{R}^{(B)} - \mathbf{R}^{(X)}$ upon Gram–Schmidt orthogonalization with respect to ζ_R [638]. The displacement vectors associated with ζ_{X-B} are shown in Figure 24.9(a). They illustrate the response of the matrix cage to the Br–Br bond elongation in the B state. In principle one could extend this procedure to account, for example, for the configuration of the outer turning point reached after Franck–Condon excitation, $\mathbf{R}^{(\mathrm{TP})}$. Numerical tests showed, however, that this geometry is already reasonably covered by the two coordinates ζ_R and ζ_{X-B}.

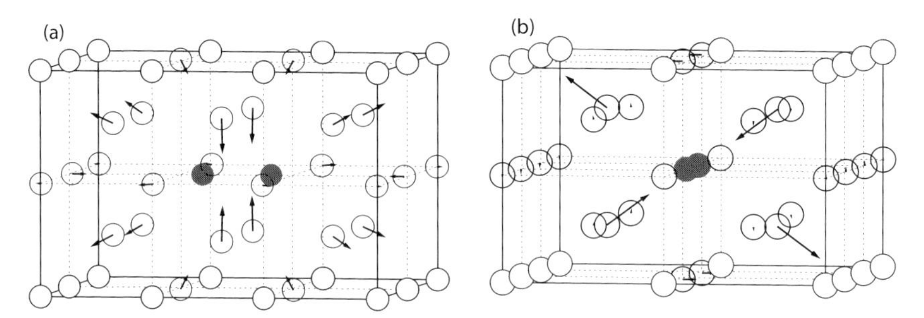

Figure 24.9 Displacement vectors of (a) the large-amplitude matrix coordinate, ζ_{X-B}, and (b) the harmonic B–C coupling mode, q_2 [638].

The PES for the B state spanned by these two large-amplitude coordinates is shown in Figure 24.10. The coupling is obvious from the tilt of the contours. Since these coordinates maintain the approximate D_{2h} symmetry of the cage, we obtain a crossing seam for the intersection with the C states (approximately given by the straight line in Figure 24.10). Hence, predissociation is negligible if only these two coordinates are considered. This gave motivation to incorporate harmonic modes coupling B and C states as is usually done in the context of vibronic coupling theory [4] (note that harmonic tuning modes coupling to the motion in a given electronic state have not been taken into account).

Inspecting the B–C coupling matrix elements, two coupling modes, q_1 and q_2, have been identified to couple most strongly to this transition. For the dominant mode, q_2, we show the displacement vectors in Figure 24.9(b). As expected, this mode lowers the local symmetry of the cage, thus opening a conical intersection at points along the crossing seam. The four-state and four-coordinate Hamiltonian thus contains the kinetic energy part ($\alpha = X, B, C$; $\omega_1/2\pi c = 60\,\mathrm{cm}^{-1}$; $\omega_2/2\pi c = 63\,\mathrm{cm}^{-1}$):

$$T = -\frac{\hbar^2}{2}\left(\frac{\partial^2}{\partial\zeta_R^2} + \frac{\partial^2}{\partial\zeta_{X-B}^2}\right) - \frac{\hbar}{2}\left(\omega_1\frac{\partial^2}{\partial q_1^2} + \omega_2\frac{\partial^2}{\partial q_2^2}\right) \tag{24.15}$$

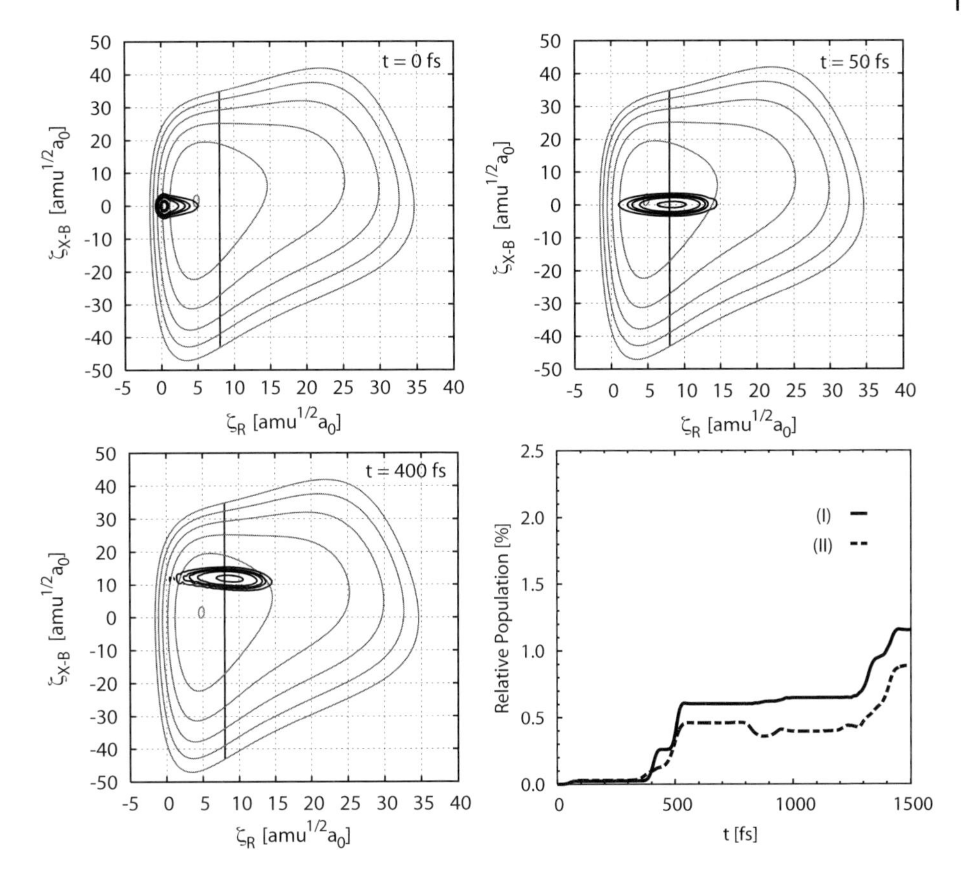

Figure 24.10 Snapshots showing the time evolution of the 2D reduced density on the diabatic B-state surface (contour levels 0.069, 0.08, 0.09, 0.10, 0.11, 0.12 E_h) after Franck–Condon excitation with a Gaussian-shaped pulse ($E_0 = 5\,\mathrm{m}E_\mathrm{h}/ea_0$ and $\tau = 30\,\mathrm{fs}$). The transition dipole moment is assumed to depend on ζ_R only. The position of the crossing seam between B and C states is indicated by a vertical line. The lower right panel shows the population of the C states relative to the total population transfer out of the ground state for the cases of vertical excitation from the vibrational ground (I) and the first excited (II) state (for propagation details, see Ref. [639]). In this simulation, the chirp rate was zero.

the diagonal potential energy part (U_α is the PES for the large-amplitude coordinates):

$$V_{\alpha\alpha} = U_\alpha(\zeta_R, \zeta_{X-B}) + \tfrac{1}{2}\hbar(\omega_1 q_1^2 + \omega_2 q_2^2) \tag{24.16}$$

and the state coupling (for both C states):

$$V_{BC} = \mathrm{i}\hbar F_{BC}^{(1)}(\zeta_{X-B})q_1 + \hbar F_{BC}^{(2)}(\zeta_{X-B})q_2 \tag{24.17}$$

The coupling matrix elements $F_{BC}^{(1)}$ and $F_{BC}^{(2)}$ have been discussed in Ref. [638]. An important point concerns their coordinate dependence along the crossing seam, that is, the coupling becomes stronger as ζ_{X-B} increases (cage compression, see Figure 24.9(a)).

The focus of the MCTDH study in Ref. [639] has been to unravel details of the predissociation dynamics. Snapshots of reduced densities on the B state after ultrafast excitation are shown in Figure 24.10. Initially, the wavepacket moves towards the cage wall, where it is reflected also along the direction of the collective matrix coordinate ζ_{X-B}. The predissociation yield can be quantified by the population of both C states relative to the total excited population, which is shown in the lower right panel of the figure. One notices that, although the wavepacket crosses the intersection seam twice during the first 400 fs, population transfer occurs only if the cage is compressed, that is, if there is an elongation along ζ_{X-B}. Apart from the energetic arguments, the reason for the effective transfer in this case can be found in the coordinate dependence of the state coupling, as mentioned above. Another scenario that has been investigated involves the case of vibrational pre-excitation of the Br–Br bond in the ground X state. Such an initial state can be prepared in principle by a pump–dump like pulse sequence. Analysing the wavepacket dynamics one finds that the population transfer is reduced (see curve (II) in Figure 24.10). Here, the reason can be traced to the fact that the wavepacket is launched deeper in the excited-state PES, where in principle it is already energetically closer to the crossing seam during the first bond elongation. However, at the same time, the magnitude of the state coupling is reduced and therefore there is less predissociation in the present model.

Experimentally, it has been shown that, for the case of dihalogens in argon, phase-locked pulse sequences can be used to control the amount of matrix excitation in the B-state wavepacket [674]. Since vibrational relaxation is important in these experiments, simulations will require the use of the flexibility of the present formulation, including in particular harmonic tuning modes.

24.4 Summary

The combination of the MCTDH wavepacket propagation and the all-Cartesian reaction surface Hamiltonian methods is ideally suited for exploration of the quantum dynamics of large systems on the basis of quantum chemical input. In light of the desirable features of an appropriate Hamiltonian, which were stated in the Introduction, the following can be concluded:

(i) Most quantum chemical methods provide a convenient means for calculating the first and second derivatives entering Equation (24.4). The restrictions in dimensionality, therefore, come from the desired accuracy where *ab initio* methods are concerned. Semi-empirical methods, on the other hand, can provide input for thousands of DOFs.

(ii) A systematic way of identifying reduced models and extending them is provided by inspecting a number of quantities such as reorganization energies, which are straightforwardly extracted from the quantum chemical input.

(iii) The Hamiltonian is of factorized form in most coordinates.

(iv) In principle, all the input for calculating spectral densities that could enter a system–bath model is available.

In the applications that have been discussed, controlled population of highly excited states of selected DOFs has been accomplished by driving with a laser field, choosing appropriate pulse forms and parameters by hand. Simulating wavepacket control with more flexible pulse forms is still a formidable task for multidimensional systems. In view of the experimental success in controlling even complex biomolecules, theoretical advances are needed. In this respect, recent success in the incorporation of optimal control theory into the MCTDH framework is very promising [72, 678].

Acknowledgments

This work has been financially supported by the Deutsche Forschungsgemeinschaft (sfb 450). I am grateful to my former PhD students H. Naundorf, M. Petković, and A. Borowski who carried out the work on PMME, SA and Br_2/Ar, respectively.

25
Polyatomic Dynamics of Dissociative Electron Attachment to Water Using MCTDH

Daniel J. Haxton, Thomas N. Rescigno and C. William McCurdy

25.1
Introduction

Dissociative electron attachment (DEA),

$$AB + e^- \to (AB^-)^* \to A^- + B \tag{25.1}$$

is the process in which a free electron attaches to the (polyatomic) molecule AB, creating a transient anion species $(AB^-)^*$, which then subsequently dissociates to ionic and neutral fragments, A^- and B. Three-body break-up may also be possible. This process may occur in a variety of environments, from plasmas, in which free electrons are pervasive, to human bodies, where harmful radiation may trigger a cascade of ionized electrons, which in turn may attach to simple molecules such as H_2O or larger molecules such as DNA bases.

Considerable theoretical and experimental effort has gone into understanding this process. Experiments have been performed since the first half of the 20th century. By the 1960s, theory had gained a footing and began to explain the experiments [679]. Key among the theoretical developments was the recognition that DEA must proceed via a resonant (metastable) electronic state, denoted $(AB^-)^*$ [679].

We are interested in the process of dissociative electron attachment to water, one of the most famous examples of DEA in a polyatomic molecule, which was studied as early as 1930 [680], and continues to attract interest to the present day [681–692]. This system exhibits considerable complexity and may be considered a challenging test for the theory of dissociative attachment in polyatomic molecules.

The fact that $(AB^-)^*$ is a metastable electronic state is what leads to much of the difficulty in treating dissociative electron attachment theoretically and computationally. First, in the fixed-nuclei picture, one must account for the electronic continuum in solving the electronic part of the problem, so that the energies and lifetimes of the metastable states of the anion can be calculated

Multidimensional Quantum Dynamics: MCTDH Theory and Applications.
Edited by Hans-Dieter Meyer, Fabien Gatti, and Graham A. Worth

ISBN: 978-3-527-32018-9

correctly as a function of the molecular geometry. There are a variety of methods for doing so. We use an implementation of the complex Kohn variational method [693–701] incorporating a fully correlated and *ab initio* expansion of the $H_2O + e^-$ system, including 37-state close coupling in the states of the target H_2O plus additional terms.

The second part of the theoretical treatment is the calculation of the nuclear dynamics of the metastable electronic states of H_2O^-. The nuclear dynamics are calculated under an approximation known as the local complex potential (LCP) model [679, 702–705], in which the effect of the decay of the resonance is accounted for by a complex-valued potential

$$V(\mathbf{q}) = E_R(\mathbf{q}) - \tfrac{1}{2}\mathrm{i}\Gamma(\mathbf{q})$$

where $\mathbf{q}$ denotes the internal nuclear coordinates of the molecule, E_R is the location of the resonance, and Γ is its width. The LCP model is an example of non-Hermitian quantum mechanics; the norm of the wavepacket decreases over time, corresponding to decay of the metastable state.

We employ the MCTDH package [13] in calculating the quantum nuclear dynamics of DEA to water. This problem is non-trivial, even for a diatomic molecule, let alone for a polyatomic one. There are several competing effects that determine the cross-section and product distributions in DEA. First, there is the tendency for the added electron to detach. This process is the inverse of the first step in Equation (25.1), the attachment of an electron to the target AB to produce AB^-. Such detachment is geometry dependent, controlled by the width $\Gamma(\mathbf{q})$, and may completely dominate survival, such that the amount of $A^- + B$ produced is only a small fraction of the AB^- initially created by the attachment process. Second, just as in other processes such as photodissociation, the shape of the potential energy curve (for a diatomic) or surface (for a polyatomic), which we denote E_R, will direct the propagating wavefunction towards products, and will determine the speed at which it gets there, the kinetic energy of the fragments and the product distribution. A large width Γ may also affect product distributions.

The problem at hand, DEA to water, is further complicated by the fact that the observed products are the result of nuclear dynamics on several Born–Oppenheimer potential surfaces that are coupled to one another and to other surfaces. We perform the simplest treatment that would be expected to reproduce the major features of the cross-section, which treatment includes both conical intersection and Renner–Teller non-adiabatic effects. Throughout this chapter, we provide full details of the parameters of the MCTDH calculations. These (half-)reactive scattering calculations have been performed in several coordinate systems, sometimes employing coupled electronic channels, and in all cases we have been able to obtain converged results with the MCTDH package.

Further information about the calculations we have performed may be found in Refs. [706–712].

25.2 Dissociative Electron Attachment to Water

Dissociative electron attachment to the water molecule [680–692] proceeds through a number of channels, each with a different energetic threshold:

$$H_2O + e^- \rightarrow \begin{cases} H + OH^- & 3.27\text{ eV} \\ H_2 + O^- & 3.56\text{ eV} \\ H^- + OH \quad (X\,^2\Pi) & 4.35\text{ eV} \\ H + H + O^- & 8.04\text{ eV} \\ H^- + OH^* \quad (^2\Sigma) & 8.38\text{ eV} \\ H^- + H + O & 8.75\text{ eV} \end{cases} \tag{25.2}$$

The production of these species occurs via three metastable Born–Oppenheimer electronic states of the H_2O^- system, whose vertical transition energies therefore determine the incident energies at which DEA occurs. Those electronic states of the anion are the $^2B_1$, $^2A_1$ and $^2B_2$ Feshbach resonances, and they are responsible for the three distinct peaks in the DEA cross-section, shown in Figure 25.1. Their potential energy surfaces contain asymptotes corresponding to the product channels listed in Equation (25.2), with the exception of the $H + OH^-$ channel; this product is a result of non-adiabatic effects [708].

The vertical transition energies of the three resonant states may be estimated from the position of the peaks in the cross-section, at approximately 6.5, 8.4 and 11.8 eV. In Figure 25.2 we show the results of approximate calculations [708] to obtain the real part of the resonance energies, E_R. This figure depicts the behaviour of the resonance energies with respect to the H–O–H bending angle θ_{HOH}, fixing the OH bond lengths at $r_1 = r_2 = 1.81\,a_0$. We see the vertical transition energies at the equilibrium geometry of the neutral, $\theta_{HOH} = 104.5°$.

There is coupling among all three resonance states, which makes the treatment of the nuclear dynamics more difficult. The $^2B_1$ and $^2A_1$ resonances are degenerate at linear geometry. This degeneracy will lead to Renner–Teller coupling [713–717] between the two states. In addition, there is a conical intersection [718–723] between the $^2A_1$ and $^2B_2$ surfaces located at approximately $\theta_{HOH} = 73°$. For a full description of the many surfaces involved in this system, see Ref. [708].

We have constructed complex-valued potential energy surfaces corresponding to the three resonance states [706, 710]. The details will be omitted here,

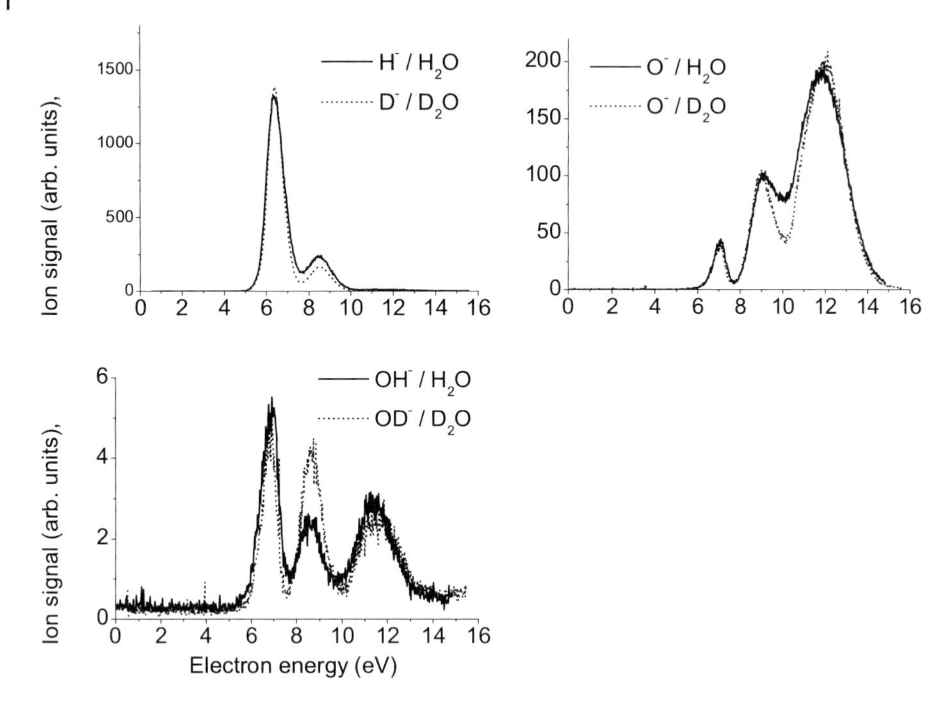

Figure 25.1 Experimental results of Fedor *et al.* [692].

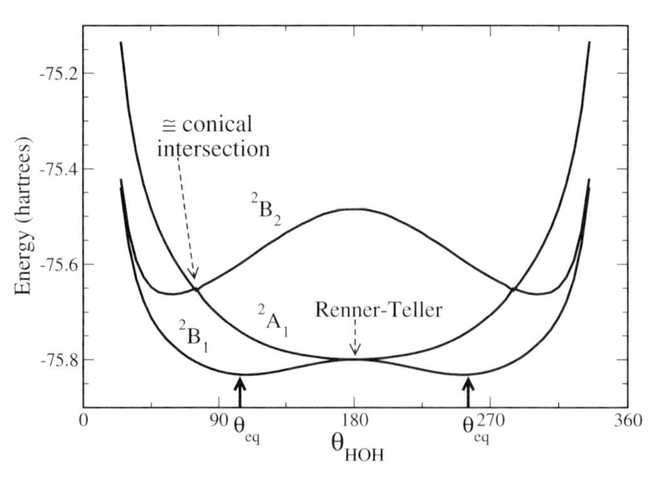

Figure 25.2 Real parts of resonance energies, in units of hartrees, for OH bond distance = 1.81 a_0 in C_{2v} geometry, plotted with respect to bending angle, in degrees [708].

but we note that the treatment of the conical intersection between the $^2A_1$ and $^2B_2$ states, which plays a major role in DEA starting from the $^2B_2$ resonance at 11.8 eV, requires a diabatization [66, 724, 725]. We have performed such a diabatization [710], which yields diabatic $^2A_1$ and $^2B_2$ surfaces, which correspond to the adiabatic surfaces at the equilibrium geometry of neutral H_2O, and a coupling term. We use these diabatic surfaces for calculations of DEA via the $^2B_2$ resonance [711]. For calculations on the Renner–Teller coupled $^2B_1$ and $1\,^2A'$ surfaces, we also use an electronically diabatic basis, that which diagonalizes the projection of electronic angular momentum about the molecular axis, $\hat{\ell}_z$.

25.3 Time-Dependent Treatment of Dissociative Attachment Within the Local Complex Potential Model

The local complex potential (LCP) model [679, 702–705], also known as the 'boomerang' model when applied to vibrational excitation, describes the nuclear dynamics in terms of the driven Schrödinger equation

$$(E - H)\xi_{\nu_i}(\mathbf{q}) = \phi_{\nu_i}(\mathbf{q}, 0) \tag{25.3}$$

in which the Hamiltonian for nuclear motion in the resonant state is

$$H = K_{\mathbf{q}} + E_R(\mathbf{q}) - \tfrac{1}{2}\mathrm{i}\Gamma(\mathbf{q}) \tag{25.4}$$

In Equations (25.3)–(25.4), the nuclear degrees of freedom are collectively denoted by $\mathbf{q}$ and the nuclear kinetic energy is denoted by $K_{\mathbf{q}}$. The energy, E, is the energy of the entire system, namely that of the target molecular state plus the kinetic energy of the incident electron,

$$E = E_{\nu_i} + \tfrac{1}{2}k^2 \tag{25.5}$$

The driving term, ϕ_{ν_i} in Equation (25.3) is defined as

$$\phi_{\nu_i}(\mathbf{q}, 0) = \sqrt{\frac{\Gamma(\mathbf{q})}{2\pi}}\,\chi_{\nu_i}(\mathbf{q}) \tag{25.6}$$

in which χ_{ν_i} is the initial vibrational wavefunction of the neutral target molecule, whose quantum numbers are collectively denoted by ν_i. The factor that multiplies χ_{ν_i}, called the 'entrance amplitude', is arrived at via certain approximations [696].

The width Γ enters the LCP equations both in the entrance amplitude and in the imaginary component of the surface. In the case of a narrow resonance, for which the imaginary component of the LCP potential $-\frac{1}{2}\mathrm{i}\Gamma$ is small and therefore for which the probability of autodetachment will be small, the magnitude of the cross-section will exhibit a linear dependence upon the width. At the other extreme, when the entrance amplitude and imaginary component of the surface are large, the magnitude of the cross-section will be determined by the survival probability, that is, the probability that dissociation will occur before autodetachment. The survival probability then exhibits an exponential dependence upon the width.

The solution of Equation (25.3) can be accomplished via time-dependent methods, as first demonstrated by McCurdy and Turner [726]. The solution $\xi_{\nu_i}(\mathbf{q})$ satisfies the boundary condition that it should contain only purely outgoing waves,

$$\xi_{\nu_i}(\mathbf{q}) = (E - H + \mathrm{i}\epsilon)^{-1}\phi_{\nu_i}(\mathbf{q}, 0) \tag{25.7}$$

By representing the Green's function, $(E - H + \mathrm{i}\epsilon)^{-1}$, by the Fourier transform of the corresponding propagator, the stationary solution $\xi_{\nu_i}(\mathbf{q})$ of Equation (25.3) can be obtained:

$$\begin{aligned} \xi_{\nu_i}(\mathbf{q}) &= \lim_{\epsilon \to 0} \mathrm{i} \int_0^\infty \mathrm{e}^{\mathrm{i}(E+\mathrm{i}\epsilon)t} \mathrm{e}^{-\mathrm{i}Ht} \phi_{\nu_i}(\mathbf{q}, 0)\, \mathrm{d}t \\ &= \lim_{\epsilon \to 0} \mathrm{i} \int_0^\infty \mathrm{e}^{\mathrm{i}(E+\mathrm{i}\epsilon)t} \phi_{\nu_i}(\mathbf{q}, t)\, \mathrm{d}t \end{aligned} \tag{25.8}$$

where we define the time-dependent nuclear wavefunction as

$$\phi_{\nu_i}(\mathbf{q}, t) = \mathrm{e}^{-\mathrm{i}Ht}\phi_{\nu_i}(\mathbf{q}, 0) \tag{25.9}$$

The present system involves three resonances with very different complex potential energy surfaces [708, 710]. The width of the $^2B_1$ state is small, of the order of 6 meV; therefore, we expect the LCP model to perform very well in this case. In contrast, both the $^2A_1$ and $^2B_2$ states have widths that may approach a quarter of an electronvolt at some geometries. In the case of the $^2A_1$ resonance, for dissociation of H^- from OH, the width increases sharply just before the resonance becomes bound, as the Feshbach resonance takes on shape resonance character [710]. This behaviour may indicate that virtual-state effects, not included in the LCP model, are important in describing the dynamics on the $^2A_1$ potential energy surface.

For the $^2B_2$ state, the width is consistently near 200 meV in the vicinity of the Franck–Condon region of the neutral. This large width may indicate a breakdown of the LCP model for this system. Additionally, the potential energy surface of the $^2B_2$ state is inherently multivalued, undergoing a branch-point degeneracy with a $^2B_2$ shape resonance [708]. We do not include this feature in the present calculations, and therefore our results on three-body break-up via the $^2B_2$ resonance are expected to be only qualitatively correct.

25.4 Coordinate Systems

The calculation of the nuclear dynamics upon these three coupled resonance surfaces using the MCTDH algorithm requires, first and foremost, the selection of appropriate coordinate systems. We choose the coordinate systems based on the product arrangements we wish to analyse. Selection of the appropriate coordinate system enables the use of FLUX84, one of the Heidelberg MCTDH package analysis programs [13], to calculate the final cross-sections, and to resolve those total cross-sections into the partial cross-sections corresponding to the rovibrational states populated. The FLUX84 program enables the discrimination of two- and three-body break-up.

For calculations of two-body break-up, we choose the Jacobi coordinate system appropriate to the diatomic fragment involved. Therefore, to analyse production of $H^- + OH$ (where the OH is in its ground electronic state, $X\,^2\Pi$, or, for DEA via the highest $^2B_2$ resonance state, may also be in the first excited state, $^2\Sigma$), we choose the Jacobi coordinate system in which the vector $\mathbf{r}$ connects the O and the H of the OH fragment, and $\mathbf{R}$ connects the second H with the OH centre of mass. To study break-up into $H_2 + O^-$, we choose the other Jacobi coordinate system in which $\mathbf{r}$ connects the two H atoms. In both cases the angle γ denotes the angle between the two Jacobi vectors.

For total rotational angular momentum $J \neq 0$, we operate within the body-fixed frame. This body-fixed frame is defined as having the z-axis collinear with the Jacobi $\mathbf{R}$ vector, and as having the molecule reside within the xz plane. As is customary, we expand the total wavefunction in a sum over normalized Wigner D-functions $\widetilde{D}^J_{MK}$,

$$\chi_{v_i}(R,r,\gamma,\alpha,\beta,\zeta) = \sum_K \widetilde{D}^J_{MK}(\alpha,\beta,\zeta)\,\frac{\chi^K_{v_i}(R,r,\gamma)}{Rr}$$
$$\widetilde{D}^J_{MK}(\alpha,\beta,\zeta) = \sqrt{\frac{2J+1}{8\pi^2}}\,D^J_{MK}(\alpha,\beta,\zeta) \qquad (25.10)$$

This coordinate system is shown in Figure 25.3.

For DEA to the middle- and highest-energy resonances, the $^2A_1$ and $^2B_2$, however, the three-body break-up channels $H + H + O^-$ and $H^- + H + O$ may be energetically open, and therefore we must consider three-body break-up. In order to do so, we cannot use the Jacobi coordinate systems, because there is not a unique dissociative coordinate corresponding to three-body break-up in the Jacobi coordinate system. Therefore, we use hyperspherical coordinates, in which there is only one dissociative coordinate, the hyperradius $\mathbf{R}$. The

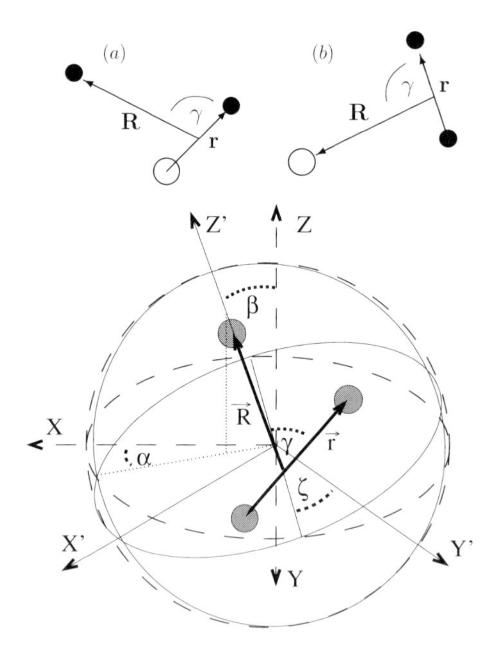

Figure 25.3 Jacobi coordinate systems used to analyse (a) the OH + H and (b) the H_2 + O arrangement channels and (below) the 'R-embedding' coordinate system with origin at the centre of mass. Primed and unprimed axes refer to body-fixed (BF) and space-fixed (SF) frames, respectively. The BF $X'Z'$ and $X'Y'$ planes are both marked with a thin circle, and the SF XZ and XY planes are marked with dashed circles. The line of nodes is also drawn. The molecule resides in the BF $X'Z'$ plane.

Delves-type coordinates [727, 728] are defined as

$$\mathbf{R} = \sqrt{R^2 + \frac{\mu_r}{\mu_R} r^2}$$

$$\theta = \tan^{-1} \sqrt{\frac{\mu_r}{\mu_R}} \frac{R}{r} \tag{25.11}$$

γ is unchanged from the Jacobi coordinate system

The coordinate $\mathbf{R}$ ranges from zero to infinity, and the coordinate θ ranges from zero to 90°.

25.5 Hamiltonians

Before discussing the primitive bases we use within the MCTDH implementation, it is necessary to discuss the Hamiltonians we employ, as these include various singularities, which must be accounted for in the numerical implementation. For the Jacobi coordinate system, the Hamiltonian reads [165,716]

$$
\begin{aligned}
H^J_{KK} &= -\frac{1}{2\mu_R}\frac{\partial^2}{\partial R^2} - \frac{1}{2\mu_r}\frac{\partial^2}{\partial r^2} + \frac{\hat{j}^2}{2\mu_r r^2} \\
&\quad + \frac{1}{2\mu_R R^2}[J(J+1) - 2K^2 + \hat{j}^2] + V(R,r,\gamma) \\
H^J_{K\pm 1,K} &= -\frac{1}{2\mu_R R^2}\sqrt{J(J+1) - K(K\pm 1)}\,\hat{j}_\pm \qquad (25.12) \\
\hat{j}^2 &= -\left(\frac{1}{\sin\gamma}\frac{\partial}{\partial\gamma}\sin\gamma\frac{\partial}{\partial\gamma} - \frac{K^2}{\sin^2\gamma}\right) \\
\hat{j}_\pm &= \mp\frac{\partial}{\partial\gamma} - K\cot\gamma
\end{aligned}
$$

For the hyperspherical version, we have only employed $J = 0$ thus far, and so write the Hamiltonian as

$$
H^0_{00} = -\frac{1}{2\mu_{\mathbf{R}}}\frac{\partial^2}{\partial \mathbf{R}^2} + \frac{1}{2\mu_{\mathbf{R}}\mathbf{R}^2}\left[-\frac{\partial^2}{\partial\theta^2} - \frac{1}{4} + \frac{1}{\sin^2\theta\cos^2\theta}\hat{j}^2\right] + V(R,r,\theta) \qquad (25.13)
$$

There are several singularities in the above Hamiltonians. One occurs due to the inverse sine-squared terms in the operator $\hat{j}^2$, becoming singular at $\gamma = 0$ or 180°. Others occur at r, R, or $\mathbf{R}$ equal to zero. Finally, in the hyperspherical coordinates, we have singularities at $\theta = 0$ or 90°.

25.6 Choice of Primitive Basis and Representation of the Hamiltonians

We employ both separable and non-separable bases implemented within MCTDH. We use the customary sine discrete variable representation (DVR) – see, for example, Appendix B of Ref. [34] – to represent all radial degrees of freedom, r, R and $\mathbf{R}$. For the bending coordinate γ, we employ either the Legendre DVR, for $J = 0$, or the extended Legendre DVR [165], for $J \neq 0$.

The extended Legendre DVR allows us to fully account for the kinetic energy singularities at $\gamma = 0$ and 180° that occur in the Hamiltonian, described above. We employ the operators $\hat{j}_+$ and $\hat{j}_-$ as implemented for the extended Legendre DVR within MCTDH. It is very important that this singularity be treated properly, for the dissociating molecule samples linear geometry for DEA via all the resonance states.

In contrast, we have found that an exact representation of the other singularities is not necessary for an accurate description of the nuclear dynamics.

That at $r = 0$ is never sampled, this geometry corresponding to an oxygen and hydrogen being on top of one another. The geometry $R = 0$ may be sampled. In particular, the dynamics via the $^2A_1$ resonance state involves a transit to linear H–O–H geometry, as one can infer from the shape of the potential energy curves shown in Figure 25.2: the $^2A_1$ potential slopes downwards to linear geometry. Therefore, the geometry $R = 0$ is relevant for the $r = r_{\mathrm{HH}}$ Jacobi coordinate system as applied to the $^2A_1$ resonance state dynamics. However, we find that we may implement the singular $1/R^2$ terms simply as a diagonal potential in the sine DVR basis, and encounter no numerical difficulties.

However, the singularity in the hyperspherical coordinate system at $\theta = 0$ and $90°$ is more problematic. For this coordinate, we also employ sine DVR, and we find that a straightforward implementation of this term does lead to difficulties with the numerical integration, as one would suspect due to large values of the singular term. Therefore, we are forced to replace $1/(\sin^2\theta\cos^2\theta)$ with $\min(1/(\sin^2\theta\cos^2\theta), A)$, where min( ) denotes the lower value of, and A is a fixed parameter. Luckily, we observe convergence of the calculation with respect to A: calculations using $A = 80$ and $A = 300$ hartree are indistinguishable, and we set the value of A at 80 hartree. Of course, 80 hartree is several orders of magnitude greater than the kinetic energy available to the dissociation, which is of the order of several electron-volts, and therefore this convergence is not unexpected.

25.7 Representation of Potential Energy Functions Using potfit

In all our calculations, we employ a separable expansion of the potential terms using the MCTDH utility *potfit* [13]. We have found the representations to be quite faithful, and we provide the potfit parameters used in Table 25.1. For brevity, we do not include the calculations done in the $r = r_{\mathrm{HH}}$ Jacobi coordinate system.

25.8 Single-Particle Function (SPF) Expansion and Mode Combinations

Proper choice of the single-particle function (SPF) expansion within the MCTDH algorithm is necessary for converged results. The current system has three internal degrees of freedom – four in the case of $J \neq 0$ calculations – and three primitive modes within MCTDH, as the K degree of freedom is combined with γ for $J \neq 0$ by the KLeg-DVR. In most cases, we were able to achieve converged results for the various calculations presented here, incorporating different final channels, surfaces and coordinate systems, using

Table 25.1 Parameters used in potfit potential fits. In all cases V_{min} and V_{max} refer to the non-separable weights used to include or discard points from the fit. The error is the root-mean-square error on these relevant points.

	Natpots					
Surface	R	r	γ	V_{min}	V_{max}	Error
Jacobi $\mathbf{r}_{OH}$						
$^2B_1$ E_R	15	15	contr	−4.67 eV	5.44 eV	32 meV
$^2B_1$ Γ	10	10	contr	0.0 eV	10.72 meV	0.026 meV
$1\,^2A'$ E_R	contr	18	16	−4.58 eV	6.81 eV	65 meV
$1\,^2A'$ Γ	10	10	contr	0.0 eV	0.005 eV	0.605 meV
$^2A_1$ E_R	15	15	contr	−3.15 eV	5.44 eV	13 meV
$^2A_1$ Γ	10	10	contr	0.0 eV	196 meV	3.83 meV
$^2B_2$ E_R	15	15	contr	−4.58 eV	6.80 eV	15.89 meV
$^2B_2$ Γ	10	10	contr	0.0 eV	272 meV	9.52 meV
Coupling E_R	15	15	contr	−2.72 eV	2.72 eV	32 meV
Coupling Γ	10	10	contr	−272 meV	272 meV	2 meV
	Natpots					
Surface	$\mathbf{R}$	θ	γ	V_{min}	V_{max}	Error
Hyperspherical						
$1\,^2A'$ E_R	18	contr	18	−5.44 eV	4.56 eV	15 meV
$1\,^2A'$ Γ	20	contr	20	0.0 eV	0.136 eV	0.326 meV
$^2A_1$ E_R	16	contr	16	−3.18 eV	6.80 eV	12.4 meV
$^2A_1$ Γ	16	contr	16	0.0 eV	198 meV	2.37 meV
$^2B_2$ E_R	16	contr	16	−4.55 eV	6.80 eV	9.97 meV
$^2B_2$ Γ	16	contr	16	0.0 eV	272 meV	2.62 meV
Coupling E_R	16	contr	16	−2.72 eV	2.72 eV	28.9 meV
Coupling Γ	16	contr	16	−272 meV	272 meV	0.97 meV

the three uncombined modes. In all cases the size of the SPF expansion is much smaller than the size of the primitive basis, demonstrating the utility and power of the MCTDH approach.

In the case of the hyperspherical calculations, analysing DEA via the uppermost $^2B_2$ state, coupled to the $^2A_1$ state via the conical intersection, we found combined modes to be appropriate. To be specific, we combined the modes θ and γ into one combined mode, and performed a two-mode MCTDH calculation with this combined mode and the hyperradius $\mathbf{R}$. We tried all mode combinations, and this one was the best performing. We see no *a priori* reason for which this particular mode combination would be superior. We interpret this superiority as indicating that the dynamics in θ and γ are more strongly correlated than that in any other pair of coordinates.

A table of the parameters of the various MCTDH calculations we have performed, including the SPF expansion, is presented in Table 25.2.

Table 25.2 Parameters used in MCTDH calculations. The parameters in the column 'CAP' are the length (in bohr) times the strength (in millihartree).

	DVR order $r \times R \times \gamma$	SPF $r \times R \times \gamma, \zeta$	Radii $r \times R$	t_f (fs)	CAP	t_{cpu} (s)
H + OH arrangement						
$^2B_1$	90×150×80	16×20×14	7×12	60	3×7	4.6E4
$1\,^2A'$	125×160×80	22×26×24	9.5×13	100	4×4	2.1E5
$1\,^2A'/$ $^2B_1$	90×150×80	18×18×18 16×16×16	7×12	40	3×7	2.1E5
$^2B_2/$ $^2A_1$	125×160×80	28×34×30 20×28×20	9.5×13	60	4×4	1.5E6
D + OD arrangement						
$^2B_1$	90×150×80	16×20×14	7×12	60	3×7	3.4E4
$1\,^2A'$	125×160×80	24×28×24	9.5×13	171	4×4	2.6E4
$^2B_2/$ $^2A_1$	125×160×80	24×28×26 22×26×18	9.5×13	75	4×4	5.6E5
O + H_2 arrangement						
$^2B_1$	90×180×80	18×40×28	7×12	80	3×7	1.0E6
$^2B_2/$ $^2A_1$	90×180×80	18×20×22 12×16×16	7×12	45	3×7	9.0E4
O + D_2 arrangement						
$^2B_1$	90×180×80	22×24×30	7×12	100	3×7	4.6E5
$^2B_2/$ $^2A_1$	90×180×80	28×26×26 20×22×16	7×12	60	3×7	4.1E5
	DVR order $\mathbf{R} \times \theta \times \gamma$	SPF $(\mathbf{R}) \times \theta \times \gamma$	Radii $\mathbf{R}$	t_f (fs)	CAP	t_{cpu} (s)
Hyperspherical H_2O						
$1\,^2A'$	150×120×64	22×36×20	12	100	4×4	1.7E5
$^2B_2/$ $^2A_1$	150×120×64	14×14 * 14×14	12	60	3×6	3.1E6
Hyperspherical D_2O						
$1\,^2A'$	200×120×64	22×36×20	16	171	4×2.5	4.2E5
$^2B_2/$ $^2A_1$	200×120×64	10×10 * 14×14	16	60	3×6	1.8E6

*For the hyperspherical calculations on the coupled $^2B_2/^2A_1$ surfaces, we contracted the degrees of freedom θ and γ into one mode.

25.9
Propagation and Natural Orbitals

For all calculations we employ the constant mean-field (CMF) scheme, using the Burlisch–Stoer (BS) integrator for the single-particle functions, and the short iterative Lanczos (SIL) integrator for the A coefficients.

The single-particle function expansion of an MCTDH calculation is deemed large enough when the lowest natural population of the SPF orbital basis falls below a certain threshold. In all calculations we aim for a maximum lowest natural population of 10^{-4} or lower, and observe that the final results are sufficiently converged.

25.10
Analysis of Flux to Calculate Cross-Sections

The dissociative attachment cross-section is given as [707]

$$\sigma_{\text{DEA}} = g\frac{4\pi^3}{k^2}\,F(E_{v_i} + \tfrac{1}{2}k^2) \tag{25.14}$$

where $F(\,)$ is the energy-resolved projected flux calculated by the FLUX84 program and g is a statistical factor. In the case of the Jacobi-coordinate calculations for two-body break-up, we calculate the projected flux to determine partial cross-sections into each bound rovibrational state $\chi_{jv}Y_{jK}(\gamma,\zeta)$ of the diatom. By doing so and summing over the bound states, we separate the two-body cross-section from the three-body component. The Jacobi coordinate system is unsuited for three-body break-up, as such break-up may occur through the CAPs in R, r, or both. In the case of the hyperspherical calculations, the final states may not be expressed as product wavefunctions in γ, θ and ζ, and therefore no projection is possible.

25.10.1
Two-Body Breakup

For the calculations on production of $H^- + OH$ via the $^2B_1$ state, we have achieved excellent agreement with experiment and are confident that we have reproduced the nuclear dynamics of dissociative attachment accurately. In our first publication on this process [706], we calculated a width of approximately 6 meV at the equilibrium geometry of the neutral. Subsequently, with a more sophisticated scattering calculation [710], we calculated a value of approximately 10 meV. Because this resonance is narrow, the cross-section exhibits a linear dependence upon this value. The results [707, 711] indicate that the former value is more accurate. In Figure 25.4 we reproduce the results of Ref. [707], showing the excellent agreement between our results and the experimental results of Belič *et al.*, with regard to both the total and fi-

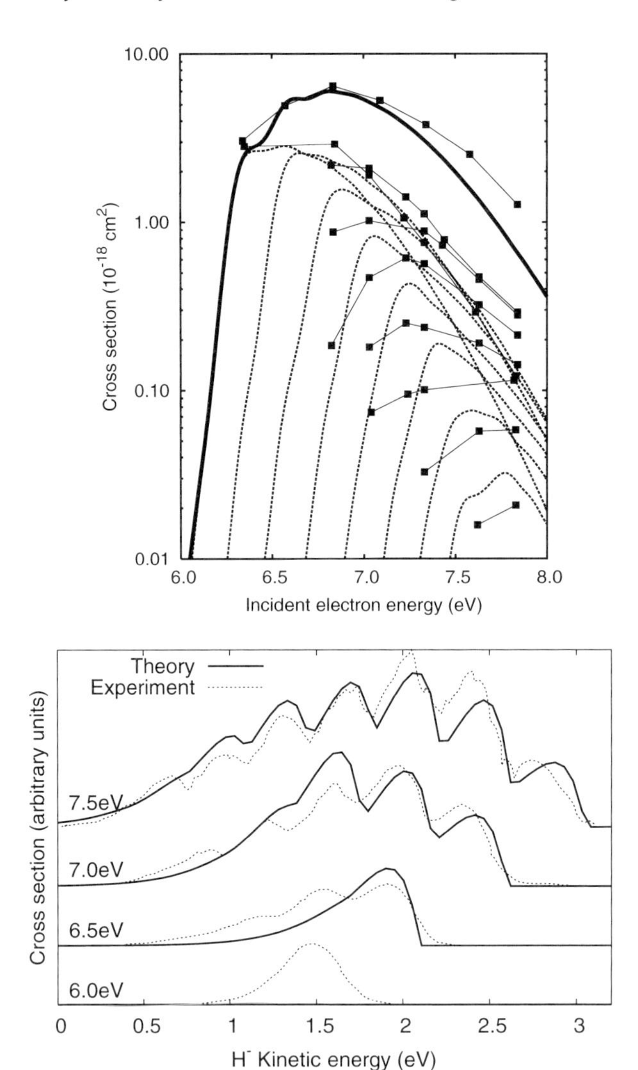

Figure 25.4 (Top panel) Cross-sections for DEA via $^2B_1$ resonance [707], from the ground rovibrational initial state: total (heavy line) and into vibrational channels $\nu = 0$ through $\nu = 7$ of OH (dotted lines, left to right), on a logarithmic scale. Also included are data from Beliĉ, Landau and Hall's [686] measurements (thin lines with squares), shifted in energy so that the maxima (present 6.81 eV, versus their value of 6.5 eV) in the total cross-section coincide. (Bottom panel) Production of $H^- + OH$ via the $^2B_1$ resonance at different incident electron energies [707], as a function of H^- fragment kinetic energy, unshifted, on arbitrary and different scales for each incident electron energy. Calculated results have been broadened using a 150 meV linewidth, consistent with the plotted experimental results from Beliĉ, Landau and Hall [686].

nal vibrational-state-resolved cross-sections. This agreement indicates that we have reproduced the major nuclear dynamics on this surface accurately.

The channel $H_2 + O^-$ is the minor channel for DEA via this resonance, as the corresponding observed cross-section is $\frac{1}{40}$th as big as that of the major channel $H^- + OH$. We calculate a cross-section for $H_2 + O^-$ that is substantially smaller than the observed cross-section, and that peaks at a higher energy [711]. Consequently, we cannot say that we have represented this channel accurately. Although the LCP model is expected to hold for DEA via such a narrow resonance, it is possible that small deviations from the LCP model may be responsible for the observed $H_2 + O^-$ cross-section. To access the $H_2 + O^-$ well, the nuclear wavepacket must traverse a large region of configuration space in which the LCP potential is essentially flat, having no significant barrier, and we regard it likely that non-local effects could affect dynamics there.

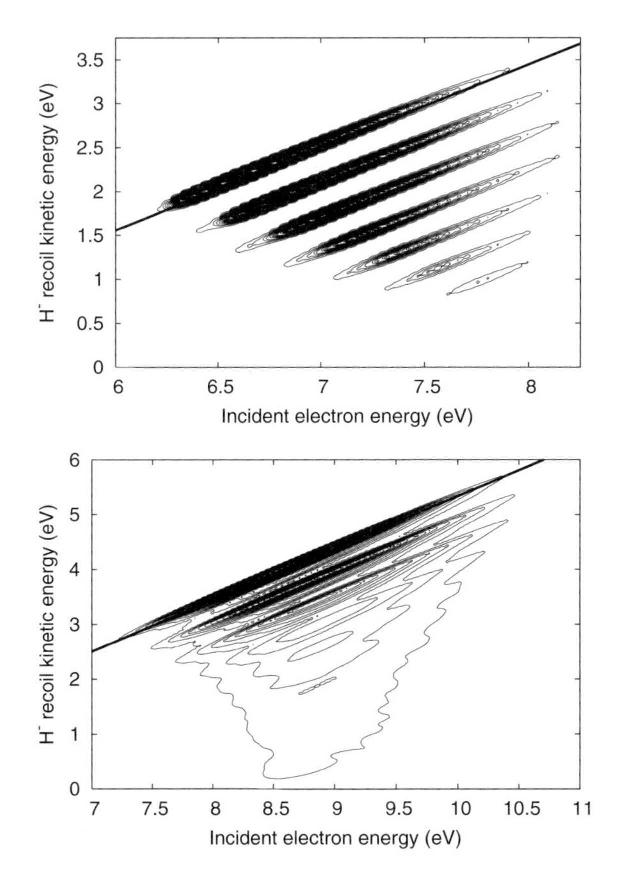

Figure 25.5 (Top panel) Cross-section for production of $H^- + OH$ from $^2B_1$ resonance as a function of incident electron energy and H^- fragment kinetic energy, unshifted, with the physical value of the maximum kinetic energy available plotted as the bold line; lowermost, deuterated. The physical value for the maximum kinetic energy is slightly lower than the value corresponding to our calculated surfaces. Contours every 2×10^{-17} cm^2 eV^{-1}. (Bottom panel) Corresponding results for $^2A_1$ state, contours every 6×10^{-18} cm^2 eV^{-1}. From [711].

The final-state projection performed by the FLUX84 utility provides a kinematically complete view of the DEA process. In Figure 25.5 we show results from calculations [711] on the $^2B_1$ and $^2A_1$ resonances. The cross-section is resolved with respect to both the incident electron energy and the kinetic energy of the recoil. Clear structure due to vibrational excitation is visible in these figures. In the case of the $^2B_1$ result, the peaks corresponding to different vibrational states are well separated, indicating that, with high enough experimental resolution, these peaks should be observed to be distinct, in contrast with the results of Beliĉ *et al.* [686] in Figure 25.4.

For DEA via the uppermost $^2B_2$ resonance, we employ a diabatic representation [710]. The $^2B_2$ and $^2A_1$ diabatic states are equal to the adiabatic states at the equilibrium geometry of the neutral. The former correlates to $H^- + OH$ ($^2\Sigma$), the $2\,^2A'$ state, and to $O^- + H_2$, the $1\,^2A'$ state, in those geometries. The latter correlates to $H^- + OH$ ($X\,^2\Pi$) and $O^- + H_2$ (triplet $\sigma_g\sigma_u$), 1 and $2\,^2A'$. The diabatic electronic coupling potential is responsible for transitions from the former to the latter.

We plot the results of calculations [712] on the two-body break-up via the $^2B_2$ resonance state in Figure 25.6. The $H^- + OH$ ($^2\Sigma$) product channel is an asymptote of the upper $2\,^2A'$ adiabatic surface; the other two asymptotes lie on the lower $1\,A'$ adiabatic surface. The $H^- + OH$ ground-state asymptote lies on the $^2A_1$ diabatic surface, whereas the others lie on the $^2B_2$ diabatic surface.

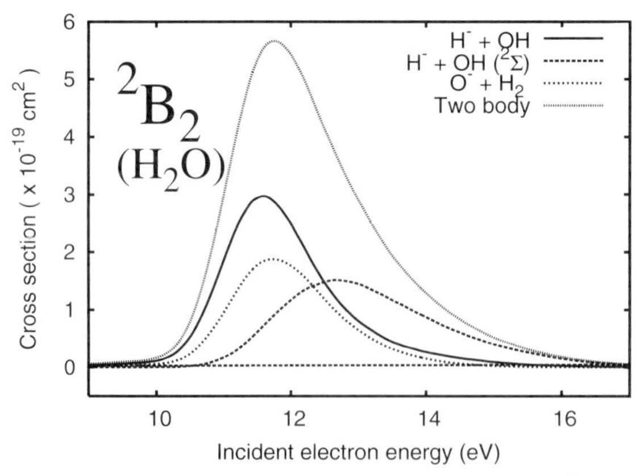

Figure 25.6 Two-body results [712] on DEA via the $^2B_2$ state, coupled to the $^2A_1$ state via the conical intersection.

Thus, we can see that, at low incident electron energy, a transition through the conical intersection is strongly favoured, whereas at high incident electron energy, the system stays on the upper $2\,^2A'$ surface. (Transition to the $^2A_1$ diabatic surface happens primarily at low incident electron energy, whereas the system may stay on the $^2B_2$ diabatic surface for all incident electron energies.)

In terms of classical trajectories, we interpret this behaviour in the following manner. At low incident electron energies, the system cannot avoid falling into the conical intersection, which lies directly downhill from the Franck–Condon region of the neutral [710]. At larger incident electron energies, the system has extra momentum, which may carry it around and avoiding the conical intersection. We will provide further details in a forthcoming publication [712].

25.10.2 Three-Body Breakup

Our calculations for three-body break-up [712], incorporating results from the hyperspherical coordinate MCTDH runs, are necessarily approximate. Even for the $^2A_1$ state, the first state for which three-body break-up is energetically open, there is an ambiguity in the three-body asymptote [708] in that both $O^- + H + H$ and $O + H^- + H$ may be considered asymptotes of the surface. There is a conical intersection between these asymptotes at infinite H–OH separation that is responsible for this ambiguity. It is plausible that transitions between these asymptotes may occur at much closer geometries and affect the cross-sections for three-body break-up and two-body break-up into high vibrational states of OH.

Owing to this ambiguity, we have set the actual asymptote of our constructed surface at 8.4 eV, equidistant between the two physical asymptotes, which lie at 8.04 and 8.75 eV, and we expect a slight discrepancy between theory and experiment for the onset of the three-body channel. The opening of this channel occurs within the $^2A_1$ dissociative attachment peak.

For dissociative attachment via the upper $^2B_2$ state, our surfaces are more approximate. As discussed above, the $^2B_2$ surface is inherently double valued, and we have constructed a single surface that interpolates between the sheets in the asymptotic region.

For both states the three-body cross-section does not, therefore, distinguish properly between $O^- + H + H$ and $O + H^- + H$. However, we find notable agreement with experiment for DEA via the $^2B_2$ state if we assign the three-body break-up to $O^- + H + H$. This result corresponds to a testable prediction: that three-body break-up is in fact dominated by production of O^- over H^-.

In Figure 25.7 we plot the overall results for this study, assigning three-body break-up to O^- production. One can see the agreement in the $^2B_2$ O^- peak. The magnitude of the cross-sections for DEA via the $^2A_1$ state is consistently overestimated by the present study. We interpret this as indicating a breakdown of the LCP model for this system, caused by the peaked behaviour of the width near the $H^- + OH$ channel and concomitant virtual state effects, leading to enhanced autodetachment in the $^2A_1$ state.

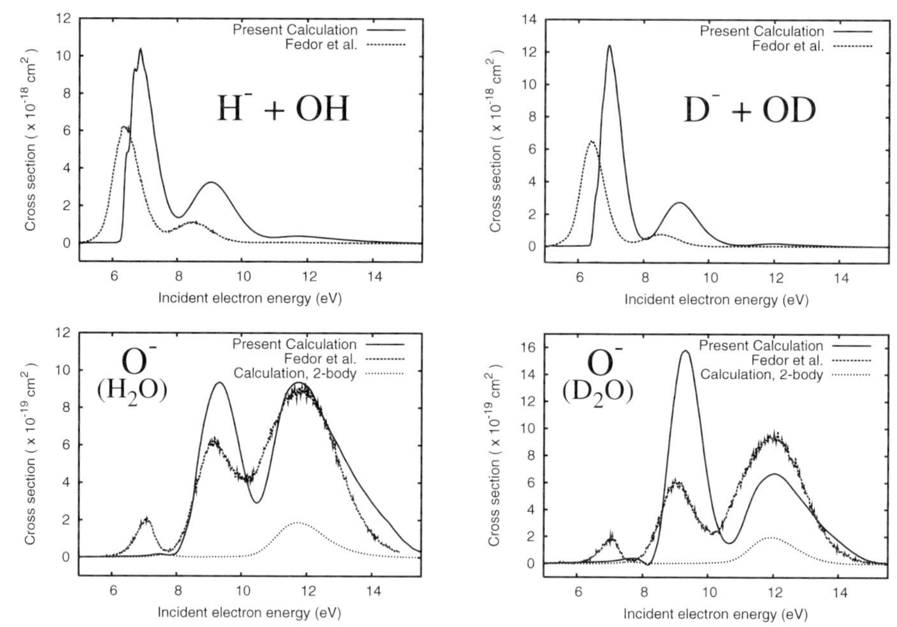

Figure 25.7 Final results of the series of studies described in the present chapter, compared with the experimental results of Fedor *et al.* [692].

As previously described [711], the discrepancies in the $^2B_1$ results are well understood. In this study [711] we have overestimated the width of this state near the Franck–Condon region of the neutral, placing it at 10 meV, not 6 meV, the physically accurate value. There is also a shift of the resonance peak of approximately 0.4 eV due to a discrepancy in the vertical excitation energy of this resonance. We have reproduced the final vibrational state distribution and isotope effects observed in experiment.

For DEA via the $^2A_1$ state, the three-body channel opens up in the middle of the experimentally observed peak. In Figure 25.8 we plot the two- and three-body cross-sections calculated. The three-body channel clearly opens at a higher incident electron energy than does the two-body channel, corresponding to the three-body asymptote of the calculated surface, 8.4 eV, in between the two physical asymptotes at 8.04 and 8.75 eV.

The three-body cross-section is calculated by subtracting the two-body, Jacobi coordinate calculation cross-section from the total cross-section calculation done in hyperspherical coordinates. It is remarkable that the MCTDH algorithm is able to represent the dynamics in both Jacobi and hyperspherical coordinate systems so accurately that such a precise cancellation of the two-body cross-sections may occur.

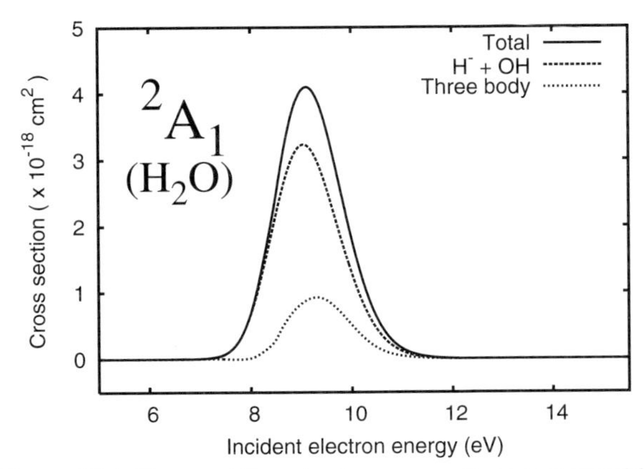

Figure 25.8 Two- and three-body cross-sections calculated for DEA via the $^2A_1$ state.

In Figure 25.9 we plot corresponding results for DEA via the $^2B_2$ state. As the three-body channel is already open, the onsets for two- and three-body break-up coincide. We see that the two- and three-body cross-sections have roughly comparable magnitudes for the full range of incident electron energy.

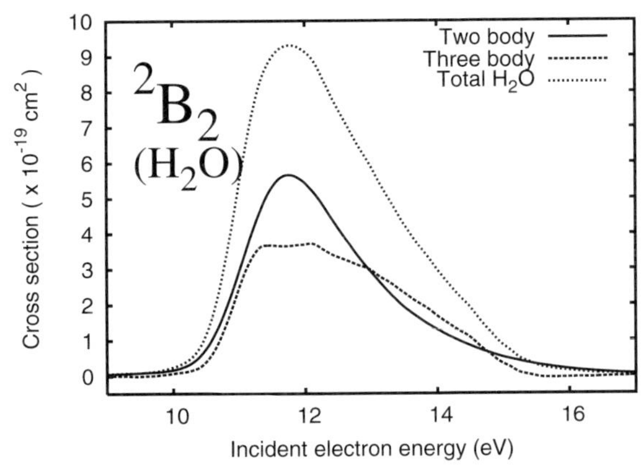

Figure 25.9 Two- and three-body cross-sections calculated for DEA via the $^2B_2$ state.

25.11 Conclusion

The MCTDH algorithm and implementation by the Heidelberg authors have enabled us to perform unprecedented calculations on dissociative attachment to a polyatomic molecule. The current series of studies represent the first such calculations incorporating the full range of nuclear motion. Although several puzzles remain for this complicated system, a substantial amount of the relevant nuclear dynamics has been elucidated by these studies. The consistency between the Jacobi coordinate system calculations and those in hyperspherical coordinates provides a powerful confirmation of the accuracy of the underlying components of the MCTDH calculation: the representation of potential energy surfaces using the potfit algorithm, the expansion of the wavefunction in the time-dependent MCTDH SPFs, the DVR primitive basis approach, and the numerical integration. We look forward to further calculations on electron–molecule scattering problems in general, and DEA in particular, with the MCTDH code.

Acknowledgments

The bulk of this work was performed under the auspices of the US Department of Energy by the University of California Lawrence Berkeley National Laboratory under contract DE-AC02-05CH11231 and was supported by the US DOE Office of Basic Energy Sciences, Division of Chemical Sciences. D.J.H. acknowledges additional support at JILA, University of Colorado, Boulder, under DOE grant number W-31-109-ENG-38 and NSF grant number ITR 0427376. C.W.M. acknowledges support from the National Science Foundation under grant number PHY-0604628. D.J.H. would like to acknowledge the Nesbitt and Kepteyn/Murnane Labs at JILA along with the rest of JILA's Center for Atomic, Molecular, and Optical Physics, funded by the NSF's Physics Frontier Center Program, which entities provided funding for the 'yotta' computer cluster at JILA. The 'yotta' cluster was used to perform the hyperspherical coordinate calculations and the recent, improved Jacobi H–OH coordinate calculations [712] on DEA via the $^2B_2$ and $^2A_1$ states.

26
Ultracold Few-Boson Systems in Traps

Sascha Zöllner, Hans-Dieter Meyer and Peter Schmelcher

26.1
Introduction

The MCTDH method as presented in Chapter 3 is designed for distinguishable particles. However, in recent years – triggered by the experimental realization of Bose–Einstein condensation – ultracold bosonic atoms have become a major research focus [260, 284]. Their high coherence and their tunability via electromagnetic fields have made it possible to study these systems at an unprecedented level of precision and control, making cold atoms a 'Rosetta stone' for many fundamental quantum effects.

While originally these ultracold quantum gases were explored in the regime of many atoms (typically, atom numbers $N \sim 10^4$), the study of few-body systems is starting to attract more and more attention. This is now possible because the ongoing miniaturization of experiments has made it almost routine to handle only a few atoms in the laboratory. Today, there is a broad range of techniques allowing for the controlled one-by-one transport and positioning of atoms [729,730], storing small ensembles on a so-called atom chip [731,732], and imaging them up to microscopic resolution [733]. This development has sparked a number of theoretical works, starting from the seminal soluble models of two atoms in a harmonic trap [734, 735], or those of bosons in a toroidal [736] or hard-wall trap [737]. Efforts to go beyond these integrable systems towards more realistic ones include the analysis of strongly correlated low-dimensional systems using quantum Monte Carlo [738] or exact-diagonalization approaches [739].

In Chapter 17, a modified MCTDH algorithm has been outlined that is tailored to ultracold bosonic systems (MCTDHB). That scheme explicitly incorporates the permutation symmetry required for *distinguishable* particles and the specific nature of the effective two-body interactions. As it stands, though, MCTDHB is designed only for larger, weakly interacting, systems (typically $N \sim 10^3$ particles), where two single-particle functions (SPFs) suffice for an

Multidimensional Quantum Dynamics: MCTDH Theory and Applications.
Edited by Hans-Dieter Meyer, Fabien Gatti, and Graham A. Worth

ISBN: 978-3-527-32018-9

accurate description. By contrast, we regard systems with *small* atom numbers. Here quantum fluctuations play a key role, and many SPFs are needed to capture the effects of strong correlations.

This chapter gives an idea of how MCTDH can be applied to few-boson systems. Section 26.2 introduces the basic model – one-dimensional (1D) trapped bosons – and lays out how it is implemented in the Heidelberg MCTDH package. Two applications are presented, which illustrate the potential for both stationary (Section 26.3) and time-dependent calculations (Section 26.4).

26.2 Model

As shown in Chapter 17, N trapped interacting bosons[1] can typically be described by the following effective Hamiltonian:

$$H = \sum_{i=1}^{N} h(p_i, x_i) + \sum_{i<j} V(x_i - x_j)$$

Here

$$h(p, x) = \frac{1}{2M} p^2 + U(x)$$

is Hamiltonian of a single particle (of mass M) in some confining potential U. [For concreteness, we focus on the case of (spinless) one-dimensional particles. The extension to higher dimensions is slightly more complicated insofar as S_+ permutes physical particles rather than degrees of freedom. To match physical and MCTDH *particles*, different degrees of freedom (DOFs) κ belonging to one and the same physical particle i have to be combined into one mode – see Chapter 3.] In one dimension, one usually employs the effective contact interaction $V(x) = g\delta(x)$, where we focus on repulsive forces: $g \geqslant 0$.

There are two peculiarities that set this problem apart from those typically tackled via MCTDH. The first issue is the requirement of bosonic permutation symmetry – that is, demanding that the true wavefunction resides in the symmetry-restricted Hilbert space $\mathbb{H}_+ = \{\Psi \in \mathbb{H} \mid S_+\Psi = \Psi\}$, where S_+ denotes the symmetrization operator over all permutations. The second issue is that the interaction potential V does not vary smoothly but rather has distribution character.

1) Be aware that here N denotes the number of atoms, *not* the number of grid points N_g.

26.2.1 Permutation Symmetry

The fact that MCTDH is designed for distinguishable particles is reflected in the MCTDH *Ansatz* for the wavefunction,

$$\Psi(\cdot,t) = \sum_J A_J(t)\Phi_J(\cdot,t), \qquad \Phi_J \equiv \varphi_{j_1} \otimes \cdots \otimes \varphi_{j_N}$$

The permutation symmetry of H clearly requires the set of single-particle functions $\{\varphi_j\}_{j=1}^n$ to be identical for each identical particle. Even so, the basis vectors Φ_J are generally not symmetric, as would be an obvious demand when dealing with bosons. (Indeed, MCTDHB employs the symmetrized version $S_+\Phi_J$, namely, number states $|n_1, n_2, \ldots\rangle_+$ in the SPF basis.) This is not a conceptual problem, though, since one may just as well keep the coefficients A_J symmetric. While this is highly redundant for $N \gg 1$, it works reasonably well for small systems. In practice, it is rarely necessary to project explicitly onto $\mathbb{H}_+$, the reason being that a symmetric initial state will not lose its symmetry under (real or imaginary) time evolution. Only numerical instabilities can destroy the symmetry. However, we have only encountered these in cases when not enough SPFs were included.

26.2.2 Modelling the Interaction

The second issue does not impose a serious restriction. In fact, while the point interaction $g\delta(x)$ is convenient as an analytic tool and for perturbative approaches, it is only one specific effective potential. At low enough energies, any model potential may be chosen so long as the low-energy scattering parameters are reproduced. Actually, for exact many-body calculations, the delta function is not an overly practicable choice, as it imposes discontinuities on the derivative of Ψ, which is an unphysical consequence of the zero-range limit. We opt to replace the delta function by a more realistic Gaussian,

$$\delta_\sigma(x) = \frac{1}{\sqrt{2\pi}\,\sigma}\,\mathrm{e}^{-x^2/2\sigma^2}$$

which converges to $\delta(x)$ in the distribution sense for ranges $\sigma \ll 2\hbar^2/M|g|$ smaller than the 1D scattering length. However, only the weaker constraint of being short-ranged compared to the average inter-particle distance is vital, $\sigma \ll L/N$ (L being the system's spatial extension). On the other hand, the range ought to be at least of the order of the grid spacing Δ_g, so that the details of V are sampled sufficiently. At the same time, the number of grid points $N_\mathrm{g} \sim L/\Delta_\mathrm{g}$ must be high enough – in our case, typically $N_\mathrm{g} \sim 150$. In addition to the high number of SPFs needed to describe very strong correlations (for our purposes, $n \sim 15$ typically suffices), this naturally limits the application of MCTDH to as few as five atoms.

26.3
Ground State: Crossover From Weak to Strong Interactions

In this section as well as the next, we shall illustrate the application of MCTDH on the example of one-dimensional bosons. Such systems nicely lend themselves to numerical studies because both limiting cases, $g = 0$ and $g \to +\infty$, are known in closed form. While the former case of non-interacting bosons is trivial, the hard-core limit of infinite repulsion permits a non-trivial mapping to non-interacting (spin-polarized) *fermions* [740]. This so-called *Bose–Fermi map* reads

$$\Psi_B = A\Psi_F, \qquad A(Q) := \prod_{i<j} \text{sgn}(x_i - x_j)$$

It is tempting to think of Pauli's exclusion principle as emulating the effect of the hard-core interaction, imposing that Ψ vanishes whenever two particles intersect ($x_i = x_j$), which is why the limit $g \to \infty$ is termed *fermionization*. The antisymmetric isometry A here serves to re-establish bosonic permutation symmetry.

While both limits are thus known by reduction to single-particle problems, the question as to how these very different limiting cases connect can be answered using numerical approaches. In this section, we want to illuminate this crossover for the ground state of $N \leqslant 5$ bosons in a harmonic trap, $U(x) = \frac{1}{2}x^2$ [281, 282]. (In what follows, we use dimensionless harmonic-oscillator units. All ground-state calculations have been done using the improved relaxation scheme – see Chapter 8.)

26.3.1
Density Profiles

To get a feeling for what happens when we go from the non-interacting case ($g = 0$) to the strongly correlated fermionization limit $g \to \infty$, let us first look at the one-body density profile $\rho(x) \equiv \langle x|\rho_1|x\rangle$, giving the probability density of finding one particle at position x. [The reduced n-body density matrix $\rho_n = \text{tr}_{n+1,\ldots,N}|\Psi\rangle\langle\Psi|$ is defined as usual by integrating out all particles $> n$. Of course, by permutation symmetry, this is independent of the numbering of the particles.]

Figure 26.1 visualizes the crossover for $N = 4, 5$ atoms. Near $g = 0$, all bosons reside in the single-particle ground state of the harmonic oscillator, $\Psi = \phi_0^{\otimes N}$, which is broadened due to repulsion. For stronger interactions, however, the profile already deviates visibly from the Gaussian shape

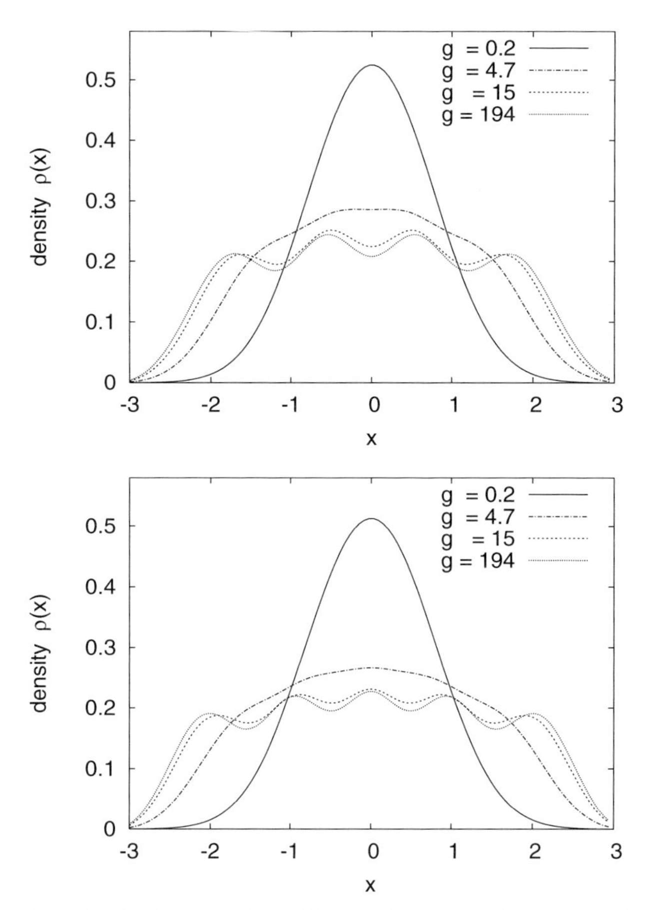

Figure 26.1 Fermionization of bosons in a harmonic trap: one-body density $\rho(x)$ for $N = 4$ (top) and $N = 5$ (bottom) for different interactions g. Note how the profile changes from a weakly interacting one ($g = 0.2$) to a flattened one due to fragmentation, and finally to a fermionized profile featuring N humps ($g \geqslant 15$).

($g = 4.7$). For very large g ($g = 15$), in turn, a structure of N peaks in the profile emerges. Physically, this means that, if we were to measure the position of a boson, we would be likely to find it at N discrete spots, and not so likely to detect it anywhere in between. This *localization* effect has a simple intuitive explanation. If the bosons repel each other very strongly, $g \to \infty$, they try to isolate each other so as to 'pay' less interaction energy. However, they cannot do that indefinitely, as they are confined in a trap. As a consequence, they tend to be pinpointed to more or less discrete positions. Note that this is the same profile that one obtains for an ideal fermion gas, in which the ground state $|n_0{=}1, \ldots, n_{N-1}{=}1\rangle_-$ is given by filling up the one-particle levels to the Fermi edge, so that the fermionized density is simply $\rho = \frac{1}{N}\sum_{a=0}^{N-1}|\phi_a|^2$. There, the seeming localization comes about because of the exclusion principle, which prevents the fermions from occupying the same point in space. By contrast, the effect here is caused by the ultrastrong repulsion.

26.3.2
Two-Body Correlations

To understand the underlying mechanism better, let us revisit the fermionization from the perspective of the two-body correlations. Figure 26.2 depicts the evolution of the two-body density $\rho_2(x_1, x_2) \equiv \langle x_1, x_2|\rho_2|x_1, x_2\rangle$, which tells us the probability density of measuring one particle at position x_1 and any second particle at x_2. In the absence of correlations, at $g = 0$, $\rho_2 = \rho_1 \otimes \rho_1$ factorizes. This leads to the symmetric Gaussian density still visible for smaller interactions $g = 0.4$ (Figure 26.2). To be sure, minor imperfections of the Gaussian shape are already anticipated here – these become even clearer when we go to higher values of $g = 4.7$. Apart from a significant broadening due to repulsion, what we see here is a *correlation hole* on the diagonal $\{x_1 = x_2\}$, signifying a depression of the two-body density. This is fairly intuitive. If the particles repel each other, it will cost a lot of energy for any two atoms to sit on top of one another, so such a configuration is avoided. This is also clear from the interaction energy

$$\mathrm{tr}(V\rho_2) \overset{\sigma\to 0}{\sim} g \int \mathrm{d}x\, \rho_2(x, x)$$

Note that this correlation hole is an inherent two-body picture. In the one-body density $\rho = \int \mathrm{d}x_2\, \rho_2(\cdot, x_2)$ it is smoothed out and is only reflected in a smeared-out profile.

When this is taken to extremes, yet another effect emerges. For $g = 15$, Figure 26.2 reveals the formation of a checkerboard pattern in ρ_2. This corresponds to the density wiggles seen in the one-body picture near fermionization. Here it has the following interpretation. Suppose we measure a first particle at, say, $x_1 \approx 2$. Then, of course, the probability of finding any second particle at $x_2 \approx x_1$ is zero, while the remaining $N - 1$ particles can be found at $N - 1$ 'discrete' spots x_2. This 'localization' shared with non-interacting fermions is a true few-body feature; for $N \gg 1$ the peaks become ever tinier modulations on the envelope density.

26.3.3
Fermionized Bosons Versus Fermions: Momentum Distribution

Up to now, one might have been under the illusion that ultrastrong interactions actually render the bosons *fermionic*. However, even though local features derived from $|\Psi_{\mathrm{B}}|^2$ are shared with their fermionic counterparts, the

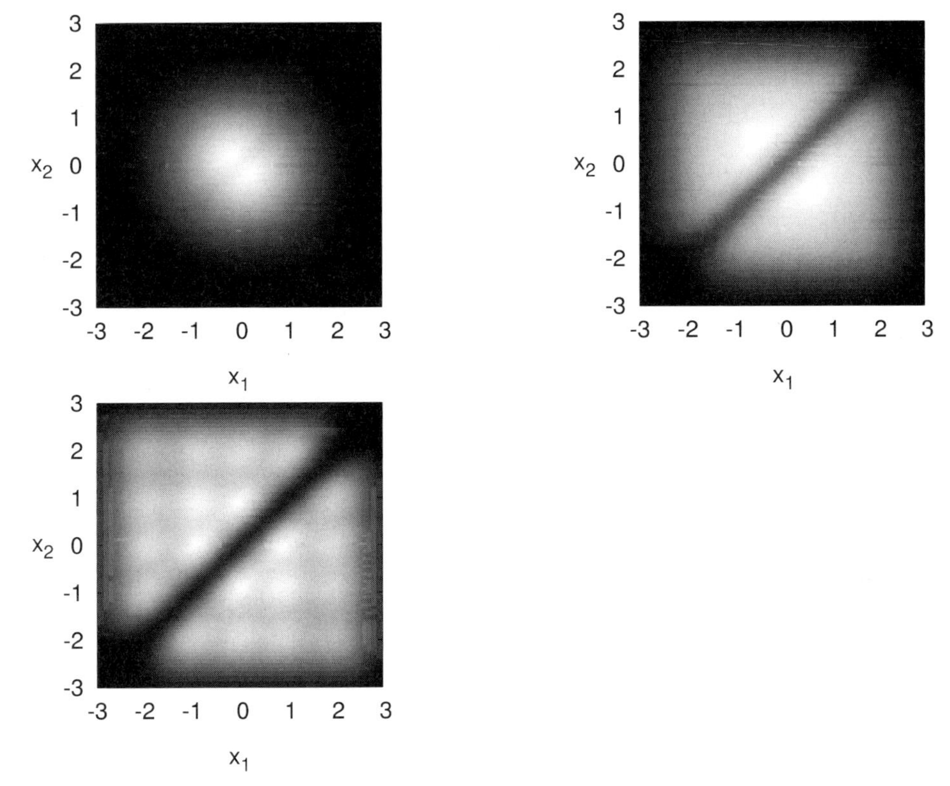

Figure 26.2 Two-body density $\rho_2(x_1, x_2)$ for $N = 5$ bosons in a harmonic trap. Shown are the interaction strengths g = 0.4 (top left), 4.7 (top right) and 15 (bottom).

bosons do retain their permutation symmetry. Their bosonic nature is reflected in off-diagonal correlations, such as the one-body momentum distribution

$$\tilde{\rho}(k) = 2\pi \langle k|\rho_1|k\rangle = \int \mathrm{d}x \int \mathrm{d}x'\, \mathrm{e}^{-\mathrm{i}k(x-x')} \langle x|\rho_1|x'\rangle$$

This quantity is not just conceptually relevant, as it relates to salient questions such as off-diagonal long-range order in ρ_1 and coherence, but it also plays a key role in cold-atom experiments.

At $g = 0$, for a harmonic trap the distribution is simply a Gaussian, $\tilde{\rho}(k)/2\pi = \pi^{-1/2}\mathrm{e}^{-k^2}$. As the interaction is switched on, we witness two effects. For one thing, the momentum peak at $k = 0$ initially sharpens, which may be understood in terms of a broadened range of the *spatial* density matrix. At the same time, the momentum spectrum also develops a long-range tail, which culminates in a decay

$$\tilde{\rho}(k) \overset{k\to\infty}{\sim} C/k^4$$

towards fermionization. This high-momentum tail reflects the strong short-range correlations for stronger interactions. Ultimately, these are also responsible for the reduction of the $k = 0$ component as $g \to \infty$ (see $g = 15, 25$ in Figure 26.3), which can be understood as a depletion of the Bose–Einstein condensed part of the system or, equivalently, the reduction of long-range order [282].

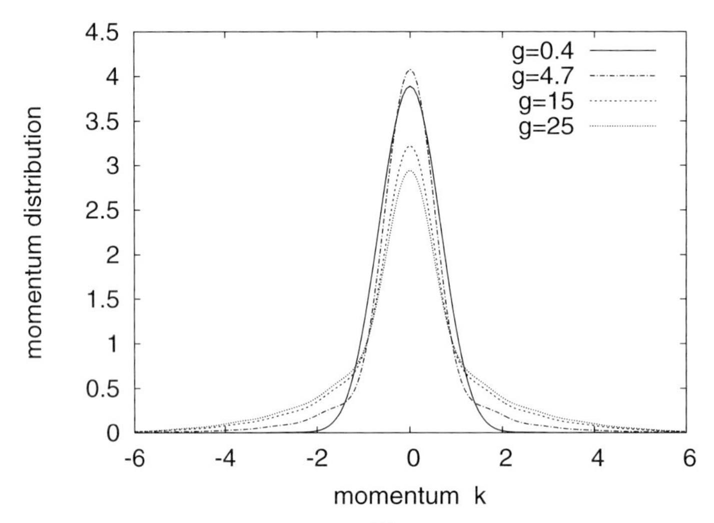

Figure 26.3 Momentum distribution $\tilde{\rho}(k)$ of $N = 5$ bosons for the interaction strengths $g = 0.4,\ 4.7,\ 15$ and 25.

Understanding these basic mechanisms of the crossover to fermionization paves the way to more detailed investigations. For instance, MCTDH allows the treatment of arbitrary trap configurations U (such as double wells), more general classes of interactions V [281, 282] and, of course, also makes excitations accessible via the improved relaxation scheme (see Chapter 8) [283].

26.4
Quantum Dynamics: Correlated Tunnelling in Double Wells

After having attained an impression of the evolution to fermionization, it is natural to ask how this affects the few-boson dynamics. A fundamental aspect we want to detail here is the tunnelling dynamics in a double-well trap, which we model as a superposition of a harmonic oscillator and a central barrier shaped as a Gaussian, $U(x) = \frac{1}{2}x^2 + h\delta_w(x)$ (we choose a width $w = 0.5$ and $h = 8$, again in harmonic oscillator units). The potential is visualized in Figure 26.4 along with the single-particle spectrum, which arranges in bands $\beta = 0, 1, \ldots$ of doublets $\epsilon_{0,1}^{(\beta)}$.

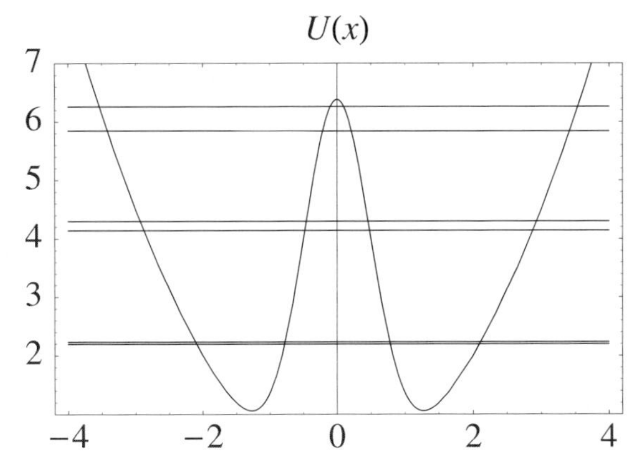

Figure 26.4 Double-well potential $U(x) = \frac{1}{2}x^2 + h\delta_w(x)$, where the horizontal lines indicate the single-particle spectrum $\{\epsilon_a\}$.

Here we want to look at the following scenario. Imagine we initially put N atoms into one (say, the right-hand) well – explicitly, we make that side energetically favourable by adding a linear external potential $-d \cdot x$ and let the system relax to its ground state $\Psi_0^{(d>0)}$. We then follow the time evolution in the *symmetric* double well by ramping the asymmetry d down to zero within some finite time τ [280, 741].

26.4.1 From Uncorrelated to Pair Tunnelling

Let us now see how the tunnelling changes as we pass from uncorrelated tunnelling ($g = 0$) to tunnelling in the presence of correlations, and finally to the fermionization limit ($g \to \infty$). It is natural to look first at the conceptually clearest situation of $N = 2$ atoms, to which we shall confine ourselves here.

In the absence of any interactions, the bosons should simply *Rabi*-oscillate back and forth between the two wells (Figure 26.5 (top)). This can be monitored by the percentage of atoms in the right-hand well

$$p_{\mathrm{R}}(t) = \langle\Theta(x)\rangle_{\Psi(t)} = \int_0^\infty \rho(x;t)\,\mathrm{d}x$$

Figure 26.5 (bottom) confirms that we obtain a harmonic oscillation at the Rabi frequency $\Delta^{(0)} = \epsilon_1^{(0)} - \epsilon_0^{(0)}$, that is, the tunnelling splitting of the lowest band. For our parameters, the corresponding period is $T = 2\pi/\Delta^{(0)} \approx 180$, which can be observed in Figure 26.5.

If we switch on repulsive interactions, see $g = 0.2$, one might naively expect the tunnelling to be enhanced. By contrast, Figure 26.5 reveals that, for short times, there is just a minute oscillation, while complete population transfer occurs on a much longer time-scale ($t \sim 300$). A look at the population dynam-

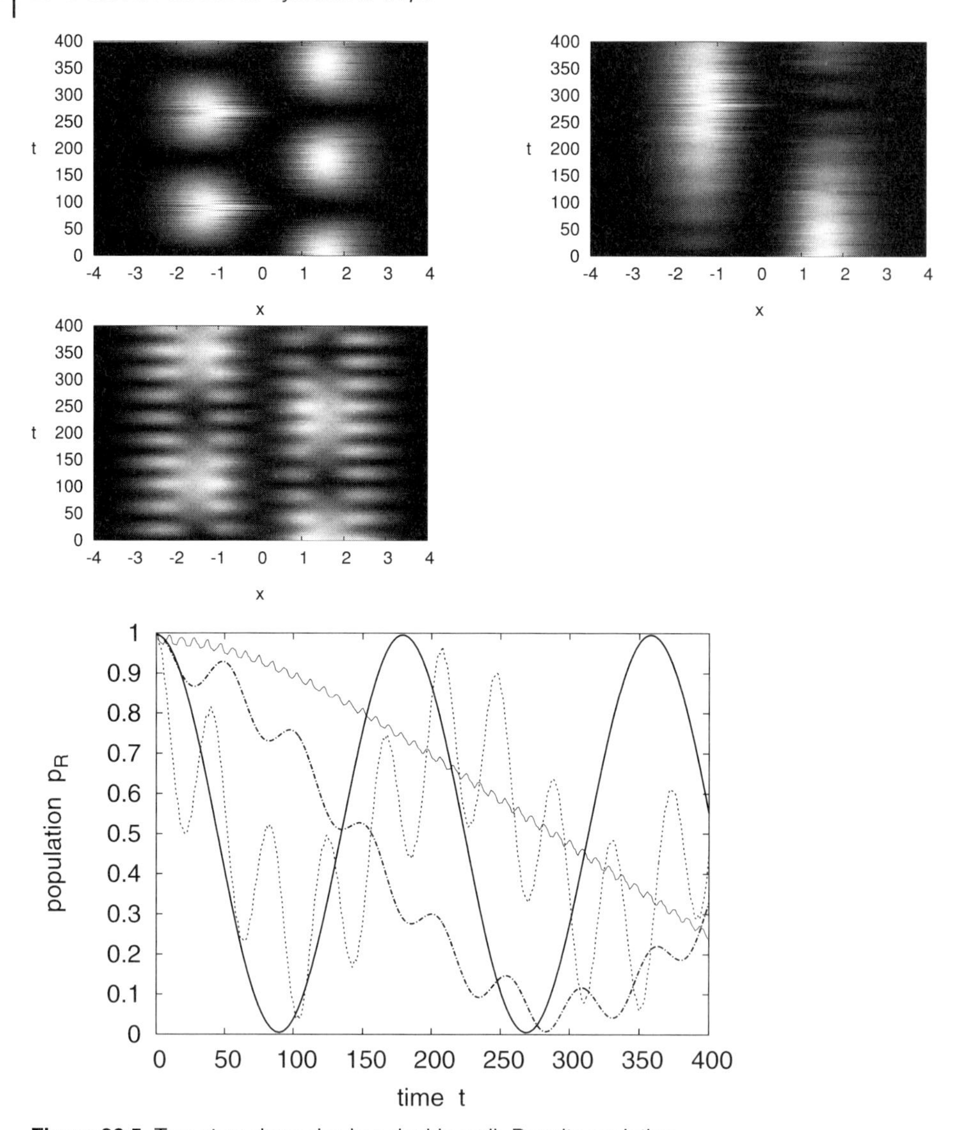

Figure 26.5 Two-atom dynamics in a double well. Density evolution $\rho(x;t)$ for g = 0 (top left), 0.2 (top right) and 25 (centre). (Bottom) Population of the right-hand well over time, $p_R(t)$, for $g = 0$ (thick full line), $g = 0.2$ (dashed-dotted line), $g = 4.7$ (thin full line) and $g = 25$ (dotted line).

ics confirms that the tunnelling oscillations have become a two-mode process. There is a fast (small-amplitude) oscillation that modulates the much slower tunnelling. As we go over to much stronger couplings, the time evolution becomes more complex, even though tunnelling is still strongly suppressed (Figure 26.5, $g = 4.7$). What is striking, though, is that, near the fermionization limit (see $g = 25$), again a simple picture emerges. The strongly correlated atom pair tunnels back and forth at about the Rabi frequency, almost like a single particle. A closer look unveils that the tunnelling is modulated by a characteristic fast, large-amplitude motion (Figure 26.5).

26.4.2 Spectral Analysis

To get an understanding of the time-scales, Figure 26.6 explores the evolution of the two-body spectrum $\{E_m(g)\}$ as g is varied. In the non-interacting case, the low-lying spectrum is given by distributing the N atoms over the lowest-band (anti)symmetric orbitals, corresponding to the lowest doublet in Figure 26.4. Thus, at $g = 0$, the few-body levels are equidistant, and only a single mode with Rabi frequency $\Delta^{(0)}$ contributes.

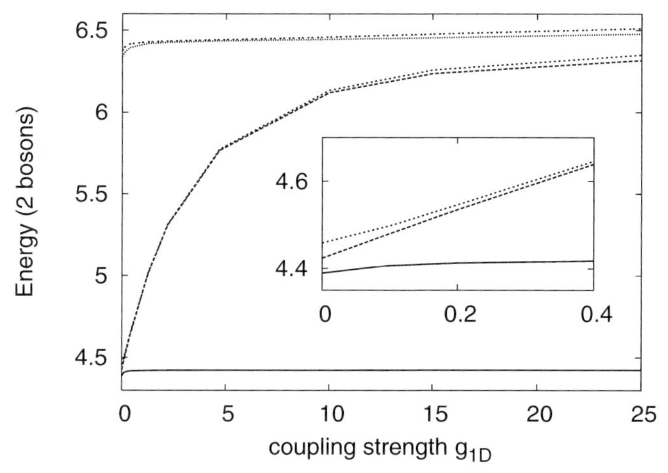

Figure 26.6 Low-lying spectrum of two bosons in a double well as a function of the interaction strength g. *Inset*: Doublet formation with increasing g.

However, as the interaction is 'switched on', the two first excited levels $E_{1,2}$ virtually glue to one another to form a doublet, whereas the gap to the ground state E_0 increases (Figure 26.6, inset). This quasi-degeneracy corresponds to the very long tunnelling times encountered for $g = 0.2$, 4.7, while the contribution from other states only makes for a fast, if tiny, modulation.

It is clear that the lowest-band scenario above breaks down for stronger interactions, when the gap to higher-lying levels in Figure 26.6 disappears. As we approach the fermionization limit (approximately realized at $g = 25$), a simple picture emerges. We are left with two doublets, each of Rabi frequency $\Delta^{(0)}$ (at $E \approx 6.3$, 6.5), separated by the (larger) upper-band tunnel splitting

$\Delta^{(1)}$ [280,741]. This makes it appealing to think of the fermionized tunnelling as that of two identical fermions hopping independently in the bands $\beta = 0$ (related to the slower tunnelling period $2\pi/\Delta^{(0)}$) and $\beta = 1$ (the faster modulation in Figure 26.5), respectively.

26.4.3 Role of Correlations

In order to unveil the physical content behind the tunnelling dynamics, let us look into the two-body correlations. Non-interacting bosons simply tunnel independently, which is reflected in the snapshots of the two-body density

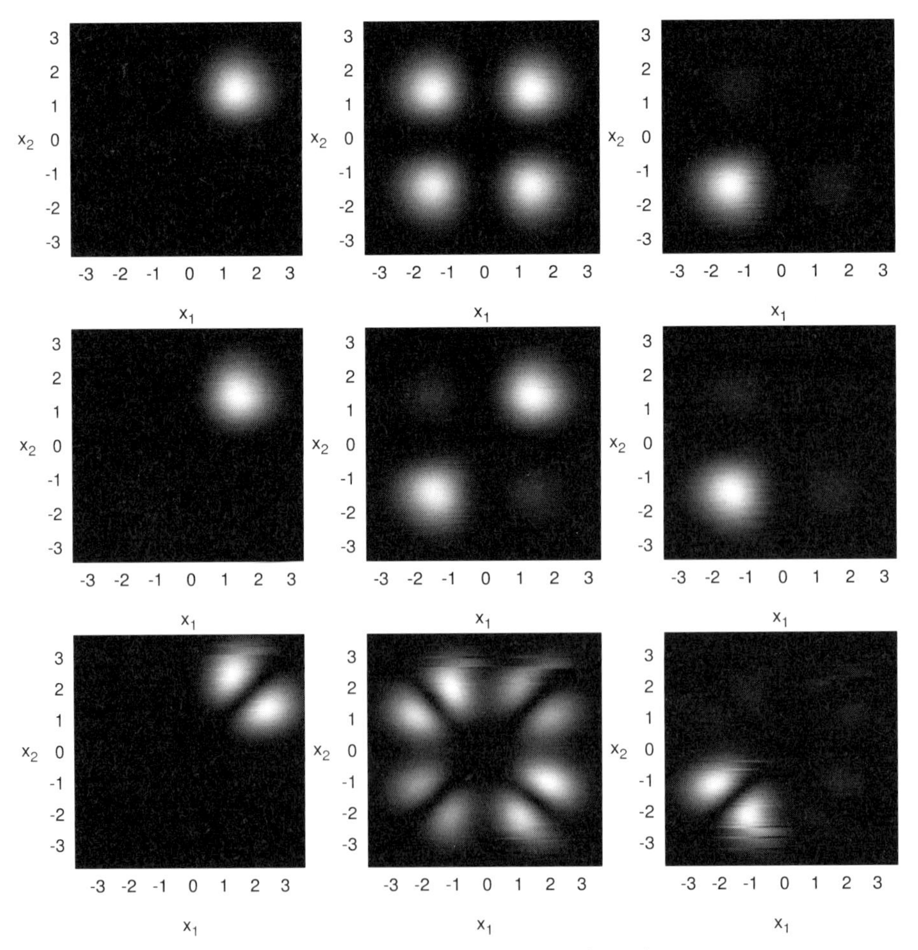

Figure 26.7 Snapshots of two-body correlation function $\rho_2(x_1, x_2)$ at $t = 0$ (left), $t = T/4$ (centre) and $t = T/2$ (right column), in terms of tunnel period T. From top to bottom: coupling strengths $g = 0,\ 0.2$ and 25.

$\rho_2(x_1, x_2)$ captured in Figure 26.7 (top row). As a consequence, if both atoms start out in the right-hand well, then in the equilibrium point of the oscillation (where $p_R = 1/2$) it will be as likely to find both atoms in the same well as in opposite ones.

As we introduce small correlations, the situation notably changes. At $g = 0.2$ (middle row), in the equilibrium point only the *diagonal* configurations have significant peaks, signifying that both atoms remain essentially in the same well in the course of tunnelling. In other words, they tunnel as pairs – which is particularly striking in view of their mutual repulsion. This reasoning also casts a new light on the fast (small-amplitude) modulations of $p_R(t)$ observed in Figure 26.5. In contrast to the slow pair tunnelling, those are linked to (suppressed) single-atom tunnelling.

It is clear that, as before, the time evolution becomes more involved as the interaction energy is raised to the fermionization limit (see $g = 25$). The two-body correlation pattern is fully fragmented not only when the pair is captured in one well (Figure 26.7, bottom left), but also when passing through the equilibrium point (centre), reflecting contributions from higher-band excited states.

The basic physics demonstrated above should provide an illustration of how the MCTDH approach may be used to study the quantum dynamics of few-boson systems. Needless to say, the range of applications even to the basic problem sketched above is nowhere near exhausted. For instance, MCTDH is predestined to treat problems with more than just two particles [280, 741]. What is more, also time-dependent Hamiltonians – as are commonly created in the lab using lasers – are handled effortlessly, offering the opportunity to consider, for example, periodically driven systems.

References

1 M. Born and R. Oppenheimer, Ann. Phys. **84**, 457 (1927).

2 M. Born and K. Huang, *Dynamical Theory of Crystal Lattices*, Oxford University Press, London, 1954.

3 L. S. Cederbaum, Born–Oppenheimer approximation and beyond, in *Conical Intersections*, edited by W. Domcke, D. R. Yarkony, and H. Köppel, pages 3–40, World Scientific, Singapore, 2004.

4 W. Domcke, D. R. Yarkony, and H. Köppel, editors, *Conical Intersections*, World Scientific, New Jersey, 2004.

5 M. J. Frisch *et al.*, Gaussian 03, Revision C.02, Gaussian, Inc., Pittsburgh, PA, 2003.

6 H.-J. Werner and P. J. Knowles, MOLPRO is a package of ab initio programs. See http://www.tc.bham.ac.uk/molpro/.

7 R. D. Levine, *Molecular Reaction Dynamics*, Cambridge University Press, Cambridge, 2005.

8 A. H. Zewail, J. Phys. Chem. A **104**, 5660 (2000).

9 P. Kukura, D. W. McCamant, S. Yoon, D. B. Wandschneider, and R. A. Mathies, Science **310**, 1006 (2005).

10 E. A. McCullough and R. E. Wyatt, J. Chem. Phys. **51**, 1253 (1969).

11 E. A. McCullough and R. E. Wyatt, J. Chem. Phys. **54**, 3578 (1971).

12 D. J. Tannor, *Introduction to Quantum Dynamics: A Time-Dependent Perspective*, University Science Books, Sausalito, CA, 2007.

13 G. A. Worth, M. H. Beck, A. Jäckle, and H.-D. Meyer, The MCTDH Package, Version 8.2, 2000. H.-D. Meyer, The MCTDH Package, Version 8.3, 2002; Version 8.4, 2007. See http://mctdh.uni-hd.de/.

14 J. M. Bowman, T. Carrington Jr., and H.-D. Meyer, Mol. Phys. **106**, *2145* (2008).

15 W. H. Miller, S. D. Schwartz, and J. W. Tromp, J. Chem. Phys. **79**, 4889 (1983).

16 E. J. Heller, J. Chem. Phys. **62**, 1544 (1975).

17 D. Kosloff and R. Kosloff, J. Comput. Phys. **52**, 35 (1983).

18 J. C. Light, I. P. Hamilton, and J. V. Lill, J. Chem. Phys. **82**, 1400 (1985).

19 J. C. Light, Discrete variable representations in quantum dynamics, in *Time-Dependent Quantum Molecular Dynamics*, edited by J. Broeckhove and L. Lathouwers, pages 185–199, Plenum, New York, 1992.

20 J. C. Light and T. Carrington Jr., Adv. Chem. Phys. **114**, 263 (2000).

21 A. D. McLachlan, Mol. Phys. **8**, 39 (1964).

22 R. B. Gerber, V. Buch, and M. A. Ratner, J. Chem. Phys. **77**, 3022 (1982).

23 R. H. Bisseling *et al.*, J. Chem. Phys. **87**, 2760 (1987).

24 P. Jungwirth and R. B. Gerber, J. Chem. Phys. **102**, 8855 (1995).

25 Z. Bihary, R. B. Gerber, and V. A. Apkarian, J. Chem. Phys. **115**, 2695 (2001).

26 N. Makri and W. H. Miller, J. Chem. Phys. **87**, 5781 (1987).

27 J. Jortner and B. Pullman, editors, *Large Finite Systems: Proceedings of the Twentieth Jerusalem Symposium of Quantum Chemistry and Biochemistry*, Reidel, Dordrecht, 1987.

28 A. D. Hammerich, R. Kosloff, and M. A. Ratner, Chem. Phys. Lett. **171**, 97 (1990).

29 B. Jackson, J. Chem. Phys. **99**, 8299 (1993).

Multidimensional Quantum Dynamics: MCTDH Theory and Applications.
Edited by Hans-Dieter Meyer, Fabien Gatti, and Graham A. Worth

ISBN: 978-3-527-32018-9

30 H.-D. Meyer, U. Manthe, and L. S. Cederbaum, Chem. Phys. Lett. **165**, 73 (1990).

31 J. V. Lill, G. A. Parker, and J. C. Light, J. Chem. Phys. **85**, 900 (1986).

32 P. A. M. Dirac, Proc. Cambridge Philos. Soc. **26**, 376 (1930).

33 J. Frenkel, *Wave Mechanics*, Clarendon Press, Oxford, 1934.

34 M. H. Beck, A. Jäckle, G. A. Worth, and H.-D. Meyer, Phys. Rep. **324**, 1 (2000).

35 J. M. Bowman, Acc. Chem. Res. **19**, 202 (1986).

36 M. A. Ratner and R. B. Gerber, J. Phys. Chem. **90**, 20 (1986).

37 U. Manthe, H.-D. Meyer, and L. S. Cederbaum, J. Chem. Phys. **97**, 3199 (1992).

38 H.-D. Meyer and G. A. Worth, Theor. Chem. Acc. **109**, 251 (2003).

39 R. Kosloff, J. Phys. Chem. **92**, 2087 (1988).

40 K. C. Kulander, editor, *Time-Dependent Methods for Quantum Dynamics*, Elsevier, Amsterdam, 1991.

41 R. Kosloff, Annu. Rev. Phys. Chem. **45**, 145 (1994).

42 G. A. Worth, J. Chem. Phys. **112**, 8322 (2000).

43 I. Burghardt, H.-D. Meyer, and L. S. Cederbaum, J. Chem. Phys. **111**, 2927 (1999).

44 C. W. Gear, *Numerical Initial Value Problems in Ordinary Differential Equations*, Prentice-Hall, Englewood Cliffs, NJ, 1971.

45 W. H. Press, S. A. Teukolsky, W. T. Vetterling, and B. P. Flannery, *Numerical Recipes*, Cambridge University Press, Cambridge, 1992.

46 C. Leforestier *et al.*, J. Comput. Phys. **94**, 59 (1991).

47 H. Tal-Ezer and R. Kosloff, J. Chem. Phys. **81**, 3967 (1984).

48 T. J. Park and J. C. Light, J. Chem. Phys. **85**, 5870 (1986).

49 U. V. Riss and H.-D. Meyer, J. Phys. B **26**, 4503 (1993).

50 M. H. Beck and H.-D. Meyer, Z. Phys. D **42**, 113 (1997).

51 U. Manthe, Chem. Phys. **329**, 168 (2006).

52 R. A. Friesner, L. S. Tuckerman, B. C. Dornblaser, and T. V. Russo, J. Sci. Comput. **4**, 327 (1989).

53 M. R. Wall and D. Neuhauser, J. Chem. Phys. **102**, 8011 (1995).

54 M. H. Beck and H.-D. Meyer, J. Chem. Phys. **109**, 3730 (1998).

55 M. H. Beck and H.-D. Meyer, J. Chem. Phys. **114**, 2036 (2001).

56 F. Gatti, M. H. Beck, G. A. Worth, and H.-D. Meyer, Phys. Chem. Chem. Phys. **3**, 1576 (2001).

57 R. Schinke, *Photodissociation Dynamics*, Cambridge University Press, Cambridge, 1993.

58 G. G. Balint-Kurti, R. N. Dixon, and C. C. Marston, J. Chem. Soc., Faraday Trans. **86**, 1741 (1990).

59 O. Vendrell, F. Gatti, and H.-D. Meyer, J. Chem. Phys. **127**, 184303 (2007).

60 U. Manthe, H.-D. Meyer, and L. S. Cederbaum, J. Chem. Phys. **97**, 9062 (1992).

61 V. Engel, Chem. Phys. Lett. **189**, 76 (1992).

62 A. Raab, G. Worth, H.-D. Meyer, and L. S. Cederbaum, J. Chem. Phys. **110**, 936 (1999).

63 S. Mahapatra, G. A. Worth, H. D. Meyer, L. S. Cederbaum, and H. Köppel, J. Phys. Chem. A **105**, 5567 (2001).

64 D. Neuhauser, J. Chem. Phys. **93**, 2611 (1990).

65 F. Richter, P. Rosmus, F. Gatti, and H.-D. Meyer, J. Chem. Phys. **120**, 6072 (2004).

66 W. Domcke and G. Stock, Adv. Chem. Phys. **100**, 1 (1997).

67 A. Stolow, Int. Rev. Phys. Chem. **22**, 377 (2003).

68 M. Seel and W. Domcke, J. Chem. Phys. **95**, 7806 (1991).

69 G. A. Worth, R. E. Carley, and H. H. Fielding, Chem. Phys. **338**, 220 (2007).

70 D. J. Tannor and S. A. Rice, Adv. Chem. Phys. **70**, 441 (1988).

71 S. Shi and H. Rabitz, J. Chem. Phys. **92**, 364 (1990).

72 L. Wang, H.-D. Meyer, and V. May, J. Chem. Phys. **125**, 014102 (2006).

73 G. A. Worth and L. S. Cederbaum, Annu. Rev. Phys. Chem. **55**, 127 (2004).

74 M. R. Brill, F. Gatti, D. Lauvergnat, and H.-D. Meyer, Chem. Phys. **338**, 186 (2007).

75 G. A. Worth, H.-D. Meyer, and L. S. Cederbaum, J. Chem. Phys. **109**, 3518 (1998).

76 D. Neuhauser and M. Baer, J. Chem. Phys. **90**, 4351 (1989).

77 U. V. Riss and H.-D. Meyer, J. Chem. Phys. **105**, 1409 (1996).

78 S. Sukiasyan and H.-D. Meyer, J. Chem. Phys. **116**, 10641 (2002).

79 A. Jäckle and H.-D. Meyer, J. Chem. Phys. **109**, 2614 (1998).

80 M.-C. Heitz and H.-D. Meyer, J. Chem. Phys. **114**, 1382 (2001).

81 F. Gatti, F. Otto, S. Sukiasyan, and H.-D. Meyer, J. Chem. Phys. **123**, 174311 (2005).

82 D. J. Tannor and D. E. Weeks, J. Chem. Phys. **98**, 3884 (1993).

83 S. Garashchuk and D. J. Tannor, J. Chem. Phys. **110**, 2761 (1999).

84 A. N. Panda, F. Otto, F. Gatti, and H.-D. Meyer, J. Chem. Phys. **127**, 114310 (2007).

85 F. Otto, F. Gatti, and H.-D. Meyer, J. Chem. Phys. **128**, 064305 (2008).

86 K. Blum, *Density Matrix Theory and Applications*, Plenum Press, New York, 1981.

87 K. Lindenberg and B. J. West, *The Non-equilibrium Statistical Mechanics of Open and Closed Systems*, VCH, New York, 1990.

88 N. G. V. Kampen, *Stochastic Processes in Physics and Chemistry*, North-Holland, Amsterdam, 1992.

89 G. Lindblad, Commun. Math. Phys. **48**, 119 (1976).

90 V. Gorini, A. Kossakowski, and E. C. G. Sudarshan, J. Math. Phys. **17**, 821 (1976).

91 E. Christensen and D. E. Evans, J. Lond. Math. Soc. **20**, 358 (1979).

92 A. G. Redfield, Adv. Magn. Reson. **1**, 1 (1965).

93 A. O. Caldeira and A. J. Leggett, Physica **121 A**, 587 (1983).

94 A. Raab, I. Burghardt, and H.-D. Meyer, J. Chem. Phys. **111**, 8759 (1999).

95 A. Raab and H.-D. Meyer, Theor. Chem. Acc. **104**, 358 (2000).

96 A. Raab and H.-D. Meyer, J. Chem. Phys. **112**, 10718 (2000).

97 C. Cattarius and H. D. Meyer, J. Chem. Phys. **121**, 9283 (2004).

98 B. Brüggemann, P. Person, H.-D. Meyer, and V. May, Chem. Phys. **347**, 152 (2008).

99 E. Fick and G. Sauermann, *The Quantum Statistics of Dynamic Processes*, Springer, Berlin, 1990.

100 R. Kosloff and H. Tal-Ezer, Chem. Phys. Lett. **127**, 223 (1986).

101 H.-D. Meyer, F. Le Quéré, C. Léonard, and F. Gatti, Chem. Phys. **329**, 179 (2006).

102 J. Hinze, J. Chem. Phys. **59**, 6424 (1973).

103 K. Drukker and S. Hammes-Schiffer, J. Chem. Phys. **107**, 363 (1997).

104 F. Culot and J. Liévin, Theor. Chem. Acc. **89**, 227 (1994).

105 F. Culot, F. Laruelle, and J. Liévin, Theor. Chem. Acc. **92**, 211 (1995).

106 E. Davidson, J. Comput. Phys. **17**, 87 (1975).

107 F. Richter, F. Gatti, C. Léonard, F. Le Quéré, and H.-D. Meyer, J. Chem. Phys. **127**, 164315 (2007).

108 E. A. Hylleraas and B. Undheim, Z. Phys. **65**, 759 (1930).

109 J. K. L. MacDonald, Phys. Rev. **43**, 830 (1933).

110 T. J. Park and J. C. Light, J. Chem. Phys. **106**, 4897 (1988).

111 U. Manthe and F. Matzkies, Chem. Phys. Lett. **252**, 71 (1996).

112 U. Manthe, J. Chem. Phys. **128**, 064108 (2008).

113 D. O. Harris, G. G. Engerholm, and G. W. Gwinn, J. Chem. Phys. **43**, 1515 (1965).

114 A. S. Dickinson and P. R. Certain, J. Chem. Phys. **49**, 4209 (1968).

115 U. Manthe, J. Chem. Phys. **105**, 6989 (1996).

116 R. van Harrevelt and U. Manthe, J. Chem. Phys. **123**, 064106 (2005).

117 A. Jäckle, Die zeitabhängige Multikonfigurations-Hartree Methode und ihre Anwendung auf reaktive Streuprozesse, PhD thesis, Universität Heidelberg, 1997.

118 R. van Harrevelt and U. Manthe, J. Chem. Phys. **121**, 5623 (2004).

119 H. Wang and M. Thoss, J. Chem. Phys. **119**, 1289 (2003).

120 R. Dawes and T. Carrington, J. Chem. Phys. **121**, 726 (2004).

121 A. Bunse-Gerstner, R. Byers, and V. Mehrmann, SIAM J. Matrix Anal. Appl. **14**, 927 (1993).

122 A. Jäckle and H.-D. Meyer, J. Chem. Phys. **104**, 7974 (1996).

123 A. Jäckle and H.-D. Meyer, J. Chem. Phys. **109**, 3772 (1998).

124 L. D. Lathauwer, B. D. Moor, and J. Vandewalle, SIAM J. Matrix Anal. Appl. **21**, 1253 (2000).

125 L. D. Lathauwer, B. D. Moor, and J. Vandewalle, SIAM J. Matrix Anal. Appl. **21**, 1324 (2000).

126 E. Schmidt, Math. Ann. **63**, 433 (1906).

127 J. M. Bowman, S. Carter, and X. Huang, Int. Rev. Phys. Chem. **22**, 533 (2003).

128 A. Chakraborty, D. G. Truhlar, J. M. Bowman, and S. Carter, J. Chem. Phys. **121**, 2071 (2004).

129 H. Rabitz and O. F. Alis, J. Math. Chem. **25**, 197 (1999).

130 O. F. Alis and H. Rabitz, J. Math. Chem. **29**, 127 (2001).

131 B. Podolsky, Phys. Rev. **32**, 812 (1928).

132 A. Nauts and X. Chapuisat, Mol. Phys. **55**, 1287 (1985).

133 B. T. Sutcliffe and J. Tennyson, Mol. Phys. **58**, 1053 (1986).

134 X. Chapuisat, A. Nauts, and J.-P. Brunet, Mol. Phys. **72**, 1 (1991).

135 M. J. Bramley and N. C. Handy, J. Chem. Phys. **98**, 1378 (1993).

136 S. Wolfram, *Mathematica, a System for Doing Mathematics by Computer*, 2nd edn., Addison-Wesley, Reading, MA, 1991.

137 M. Menou and X. Chapuisat, J. Mol. Spectrosc. **159**, 300 (1993).

138 D. Lauvergnat, A. Nauts, Y. Justum, and X. Chapuisat, J. Chem. Phys. **114**, 6592 (2001).

139 D. Lauvergnat and A. Nauts, J. Chem. Phys. **116**, 8560 (2002).

140 D. Lauvergnat and A. Nauts, Chem. Phys. **305**, 105 (2004).

141 X. Chapuisat and C. Iung, Phys. Rev. A **45**, 6217 (1992).

142 F. Gatti, Y. Justum, M. Menou, A. Nauts, and X. Chapuisat, J. Mol. Spectrosc. **181**, 403 (1997).

143 F. Gatti *et al.*, J. Chem. Phys. **108**, 8804 (1998).

144 F. Gatti, C. Iung, M. Menou, and X. Chapuisat, J. Chem. Phys. **108**, 8821 (1998).

145 F. Gatti, J. Chem. Phys. **111**, 7225 (1999).

146 C. Iung, F. Gatti, A. Viel, and X. Chapuisat, Phys. Chem. Chem. Phys. **1**, 3377 (1999).

147 M. Mladenović, J. Chem. Phys. **112**, 1070 (2000).

148 F. Gatti, C. Munoz, and C. Iung, J. Chem. Phys. **114**, 8275 (2001).

149 F. Gatti and A. Nauts, Chem. Phys. **295**, 167 (2003).

150 F. Gatti and C. Iung, J. Theor. Comput. Chem. **2**, 507 (2003).

151 C. Iung and F. Gatti, Int. J. Quant. Chem. **106**, 130 (2006).

152 F. Gatti and H.-D. Meyer, Chem. Phys. **304**, 3 (2004).

153 B.Pouilly, M. Monnerville, F. Gatti, and H.-D. Meyer, J. Chem. Phys. **122**, 184313 (2005).

154 G. Pasin, F. Gatti, C. Iung, and H.-D. Meyer, J. Chem. Phys. **124**, 194304 (2006).

155 G. Pasin, C. Iung, F. Gatti, and H.-D. Meyer, J. Chem. Phys. **126**, 024302 (2007).

156 O. Vendrell, F. Gatti, D. Lauvergnat, and H.-D. Meyer, J. Chem. Phys. **127**, 184302 (2007).

157 A. Messiah, *Quantum Mechanics*, vol. 1, John Wiley & Sons, New York, 1962.

158 R. N. Zare, *Angular Momentum*, John Wiley & Sons, New York, 1988.

159 C. Eckart, Phys. Rev. **47**, 552 (1935).

160 J. H. Van Vleck, Rev. Mod. Phys. **23**, 213 (1951).

161 O. Klein, Z. Phys. **58**, 730 (1929).

162 L. C. Biendenharn and J. D. Louck, *Angular Momentum in Quantum Mechanics*, Addison-Wesley, Reading, MA, 1981.

163 G. C. Corey, J. W. Tromp, and D. Lemoine, Fast pseudospectral algorithm in curvilinear coordinates, in *Numerical Grid Methods and Their Application to Schrödinger's Equation*, edited by C. Cerjan, pages 1–23, Kluwer Academic, Dordrecht, 1993.

164 G. C. Corey and D. Lemoine, J. Chem. Phys. **97**, 4115 (1992).

165 S. Sukiasyan and H.-D. Meyer, J. Phys. Chem. A **105**, 2604 (2001).

166 A. Nauts and X. Chapuisat, Chem. Phys. Lett. **136**, 164 (1987).

167 C. Leforestier, A. Viel, F. Gatti, C. Munoz, and C. Iung, J. Chem. Phys. **114**, 2099 (2001).

168 F. Gatti, Chem. Phys. Lett. **373**, 146 (2003).

169 C. Iung, F. Gatti, J.-M. Ortiz, and H.-D. Meyer, in preparation.

170 M. Klessinger and J. Michl, *Excited States and Photochemistry of Organic Molecules*, Wiley-VCH, New York, 1994.

171 M. A. Robb, F. Bernardi, and M. Olivucci, Pure Appl. Chem. **67**, 783 (1995).

172 G. A. Worth, M. A. Robb, and M. J. Bearpark, Semiclassical nonadiabatic trajectory computations in photochemistry: Is the reaction path enough to understand a photochemical reaction mechanism?, in *Computational Photochemistry*, edited by M. Olivucci, pages 171–190, Elsevier, Amsterdam, 2005.

173 H. Köppel, W. Domcke, and L. S. Cederbaum, Adv. Chem. Phys. **57**, 59 (1984).

174 G. Worth, H.-D. Meyer, and L. Cederbaum, Multidimensional dynamics involving a conical intersection: Wavepacket calculations using the MCTDH method, in *Conical Intersections: Electronic Structure, Dynamics and Spectroscopy*, edited by W. Domcke, D. Yarkony, and H. Köppel, pages 583–617, World Scientific, Singapore, 2004.

175 J. C. Tully and R. K. Preston, J. Chem. Phys. **55**, 562 (1971).

176 J. C. Tully, Faraday Discuss. **110**, 407 (1998).

177 M. D. Hack and D. G. Truhlar, J. Phys. Chem. A **104**, 7917 (2000).

178 Y. L. Volobuev, M. D. Hack, M. S. Topaler, and D. G. Truhlar, J. Chem. Phys. **112**, 9716 (2000).

179 Y. L. Volobuev, M. D. Hack, and D. G. Truhlar, J. Phys. Chem. A **103**, 6225 (1999).

180 M. D. Hack, A. W. Jasper, Y. L. Volobuev, D. W. Schwenke, and D. G. Truhlar, J. Phys. Chem. A **104**, 217 (2000).

181 B. R. Smith, M. J. Bearpark, M. A. Robb, F. Bernardi, and M. Olivucci, Chem. Phys. Lett. **242**, 27 (1995).

182 T. Vreven *et al.*, J. Am. Chem. Soc. **119**, 12687 (1997).

183 A. Sanchez-Galvez *et al.*, J. Am. Chem. Soc. **122**, 2911 (2000).

184 F. Jolibois, M. J. Bearpark, S. Klein, M. Olivucci, and M. A. Robb, J. Am. Chem. Soc. **122**, 5801 (2000).

185 M. J. Patterson, P. A. Hunt, M. A. Robb, and O. Takahashi, J. Phys. Chem. A **106**, 10494 (2002).

186 G. A. Worth and M. A. Robb, Adv. Chem. Phys. **124**, 355 (2002).

187 G. A. Worth, M. A. Robb, and B. Lasorne, Mol. Phys. **106**, *2077* (2008).

188 M. Ben-Nun and T. J. Martínez, Chem. Phys. **259**, 237 (2000).

189 T. J. Martínez, Acc. Chem. Res. **39**, 119 (2006).

190 M. Ben-Nun and T. J. Martínez, Adv. Chem. Phys. **121**, 439 (2002).

191 M. Ben-Nun, J. Quenneville, and T. J. Martínez, J. Phys. Chem. A **104**, 5161 (2000).

192 D. Lauvergnat, E. Baloïtcha, G. Dive, and M. Desouter-Lecomte, Chem. Phys. **326**, 500 (2006).

193 G. Worth, P. Hunt, and M. Robb, J. Phys. Chem. A **107**, 621 (2003).

194 I. Burghardt, K. Giri, and G. A. Worth, J. Chem. Phys. **129**, *174104* (2008).

195 G. Worth and I. Burghardt, Chem. Phys. Lett. **368**, 502 (2003).

196 G. A. Worth, M. A. Robb, and I. Burghardt, Faraday Discuss. **127**, 307 (2004).

197 S.-Y. Lee and E. J. Heller, J. Chem. Phys. **76**, 3035 (1982).

198 B. Lasorne, M. A. Robb, and G. A. Worth, Phys. Chem. Chem. Phys. **9**, 3210 (2007).

199 B. Lasorne, M. J. Bearpark, M. A. Robb, and G. A. Worth, J. Phys. Chem. A **12**, *13017* (2008), Submitted.

200 B. Lasorne, M. J. Bearpark, M. A. Robb, and G. A. Worth, Chem. Phys. Lett. **432**, 604 (2006).

201 H. Köppel, J. Gronki, and S. Mahapatra, J. Chem. Phys. **115**, 2377 (2001).

202 H. Köppel and B. Schubert, Mol. Phys. **104**, 1069 (2006).

203 M. J. Frisch *et al.*, Gaussian, Development Version, Revision G.01, Gaussian, Inc., Wallingford, CT, 2007.

204 R. P. A. Bettens and M. A. Collins, J. Chem. Phys. **111**, 816 (1999).

205 L. S. Cederbaum, W. Domcke, H. Köppel, and W. von Niessen, Chem. Phys. **26**, 169 (1977).

206 C. Cattarius, G. A. Worth, H.-D. Meyer, and L. S. Cederbaum, J. Chem. Phys. **115**, 2088 (2001).

207 J. C. Tully, J. Chem. Phys. **93**, 1061 (1990).

208 A. Untch, K. Weide, and R. Schinke, J. Chem. Phys. **95**, 6496 (1991).

209 B. Lasorne *et al.*, J. Chem. Phys. **128**, 124307 (2008).

210 X. Huang, B. J. Braams, and J. M. Bowman, J. Chem. Phys. **122**, 044308 (2005).

211 A. Brown, A. B. McCoy, B. J. Braams, Z. Jin, and J. M. Bowman, J. Chem. Phys. **121**, 4150 (2004).

212 M. A. Collins, Adv. Chem. Phys. **93**, 389 (1996).

213 K. C. Thompson, M. J. T. Jordan, and M. A. Collins, J. Chem. Phys. **108**, 564 (1998).

214 M. A. Collins and D. H. Zhang, J. Chem. Phys. **111**, 9924 (1999).

215 G. A. Worth, H.-D. Meyer, and L. S. Cederbaum, J. Chem. Phys. **105**, 4412 (1996).

216 H. Wang, J. Chem. Phys. **113**, 9948 (2000).

217 F. Huarte-Larrañaga and U. Manthe, J. Chem. Phys. **113**, 5115 (2000).

218 H. Wang, M. Thoss, and W. Miller, J. Chem. Phys. **115**, 2979 (2001).

219 M. Thoss, H. Wang, and W. H. Miller, J. Chem. Phys. **115**, 2991 (2001).

220 H. Köppel, M. Döscher, I. Baldea, H.-D. Meyer, and P. G. Szalay, J. Chem. Phys. **117**, 2657 (2002).

221 M. Nest and H.-D. Meyer, J. Chem. Phys. **117**, 10499 (2002).

222 F. Huarte-Larrañaga and U. Manthe, J. Chem. Phys. **117**, 4635 (2002).

223 M. Thoss and H. Wang, Chem. Phys. Lett. **358**, 298 (2002).

224 H. Wang and M. Thoss, J. Phys. Chem. A **107**, 2126 (2003).

225 T. Wu, H.-J. Werner, and U. Manthe, Science **306**, 2227 (2004).

226 M. D. Coutinho-Neto, A. Viel, and U. Manthe, J. Chem. Phys. **121**, 9207 (2004).

227 Z. Bačić and J. C. Light, J. Chem. Phys. **85**, 4594 (1986).

228 H. Wang and M. Thoss, J. Chem. Phys. **124**, 034114 (2006).

229 I. Kondov, M. Thoss, and H. Wang, J. Phys. Chem. A **110**, 1364 (2006).

230 H. Wang, D. E. Skinner, and M. Thoss, J. Chem. Phys. **125**, 174502 (2006).

231 M. Thoss and H. Wang, Chem. Phys. **322**, 210 (2006).

232 I. Kondov, M. Thoss, and H. Wang, J. Phys. Chem. C **111**, 11970 (2007).

233 H. Wang and M. Thoss, J. Phys. Chem. A **111**, 10369 (2007).

234 I. R. Craig, H. Wang, and M. Thoss, J. Chem. Phys. **127**, 144503 (2007).

235 M. Thoss, I. Kondov, and H. Wang, Phys. Rev. B **76**, 153331 (2007).

236 H. Wang and M. Thoss, Chem. Phys. **347**, 139 (2008).

237 C. Mak and D. Chandler, Phys. Rev. A **44**, 2352 (1991).

238 R. Egger and U. Weiss, Z. Phys. B **89**, 97 (1992).

239 N. Makri and D. Makarov, J. Chem. Phys. **102**, 600 (1995).

240 L. Mühlbacher and R. Egger, J. Chem. Phys. **118**, 179 (2003).

241 J. Caillat, J. Zanghellini, and A. Scrinzi, AURORA Tech. Rep. **19**, 1 (2004).

242 J. Zanghellini, M. Kitzler, C. Fabian, T. Brabec, and A. Scrinzi, Laser Phys. **13**, 1064 (2003).

243 J. Caillat *et al.*, Phys. Rev. A **71**, 012712 (2005).

244 C. Ede, G. Jordan, and A. Scrinzi, Comput. Phys. Commun., in preparation.

245 A. Scrinzi and N. Elander, J. Chem. Phys. **98**, 3866 (1993).

246 T. N. Rescigno and C. W. McCurdy, Phys. Rev. A **62**, 032706 (2000).

247 T. Kato and H. Kono, Chem. Phys. Lett. **392**, 533 (2004).

248 M. Nest, R. Padmanaban, and P. Saalfrank, J. Chem. Phys. **126**, 214106 (2007).

249 E. Schmidt, Math. Ann. **63**, 433 (1907).

250 W. Hackbusch, Computing **62**, 89 (1999).

251 G. Jordan, J. Caillat, C. Ede, and A. Scrinzi, J. Phys. B **39**, 341 (2006).

252 M. Lewenstein, P. Balcou, M. Y. Ivanov, A. L'Huillier, and P. B. Corkum, Phys. Rev. A **49**, 2117 (1994).

253 R. Santra and A. Gordon, Phys. Rev. Lett. **96**, 073906 (2006).

254 S. Patchkovskii, Z. Zhao, T. Brabec, and D. M. Villeneuve, Phys. Rev. Lett. **97**, 123003 (2006).

255 G. Jordan and A. Scrinzi, New J. Phys. **10**, 025035 (2008).

256 I. Lesanovsky *et al.*, Phys. Rev. A **73**, 033619 (2006).

257 P. Ring and P. Schuck, *The Nuclear Many-Body Problem*, Springer, Berlin, 2000.

258 J. H. McGuire, *Electron Correlation Dynamics in Atomic Collisions*, Cambridge University Press, Cambridge, 1996.

259 N. B. Kopnin, *Theory of Nonequilibrium Superconductivity*, Oxford University Press, Oxford, 2001.

260 L. Pitaevskii and S. Stringari, *Bose–Einstein Condensation*, Oxford University Press, Oxford, 2003.

261 A. I. Streltsov, O. E. Alon, and L. S. Cederbaum, Phys. Rev. A **73**, 063626 (2006).

262 P.-O. Löwdin, Phys. Rev. **97**, 1474 (1955).

263 A. J. Coleman and V. I. Yukalov, *Reduced Density Matrices: Coulson's Challenge*, Springer, New York, 2000.

264 D. A. Mazziotti, editor, *Reduced-Density-Matrix Mechanics: With Application to Many-Electron Atoms and Molecules*, Adv. Chem. Phys., vol. 134, John Wiley & Sons, New York, 2007.

265 A. I. Streltsov, O. E. Alon, and L. S. Cederbaum, Phys. Rev. Lett. **99**, 030402 (2007).

266 O. E. Alon, A. I. Streltsov, and L. S. Cederbaum, Phys. Rev. A **77**, 033613 (2008).

267 M. Nest, T. Klamroth, and P. Saalfrank, J. Chem. Phys. **122**, 124102 (2005).

268 O. E. Alon, A. I. Streltsov, and L. S. Cederbaum, J. Chem. Phys. **127**, 154103 (2007).

269 P. Kramer and M. Saraceno, *Geometry of the Time-Dependent Variational Principle*, Springer, Berlin, 1981.

270 H.-J. Kull and D. Pfirsch, Phys. Rev. E **61**, 5940 (2000).

271 D. R. Yarkony, editor, *Modern Electronic Structure Theory*, Adv. Ser. Phys. Chem., vol. 2, World Scientific, Singapore, 1995.

272 A. Szabo and N. S. Ostlund, *Modern Quantum Chemistry*, Dover, Mineola, NY, 1996.

273 O. E. Alon, A. I. Streltsov, and L. S. Cederbaum, Phys. Rev. A **76**, 062501 (2007).

274 J. Carlson, V. R. Pandharipande, and R. B. Wiringa, Nucl. Phys. **A401**, 59 (1983).

275 H. P. Büchler, A. Micheli, and P. Zoller, Nature Phys. **3**, 726 (2007).

276 J.-P. Blaizot and G. Ripka, *Quantum Theory of Finite Systems*, MIT Press, Cambridge, MA, 1986.

277 A. I. Streltsov, O. E. Alon, and L. S. Cederbaum, Phys. Rev. Lett. **100**, 130401 (2008).

278 O. E. Alon, A. I. Streltsov, and L. S. Cederbaum, Phys. Lett. A **373**, 301 (2009).

279 C. Matthies, S. Zöllner, H.-D. Meyer, and P. Schmelcher, Phys. Rev. A **76**, 023602 (2007).

280 S. Zöllner, H.-D. Meyer, and P. Schmelcher, Phys. Rev. Lett. **100**, 040401 (2008).

281 S. Zöllner, H.-D. Meyer, and P. Schmelcher, Phys. Rev. A **74**, 053612 (2006).

282 S. Zöllner, H.-D. Meyer, and P. Schmelcher, Phys. Rev. A **74**, 063611 (2006).

283 S. Zöllner, H.-D. Meyer, and P. Schmelcher, Phys. Rev. A **75**, 043608 (2007).

284 C. J. Pethick and H. Smith, *Bose–Einstein Condensation in Dilute Gases*, Cambridge University Press, Cambridge, 2002.

285 C. Menotti, J. R. Anglin, J. I. Cirac, and P. Zoller, Phys. Rev. A **63**, 023601 (2001).

286 L. S. Cederbaum, O. E. Alon, and A. I. Streltsov, Phys. Rev. A **73**, 043609 (2006).

287 L. S. Cederbaum and A. I. Streltsov, Phys. Lett. A **318**, 564 (2003).

288 O. E. Alon, A. I. Streltsov, and L. S. Cederbaum, Phys. Lett. A **347**, 88 (2005).

289 O. E. Alon, A. I. Streltsov, and L. S. Cederbaum, Phys. Lett. A **362**, 453 (2007).

290 A. I. Streltsov, L. S. Cederbaum, and N. Moiseyev, Phys. Rev. A **70**, 053607 (2004).

291 O. E. Alon, A. I. Streltsov, and L. S. Cederbaum, Phys. Rev. Lett. **95**, 030405 (2005).

292 O. E. Alon and L. S. Cederbaum, Phys. Rev. Lett. **95**, 140402 (2005).

293 O. E. Alon, A. I. Streltsov, and L. S. Cederbaum, Phys. Rev. Lett. **97**, 230403 (2006).

294 L. S. Cederbaum, A. I. Streltsov, Y. B. Band, and O. E. Alon, Phys. Rev. Lett. **98**, 110405 (2007).

295 L. S. Cederbaum, A. I. Streltsov, and O. E. Alon, Phys. Rev. Lett. **100**, 040402 (2008).

296 W. Domcke, H. Köppel, and L. S. Cederbaum, Mol. Phys. **43**, 851 (1981).

297 M. Klessinger, Angew. Chem. **107**, 597 (1995).

298 S. Mahapatra and H. Köppel, J. Chem. Phys. **109**, 1721 (1998).

299 H. Köppel, M. Döscher, and S. Mahapatra, Int. J. Quant. Chem. **80**, 942 (2000).

300 M. Garavelli, P. Celani, F. Bernardi, M. R. Robb, and M. Olivucci, J. Am. Chem. Soc. **119**, 6891 (1997).

301 S. Hahn and G. Stock, J. Phys. Chem. B **104**, 1146 (2000).

302 A. L. Sobolewski, W. Domcke, C. Dedonder-Lardeux, and C. Jouvet, Phys. Chem. Chem. Phys. **4**, 1093 (2002).

303 A. L. Sobolewski, W. Domcke, and C. Hättig, Proc. Natl. Acad. Sci. USA **102**, 17903 (2005).

304 M. Baer, Chem. Phys. Lett. **35**, 112 (1975).

305 V. Sidis, Adv. Chem. Phys. **82**, 73 (1992).

306 T. Pacher, L. S. Cederbaum, and H. Köppel, Adv. Chem. Phys. **84**, 293 (1993).

307 C. A. Mead and D. G. Truhlar, J. Chem. Phys. **77**, 6090 (1982).

308 R. Englman, *The Jahn–Teller Effect in Molecules and Crystals*, John Wiley & Sons, New York, 1972.

309 I. B. Bersuker and V. Z. Polinger, *Vibronic Interactions in Molecules and Crystals*, Springer, Heidelberg, 1989.

310 M. Döscher, H. Köppel, and P. G. Szalay, J. Chem. Phys. **117**, 2645 (2002).

311 H. Köppel, B. Schubert, and H. Lischka, Chem. Phys. **343**, 319 (2008).

312 I. Yamazaki, T. Murao, T. Yamanaka, and K. Yoshihara, Faraday Discuss. Chem. Soc. **75**, 395 (1983).

313 L. Seidner, G. Stock, A. L. Sobolewski, and W. Domcke, J. Chem. Phys. **96**, 5298 (1992).

314 P. Baltzer *et al.*, Chem. Phys. **196**, 551 (1995).

315 C. Woywood and W. Domcke, Chem. Phys. **162**, 349 (1992).

316 S. Mahapatra, L. S. Cederbaum, and H. Köppel, J. Chem. Phys. **111**, 10452 (1999).

317 G. Worth and L. Cederbaum, Chem. Phys. Lett. **348**, 477 (2001).

318 A. Markmann, G. A. Worth, and L. S. Cederbaum, J. Chem. Phys. **122**, 144320 (2005).

319 C. Cattarius, A. Markmann, and G. A. Worth, The VCHAM program, 2007. See http://mctdh.uni-hd.de/.

320 G. A. Worth, J. Photochem. Photobiol. A **190**, 190 (2007).

321 A. Markmann *et al.*, J. Chem. Phys. **123**, 204310 (2005).

322 G. A. Worth, G. Welch, and M. J. Paterson, Mol. Phys. **104**, 1095 (2006).

323 J. Schön and H. Köppel, J. Chem. Phys. **103**, 9292 (1995).

324 S. A. Trushin, W. Fuß, and W. E. Schmid, Chem. Phys. **259**, 313 (2000).

325 P. Baltzer *et al.*, Chem. Phys. **224**, 95 (1997).

326 I. Baldea and H. Köppel, J. Chem. Phys. **124**, 064101 (2006).

327 I. Baldea, J. Franz, P. G. Szalay, and H. Köppel, Chem. Phys. **329**, 65 (2006).

328 E. Gindensperger, I. Baldea, J. Franz, and H. Köppel, Chem. Phys. **338**, 207 (2007).

329 S. Faraji, H.-D. Meyer, and H. Köppel, J. Chem. Phys. **129**, 074311 (2008).

330 J. R. Fletcher, M. C. M. O'Brien, and S. N. Evangelou, J. Phys. A **13**, 2035 (1980).

331 L. S. Cederbaum, E. Haller, and W. Domcke, Solid State Commun. **35**, 879 (1980).

332 L. S. Cederbaum, E. Gindensperger, and I. Burghardt, Phys. Rev. Lett. **94**, 113003 (2005).

333 E. Gindensperger, I. Burghardt, and L. S. Cederbaum, J. Chem. Phys. **124**, 144103 (2006).

334 E. Gindensperger, H. Köppel, and L. S. Cederbaum, J. Chem. Phys. **126**, 034107 (2007).

335 E. Gindensperger and L. S. Cederbaum, J. Chem. Phys. **127**, 124107 (2007).

336 I. Burghardt, E. Gindensperger, and L. S. Cederbaum, Mol. Phys. **104**, 1081 (2006).

337 H. Tamura, E. R. Bittner, and I. Burghardt, J. Chem. Phys. **126**, 021103 (2007).

338 H. Tamura, E. R. Bittner, and I. Burghardt, J. Chem. Phys. **127**, 034706 (2007).

339 H. Tamura, A. G. S. Ramon, E. R. Bittner, and I. Burghardt, J. Phys. Chem. B **112**, 495 (2008).

340 E. Gindensperger, I. Burghardt, and L. S. Cederbaum, J. Chem. Phys. **124**, 144104 (2006).

341 I. Burghardt, J. T. Hynes, E. Gindensperger, and L. S. Cederbaum, Phys. Scr. **73**, C42 (2006).

342 E. V. Gromov *et al.*, J. Chem. Phys. **121**, 4585 (2004).

343 S. Saddique and G. A. Worth, Chem. Phys. **329**, 99 (2006).

344 T. S. Venkatesan, S. Mahapatra, H.-D. Meyer, H. Köppel, and L. S. Cederbaum, J. Phys. Chem. A **111**, 1746 (2007).

345 U. Manthe, T. Seideman, and W. H. Miller, J. Chem. Phys. **99**, 10078 (1993).

346 D. H. Zhang, J. Z. H. Zhang, Y. Zhang, D. Wang, and Q. Zhang, J. Chem. Phys. **102**, 7400 (1995).

347 D. Neuhauser, J. Chem. Phys. **100**, 9272 (1994).

348 W. Zhu, J. Dai, J. Z. H. Zhang, and D. H. Zhang, J. Chem. Phys. **105**, 4881 (1996).

349 U. Manthe and F. Matzkies, J. Chem. Phys. **113**, 5725 (2000).

350 D. Wang, J. Chem. Phys. **124**, 201105 (2006).

351 T. Yamamoto, J. Chem. Phys. **33**, 281 (1960).

352 W. H. Miller, J. Chem. Phys. **61**, 1823 (1974).

353 F. Huarte-Larrañaga and U. Manthe, Z. Phys. Chem. **221**, 171 (2007).

354 F. Matzkies and U. Manthe, J. Chem. Phys. **108**, 4828 (1998).

355 U. Manthe, Chem. Phys. Lett. **241**, 497 (1995).

356 D. H. Zhang and J. C. Light, J. Chem. Phys. **104**, 6184 (1996).

357 F. Huarte-Larrañaga and U. Manthe, J. Chem. Phys. **123**, 204114 (2005).

358 U. Manthe, W. Bian, and H.-J. Werner, Chem. Phys. Lett. **313**, 647 (1999).

359 U. Manthe and F. Matzkies, Chem. Phys. Lett. **282**, 442 (1998).

360 F. Huarte-Larrañaga and U. Manthe, J. Chem. Phys. **116**, 2863 (2002).

361 F. Matzkies and U. Manthe, J. Chem. Phys. **110**, 88 (1999).

362 F. Matzkies and U. Manthe, J. Chem. Phys. **112**, 130 (2000).

363 F. Huarte-Larrañaga and U. Manthe, J. Phys. Chem. A **105**, 2522 (2001).

364 J. M. Bowman, J. Phys. Chem. **95**, 4960 (1991).

365 M. J. T. Jordan and R. G. Gilbert, J. Chem. Phys. **102**, 5669 (1995).

366 T. Wu, H.-J. Werner, and U. Manthe, J. Chem. Phys. **124**, 164307 (2006).

367 T. Wu and U. Manthe, J. Chem. Phys. **119**, 14 (2003).

368 D. L. Baulch *et al.*, J. Chem. Phys. Ref. Data **21**, 411 (1992).

369 P.-M. Marquaire, A. G. Dastidar, K. C. Manthorne, and P. D. Pacey, Can. J. Chem. **72**, 600 (1994).

370 H. G. Yu and G. Nyman, J. Chem. Phys. **111**, 3508 (1999).

371 M. L. Wang, Y. Li, J. Z. H. Zhang, and D. H. Zhang, J. Chem. Phys. **113**, 1802 (2000).

372 D. Y. Wang and J. M. Bowman, J. Chem. Phys. **115**, 2055 (2001).

373 L. Zhang, Y. Lu, S. Y. Lee, and D. H. Zhang, J. Chem. Phys. **127**, 234313 (2007).

374 J. Z. Pu, J. C. Corchado, and D. G. Truhlar, J. Chem. Phys. **115**, 6266 (2001).

375 F. Huarte-Larrañaga and U. Manthe, J. Chem. Phys. **118**, 8261 (2003).

376 G. Ertl, J. Vac. Sci. Technol. A **1**, 1247 (1983).

377 J. P. Van Hook, Catal. Rev. Sci. Eng **21**, 1 (1980).

378 I. Estermann and O. Stern, Z. Phys. **61**, 95 (1930).

379 A. W. Kleyn and T. C. M. Horn, Phys. Rep. **199**, 192 (1991).

380 J. D. White, J. Chen, D. Matsiev, D. J. Auerbach, and A. M. Wodtke, Nature **433**, 503 (2005).

381 G. Boato, P. Cantini, and L. Mattera, Jpn. J. Appl. Phys. Suppl. **2**, 553 (1974).

382 R. G. Rowe and G. Ehrlich, J. Chem. Phys. **62**, 735 (1975).

383 C. T. Rettner, F. Fabre, J. Kimman, and D. J. Auerbach, Phys. Rev. Lett. **55**, 1904 (1985).

384 H. A. Michelsen, C. T. Rettner, and D. J. Auerbach, Surf. Sci. **272**, 65 (1992).

385 M. J. Murphy and A. Hodgson, J. Chem. Phys. **108**, 4199 (1998).

386 M. Dohle and P. Saalfrank, Surf. Sci. **373**, 95 (1997).

387 P. Nieto *et al.*, Science **312**, 86 (2006).

388 A. Gross, S. Wilke, and M. Scheffler, Phys. Rev. Lett. **75**, 2718 (1995).

389 G. J. Kroes, E. J. Baerends, and R. C. Mowrey, Phys. Rev. Lett. **78**, 3583 (1997).

390 J. Dai and J. C. Light, J. Chem. Phys. **107**, 1676 (1997).

391 C. T. Rettner, H. A. Michelsen, and D. J. Auerbach, J. Chem. Phys. **102**, 4625 (1995).

392 H. Hou, S. J. Gulding, C. T. Rettner, A. M. Wodtke, and D. J. Auerbach, Science **277**, 80 (1997).

393 G. Anger, A. Winkler, and K. D. Rendulic, Surf. Sci. **220**, 1 (1989).

394 G. Comsa and R. David, Surf. Sci. **117**, 77 (1982).

395 E. Watts and G. O. Sitz, J. Chem. Phys. **114**, 4171 (2001).

396 A. C. Luntz, J. K. Brown, and M. D. Williams, J. Chem. Phys. **93**, 5240 (1990).

397 C. Díaz *et al.*, Phys. Rev. Lett. **96**, 096102 (2006).

398 L. B. F. Juurlink, P. R. McCabe, R. R. Smith, C. L. DiCologero, and A. L. Utz, Phys. Rev. Lett. **83**, 868 (1999).

399 R. Beck *et al.*, Science **302**, 98 (2003).

400 A. Capellini and A. P. J. Jansen, J. Chem. Phys. **104**, 3366 (1996).

401 M. Ehara, H.-D. Meyer, and L. S. Cederbaum, J. Chem. Phys. **105**, 8865 (1996).

402 E. Gindesperger, C. Meier, J. A. Beswick, and M.-C. Heitz, J. Chem. Phys. **116**, 10051 (2002).

403 A. P. J. Jansen, J. Chem. Phys. **99**, 4055 (1993).

404 R. van Harrevelt and U. Manthe, J. Chem. Phys. **121**, 3829 (2004).

405 R. van Harrevelt and U. Manthe, J. Chem. Phys. **123**, 124706 (2005).

406 C. Crespos, H.-D. Meyer, R. C. Mowrey, and G. J. Kroes, J. Chem. Phys. **124**, 074706 (2006).

407 R. van Harrevelt and U. Manthe, J. Chem. Phys. **122**, 234702 (2005).

408 A. P. J. Jansen and H. Burghgraef, Surf. Sci. **344**, 149 (1995).

409 R. Milot and A. P. J. Jansen, J. Chem. Phys. **109**, 1966 (1998).

410 R. Milot and A. P. J. Jansen, Surf. Sci. **452**, 179 (2000).

411 R. Milot and A. P. J. Jansen, Phys. Rev. B **61**, 15657 (2000).

412 J.-Y. Fang and H. Guo, J. Chem. Phys. **101**, 5831 (1994).

413 J.-Y. Fang and H. Guo, Chem. Phys. Lett. **235**, 341 (1995).

414 J.-Y. Fang and H. Guo, J. Mol. Struct. (Theochem) **341**, 201 (1995).

415 S. Woittequand *et al.*, Chem. Phys. Lett. **406**, 202 (2005).

416 S. Woittequand *et al.*, Surf. Sci. **601**, 3034 (2007).

417 I. Andrianov and P. Saalfrank, Chem. Phys. Lett. **433**, 91 (2006).

418 G. K. Paramanov, I. Andrianov, and P. Saalfrank, J. Phys. Chem. C **111**, 5432 (2007).

419 A. Markmann, J. L. Gavartin, and A. L. Schluger, Chem. Phys. **330**, 253 (2006).

420 J. Campos-Martínez and R. D. Coalson, J. Chem. Phys. **99**, 9629 (1993).

421 G. J. Kroes, Prog. Surf. Sci. **60**, 1 (1999).

422 D. Neuhauser, Chem. Phys. Lett. **200**, 173 (1992).

423 G. Wolken, J. Chem. Phys. **59**, 1159 (1973).

424 G. J. Kroes and R. C. Mowrey, J. Chem. Phys. **101**, 805 (1994).

425 R. B. Gerber, L. H. Beard, and D. J. Kouri, J. Chem. Phys. **74**, 4709 (1981).

426 H. F. Bowen, D. J. Kouri, R. C. Mowrey, A. T. Yinnon, and R. B. Gerber, J. Chem. Phys. **99**, 704 (1993).

427 D. Halstead and S. Holloway, J. Chem. Phys. **93**, 2859 (1990).

428 D. A. McCormack *et al.*, Faraday Discuss. **117**, 109 (2000).

429 M. F. Somers *et al.*, J. Chem. Phys. **117**, 6673 (2002).

430 G. R. Darling and S. Holloway, Surf. Sci. **304**, L461 (1994).

431 H. A. Michelsen and D. J. Auerbach, J. Chem. Phys. **94**, 7502 (1991).

432 R. A. Olsen *et al.*, J. Chem. Phys. **116**, 3841 (2002).

433 E. Pijper, G. J. Kroes, R. A. Olsen, and E. J. Baerends, J. Chem. Phys. **116**, 9435 (2002).

434 A. Logadottir and J. K. Nørskov, J. Catal. **220**, 273 (2003).

435 G. Haase, M. Asscher, and R. Kosloff, J. Chem. Phys. **90**, 3346 (1989).

436 U. Manthe, J. Theor. Comput. Chem. **1**, 153 (2002).

437 S. Dahl *et al.*, Phys. Rev. Lett. **83**, 1814 (1999).

438 P. M. Holmblad, J. Wambach, and I. Chorkendorff, J. Chem. Phys. **102**, 8255 (1995).

439 P. Maroni *et al.*, Phys. Rev. Lett. **94**, 246104 (2005).

440 G. J. Kroes and D. C. Clary, J. Phys. Chem. **96**, 7079 (1992).

441 M. Nest and H.-D. Meyer, J. Chem. Phys. **119**, 24 (2003).

442 I. Andrianov and P. Saalfrank, J. Chem. Phys. **124**, 034710 (2006).

443 D. Newns, Phys. Rev. B **178**, 1123 (1969).

444 A. Petersson, M. Ratner, and H. Karlsson, J. Phys. Chem. B **104**, 8498 (2000).

445 W. Domcke, Phys. Rep. **208**, 97 (1991).

446 M. F. Somers, R. A. Olsen, H. F. Busnengo, E. J. Baerends, and G. J. Kroes, J. Chem. Phys. **121**, 11379 (2004).

447 H. F. Busnengo, A. Salin, and W. Dong, J. Chem. Phys. **112**, 7641 (2000).

448 J. Ischtwan and M. A. Collins, J. Chem. Phys. **100**, 8080 (1994).

449 V. A. Ukraintsev and I. Harrison, J. Chem. Phys. **101**, 1564 (1994).

450 R. R. Smith, D. R. Killelea, D. F. DelSesto, and A. L. Utz, Science **304**, 992 (2004).

451 L. B. F. Juurlink, P. R. McCabe, R. R. Smith, D. R. Killelea, and A. L. Utz, Phys. Rev. Lett. **94**, 208303 (2005).

452 Y. Xiang, J. Z. H. Zhang, and D. Y. Wang, J. Chem. Phys. **117**, 7698 (2002).

453 S. Nave and B. Jackson, Phys. Rev. Lett. **98**, 173003 (2007).

454 G.-J. Kroes and H.-D. Meyer, Chem. Phys. Lett. **440**, 334 (2007).

455 Y. H. Huang, C. T. Rettner, D. J. Auerbach, and A. M. Wodtke, Science **290**, 111 (2000).

456 R. E. Wyatt and C. Iung, Quantum mechanical studies of molecular sepctra and dynamics, in *Dynamics of Molecules and Chemical Reactions*, edited by R. E. Wyatt and J. Z. H. Zhang, pages 59–122, Marcel Dekker, New York, 1996.

457 F. Remacle and R. D. Levine, Spectra, rates, and intramolecular dynamics, in *Dynamics of Molecules and Chemical Reactions*, edited by R. E. Wyatt and J. Z. H. Zhang, pages 1–58, Marcel Dekker, New York, 1996.

458 J. T. Hynes and R. Rey, *Ultrafast Raman and Infrared Spectroscopy*, edited by M. Fayer, Marcel Dekker, New York, 2001.

459 E. T. J. Nibbering and T. Elsaesser, Chem. Rev. **104**, 1887 (2004).

460 R. Marquardt and M. Quack, in *Encyclopedia of Chemical Physics and Physical Chemistry*, edited by J. Moore and N. Spencer, Chap. A13.13, IOP, Bristol, 2001.

461 R. A. Marcus, J. Chem. Phys. **20**, 359 (1952).

462 T. Brixner, N. H. Damreuer, P. Niklaus, and G. Gerber, Nature **414**, 57 (2001).

463 R. J. Levis, G. M. Menkir, and H. Rabitz, Science **292**, 709 (2001).

464 K. K. Lehmann, G. Scoles, and B. H. Pate, Annu. Rev. Phys. Chem. **45**, 241 (1994).

465 T. Uzer and W. H. Miller, Phys. Rep. **199**, 73 (1991).

466 M. Quack, Annu. Rev. Phys. Chem. **41**, 839 (1990).

467 M. Hippler and M. Quack, in *Isotope Effects in Chemistry and Biology*, Part III *Isotope Effects in Chemical Dynamics*, edited by A. Kohen and H.-H. Limbach, Chap. 11, Marcel Dekker, New York, 2005.

468 A. H. Zewail, *Femtochemistry*, edited by F. C. De Schryver, S. De Feyter and G. Schweitzer, Wiley-VCH, Weinheim, 2001.

469 M. Dantus and A. H. Zewail, Chem. Rev. **104**, 1717 (2004).

470 H. Rabitz, R. de Vivie-Riedle, M. Motkuz, and K. Kompa, Science **288**, 824 (2000).

471 T. Brixner and G. Gerber, Comput. Phys. Commun. **4**, 418 (2003).

472 M. Dantus and V. Lozovoy, Chem. Rev. **104**, 1813 (2004).

473 C. Bloch, Nucl. Phys. **6**, 329 (1958).

474 P. Durand and J. P. Malrieu, *Ab initio Methods in Quantum Chemistry*, edited by K. P. Lawley, John Wiley & Sons, New York, 1987.

475 R. E. Wyatt, C. Iung, and C. Leforestier, J. Chem. Phys. **97**, 3458 (1992).

476 R. E. Wyatt, C. Iung, and C. Leforestier, J. Chem. Phys. **97**, 3477 (1992).

477 C. Iung, C. Leforestier, and R. E. Wyatt, J. Chem. Phys. **98**, 6722 (1993).

478 R. E. Wyatt and C. Iung, J. Chem. Phys. **98**, 6758 (1993).

479 R. E. Wyatt and C. Iung, J. Chem. Phys. **101**, 3671 (1994).

480 C. Iung and R. E. Wyatt, J. Chem. Phys. **99**, 2261 (1993).

481 R. E. Wyatt, C. Iung, and C. Leforestier, Acc. Chem. Res. **28**, 423 (1995).

482 A. Maynard, R. E. Wyatt, and C. Iung, J. Chem. Phys. **106**, 9483 (1997).

483 T. J. Minehardt and R. E. Wyatt, J. Chem. Phys. **109**, 8330 (1998).

484 B. H. Pate, K. K. Lehmann, and G. Scoles, J. Chem. Phys. **95**, 3891 (1991).

485 V. A. Mandelshtam and H. S. Taylor, Phys. Rev. Lett. **78**, 3274 (1997).

486 O. Vendrell, F. Gatti, and H.-D. Meyer, Angew. Chem. Int. Edn. **46**, 6918 (2007).

487 A. Campargue and F. Stoeckel, J. Chem. Phys. **85**, 1220 (1986).

488 O. Boyarkin, M. Kowalszyk, and T. Rizzo, J. Chem. Phys. **118**, 93 (2003).

489 O. Boyarkin and T. Rizzo, J. Chem. Phys. **103**, 1985 (1995).

490 O. Boyarkin and T. Rizzo, J. Chem. Phys. **105**, 6285 (1996).

491 O. Boyarkin, R. Settle, and T. Rizzo, Ber. Bunsenges. Phys. Chem. **99**, 504 (1995).

492 M. Coffey, H. Berghout, E. W. III, and F. Crim, J. Chem. Phys. **110**, 10850 (1999).

493 A. Callegari, J. Rebstein, R. Jost, and T. Rizzo, J. Chem. Phys. **111**, 7359 (1999).

494 O. V. Boyarkin, T. R. Rizzo, and D. S. Perry, J. Chem. Phys. **110**, 11346 (1999).

495 B. Kuhn and T. R. Rizzo, J. Chem. Phys. **112**, 7461 (2000).

496 Y. T. L. R. H. Page, Y. R. Shen, J. Chem. Phys. **88**, 4621 (1988).

497 A. Callegari *et al.*, J. Chem. Phys. **106**, 432 (1997).

498 A. Callegari *et al.*, J. Chem. Phys. **113**, 10583 (2000).

499 C. Iung, F. Gatti, and H.-D. Meyer, J. Chem. Phys. **120**, 6992 (2004).

500 H. J. Bernstein and G. Herzberg, J. Phys. Chem. **16**, 4765 (1948).

501 D. Romanini and A. Campargue, Chem. Phys. Lett. **254**, 52 (1996).

502 H. R. Dubal and M. Quack, J. Chem. Phys. **81**, 3779 (1984).

503 J. Segall, R. N. Zare, H. R. Dubal, M. Lewerentz, and M. Quack, J. Chem. Phys. **86**, 634 (1987).

504 J. Wong, W. H. Green, C. Cheng, and C. B. Moore, J. Chem. Phys. **86**, 5994 (1987).

505 Z. Lin, K. Boraas, and J.-P. Reilly, Chem. Phys. Lett. **217**, 239 (1994).

506 A. Maynard, R. E. Wyatt, and C. Iung, J. Chem. Phys. **103**, 8372 (1995).

507 V. A. Mandelshtam and H. S. Taylor, J. Chem. Phys. **107**, 6756 (1997).

508 S. G. Ramesh and E. L. Sibert, Mol. Phys. **103**, 149 (2005).

509 M. M. Sprague, S. G. Ramesh, and E. S. Sibert, J. Chem. Phys. **124**, 114307 (2006).

510 K. M. Gough and B. R. Henry, J. Phys. Chem. **88**, 1298 (1984).

511 M. G. Sowa and B. R. Henry, J. Chem. Phys. **95**, 3040 (1991).

512 L. Anastasakos and T. A. Wildman, J. Chem. Phys. **99**, 9453 (1993).

513 H. G. Kjaergaard, D. M. Turnbull, and B. R. Henry, J. Phys. Chem. **101**, 2589 (1997).

514 C. Zhu, H. G. Kjaergaard, and B. R. Henry, J. Chem. Phys. **107**, 691 (1997).

515 H. G. Kjaergaard, D. M. Turnbull, and B. R. Henry, J. Phys. Chem. **102**, 6095 (1998).

516 H. G. Kjaergaard, Z. Zong, A. J. McAlees, D. Horward, and B. R. Henry, J. Phys. Chem. A **104**, 6398 (2000).

517 C. Lapouge and D. Cavagnat, J. Phys. Chem. A **102**, 8393 (1998).

518 D. Cavagnat and L. Lespade, J. Chem. Phys. **114**, 6030 (2001).

519 Y. S. Choi and C. B. Moore, J. Chem. Phys. **94**, 5414 (1991).

520 Y. S. Choi and C. B. Moore, J. Chem. Phys. **97**, 1010 (1992).

521 Y. S. Choi and C. B. Moore, J. Chem. Phys. **103**, 9981 (1995).

522 J. C. Crane *et al.*, J. Mol. Spectrosc. **183**, 273 (1997).

523 J. C. Crane *et al.*, J. Phys. Chem. A **102**, 9433 (1998).

524 T. Yamamoto and S. Kato, J. Chem. Phys. **107**, 6114 (1997).

525 C. Iung and G. Pasin, J. Phys. Chem. **111**, 10426 (2007).

526 F. Ribeiro, C. Iung, and C. Leforestier, Chem. Phys. Lett. **362**, 199 (2002).

527 F. Ribeiro, C. Iung, and C. Leforestier, J. Chem. Phys. **123**, 054106 (2005).

528 C. Iung and F. Ribeiro, J. Chem. Phys. **121**, 174105 (2005).

529 C. Iung, F. Ribeiro, and E. L. Sibert, J. Phys. Chem. A **110**, 5420 (2006).

530 F. Ribeiro, C. Iung, and C. Leforestier, J. Theor. Comput. Chem. **2**, 609 (2003).

531 G. Pasin, C. Iung, F. Gatti, F. Richter, C. Léonard and H.-D. Meyer, J. Chem. Phys. **129**, *144304* (2008).

532 S. Carter and N. C. Handy, J. Mol. Spectrosc. **192**, 263 (1998).

533 Léonard, G. Chambaud, P. Rosmus, S. Carter, and N. C. Handy, Phys. Chem. Chem. Phys. **3**, 508 (2001).

534 R. Loudon, *The Quantum Theory of Light*, Oxford University Press, Oxford, 1983.

535 E. A. Donley *et al.*, Mol. Phys. **99**, 1275 (2001).

536 R. Marquardt, M. Quack, I. Thanopoulos, and D. Luckhaus, J. Chem. Phys. **118**, 643 (2003).

537 F. Richter, M. Hochlaf, P. Rosmus, F. Gatti, and H.-D. Meyer, J. Chem. Phys. **120**, 1306 (2004).

538 J. M. Guilmot, F. Mélen, and M. Herman, J. Mol. Spectrosc. **160**, 401 (1993).

539 T. J. Lee and A. P. Rendell, J. Chem. Phys. **94**, 6229 (1991).

540 L. Khriatchev, J. Lundell, E. Isoniemi, and M. Räsänen, J. Chem. Phys. **113**, 4265 (2000).

541 D. Luckhaus, J. Chem. Phys. **118**, 8797 (2003).

542 P. A. McDonald and J. S. Shirk, J. Chem. Phys. **77**, 2355 (1982).

543 A. E. Shirk and J. S. Shirk, Chem. Phys. Lett. **97**, 549 (1983).

544 J. M. Guilmot, M. Carleer, M. Godefroid, and M. Herman, J. Mol. Spectrosc. **143**, 81 (1990).

545 J. M. Guilmot, M. Godefroid, and M. Herman, J. Mol. Spectrosc. **160**, 387 (1993).

546 R. Schanz, V. Botan, and P. Hamm, J. Chem. Phys. **122**, 044509 (2005).

547 V. Botan, R. Schanz, and P. Hamm, J. Chem. Phys. **124**, 234511 (2006).

548 Y. Guan, G. C. Lynch, and D. L. Thompson, J. Chem. Phys. **87**, 6957 (1987).

549 Y. Guan and D. L. Thompson, Chem. Phys. **139**, 147 (1989).

550 Y. Qin and D. L. Thompson, J. Chem. Phys. **96**, 1992 (1992).

551 Y. Qin and D. L. Thompson, J. Chem. Phys. **100**, 6445 (1994).

552 P. M. Agrawal, D. L. Thompson, and L. M. Raff, J. Chem. Phys. **102**, 7000 (1995).

553 Y. Guo and D. L. Thompson, J. Chem. Phys. **118**, 1673 (2003).

554 P. A. Hunt and M. A. Robb, J. Am. Chem. Soc. **127**, 5720 (2005).

555 W. H. Louisell, *Quantum Statistical Properties of Radiation*, John Wiley & Sons, New York, 1990.

556 R. Alicki and K. Lendi, *Quantum Dynamical Semigroups and Applications*, Springer, Heidelberg, 1987.

557 A. Redfield, Phys. Rev. **98**, 1787 (1955).

558 A. Redfield, IBM J. **1**, 19 (1957).

559 A. Caldeira and A. Leggett, Phys. Rev. A **31**, 1059 (1985).

560 F. Haake, H. Risken, C. Savage, and D. Walls, Phys. Rev. A **34**, 3969 (1986).

561 L. Diósi, Europhys. Lett. **22**, 1 (1993).

562 L. Diósi, Physica A **199**, 517 (1993).

563 M. Nest and P. Saalfrank, Chem. Phys. **268**, 65 (2001).

564 G. A. Worth, H.-D. Meyer, and L. S. Cederbaum, Chem. Phys. Lett. **299**, 451 (1999).

565 R. Baer and R. Kosloff, Chem. Phys. **106**, 8862 (1997).

566 D. Gelman and R. Kosloff, Chem. Phys. Lett. **381**, 129 (2003).

567 H.-P. Breuer and F. Petruccione, *The Theory of Open Quantum Systems*, Oxford University Press, Oxford, 2002.

568 U. Weiss, *Quantum Dissipative Systems*, World Scientific, Singapore, 1992.

569 G. Ford, M. Kac, and P. Mazur, J. Math. Phys. **6**, 504 (1965).

570 W. Lamb and R. Retherford, Phys. Rev. **72**, 241 (1947).

571 B. Bransden and C. Joachain, *Introduction to Quantum Mechanics*, Longman Scientific & Technical, Harlow, 1989.

572 W. Demtröder, *Laserspektroskopie*, Springer, Berlin, 2000.

573 F. Jensen, *Introduction to Computational Chemistry*, John Wiley & Sons, Chichester, 1999.

574 G. Lindblad, Commun. Math. Phys. **40**, 147 (1975).

575 R. Kosloff, M. Ratner, and W. Davis, J. Chem. Phys. **106**, 7036 (1997).

576 G. Ertl and H.-J. Freund, Phys. Today **52**, 32 (1999).

577 G. Ertl, Chem. Record **1**, 33 (2001).

578 S. Dittrich, H.-J. Freund, C. Koch, R. Kosloff, and T. Klüner, J. Chem. Phys. **124**, 024702 (2006).

579 M. Bonfanti, R. Martinazzo, G. F. Tantardini, and A. Ponti, J. Phys. Chem. C **111**, 5825 (2007).

580 T. Klamroth and P. Saalfrank, J. Chem. Phys. **112**, 10571 (2000).

581 M. Sage, Chem. Phys. **35**, 375 (1978).

582 M. Nieto and L. Simmons, Phys. Rev. A **19**, 438 (1979).

583 S. Efrima, C. Jedrzejek, K. Freed, E. Hood, and H. Metiu, J. Chem. Phys. **79**, 2436 (1983).

584 L. Infeld and T. Hull, Rev. Mod. Phys. **23**, 21 (1951).

585 F. Cooper, A. Khare, and U. Sukhatme, Phys. Rep. **251**, 269 (1995).

586 M. Nest and R. Kosloff, J. Chem. Phys. **127**, 134711 (2007).

587 U. Manthe and F. Huarte-Larrañaga, Chem. Phys. Lett. **349**, 321 (2001).

588 G. C. U. Manthe and H.-J. Werner, Phys. Chem. Chem. Phys. **6**, 5026 (2004).

589 F. Huarte-Larrañaga and U. Manthe, J. Chem. Phys. **116**, 2863 (2002).

590 P. Gaspard and M. Nagaoka, J. Chem. Phys. **111**, 5668 (1999).

591 F. Haake and R. Reibold, Phys. Rev. A **32**, 2462 (1985).

592 N. Erez, G. Gordon, M. Nest, and G. Kurizki, Nature, **452**, 724 (2008).

593 R. Martinazzo, M. Nest, P. Saalfrank, and G. Tantardini, J. Chem. Phys. **125**, 194102 (2006).

594 R. Pomes and B. Roux, Biophys. J. **82**, 2304 (2002).

595 A. Smondyrev and G. Voth, Biophys. J. **83**, 1987 (2002).

596 A. Burykin and A. Warshel, Biophys. J. **85**, 3696 (2003).

597 K.-D. Kreuer, Chem. Mater. **8**, 610 (1996).

598 K. D. Kreuer, J. Membrane Sci. **185**, 29 (2001).

599 N. Agmon, Biophys. J. **88**, 2452 (2005).

600 H. S. Mei, M. E. Tuckerman, D. E. Sagnella, and M. L. Klein, J. Phys. Chem. B **102**, 10446 (1998).

601 N. Agmon, Isr. J. Chem. **39**, 493 (1999).

602 D. Marx, M. Tuckerman, J. Hutter, and M. Parrinello, Nature **397**, 601 (1999).

603 G. Granucci, J. T. Hynes, P. Millié, and T.-H. Tran-Thi, J. Am. Chem. Soc. **122**, 12243 (2000).

604 C. Tanner, C. Manca, and S. Leutwyler, Science **302**, 1736 (2003).

605 W. Domcke and A. L. Sobolewski, Science **302**, 1693 (2003).

606 C. Manca, C. Tanner, S. Coussan, A. Bach, and S. Leutwyler, J. Chem. Phys. **121**, 2578 (2004).

607 O. F. Mohammed, D. Pines, J. Dreyer, E. Pines, and E. T. J. Nibbering, Science **310**, 83 (2005).

608 H. Decornez, K. Drukker, and S. Hammes-Schiffer, J. Phys. Chem. A **103**, 2891 (1999).

609 K. Giese, H. Ushiyama, K. Takatsuka, and O. Kühn, J. Chem. Phys. **122**, 124307 (2005).

610 M. Petkovic and O. Kühn, J. Phys. Chem. A **107**, 8458 (2003).

611 H. Naundorf, G. A. Worth, H.-D. Meyer, and O. Kühn, J. Phys. Chem. A **106**, 719 (2002).

612 J. Ortiz-Sanchez, R. Gelabert, M. Moreno, and J. Lluch, J. Phys. Chem. A **110**, 4649 (2006).

613 K. Giese and O. Kühn, J. Chem. Phys. **123**, 054315 (2005).

614 O. Vendrell and H.-D. Meyer, J. Chem. Phys. **122**, 104505 (2005).

615 R. Pomes and B. Roux, Biophys. J. **71**, 19 (1996).

616 D. J. Mann and M. D. Halls, Phys. Rev. Lett. **90**, 195503 (2003).

617 K. R. Asmis *et al.*, Science **299**, 1375 (2003).

618 T. D. Fridgen, T. B. McMahon, L. MacAleese, J. Lemaire, and P. Maitre, J. Phys. Chem. A **108**, 9008 (2004).

619 J. M. Headrick, J. C. Bopp, and M. A. Johnson, J. Chem. Phys. **121**, 11523 (2004).

620 N. I. Hammer *et al.*, J. Chem. Phys. **122**, 244301 (2005).

621 L. McCunn, J. Roscioli, M. Johnson, and A. McCoy, J. Phys. Chem. B **112**, 321 (2008).

622 M. V. Vener, O. Kühn, and J. Sauer, J. Chem. Phys. **114**, 240 (2001).

623 J. Dai, Z. Bacic, X. C. Huang, S. Carter, and J. M. Bowman, J. Chem. Phys. **119**, 6571 (2003).

624 J. Sauer and J. Dobler, Chem. Phys. Chem. **6**, 1706 (2005).

625 M. Kaledin, A. L. Kaledin, and J. M. Bowman, J. Phys. Chem. A **110**, 2933 (2006).

626 S. Carter, S. J. Culik, and J. M. Bowman, J. Chem. Phys. **107**, 10458 (1997).

627 D. J. Wales, J. Chem. Phys. **110**, 10403 (1999).

628 A. B. McCoy, X. Huang, S. Carter, M. Y. Landeweer, and J. M. Bowman, J. Chem. Phys. **122**, 061101 (2005).

629 O. Kühn and L. Wöste, editors, *Analysis and Control of Ultrafast Photoinduced Reactions*, Springer Ser. Chem. Phys., vol. 87, Springer, Heidelberg, 2007.

630 V. May and O. Kühn, *Charge and Energy Transfer Dynamics in Molecular Systems*, Wiley-VCH, Weinheim, 2004.

631 O. Kühn, Chem. Phys. Lett. **402**, 48 (2005).

632 G. K. Paramonov, H. Naundorf, and O. Kühn, Eur. J. Phys. D **14**, 205 (2001).

633 H. Naundorf and O. Kühn, Ultrafast laser-driven hydrogen bond dynamics, in *Femtochemistry and Femtobiology*, edited by A. Douhal and J. Santamaria, page 438, World Scientific, Singapore, 2002.

634 M. Petković and O. Kühn, Chem. Phys. **304**, 91 (2004).

635 M. Petković and O. Kühn, Isotope effect on the IVR dynamics after ultrafast IR excitation of the hydrogen bond in salicylaldimine, in *Ultrafast Molecular Events in Chemistry and Biology*, edited by M. M. Martin and J. T. Hynes, page 181, Elsevier, Amsterdam, 2004.

636 K. Giese, M. Petković, H. Naundorf, and O. Kühn, Phys. Rep. **430**, 211 (2006).

637 A. Borowski and O. Kühn, Theor. Chem. Acc. **117**, 521 (2007).

638 A. Borowski and O. Kühn, J. Photochem. Photobiol. A: Chem. **190**, 169 (2007).

639 A. Borowski and O. Kühn, Chem. Phys. **347**, 523 (2008).

640 B. A. Ruf and W. H. Miller, J. Chem. Soc. Faraday. Trans. **84**, 1523 (1988).

641 K. G. D. Lahav and O. Kühn, J. Theor. Comput. Chem. **3**, 567 (2004).

642 P. Balling, D. Maas, and L. Noordam, Phys. Rev. A **50**, 4276 (1994).

643 F. G. Parak and G. U. Nienhaus, ChemPhysChem **3**, 249 (2002).

644 G. S. Kachalova, A. N. Popov, and H. D. Bartunik, Science **284**, 473 (1999).

645 F. Schotte *et al.*, Science **300**, 1944 (2003).

646 G. Hummer, F. Schotte, and P. A. Anfinrud, Proc. Natl. Acad. Sci. USA **101**, 15330 (2004).

647 T. Polack *et al.*, Phys. Rev. Lett. **93**, 018102 (2004).

648 J. Deak, H.-L. Chiu, C. Lewis, and R. Miller, J. Phys. Chem. B **102**, 6621 (1998).

649 Y. Mizutani and T. Kitagawa, Chem. Record **1**, 258 (2001).

650 M.-L. Groot *et al.*, Proc. Natl. Acad. Sci. USA **99**, 1323 (2002).

651 F. Rosca *et al.*, J. Phys. Chem. A **106**, 3540 (2002).

652 D. E. Sagnella, J. E. Straub, T. A. Jackson, M. Lim, and P. A. Anfinrud, Proc. Natl. Acad. Sci. USA **96**, 14324 (1999).

653 J. R. Hill *et al.*, J. Phys. Chem. **98**, 11213 (1994).

654 J. R. Hill *et al.*, J. Phys. Chem. **100**, 12100 (1996).

655 K. Rector, J. Jiang, M. Berg, and M. Fayer, J. Phys. Chem. B **105**, 1081 (2001).

656 N. Alberding *et al.*, Science **192**, 1002 (1976).

657 J. N. Harvey, J. Am. Chem. Soc. **122**, 12401 (2000).

658 S. Chelkowski, A. D. Bandrauk, and P. B. Corkum, Phys. Rev. Lett. **65**, 2355 (1990).

659 C. Ventalon *et al.*, Proc. Natl. Acad. Sci. USA **101**, 13216 (2004).

660 R. J. D. Miller, Acc. Chem. Res. **27**, 145 (1994).

661 A. Dreuw, B. D. Dunietz, and M. Head-Gordon, J. Am. Chem. Soc. **124**, 12070 (2002).

662 C. Meier and M.-C. Heitz, J. Chem. Phys. **123**, 044504 (2005).

663 D. D. Klug *et al.*, Proc. Natl. Acad. Sci. USA **99**, 12526 (2002).

664 M. J. Frisch *et al.*, Gaussian 98, Revision A.7, Gaussian, Inc., Pittsburgh, PA, 1998.

665 E. T. J. Nibbering *et al.*, Vibrational dynamics of hydrogen bonds, in *Analysis and Control of Ultrafast Photoinduced Reactions*, Springer Ser. Chem. Phys., vol. 87, edited by O. Kühn and L. Wöste, page 619, Springer, Heidelberg, 2007.

666 J. Stenger *et al.*, J. Phys. Chem. A **105**, 2929 (2001).

667 K. Heyne, E. T. J. Nibbering, T. Elsaesser, M. Petković, and O. Kühn, J. Phys. Chem. A **108**, 6083 (2004).

668 V. Apkarian and N. Schwentner, Chem. Rev. **99**, 1481 (1999).

669 P. Jungwirth and R. B. Gerber, Chem. Rev. **99**, 1583 (1999).

670 V. S. Batista and D. F. Coker, J. Chem. Phys. **106**, 6923 (1997).

671 G. Chaban *et al.*, J. Phys. Chem. A **105**, 2770 (2001).

672 F. O. Ellison, J. Am. Chem. Soc. **85**, 3540 (1963).

673 M. Gühr, M. Bargheer, M. Fushitani, T. Kiljunen, and N. Schwentner, Phys. Chem. Chem. Phys. **9**, 779 (2007).

674 M. Bargheer *et al.*, Coherence and control of molecular dynamics in rare gas matrices, in *Analysis and Control of Ultrafast Photoinduced Reactions*, Springer Ser. Chem. Phys., vol. 87, edited by O. Kühn and L. Wöste, page 257, Springer, Heidelberg, 2007.

675 M. V. Korolkov and J. Manz, Z. Phys. Chem. **217**, 115 (2003).

676 M. Bargheer *et al.*, Phys. Chem. Chem. Phys. **4**, 5554 (2002).

677 A. B. Alekseyev, M. V. Korolkov, O. Kühn, J. Manz, and M. Schröder, J. Photochem. Photobiol. A: Chem. **180**, 262 (2006).

678 M. Schröder, J.-L. Carreon-Macedo, and A. Brown, Phys. Chem. Chem. Phys. **10**, 850 (2008).

679 T. F. O'Malley, Phys. Rev. **150**, 14 (1966).

680 W. N. Lozier, Phys. Rev. **36**, 1417 (1930).

681 I. S. Buchel'nikova, Zh. Eksperim. Teor. Fiz. **35**, 1119 (1959).

682 R. N. Compton and L. G. Christophorou, Phys. Rev. **154**, 110 (1967).

683 C. E. Melton, J. Chem. Phys. **57**, 4218 (1972).

684 L. Sanche and G. J. Schultz, J. Chem. Phys. **58**, 479 (1972).

685 S. Trajmar and R. I. Hall, J. Phys. B **7**, L458 (1974).

686 D. S. Beliĉ, M. Landau, and R. I. Hall, J. Phys. B **14**, 175 (1981).

687 M. G. Curtis and I. C. Walker, J. Chem. Soc. Faraday Trans. **88**, 2805 (1992).

688 C. R. Claydon, G. A. Segal, and H. S. Taylor, J. Chem. Phys. **54**, 3799 (1971).

689 M. Jungen, J. Vogt, and V. Staemmler, Chem. Phys. **37**, 49 (1979).

690 T. J. Gil, T. N. Rescigno, C. W. McCurdy, and B. H. Lengsfield III, Phys. Rev. A **49**, 2642 (1994).

691 J. D. Gorfinkel, L. A. Morgan, and J. Tennyson, J. Phys. B **35**, 543 (2002).

692 J. Fedor *et al.*, J. Phys. B **39**, 3935 (2006).

693 W. Kohn, Phys. Rev. **74**, 1763 (1948).

694 R. K. Nesbet, Phys. Rev. **175**, 134 (1968).

695 R. K. Nesbet, Phys. Rev. **179**, 60 (1969).

696 A. U. Hazi, T. Rescigno, and M. Kurilla, Phys. Rev. A **23**, 1089 (1981).

697 W. H. Miller and B. M. D. D. Jansen op de Haar, J. Chem. Phys. **86**, 6213 (1987).

698 B. I. Schneider and T. N. Rescigno, Phys. Rev. A **37**, 3749 (1988).

699 B. H. Lengsfield III and T. N. Rescigno, Phys. Rev. A **44**, 2913 (1991).

700 T. N. Rescigno, C. W. McCurdy, A. E. Orel, and B. H. Lengsfield III, Computational methods for electron–molecule collisions, in *Computational Methods for Electron–Molecule Collisions*, edited by W. M. Huo and F. A. Gianturco, pages 1–44, Plenum, New York, 1995.

701 T. N. Rescigno, B. H. Lengsfield III, and C. W. McCurdy, The incorporation of modern electronic structure methods in electron–molecule collision problems: Variation calculations using the complex Kohn method, in *Modern Electronic Structure Theory*, edited by D. R. Yarkony, pages 501–588, World Scientific, Singapore, 1995.

702 D. T. Birtwistle and A. Herzenberg, J. Phys. B **4**, 53 (1971).

703 L. Dube and A. Herzenberg, Phys. Rev. A **20**, 194 (1979).

704 J. N. Bardsley and J. M. Wadehra, J. Chem. Phys. **78**, 7227 (1983).

705 T. F. O'Malley and H. S. Taylor, Phys. Rev. **176**, 207 (1968).

706 D. J. Haxton, Z. Zhang, C. W. McCurdy, and T. N. Rescigno, Phys. Rev. A **69**, 062713 (2004).

707 D. J. Haxton, Z. Zhang, H.-D. Meyer, T. N. Rescigno, and C. W. McCurdy, Phys. Rev. A **69**, 062714 (2004).

708 D. J. Haxton, T. N. Rescigno, and C. W. McCurdy, Phys. Rev. A **72**, 022705 (2005).

709 D. J. Haxton, C. W. McCurdy, and T. N. Rescigno, Phys. Rev. A **73**, 062724 (2006).

710 D. J. Haxton, C. W. McCurdy, and T. N. Rescigno, Phys. Rev. A **75**, 012710 (2007).

711 D. J. Haxton, T. N. Rescigno, and C. W. McCurdy, Phys. Rev. A **75**, 012711 (2007).

712 D. J. Haxton, T. N. Rescigno, and C. W. McCurdy, Phys. Rev. A **78**, *040702* (R) (2008).

713 R. Renner, Z. Phys. **92**, 172 (1934).

714 C. Jungen and A. J. Merer, Mol. Phys. **40**, 1 (1980).

715 S. Carter and N. C. Handy, Mol. Phys. **52**, 1367 (1984).

716 C. Petrongolo, J. Chem. Phys. **89**, 1297 (1988).

717 A. Loettgers *et al.*, J. Chem. Phys. **106**, 3186 (1997).

718 D. R. Yarkony, Rev. Mod. Phys. **68**, 985 (1996).

719 D. R. Yarkony, J. Chem. Phys. **105**, 10456 (1996).

720 D. R. Yarkony, J. Chem. Phys. **105**, 6277 (2001).

721 C. A. Mead, Rev. Mod. Phys. **64**, 51 (1992).

722 M. Baer, Phys. Rep. **358**, 75 (2002).

723 M. Baer, J. Phys. Chem. A **104**, 3181 (2000).

724 A. Macias and A. Riera, J. Phys. B **11**, L489 (1978).

725 H.-J. Werner and W. Meyer, J. Chem. Phys. **74**, 5802 (1981).

726 C. W. McCurdy and J. L. Turner, J. Chem. Phys. **78**, 6773 (1983).

727 L. M. Delves, Nucl. Phys. **9**, 391 (1959).

728 L. M. Delves, Nucl. Phys. **20**, 275 (1960).

729 S. Kuhr *et al.*, Science **293**, 278 (2001).

730 S. Nussmann *et al.*, Phys. Rev. Lett. **95**, 173602 (2005).

731 R. Folman, P. Krüger, J. Schmiedmayer, J. Denschlag, and C. Henkel, Adv. At. Mol. Opt. Phys. **48**, 263 (2002).

732 J. Fortágh and C. Zimmermann, Rev. Mod. Phys. **79**, 235 (2007).

733 S. Fölling *et al.*, Nature **448**, 1029 (2007).

734 T. Busch, B. G. Englert, K. Rzazewski, and M. Wilkens, Found. Phys. **28**, 549 (1998).

735 Z. Idziaszek and T. Calarco, Phys. Rev. A **71**, 050701 (2005).

736 K. Sakmann, A. I. Streltsov, O. E. Alon, and L. S. Cederbaum, Phys. Rev. A **72**, 033613 (2005).

737 Y. Hao, Y. Zhang, J. Q. Liang, and S. Chen, Phys. Rev. A **73**, 063617 (2006).

738 D. Blume, Phys. Rev. A **66**, 053613 (2002).

739 F. Deuretzbacher, K. Bongs, K. Sengstock, and D. Pfannkuche, Phys. Rev. A **75**, 013614 (2007).

740 M. Girardeau, J. Math. Phys. **1**, 516 (1960).

741 S. Zöllner, H.-D. Meyer, and P. Schmelcher, Phys. Rev. A **78**, 013621 (2008).

Index

Multidimensional Quantum Dynamics: MCTDH Theory and Applications.
Edited by Hans-Dieter Meyer, Fabien Gatti, and Graham A. Worth

ISBN: 978-3-527-32018-9

t